SEMICONDUCTOR DEVICES

RAJENDRA P. NANAVATI
Syracuse University

SEMICONDUCTOR DEVICES

BJTS, JFETS, MOSFETS, and Integrated Circuits

INTEXT
HARPER & ROW, PUBLISHERS
New York Hagerstown Philadelphia San Francisco London

SEMICONDUCTOR DEVICES: BJTS, JFETS, MOSFETS, and Integrated Circuits

Library of Congress Cataloging in Publication Data

Nanavati, Rajendra P
Semiconductor devices.

(The Intext Educational Publishers series in circuits, systems, communications, and computers.)
Companion vol. to the author's Network theory, models, integrated circuits, and electronics.
Includes bibliographies.
1. Semiconductors. I. Title.
TK7871.85.N32 621.3815'2 74-10863
ISBN 0-7002-2463-7

To my youngest daughter Asha and the hope that she may spend her entire life in a peaceful world.

CONTENTS

PREFACE

Since the interval between the discovery of new phenomena and their engineering applications is rapidly diminishing, knowledge of fundamental theory is essential to understanding not only electronic devices of today but also those that will be developed in the future. Therefore I have tried to balance the discussion of semiconductor devices in this book with a presentation of theory in sufficient depth. Basic assumptions are stated, as are the approximations used to obtain the final results. Moreover, the presentation is given in enough depth so that the theories developed here may be extended as necessary. Because of this emphasis on principles, almost all the material in this book has been covered satisfactorily by juniors in the electrical engineering department at Syracuse University.

Included are all the important semiconductor devices, including the bipolar junction transistor (BJT), the junction field effect transistor (FET), and the metal-oxide semiconductor field effect transistor (MOSFET). Since the device models are, by and large, equivalent circuits, each element may be expressed in terms of functional parameters, and desirable changes in the parameters may be readily related to the necessary performance of the device in a given circuit application. In certain cases more sophisticated models need to be considered. To predict the time taken by a transistor to turn off in a given circuit, for example, it is necessary to know the detailed distribution of minority carriers in the base region, as shown in Chapter 6.

Integrated circuits have revolutionized the electronics world, and their impact will be felt not only in electrical engineering education but also in medical practice, international communication, computers, banks, governmental operations, outer space exploration, and a host of other areas. There is sufficient material in the areas of semiconductor device physics and fabrication technology to enable the student to grasp the essentials of integrated circuits.

Chapter 1 covers the ideas of modern physics, assuming no prior knowledge, and Chapter 2 describes band theory and semiconductor processes. These two chapters constitute a broad foundation for the theories developed in the remainder of the book. Chapters 3 through 7 deal with the device theory of *p-n* junctions and the bipolar junction transistor. In Chapter 6 both the wide band steady state is analysed and transient analysis of junction transistors is carried out.

Chapters 8 and 9 develop the theory of FETS and MOSFETS respectively. Chapter 11 goes into the details of modern fabrication technology. The integrated circuit and the technology behind it are also introduced in this chapter. Extensions of the junction theory are considered in Chapter 12. These go somewhat beyond the junior level, but are included in the interest of completeness.

In Chapter 10 a number of additional semiconductor devices are considered, demonstrating that the material in the first 9 chapters enables the reader to understand many present day devices and those that may be developed in the future.

Chapter 11 can be covered as early as after Chapter 3. Alternatively, it can be introduced after Chapter 7, 8, or 9. Vacuum tube physics is covered in an extensive appendix E.

The reader will note that some chapters have equations numbering in several hundreds. This is done to aid clarity in the presentation as well as to make it easier to refer to specific steps in the development. Equations are numbered separately in each worked out example beginning with one. Only those results obtained in the examples which are generally true and valuable are given a number prefixed by the chapter number.

The photo-micrographs of integrated circuits provided by Fairchild Semiconductor, Mountain View, California; and by the Texas Instruments Corporation add considerably to the book. The author is deeply grateful to them.

Without the patience and encouragement of my wife and the forebearance of my two daughters over many weekends and countless evenings, this book would have been impossible to write.

Most of the manuscript typing was done by Mrs. Stephanie Micale, Mrs. Elka Bolich, Mrs. Linda Lopez and Mrs. Sue Garner whose help deserves special mention.

The many stimulating questions and comments of the Juniors in the electrical engineering department at Syracuse University have been most helpful during the preparation of this manuscript.

LIST OF SYMBOLS

A	area
A_b	cross-sectional area through which the base current flows
A_c	cross-sectional area of the collection junction
A_e	cross-sectional area of the emitter junction
A_i	current gain
A_p	power gain
A_s	area over which surface recombination takes place
A_v	voltage gain
A_{v0}	low frequency voltage gain
a	lattice constant; or acceleration in Chapter 2; doping gradient in the transition region of linearly graded junctions in Chapter 3; half channel width in FET shown in Fig. 8–1.
$a_{11}, a_{12}, a_{21}, a_{22}$	defined by Eqs. (4–76) and (4–77)
B	magnetic field density Chapters 1, 2; volume of the charge depleted transition region
B_c	volume of the charge neutral channel region
b	defined by Eq. (12–104); channel width defined on Fig. 8–1
b_t	tunneling barrier width in a tunnel diode
c	velocity of light
c_1	a numerical factor defined in Eq. (12–67)
C	volume density of chemical impurities Chapter 11; input gate capacitance of FET (Chapter 8)
C_{av}	average linear capacitance which displaces the same amount of charge as the nonlinear junction capacitance

C_{cp}	defined by Eq. (6–18)
C_{ep}	defined by Eq. (6–19)
C_d	diffusion capacitance
C_{dc}	defined by Eq. (6–211)
C_{de}	defined by Eq. (6–210)
C_{ds}	capacitance between drain and source
C_{gd}	capacitance between gate and drain
C_{gch}	gate to channel capacitance
C_{gs}	capacitance between gate and source
C_i	input capacitance Eq. (9–82)
C_j	junction capacitance
C_{jd}	junction capacitance between drain and substrate
C_{js}	junction capacitance between source and substrate
C_ℓ	linear stray capacitance of the *p-n* junction
$C_{\ell c}$	linear stray capacitance associated with the collector junction
$C_{\ell e}$	linear stray capacitance associated with the emitter junction
C_n	defined by Eq. (5–8)
C_{nc}	collector junction capacitance
C_{ne}	emitter junction capacitance
C_o	output capacitance Eq. (9–83)
C_{ox}	oxide capacitance
c_{ox}	oxide capacitance per unit area
C_p	defined by Eq. (5–7)
C_s	surface concentration of doping impurities
C_{tc}	collector junction capacitance
C_{te}	emitter junction capacitance
D	diffusion constant
D_n	electron diffusion constant
D_p	hole diffusion constant
$D_{p,\,\mathrm{eff}}$	effective diffusion constant for holes taking account of high injection affects
d_1	channel width in MOSFETs
E	energy
E_a	energy of acceptor levels
E_c	energy at the bottom of the conduction band
E_{cn}	conduction band edge energy on the *n*-side
E_{cp}	conduction band edge energy on the *p*-side
$\mathscr{E}$	electric field
$\mathscr{E}_b$	built-in field

$\mathscr{E}_e$	electric field in the epitaxial region
$\mathscr{E}_{ox}$	electric field in the oxide
$\mathscr{E}_{TL}$	electric field defined by Eq. (9–56)
E_f	Fermi level
E_{fm}	Fermi level in a metal
E_{fs}	Fermi level in a semi-conductor
E_g	band gap
E_n	quantized or discrete energy level
E_i	intrinsic energy
E_k	kinetic energy
E_r	recombination center energy
E_v	energy at the top of the valence band
E_{vn}	valence band edge on the n-side
E_{vp}	valence band edge on the p-side
E_t	trap energy
E_x	energy gap through which the electron tunnels
E_1	defined by Eq. (10–45)
F	force
$F_{\frac{1}{2}}$	Fermi–Dirac integral
F_{ext}	external force
F_{in}	internal force
f	field factor defined by Eq. (7–19)
$f(E)$	Fermi–Dirac distribution function
$f(b)$	defined by Eq. (12–103)
f_m	maximum frequency of oscillation for transistor
f_{maxd}	defined by Eq. (7–120)
f_T	defined by Eq. (7–111)
f_T^*	defined by Eq. (7–118)
f_t	fraction of traps occupied by electrons
f_{tc}	cut-off frequency due to transit time effect in FET
f_{tp}	fraction of traps occupied by holes
G_c	generation rate of carriers
G_e	defined by Eq. (2–63)
G_n	generation rate of electrons
G_0	channel conductance with gate to source voltage zero in FET.
G_p	generation rate of holes
$(GB)_v$	voltage gain-bandwidth product
$(GB)_{ic}$	current gain-bandwidth product for cascaded current gain
g	defined by Eq. (7–47)

g_{cc}	defined by Eq. (6–199)
g_D	channel conductance, Chapters 7, 8
g_d	diffusion conductance
$g'_{d\ell}$	leakage conductance between gate to drain
g_{ee}	defined by Eq. (6–198)
g'_ℓ	leakage conductance
$g'_{\ell c}$	leakage conductance of the collector junction
$g'_{\ell e}$	leakage conductance of the emitter junction
$g'_{\ell s}$	leakage conductance between the gate and source
g_m	defined by Eq. (6–257); tranconductance of a MOSFET in Chapter 9
g_h	defined by Eq. (10–8)
g_n	defined by Eq. (12–213)
$g_n(Z_n)$	defined by Eq. (12–213)
$g_p(Z_p)$	defined by Eq. (12–210)
h	Planck's constant; channel depth in Chapters 8, 9
$h_{11e}, h_{12e}, h_{21e}, h_{22e}$	common emitter h–parameters
$\hbar$	h/π
h_{fb}	forward short circuit current gain in the common base circuit
h_{fc}	forward short circuit current gain in the common emitter circuit
I	current
I_B	d-c base current
I_C	d-c collector current
I_{CM}	maximum collector current
I_{CO}	collector current when the emitter current is zero
I_{DO}	drain current in saturation
I_D	photo-diode current Chapter 10, drain current for Chapter 7, Chapter 8; given by Eq. (7–57)
I_E	d-c emitter current
I_{EM}	maximum emitter current
I_{EPD}	defined by Eq. (7–55)
I_{CPD}	defined by Eq. (7–56)
I_P	peak current
I_{PH}	photon generated diode current
I_S	back saturation current for *p-n* junction diodes
I_{SC}	photodiode short circuit current
I_V	valley current
I_c	amplitude of the a-c collector current
I_e	amplitude of the a-c emitter current

I_f	feedback current
I_{bs}	base current which just saturates the transistor
I_{bx}	base current in excess of I_{bs}
I_{nC}	d-c electron current crossing the collector junction
I_{nE}	d-c electron current crossing the emitter junction
I_{ne}	a-c electron current crossing the emitter junction
I_{n0}	d-c electron current crossing the *p-n* junction
I_{cf}	forward collector current
I_{cr}	reverse collector current
I_{cf}	forward emitter current
I_{er}	reverse emitter current
I_m	defined by Eq. (8–67)
I_{es}	emitter current to just saturate the transistor
I_p	hole current
I_{pC}	d-c hole current crossing the collector junction
I_{pE}	d-c hole current crossing the emitter junction
I_{pc}	a-c hole current crossing the emitter junction
I_{p0}	d-c hole current crossing the *p-n* junction
I_{SE}	back saturation current of the emitter junction
I_{SR}	surface recombination current
I_{VR}	volume recombination current
I_{tc}	total a-c current crossing the collector junction
I_{te}	total a-c current crossing the emitter junction
J	current density
J_c	collector current which equals the current in the epitaxial region
J_{pC}	d-c hole current density crossing the collector junction
J_{pE}	d-c hole current density crossing the emitter junction
J_{ptc}	total a-c current density crossing the collector junction
J_{pte}	total a-c current density crossing the emitter junction
J_0	defined by Eq. (12–236)
j	$\sqrt{-1}$
j_n	electron current density
j_p	hole current density
j_{p1}	unperturbed a-c minority carrier current density in the base region
j_{pt}	total hole current density including the contribution of the perturbed solution
j_{p0}	d-c minority carrier current density in the base due to d-c voltages across the emitter and collector junctions
k	Boltzmann's constant

k_a	defined by Eq. (3–94)
k_{at}	defined by Eq. (3–99)
$\mathbf{k}_c$	wave vector
k_c	wave number
k_ℓ	defined by Eq. (3–105)
k_v	defined by Eq. (8–65)
L	diffusion length; channel length in Chapters 8, 9
L_d	distance over which the transition regions meet at the drain end Chapters 8, 9; distance constant associated with the impurity grading in the base region Chapter 7
L_n	diffusion length for electrons
L_{cn}	diffusion length for electrons in the type collector
L_{ne}	diffusion length for electrons in the *p*-type emitter
L_p	diffusion length for holes
L_{pc}	diffusion length for holes in the *n*-type collector
L_{pe}	diffusion length for holes in the *n*-type emitter
L_s	effective channel length
ℓ	a quantum number in Chapter 1; length of a resistive path
M	multiplication factor
MAG	maximum available power gain
m	mass; quantum number in Eq. (1–56)
m_e	electron mass
m^*	effective mass
m_e^*	effective mass of electrons
m_h^*	effective mass of holes
m_ℓ^*	longitudinal effective mass
m_t^*	transverse effective mass
m_0	rest mass of electron
$N(E)$	density of states function
$N_c(E)$	density of states function in the conduction band
N_c	defined by Eq. (2–74)
N_a	volume density of acceptors
N_d	volume density of donor atoms
N_0	defined by Eq. (2–62)
N_{eff}	effective net doping density
$N_v(E)$	density of states function in the valence band
N_v	defined by Eq. (2–78)
N_x	density of occupied band gap states above the top of valence band

n_1	density of electrons in the conduction band when the Fermi level is at the trap level
n	integer and quantum number corresponding to energy; also electron volume density; optimum transformer turns ratio in Chapter 6
n_a	density of electrons in the acceptor states
n_d	density of electron in the donor states
n_i	intrinsic volume density
n_{inj}	injected electrons
n_n	thermal equilibrium electron density in an *n*-type crystal
n_p	thermal equilibrium density of electron density in a *p*-type crystal
n_{pc}	thermal equilibrium electron density in the *p*-type collector
n_{pe}	thermal equilibrium hole density in the *p*-type emitter
n_{xtn}	electron density at x_{tn} density in the *p*-type material
P	momentum in Chapters 1, 2; hole density in other chapters except in Chapter 8 where it is defined by Eq. (8–68)
$\tilde{p}$	perturbed solution
P	probability
P_x	probability of electrons tunneling from band gap states to the valence band
$p_0(x)$	steady state hole density
$p_1(x)$	a-c amplitude of the sinusoidally varying hole density
p_1	hole density in the valence bands when the Fermi level is at the trap level
p_i	intrinsic hole density
p_{inj}	injected holes density
p_n	thermal equilibrium density of holes in an *n*-type crystal
p_p	thermal equilibrium density of holes in a *p*-type crystal
p_{pe}	thermal equilibrium density of holes in the *p*-type emitter
p_{pc}	thermal equilibrium hole density in the *p*-type collector
P_t	total hole density solution including the perturbed solution
p_C	defined by Eq. (4–3)
p_{xtn}	hole density at x_{tn}
p_E	defined by Eq. (4–2)
p_e	a-c emitter boundary condition

P_{in}	input power
P_0	defined by Eq. (7–42)
P_{out}	output power
Q	charge
Q_b	excess minority carrier base charge
Q_{bi}	initial base charge
Q_{bss}	steady-state base charge
Q_{bs}	base minority carrier charge just at saturation
Q_{bx}	base minority carrier charge in excess of Q_{bs}
Q_n	free electron charge in the inverted region
Q_{tot}	total charge
q	magnitude of the electronic charge
s	spin quantum number Chapters 1, 2; surface recombination velocity
T_b	defined by Eq. (6–425)
T_{bv}	defined by Eq. (6–444)
T_c	defined by Eq. (6–427)
T_d	time for which the base is exposed to doping impurities from infinite source
T_e	defined by Eq. (6–426)
T_{fd}	diode fall time
T_p	normalized time, Eq. (6–427)
T_s	defined by Eq. (6–417)
T_s^*	defined by Eq. (6–416)
T_{sd}	normalized diode storage time
T	absolute temperature; time constant; duration of "on" pulse in Chapter 6.
T_0	defined by Eq. (5–30) low frequency of mid-band gain
$\bar{t}$	mean free time
$\bar{t}_n$	mean free time for electrons
$\bar{t}_p$	mean free time for holes
t_d	delay time
t_{dc}	delay time due to capacitor charge readjustment time
t_{dcb}	delay time due to capacitor charge readjustment time in common base circuit
t_{dce}	delay time due to capacitor charge readjustment time in common emitter circuit
t_{dp}	propagation delay time
t_f	fall time
t_{fb}	common base circuit fall time

t_{fc}	common collector circuit fall time
t_{fe}	common emitter circuit fall time
$t_{f\min}$	minimum fall time
t_{ox}	oxide thickness
t_r	rise time
t_{rb}	common base circuit rise time
t_{rc}	common collector circuit rise time
t_{re}	common emitter circuit rise time
t_s	storage time
t_{sb}	common base circuit storage circuit
t_{sc}	common collector circuit storage time
t_{se}	common emitter circuit storage time
U	recombination rate
U_{cn}	rate of recombination of electrons
U_{cp}	rate of recombination of holes
U_{ct}	capture rate of conduction band electrons by available traps
U_r	recombination rate in a reverse biased *p-n* junction
U_{tc}	rate of electron emission from traps to conduction band
U_{tv}	rate of hole emission from traps to valence band
U_{vt}	rate of arrival of holes from the valence band to the traps
u	velocity
u_n	electron velocity
u_p	hole velocity
V	potential energy in Chapter 1; voltage in other chapters
V_B	d-c base voltage
V_{BD}	breakdown voltage
V_{BI}	built-in voltage
V_{BIC}	built-in voltage of collector junction
V_{BIE}	built-in voltage of emitter junction
V_{BO}	defined by Eq. (10–2)
V_C	d-c collector voltage with respect to the base due to an external source
V_c	a-c amplitude of collector to base voltage
V_{DS}	d-c drain to source voltage
V_{DSO}	d-c drain to source voltage at the onset of saturation
V_E	d-c emitter voltage with respect to the base due to an external source
V_e	a-c amplitude of emitter to base voltage
V_F	final voltage

$V_{f\max}$	maximum between V_{fn} and V_{fp}
$V_{f\min}$	minimum between V_{fn} and V_{fp}
V_{fn}	voltage difference between the Fermi level and conduction band edge
V_{fp}	voltage difference between the valence band edge and Fermi level
V_A	avalanche breakdown voltage
V_{D1}	defined by Eq. (12–138)
V_G	d-c gate voltage
V_{GD}	d-c gate to drain voltage
V_{GS}	d-c gate to source voltage
$V_{GS\text{eff}}$	defined by Eq. (9–62)
V_I	initial voltage
V_d	defined by Eq. (7–5)
V_m	defined by Eq. (12–23a)
V_0	defined by Eq. (12–139)
V_{OC}	photo-diode open circuit voltage
V_P	peak voltage in Chapter 10; pinch-off voltage defined by Eq. (8–9) in FET
V_T	threshold voltage
V_V	valley voltage
V_t	total voltage across the transition region
v	volume in Chapter 1; voltage in other chapters
v_1	amplitude of a-c voltage
v_0	d-c voltage
v_c	a-c collector to base voltage
v_C	total collector junction voltage
v_e	a-c emitter to base voltage
v_g	voltage corresponding to the band gap energy
x	position coordinate
x_o	tunneling barrier width in Chapter 1. Oxide thickness in Chapter 9
x_t	transition region width
x_{tc}	collector transition region
x_{te}	emitter transition region
$x_{t\max}$	maximum value of transition region width in MOS-FETS
x_{tn}	transition region on the *n*-side of the junction
x_{tp}	transition region edge on the *p*-side of the junction
Y	admittance
$Y_{11}, Y_{12}, Y_{21}, Y_{22}$	admittance parameters

Y^i	intrinsic admittance
Y^i_{cc}, Y^i_{ce}, Y^i_{ec}, Y^i_{ee}	common base intrinsic admittances
Y^i_{ccd}, Y^i_{ced}, Y^i_{ecd}, Y^i_{eed}	common base intrinsic admittances for drift transistors
Y_{cc}	defined by Eq. (6–163)
$Y_{\ell c}$	leakage admittance of the collector junction
$Y_{\ell e}$	leakage admittance of the emitter junction
Y_m	defined by Eq. (7–93)
Z	impedance
Z'_b	extrinsic base impedance
Z_c	collector impedance
Z_e	emitter impedance
Z_n	high injection parameter for *n-p-n* transistors
Z_p	high injection parameter Eq. (12–20) for *p-n-p* transistors
z_1, z_2	defined by Eq. (12–105)
α	short-circuit current gain for common-base circuit
α^*	collector multiplication factor
α_d	short-circuit current gain in a drift transistor operated in the common-base mode
α_{d0}	low-frequency value of α_d
α_i	ionization rate in the transition region of the *p-n* junction
α_N, α_0	low-frequency value of α for a normal transistor
$\alpha_i(j\omega)$	short-circuit common-base current gain for a transistor operated inversely
α_I	low-frequency value of $\alpha_i(j\omega)$
β	short-circuit current gain for common-emitter mode of transistor operation
β^*	base transport factor
β_0	low-frequency value of β
β_0^*	low-frequency value of β^*
β_g	constant defined by Eq. (2–102)
γ	injection efficiency
γ_0	low frequency value of the injection efficiency
γ_{ss}	charge per unit area at the oxide–semiconductor interface
γ_u	injection efficiency for the unijunction transistor

ε	permittivity
ε_d	defined by Eq. (7–44)
ε_g	defined by Eq. (7–65)
ε_{ox}	permittivity of the oxide
ε_g	defined by Eq. (7–66)
η_c	normalized conduction band edge
η	E/kT normalized energy defined by Eq. (2–64) in Chapter 2; capacitance ratio defined by Eq. (6–325)
η_f	E_f/kT normalized Fermi level (Chapter 2)
η_a	E_a/kT normalized acceptor energy (Chapter 2)
η_d	E_d/kT normalized donor energy (Chapter 2)
η_g	E_g/kT (Chapter 2)
θ_h	Hall angle
θ_n	Hall angle in an *n*-type semiconductor
θ_p	Hall angle in a *p*-type semiconductor
λ	wavelength
μ	mobility
μ_{d1}	drift mobility of holes in valence band 1
μ_{d2}	drift mobility of holes in valance band 2
μ_{dn}	drift mobility of electrons
μ_{dp}	drift mobility of holes
μ_{hn}	Hall mobility of electrons
σ	defined by Eq. (7–49)
σ_0	defined by Eq. (7–48)
σ_B	base conductivity
σ_C	collector conductivity
σ_E	conductivity of the emitter
σ_i	conductivity of the intrinsic semiconductor
σ_n	conductivity due to electrons
σ_p	conductivity due to holes
σ_{nc}	conductivity of *p*-type collector due to only electrons
σ_{pc}	conductivity of *p*-type collector due to only holes
τ	base transit time
τ_B	defined by Eq. (7–113)
τ_C	defined by Eq. (7–114)
τ_{CB}	defined by Eq. (7–115)
τ_E	defined by Eq. (7–112)

τ_{eff}	effective time constant
τ_n	lifetime of electrons
τ_p	lifetime of holes
τ_{cn}	lifetime of electrons in the *p*-type collector
τ_{en}	lifetime of electrons in the *p*-type emitter
τ_{ec}	emitter to collector base delay time
τ_s	surface lifetime
τ_t	transit time
φ_n	quasi Fermi potential for electrons
φ_p	quasi Fermi potential for holes
$\varphi_{\alpha d}$	phase angle of α_d at its cut-off frequency
Ψ	Schrödinger wave function as a function of space coordinates and time
ψ	wavefunction as a function only of space coordinates
ψ_i	intrinsic potential
ψ_n	intrinsic potential in *n*-type semiconductor
ψ_p	intrinsic potential in *p*-type semiconductor
ω	radian frequency
ω_α, ω_N	α cut-off frequency for a normally operated transistor
$\omega_{\alpha d}$	cut-off frequency of α_d
ω_β	β cut-off frequency
ω_{cv}	cut-off frequency for voltage gain
ω_I	α cut-off frequency for the common-base short-circuit current gain of an inversely operated transistor

CHAPTER 1

SOME KEY CONCEPTS OF MODERN PHYSICS

1-1 INTRODUCTION

The main theme of this book is the modeling of electronic devices. We shall consider both *active* and *passive* devices, though the major emphasis will be on the active devices. For the present an active device is one which is capable of providing power gain (output power divided by input power) greater than unity. Examples of active devices are transistors, tubes, metal-oxide-semiconductor field effect transistors, tunnel diodes, etc. Passive devices are those whose power gain does not exceed unity. Examples of passive devices are resistances, capacitances, inductances, transformers, *p-n* junction diodes, etc. Only active devices can enable a circuit to provide overall power gain. In the transmission of signal information, and power from one location to another there are inevitable power losses. Unless these losses are made up, the signal being transmitted will diminish. Active circuits utilizing active devices, through linear amplification which does not distort the signal, allow the signal to reach the desired destination in adequate strength and undistorted. This is one of the reasons why it is important to study active devices.

In order to intelligently use a device in an electric circuit or system it is essential to characterize the device in terms of its behavior at its terminals. Any description or characterization which is helpful in predicting the behavior of the device constitutes a *model.* There are many different ways of obtaining models for devices. Two fundamentally different approaches to the problem are in widespread use. One approach starts with the emperical study of the *terminal* behavior of the device. An electrical circuit is then devised which behaves at its terminals similar to the actual device to be modeled. This approach has some major disadvantages associated with it. Suppose the terminal measurements, for example current versus voltage, are performed at a given temperature. If the terminal performance of the device depends on temperature, it will be necessary to retake the data under various temperatures. If the terminal behavior depends not only on voltage and current but also on their rates of change with time, the measurements would need to be made under dynamic conditions. If more than one environmental factor can materially affect the terminal behavior, it is not hard to see that an enormous catalogue of data would need to be kept. Furthermore, because of lack of coherence or structure to the data, no easy way would exist to extrapolate them to different domains of intended use for the device.

Because of the above mentioned overwhelming difficulties a second approach to modeling is preferable wherever possible. This second approach starts with the understanding of the internal physics of the device. This understanding permits the writing of precise mathematical relationships between the important parameters of the device. Some structural parameters, depending on how the device is constructed, as well as some internal parameters of the device will be necessarily involved in the mathematical formulations. Solutions of mathematical equations lead to the terminal current–voltage relationships. If temperature is an important parameter of the device, the resulting equations will show that fact. These theoretical solutions can provide both coherence and structure to data taken at the terminals. Moreover, this understanding may pinpoint the critical measurements that ought to be made so that accurate prediction of behavior over a wider range of applications becomes possible. Indeed the deeper the understanding of the internal functioning of the device, the greater our ability to predict the behavior of the device over an extended range of applications.

For these reasons we shall follow this second approach to modeling devices. Because so many present-day electronic devices—and many yet to come—will probably depend on quantum mechanical phenomena, in this chapter we take a panoramic view of *quantum theory.* For example, there is no classical way to describe the phenomenon of *quantum mechanical tunneling* on which the intriguing negative resistance characteristics of the tunnel diode is dependent. In Chapter 2 we take a look at the fundamental processes that go on in semiconductors that will be of major importance to our understanding of the semiconductor devices we shall study in this book. The vacuum tube physics is covered in Appendix E. Because of the pre-eminence of semiconductor devices among electronic devices, this book is almost wholly devoted to them. The broad foundation of the first two chapters will enable us to study many semiconductor devices. The *p-n* junction diode and then the bipolar junction transistors are studied in Chapters 3 through 7, 10, and 12. The semiconductor fabrication processes and the important subject of integrated circuits are covered in Chapter 11. Integrated circuits have drastically altered the electronics world and we shall see an ever increasing use of them in many phases of everyday life in the 1970s. In order to understand integrated circuits the simultaneous knowledge of device fabrication technology and device physics is absolutely essential. The fabrication processes provide both the structural parameters as well as the limits on parameter values of devices. Major extensions of the junction theory are considered in Chapter 12.

Field effect devices are considered in Chapters 8 and 9. The metal–oxide–semiconductor field effect transistor (MOSFETS) is destined to see much more widespread use in the future.

The models considered in this book are by and large electrical equivalent circuits, to the extent that the equivalent circuits are valid models of the devices considered. They allow us to use the powerful and highly developed tools of network theory to predict the behavior of these devices embedded in electrical networks and other systems. All models have limitations in the range of their validity, and it is important therefore to be aware of these limitations before models are used. Unfortunately it is not hard to find examples where a model has been used and conclusions drawn which were not really valid as the model was used outside the range of its validity.

Generally the complexity of a model grows with the demand for increasing accuracy. The intended use of the model should determine the model that should be chosen. If the device is to be used in a limited set of circumstances a simple model would be adequate. If the model is to be valid under widely varying circumstances a much more complicated model would be necessary. The analysis of such a model would also be correspondingly complicated. It would be clearly impractical to demand an accuracy of 1 per cent from the model if the accuracy of other components in the rest of the system is 20 per cent. Thus most models have a limited range of validity and their user must be cognizant of the limitations. For this reason all models derived in this book will also be accompanied by a careful listing of conditions under which they are valid. In any given application that model is best which (a) is valid under the conditions in question, and (b) is no more complex than necessary.

1-2 BLACKBODY RADIATION

Quantum theory may be said to have been born with Planck's proposed theory for *blackbody* radiation. To be precise a blackbody is any body which completely absorbs all incident radiation without *reflecting* any of it. While no reflection occurs, the blackbody does radiate power. Two practical examples of blackbody defined above are *lamp black* and a large enclosure having a very small pin-hole opening. Note that the latter does satisfy the definition of blackbody because all incident radiation entering through the pin-hole will suffer multiple internal reflections with negligible chance of it leaving the body by the only opening which is itself quite small.

Figure 1–1 shows how the radiated power density varies as a function of wavelength of the radiation. These curves were well known experimentally, but lacked a satisfactory theory to explain them. Two different attempts were made through classical theories to explain these curves of blackbody radiation. These attempts are known as *Wien's law* and *Rayleigh-Jean's law*. The former agreed with experimental results reasonably well for short wavelengths and the latter agreed fairly well for long wavelengths. Neither theory explained the curves of Fig. 1–1 over the entire range. As the above theories were well grounded in classically well-accepted theories it was clear that some new theory was necessary to explain the experimental results.

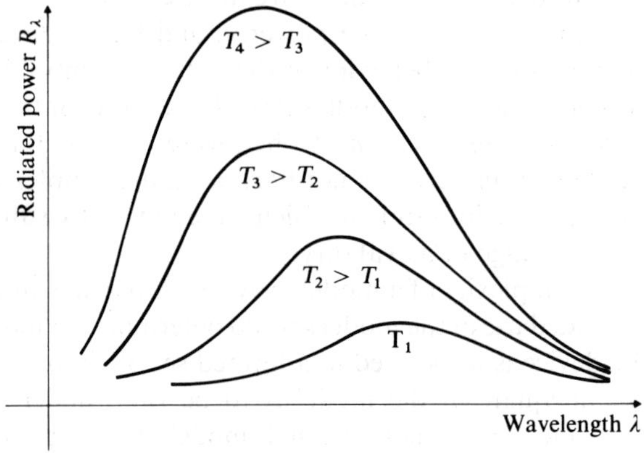

Fig. 1–1 Experimental curves for blackbody radiation.

Planck provided the new theory of blackbody radiation. He emperically obtained the analytical expression

$$R_\lambda = \frac{2\pi c^2 h}{\lambda^5} \cdot \frac{1}{(e^{h\nu/kT} - 1)}, \tag{1–1}$$

where R_λ = total emissive power, watts per square centimeter per angstrom† (Å),
λ = wavelength, A,
ν = frequency, cps,
T = absolute temperature, T°K of the black body
k = Boltzmann constant = 1.38×10^{-16} erg/°K,
c = velocity of light = 2.997930×10^8 m/sec,
h = Planck's constant = 6.6252×10^{-34} joule-sec.

Planck's constant h is a universal constant. All key results of quantum theory have this constant in them. The final successful model for blackbody radiation which was consistent with Eq. (1–1) was as follows. The sources of radiation from the blackbody were a group of oscillators. These oscillators satisfied the following conditions.

† Angstrom is 10^{-10} meters.

1. Each oscillator could have *only* a discrete set of energy levels E_n given by

$$E_n = nh\nu, \tag{1–2}$$

where $n =$ positive integer, for example 1, 2, 3, ...

2. Each oscillator could absorb or radiate energy in an indivisible packet of energy $h\nu$ called *quantum* of energy. This process is accompanied by a sudden jump in energy of an oscillator from one to another of the permissible levels of energy given by Eq. (1–2).
3. The ratio of the probability of nonemission to that of emission is proportional to the intensity of the incident radiation.

Each of the above three parts of the proposed theory was a revolutionary departure from the classical theories. There was nothing in classical electromagnetic theory that suggested the existence of only discrete allowable energies that oscillators could have given by Eq. (1–2) and no other allowed energies. Neither did the classical theories allow for the existence of indivisible quantum of energy. Certainly the most drastic change from the old theories is the third condition. Up to this point scientific theories were essentially deterministic. For example, Newtonian mechanics says that if the initial velocity and position of a particle are known together with all the forces acting on the particle, then the trajectory of the particle is uniquely determined for all time. So the probabilistic description of nature was accepted with the greatest reluctance. *Indeed it turns out that the probabilistic description is a fundamental part of modern quantum theory.* Elements of each of the three parts of Planck's theory are also accepted in the most modern theories of physics.

1-3 PARTICLE NATURE OF WAVES

As already pointed out Planck's theory for blackbody radiation represented quite a serious challenge to the well-established classical theories. Physicists were reluctant to fully accept the new ideas without further evidence. Before long several other experimental findings needed new theories to explain them.

One such experimental result was the *photo-electric effect.* In this experiment a vacuum photocell having two electrodes, a photosensitive emitter, and a collector, was connected externally to a variable voltage source V_s. Monochromatic light of a given intensity was made to shine on the photosensitive emitter. In these circumstances a current† is found to flow *in the external circuit* from the emitter to the collector through the voltage source V_s as shown in Fig. 1–2. The value of the voltage source which stops all current flow is noted. This value is called the *stopping voltage* and is given the symbol V_0.

† It is useful to remember that positive conventional current flow is defined to be opposite the direction of electron flow.

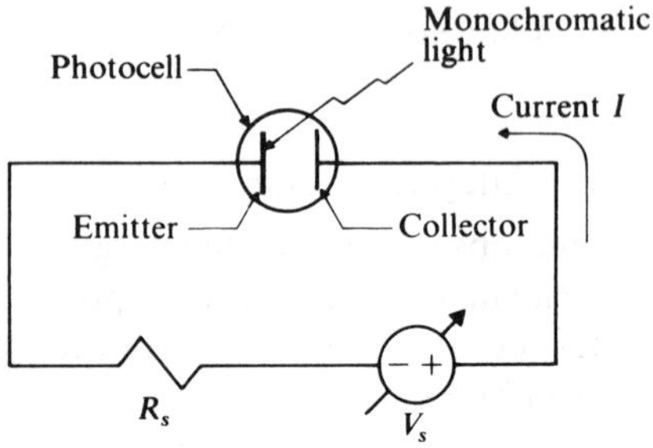

Fig. 1–2 Photo-electric experiment.

According to classical theories it would be expected that the stopping voltage V_0 should be a function of the intensity of light. Contrary to this expectation, experiments showed that the stopping voltage was *not* a function of intensity as seen in Fig. 1–3. It was Einstein who developed a simple though revolutionary theory to explain these experimental findings. His theory asserted that the *incoming light* had to be thought of as consisting of indivisible packets or *quanta of energy* of light called *photons* having energy $h\nu$. Thus incoming light is said to be *quantized.* Note that the energy of each photon is proportional to frequency ν alone; the proportionality constant being Planck's constant h. Einstein's theory says that when the magnitude of the energy of incoming light quanta is sufficient to overcome *both* the stopping potential and the work function, then the electron current will start flowing. *Work function* $q\phi$ is the energy necessary to supply to the most energetic electron in a solid to just overcome the *binding energy,* which keeps it tied to the solid, and hence frees the electron. So the energy of the

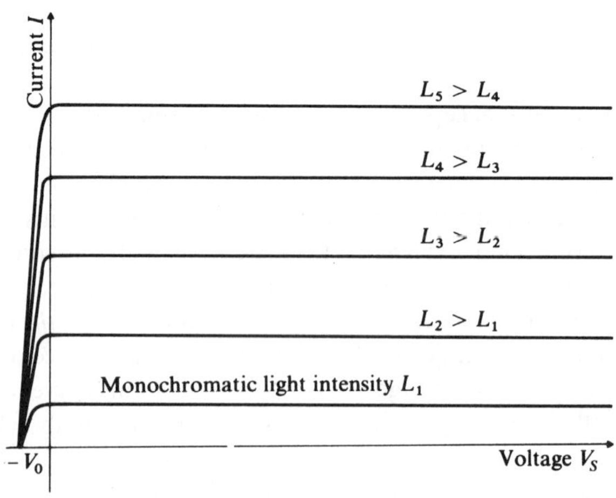

Fig. 1–3 Experimental results of the photoelectric effect.

incoming photon is related to $q\phi$ and V_0 by the following simple equation

$$qV_0 + q\phi = h\nu \tag{1–3}$$†

or

$$V_0 = \frac{h\nu}{q} - \phi. \tag{1–4}$$

An additional confirmation of Einstein's proposed theory was provided when the stopping voltage was plotted as a function of the frequency of incoming light. First, such a plot was indeed linear, and second, the slope of the plot was in fact h/q as predicted by Eq. (1–4). Classically there is no reason to expect a simple relationship to exist between stopping voltage and frequency of incoming light.

The success of Einstein's theory ironically raised some uncomfortable questions. For example, how could this new theory which considers light to be made up of photons (which is essentially a particle description) be reconciled with Huygen's highly successful wave theory of light? We shall resolve this question a little later in this chapter. What is to be noted here is that a well-known wave phenomenon seems to have particle nature which needs to be taken into account.

Another example where the particle nature of wave phenomenon needs to be taken into account is vibrations of atomic cores in a crystal. Just as the photon is the name for the particle description of light energy, *phonon* is the particle description for crystal lattice vibrational energy. For example, the interactions between an electron and lattice vibrations can be viewed as collisions between phonons and electrons. This procedure is indeed followed in understanding a number of phenomena in bulk semiconductors and semiconductor device behavior.

The *Compton effect* provides yet another case where the particle nature of waves necessarily needs to be taken into account. It is found that when X rays are scattered by atoms, not only is the original frequency ν_0 of the incident X rays present, but also a new frequency υ_δ *not present* in the original X rays. The presence of υ_δ is quite contrary to the usual observation that when waves meet physical obstacles in their path, their original frequency and wavelength do not change. Compton was able to explain the findings on X rays scattered by atoms. He did this by assuming that the scattering of the X ray could be represented by an elastic collision‡ between an X ray photon and an electron. Relativistic expressions for energy and momentum had to be used because of the high velocity of X rays. Such a calculation gives the following result:

$$\lambda_\delta - \lambda_0 = \frac{h(1\text{-}\cos\delta)}{m_0 c}, \tag{1–5}$$

† The symbol q will always stand for the *magnitude* of the electronic charge. The actual electronic charge is hence $-q$.

‡ An elastic collision is one in which total energy and momentum are both conserved.

where λ_δ, and λ_0 = wavelengths corresponding to ν_δ and ν_0 respectively,

c = velocity of light,

m_0 = rest mass of electron,

δ = angle between the final (i.e. after collision) and original momentum of the X ray photon.

Equation (1–5) not only indicates that there is indeed a new wavelength λ_δ and hence a new frequency ν_δ which is anticipated, but also what its value is.

1–4 WAVE NATURE OF PARTICLES

Frequently important break-throughs in science have followed the asking of a significant question not asked before. DeBroglie asked that if waves could have particle nature, could it be that particles have wave nature?

Assuming that the answer to the above question is "yes," it would be reasonable to study the photon which is the particle description of light which had already been accepted as truly a wave phenomenon. Let us see how the particle nature and the wave nature of light are related one to the other by noting that the photon travels at the velocity of light c. The momentum p of the photon can be expressed in terms of its energy $h\nu$ and velocity c as follows:

$$p = \frac{h\nu}{c}. \tag{1–6}$$

The wavelength λ is related to the frequency ν by the following:

$$\nu = \frac{c}{\lambda}. \tag{1–7}$$

Substituting Eq. (1–7) into Eq. (1–6) we obtain

$$p = \frac{h\nu}{c} = \frac{hc}{c\lambda} = \frac{h}{\lambda}. \tag{1–8}$$

Thus the wavelength of the photon can be related to its momentum simply by

$$\lambda = \frac{h}{p}. \tag{1–9}$$

DeBroglie hypothesized that Eq. (1–9), though derived for the photon, is valid in general. That is, if the momentum of a particle is p, then the wavelength corresponding to its wave nature is given by Eq. (1–9).

Subsequent to the statement of the *DeBroglie* hypothesis experiments were designed by *Davisson* and *Germer* to test this hypothesis. Electrons, and neutrons, which had been accepted as particles, were confronted with a diffraction grating of atomic size, e.g. orderly atomic planes of a crystal. It was observed that, based on the wavelength computed by Eq. (1–9), electrons and neutrons diffracted exactly as would be expected of waves. This was indeed a dramatic experimental verification of the DeBroglie hypothesis.

1-5 WAVE-PARTICLE DUALITY

In the preceding two sections we saw that waves have a particle nature; and by the DeBroglie hypothesis particles have a wave nature. There thus emerges a certain *wave–particle duality*. The following questions demand answers. Is an electron a wave or a particle? When can it be considered a particle, and when a wave? Is light a wave phenomenon or a particle phenomenon?

We shall soon see from the general formulation of quantum theory that wave nature and particle nature are but two sides of the same coin, or two aspects of the same truth. Each description is the compliment of the other.

1-6 DISCRETENESS IN NATURE

We have already seen that in the new theories considered so far, one of the distinguishing characteristics is the presence of discreteness of certain quantities that had been classically considered to be continuously variable. In Planck's blackbody radiation hypothesis, not only did oscillators have to have discrete energies, given by Eq. (1–2), but also the incident radiation had to be considered quantized and hence essentially discrete. Einstein's photo-electric hypothesis necessitated considering the light to consist of photons which are a manifestation of discreteness of incoming light energy.

There are many other instances where discreteness is essential in explaining the experimental observations. The most important of these are the permissible energy levels in an atom. To see why it was necessary to postulate these levels let us look at the details of the reasoning that led up to it.

1. It was known that all matter was made up of tiny particles called *atoms*, and that these were neutral.
2. In a celebrated experiment Rutherford used α particles (doubly charged helium atoms having positive charge) with which to bombard gold leaves. Rutherford showed that he could correctly predict the angle of deflection if it is hypothesized that at the center (nucleus) of the atom there exists a positive charge equal in magnitude to the electronic charge times the atomic number, and that Coulomb's law can be used for this calculation.
3. If the atom had a positive charge and yet had to be neutral, it is necessary to say that there must also be negative charges (equal in magnitude to the positive charge).

   The acceptance of positive and negative charges in the atom raised serious questions about the separate though *stable* arrangement for these charges. They could not continue to exist in a *static* arrangement because Coulomb's law would draw the positive and negative charges together. A dynamic arrangement, for example, of the negative charges moving at constant angular velocity around the positive charges in the center, might perhaps be possible. Even this idea, however, runs into trouble. Though the negative charges are rotating at constant angular velocity, they experience radial acceleration. Classical electromagnetic theory predicts that an accelerating charge radiates energy. This energy loss manifests itself in the progressive reduction of the radius. A rather crude calculation in the case of the hydrogen atom leads to the prediction that it requires approximately 10^{-8} seconds for the negative charge to fall into the nucleus.

4. Thus we conclude that neither the static nor the dynamic arrangement of positive and negative charges is possible in a stable arrangement according to classical theories which are consistent with the experimental findings of Rutherford. It was clear that a new theory was necessary to explain how a stable configuration of positive and negative charges in an atom is possible.

Bohr came up with the theory which explained the stable arrangement of charges in the atom. First, he postulated that only a discrete set of energy levels existed in an atom at which electrons could exist. These energy levels corresponded to the angular momentum of the atom equaling integral multiples of $\hbar$.† Corresponding to each energy level is a distinct orbit in which the electron revolves around the nucleus. Second, so long as an electron remained in a given energy state and hence in a fixed orbit, it did not lose energy even though radial acceleration was experienced by the electron. It should be noticed that this part of the Bohr hypothesis is in direct contradiction to classical electromagnetic field theory as already explained. The atom could radiate or accept energy only when an electron made a sudden jump from higher to lower energy state, or lower to higher energy state respectively. Third, the transitions of an electron going from one of the allowed discrete energy levels to another corresponded to a *single photon being either given up or absorbed.* Thus, if E_2 and E_1 are two of the permissible energy levels for an atom, where $E_2 > E_1$, transition from E_1 to E_2 is related to the photon energy $h\nu$ by the following relationship

$$E_2 - E_1 = h\nu. \tag{1–10}$$

Using the relationship that angular momentum equals $n\hbar$ (where n is an integer) and combining this with equations of total energy and force in dynamic equilibrium, Bohr obtained for the permissible energy levels in hydrogen atom the following equation

$$E_n = -\frac{13.6}{n^2} \quad \text{(electron volts)}, \tag{1–11}$$

where n is an integer.

---

†$\hbar \triangleq \dfrac{h}{2\pi}$ pronounced "h-bar."

Corresponding to these discrete energy levels is the following expression for radius

$$r_n = \frac{n^2\hbar^2}{m_e q^2}\,(\text{cm}), \tag{1–12}$$

where m_e is the electronic mass and n is an integer.

Energy E_1 and radius r_1 in particular are called *ground state energy* and *first Bohr radius* respectively. In Eq. (1–11) the number 13.6 comes from a combination of m_e, q, and $\hbar$.

Figure 1–4 shows a rough plot of the permissible energy levels for hydrogen atoms, and some sets of transitions that an electron could make. For example, the transitions from energies corresponding to $n > 1$ to $n = 1$ (note that the number n identifies the energy through Eq. (1–11) and is called a *quantum number*) correspond to the emission of photons of energy corresponding to the combined results of Eqs. (1–10) and (1–11). Corresponding to these transitions one expects a radiated spectrum which can be calculated and which had been observed before Bohr's theory was stated. The spectrum of these transitions is called the *Lyman series.* Other such series shown in Fig. 1–4 are *Balmer series* and *Paschen series.* Equations (1–10) and (1–11) give a rational explanation for these observed series which were known before, but no reason for their existence was given prior to

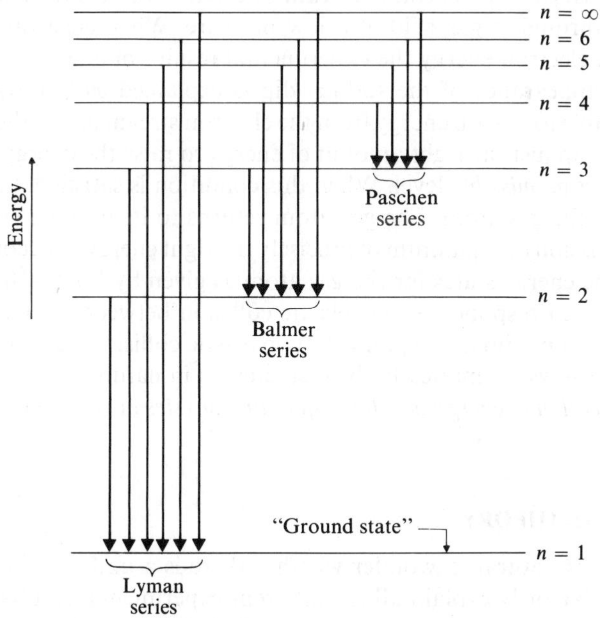

Fig. 1–4 Different permissible energy states according to Bohr theory for hydrogen atom. Several spectral series obtained from sets of transitions between different energy states are also shown.

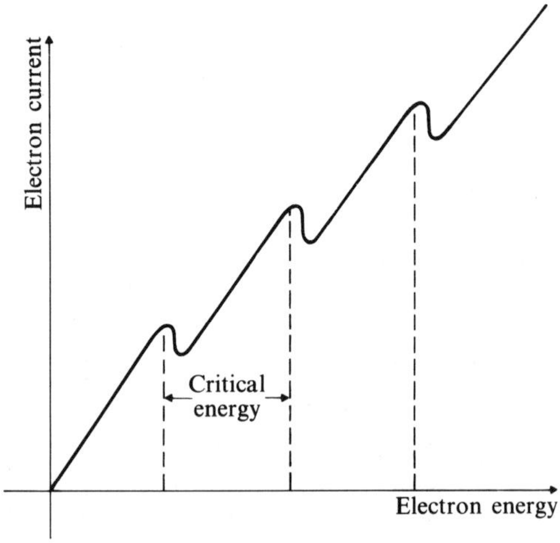

Fig. 1–5 Experimental results of Franck–Hertz experiment.

Bohr's theory. These experimental results also served as a check on Bohr's theory.

Another remarkable piece of evidence of discreteness in nature is exposed by the *Franck–Hertz experiment.* A stream of electrons of controlled energy is made to pass through a gas held at a low pressure. When electron current is plotted against electron energy the experimental result obtained is as shown in Fig. 1–5. The appearance of the sudden dip is explained as follows. The gas atoms are not able to absorb energy from the electron stream unless the electrons are able to give up just the right amount of energy to raise the energy of the gas atom to the *next permissible* level. When this condition is satisfied, the electron loses energy to the gas atom. The gas atom returns to its original state with a consequent emission of a quantum of precisely the right energy to account for the difference in the energy states for the gas atom as given by Eq. (1–10). The first dip in the curve corresponds to one elastic collision between electron and gas atom, and the second dip corresponds to two elastic collisions, etc. This experiment not only shows dramatically the discreteness in nature, but it also shows that *the conservation of energy is valid in quantum physics in just the way as it is in classical theories.*

1–7 A GENERAL THEORY

At this point it is natural to wonder whether there is a unified general theory which can satisfactorily explain all the different experimental results discussed so far. If such a general theory exists, not only is it expected to be a simpler formulation than the sum total of individual theories, but it is also most important that it can be used to predict new results.

It turns out that a general theory is indeed possible. In fact there are two different formulations of the general quantum theory. One is *Heisenberg's formulation*, and the other is *Schrödinger's formulation*. We shall concentrate on the latter as it requires the knowledge of classical differential equations, whereas the former requires more advanced mathematics. There is no loss in generality involved in concentrating on Schrödinger's formulation as it is just as valid as Heisenberg's formulation.

We expect that the general theory must be capable of giving the correct prediction to all major known results. If a single differential equation is to be adequate the DeBroglie hypothesis leads us to the wave equation. A general form of the one-dimensional solution $\psi(x)$ to wave equation has the following form

$$\psi(x,t) = Ae^{-2\pi j(vt - x/\lambda)} + Be^{-2\pi j(vt + x/\lambda)}, \tag{1–13}$$

where $\psi(x,t)$ = wave function,

A, B = arbitrary coefficients,

λ = wavelength,

v = frequency, and

$j = \sqrt{-1}$.

The first term on the right-hand side of Eq. (1–13) is a wave traveling to the right, and the second term is a wave traveling to the left. We must now build into this solution as many of the key features of the known results as possible. Two of these are the energy of the fundamental indivisible packet of energy

$$E = hv \tag{1–14}$$

and the DeBroglie wavelength

$$\lambda = \frac{h}{p}. \tag{1–9}$$

We can solve for v from Eq. (1–14) and substitute this expression together with Eq. (1–9) into Eq. (1–13) to obtain

$$\psi(x,t) = Ae^{-2\pi j[E(t/h) - p(x/h)]} + Be^{-2\pi j[E(t/h) + p(x/h)]}. \tag{1–15}$$

In Eq. (1–14) E stands for *total* energy which can be written as the sum of kinetic and potential energies as follows

$$E = \frac{p^2}{2m} + V, \tag{1–16}$$

where p = momentum of the particle,

m = mass of the particle,

V = potential energy.

The first term on the right side of Eq. (1–16) is kinetic energy E_k as can be easily seen from the following

$$E_k = \frac{1}{2}mu^2 = \frac{1}{2}\frac{m^2u^2}{m} = \frac{1}{2}\frac{(mu)^2}{2m} = \frac{p^2}{2m},$$

where E_k = kinetic energy, u = velocity.

Eq. (1–16) can be solved for p to obtain

$$p = \sqrt{(E - V)2m}. \tag{1–17}$$

Eq. (1–15) can be seen to be the product of two functions $\psi(x)$ and $T(t)$. The latter function is seen to be

$$T(t) = \mathrm{e}^{-2\pi j(Et/h)}. \tag{1–18}$$

This solution is a pure exponential, and implies the following differential equation

$$\frac{dT}{dt} + \left(\frac{2\pi jE}{h}\right)T = 0 \tag{1–19}†$$

or

$$\frac{h}{2\pi jT}\frac{dT}{dt} = -E. \tag{1–20}$$

The function $\psi(x)$ is

$$\psi(x) = A\exp[(2\pi j/h)\sqrt{2m(E - V)}x] + B\exp[-(2\pi j/h)\sqrt{2m(E - V)}x], \tag{1–21}$$

where A, and B are arbitrary coefficients. Equation (1–16) has been utilized in obtaining Eq. (1–21). The solution represented by Eq. (1–21) can be seen to be the solution of the following ordinary differential equation

$$\frac{d^2\psi}{dx^2} - \frac{8\pi^2m}{h^2}(V - E)\psi = 0. \tag{1–22}‡$$

Equation (1–22) can be rearranged as follows

$$\frac{h^2}{8\pi^2m\psi}\frac{d^2\psi}{dx^2} - V = -E. \tag{1–23}$$

To interpret the results obtained so far, we shall find it useful to recall that the wave equation we are looking for is a partial differential equation of the second order. Whenever the solution to a partial differential equation can be expressed as the product of functions each of which is a function of only one variable,

† For brevity $T(t)$ is being replaced by simple T.
‡ For brevity $\psi(x)$ is being replaced by simple ψ.

the original partial differential equation separates into several *ordinary* differential equations. This separation usually involves the use of *separation constants.* Equations (1–20) and (1–23) should be recognized as the two ordinary differential equations which arise from the same partial differential parent equation. The separation constant is seen to be $-E$ (i.e. negative total energy of the system).

The left hand sides of Eqs. (1–20) and (1–23) can be set equal to each other as the right hand sides are equal to the identical quantity $(-E)$. Thus

$$\frac{h^2}{8\pi^2 m\psi}\frac{d^2\psi}{dx^2} - V = \frac{h}{2\pi jT}\frac{dT}{dt}. \tag{1–24}$$

We can multiply both sides of Eq. (1–24) by ψ to obtain

$$\frac{h^2}{8\pi^2 m}\frac{d^2\psi}{dx^2} - V\psi = \frac{h\psi}{2\pi jT}\frac{dT}{dt}. \tag{1–25}$$

Both sides of Eq. (1–25) can be multiplied by T to obtain

$$\frac{h^2 T}{8\pi^2 m}\frac{d^2\psi}{dx^2} - V\psi T = \frac{h\psi}{2\pi j}\cdot\frac{dT}{dt}. \tag{1–26}$$

By definition

$$\Psi(x, t) \triangleq \psi(x)\, T(t). \tag{1–27}†$$

Use can be made of Eq. (1–27) to rewrite Eq. (1–26) as follows

$$\frac{\hbar^2}{2m}\frac{\partial^2\Psi}{\partial x^2} - V\Psi = \frac{\hbar}{j}\frac{\partial\Psi}{\partial t}. \tag{1–28}$$

In the steps leading to Eq. (1–28), we had considered only one position coordinate, namely x. Had we considered the general three-dimensional case, the following general equation would have been obtained

$$\frac{\hbar^2}{2m}\nabla^2\Psi - V\Psi = \frac{\hbar}{j}\frac{\partial\Psi}{\partial t}. \tag{1–29}$$

Here ∇^2 is the *Laplacian operator.* This operator takes the following form in Cartesian coordinates

$$\nabla^2 = \frac{\partial^2}{\partial x^2} + \frac{\partial^2}{\partial y^2} + \frac{\partial^2}{\partial z^2}. \tag{1–30}$$

Equation (1–29) is *Schrödinger's equation.* It is a very general formulation of quantum theory. With its help we can predict all the known results of quantum theory.

† Recall that definitions are denoted by a triangle on top of an equal sign.

We may well wonder what physical meaning $\Psi(x, t)$ could possibly have since, in general, it is a complex quantity as can be seen from Eq. (1–21). Understandably $\Psi(x, t)$ *does not itself have any physical meaning*. It is, however, closely related to the following *real* quantity which can be given physical meaning

$$\Psi\Psi^* \, dv = |\Psi|^2 \, dv, \tag{1-31}$$

where dv = infinitesimal volume element of space,

Ψ^* = the complex conjugate of Ψ.

The quantity of Eq. (1–31) is proportional to the probability of finding the particle whose mass is m (as given in Schrödinger's Eq. 1–29) in the volume dv. If $\psi(x, t)$ is known, the *probability* P of finding the particle in a finite volume Δv is given by

$$P = \int_{\Delta v} \Psi\,\Psi^* \, dv. \tag{1-32}$$

A look at Eqs. (1–29) and (1–32) enables us to put down the following reasonable requirements for physically meaningful problems. (a) $\psi(x, t)$ must be single valued and continuous. (b) First derivatives must be continuous,

$$\text{(c)} \quad \int_{\text{all space}} \Psi\,\Psi^* \, dv = 1. \tag{1-33}$$

The requirement of Ψ and the first derivatives having to be continuous is reasonable in view of the fact that existence of the second derivative, which is an integral part of Eq. (1–29), requires (a) and (b) to be valid. The *uniqueness* or single valuedness requirement of (a) assures prevention of ambiguous predictions. The requirement of (c) is to assure normalizability of $\Psi\,\Psi^*$. Normalizability is essential to the validity of Eq. (1–32) as probability necessarily is a number in the range

$$0 \leqslant P \leqslant 1. \tag{1-34}^\dagger$$

If the probability of an event is one, it means that *it will certainly* happen. If, however, the probability of an event is zero, it implies that *it will certainly not happen*. Equation (1–33) is merely saying that the particle has got to be somewhere in all of space!

The above interpretation of Ψ was historically arrived at reluctantly. The experimental results forced this interpretation on quantum physicists. The solution to a problem is thought to be complete when the wave function Ψ is found. The above interpretation of Ψ makes the following question meaning-

† In some instances normalization is not possible. In such a case the ratio of $|\Psi|^2$ in different regions carries more information. See Section 10–9 for such a case.

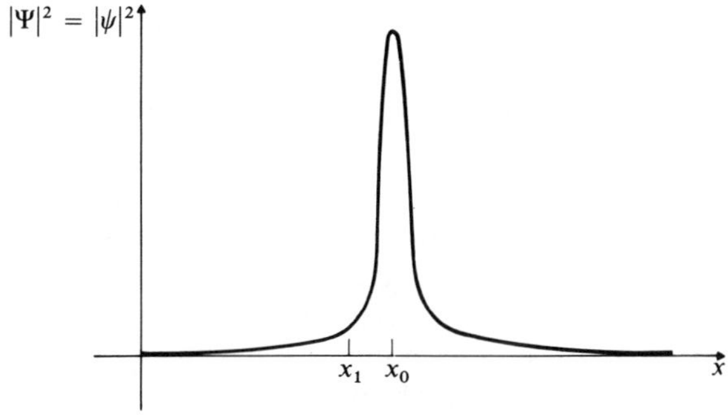

Fig. 1–6 One particular wave function. The particle is most likely to be at x_0; it is not impossible for it to be at x_1.

less "where *exactly* is the particle?" The most that can be said is that the place where $|\psi|^2$ is a maximum is where the particle is most likely to be. Note that it does not mean that the location of the particle could not be at a value other than that of the maximum value of $|\Psi|^2$. For example, Fig. 1–6 shows one possible distribution of $|\Psi|^2$. It is clear that the most likely place for the particle to be is x_0. Note that $|\Psi|^2$ is not zero at x_1, so that it is possible to find the particle at x_1 though the probability of finding it at x_1 is decidedly less than that at x_0. For example, calculation for the radius of an atom is done by computing the maximum of $|\Psi|^2$ for the electron of the outermost shell. The electron is thought to exist in a smeared out region around the maximum of $|\Psi|^2$. In many cases the region where $|\Psi|^2$ is different from zero is quite small. Such a Ψ function is called a *wave packet*. Figure 1–7 shows the picture of a typical wave packet.

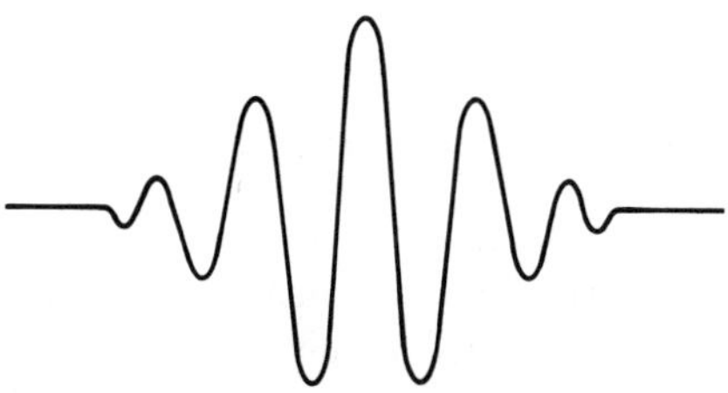

Fig. 1–7 Wave packet.

We are now ready to answer the questions asked in Section 1–5. If the wave packet is relatively compact, i.e. well localized, it may be considered as a particle. If the wave packet is well spread out, i.e. $|\Psi|^2$ has appreciable value over large distances, it may be considered more as a wave. The expressions "large distances" and "localized" are somewhat vague. More meaningful statements can be made when a comparison is made with the things with which the wave packet interacts. As a specific example let us consider the electron. We say that the electron is guided in its movement by its wave packet. In the interaction of electrons with the atoms in a crystal, the spatial magnitude of the wave packet of electrons is fairly extensive compared with the inter atomic distances, and hence the wave nature of electrons dominates in this case. In the Compton effect the electron wave packet is quite small and the interaction with X-ray photons can be considered as collisions between two particles. So we see that the wave particle descriptions are merely approximations of the descriptions in terms of Ψ function which is much more accurate and general.

Before closing this section it is worthwhile noting that because

$$|T(t)| = 1 \tag{1–35}$$

it follows that

$$|\Psi(x, t)|^2 = |\psi(x)|^2. \tag{1–36}$$

As $T(t)$ is given in general by Eq. (1–18) the problem of solving for $\Psi(x, t)$ reduces to that of solving for $\psi(x)$.

So far we have not mentioned the existence of discrete energy levels and discrete frequency spectra emitted from substances from the standpoint of Schrödinger's theory. Equations (1–20) and (1–23) show the separation constant $-E$. It is well known from partial differential equation theory that the *separation constants can have a discrete set of allowed values to be consistent with the boundary conditions of the problem.* As one of the separation constants is indeed the negative of the total energy E, it is clear that it is possible to obtain discrete values of energies (as found in Bohr's theory) from this general formulation. Through Eq. (1–10) the discrete frequency spectra can be obtained from the discrete energy levels.

1-8 HEISENBERG'S UNCERTAINTY PRINCIPLE

The general formulation of quantum theory was obtained in the last section. The final result may seem somewhat disheartening, since the best answer can be only "this is the place the particle is most likely to be." The definitive Newtonian predictions of the trajectory of a particle seem to be replaced by the far less satisfactory probabilistic prediction referred to above. Is nature all that unpredictable? The answer, alas, is yes!

Heisenberg has shown that *certain pairs of variables, or measurable quantities,*† *cannot be simultaneously measured to arbitrary degrees of accuracy.*

---

† Also referred to as *observables* in the literature on quantum mechanics.

Two examples of such sets of variables are the following

$$\Delta x \, \Delta p_x \geqslant (\sim \hbar), \tag{1-37}†$$

$$\Delta E \, \Delta t \geqslant (\sim \hbar), \tag{1-38}†$$

where Δx = uncertainty in the measured value of x,

Δp_x = uncertainty in the measured value of the x-component of momentum,

ΔE = uncertainty in the measured value of energy,

Δt = uncertainty in the measured value of time a given event occurred.

Inequalities analogous to Eq. (1–37) for the y- and z-components of momentum are also valid. Equation (1–37) shows that if a *simultaneous* measurement is made of *both* momentum and position of a particle, that no matter how the measurement is made, the product of the uncertainty in the knowledge of x and p_x has a fundamental lower bound. If we wish to measure the time at which a photon is emitted as well as its energy, Eq. (1–38) shows that the product of the uncertainty in the knowledge of energy and time has a lower bound given by the right hand side of Eq. (1–38). The uncertainty in the knowledge of certain quantities predicted by Eqs. (1–37) and (1–38) is called the *uncertainty principle* (also called *indeterminacy principle*). This uncertainty in the knowledge of certain variables is fundamental, and is *not* something which we may some day improve upon by the use of better and more accurate measuring instruments of the future. The uncertainty principle is a *unique* and *key concept* of quantum physics. It is unlike the use of statistics in classical thermodynamics, where if one could follow, for example, a gas molecule it would be expected to follow Newton's laws; and we use statistics because individual molecules cannot be followed in detail. According to quantum physics the lower bound on the accuracy of making certain sets of observations is not due to inconvenience, nor the dumb state of our present knowledge, but rather due to the impossibility of making such measurements. The reason is that the process of making measurements itself disturbs the system in unpredictable ways because the laws of quantum physics also apply to the *process of measurement itself*.

It is important to note that the uncertainty principle does *not* say that you cannot in principle measure, for example, position exactly. If position were determined *exactly*, that is $\Delta x = 0$, Eq. (1–37) would give

$$\Delta p_x = \frac{\hbar}{\Delta x} = \frac{\hbar}{0} = \infty$$

or there would be infinite uncertainty in the knowledge of momentum p_x. If

†Here the notation $\sim \hbar$ means of the order of $\hbar$. Different derivations of this result give somewhat different answers. These are always within an order of magnitude of each other.

momentum p_x is known exactly, this can only be at the cost of all knowledge of position x because $\Delta x = \infty$.

It may be said that qualitatively correct predictions can be made if theories of classical physics are used subject to the limitations of the uncertainty principle. In at least one specific case, even a quantitatively correct result is obtained by following this idea in the case of the hydrogen atom in obtaining Bohr's first radius.

Uncertainty principle predicts that it takes momentum given by

$$p \cong \frac{\hbar}{\Delta x} \tag{1}$$

to localize the electron within Δx of the nucleus. We can compute the kinetic energy E_k corresponding to this momentum as follows

$$E_k = \frac{p^2}{2m} = \frac{\hbar^2}{2m(\Delta x)^2}. \tag{2}$$

The potential energy due to Coulomb's law is

$$V \cong -\frac{q^2}{\Delta x}. \tag{3}$$

We can now impose the condition that the total energy, i.e. kinetic plus potential energy, for the system be a minimum. Thus

$$E_{\text{total}} = E_k + V \cong \frac{\hbar^2}{2m(\Delta x)^2} - \frac{q^2}{\Delta x} \tag{4}$$

minimum total energy occurs where the derivative of E_{total} with respect to Δx is zero.

$$\frac{\partial E_{\text{total}}}{\partial(\Delta x)} = -\frac{\hbar^2}{m(\Delta x)^3} + \frac{q^2}{(\Delta x)^2} = 0 \tag{5}$$

or

$$-\frac{\hbar^2}{m} + q^2\,\Delta x = 0. \tag{6}$$

finally

$$\Delta x \cong \frac{\hbar^2}{mq^2}. \tag{7}$$

The last result Eq. (7) can be seen, from Eq. (1–12), to be the first Bohr radius.

The fact that Heisenberg's uncertainty principle is true allows us to accept a little more gracefully that the answers of quantum mechanics are probabilistic and not deterministic.

Newton's laws are reinforced by our ordinary everyday experiences. For example it is eminently reasonable to expect that a ball thrown with a force $2F$ will accelerate twice as much as one which is thrown with a force F. This is a reasonable expectation because we have all experienced the event described. It is natural to wonder therefore if there are any human experiences which will make the quantum concepts seem more reasonable. I believe that there are. The human thought processes have many aspects similar to quantum processes. The uncertainty principle for example is readily verified by thought processes. If you concentrate hard on what precise thought you are thinking at any given moment, it will be hard to keep track of the direction it was headed. Furthermore, if you concentrate instead on the precise *direction* your thought is taking at any moment, it will be hard to know the thought itself at that moment. Thus, there seems to exist a limit to the accuracy with which the thought *and* the direction it is going can be determined. (You may find it interesting to try this experiment yourself to see if you agree with the above assertions.) Here, the analogy to the uncertainty principle applied to momentum and position suggests itself strongly. The learning process is often discontinuous. Very often one proceeds from one level of understanding to a distinctly different level. The "aha!" experience that occurs when we develop a sudden insight in the solution to a problem is analogous to the *quantum jump*.

1–9 QUANTUM-MECHANICAL TUNNELING

One of the new predictions based on the general formulation of quantum theory of Section 1–7 is quantum mechanical *tunneling*. The solution to this problem is unobtainable through classical mechanics.

An example of the problem we wish to consider is portrayed in Fig. 1–8. This figure shows a bowling alley with pins on the left, and the railing responsible for transporting the ball back to the bowler. The normal practice after a ball has been bowled is the return of the ball to the bowler by putting the ball at point A, which is higher than point B, and just letting it go by virtue of gravity. In Fig. 1–8a it is clear that even with zero initial velocity of the ball, it indeed will be collected by the bowler at point B. If the path along which the ball slides is truly frictionless, the ball could be collected at B even if point A is as high as B. Usually A is made higher than B to take account of energy loss due

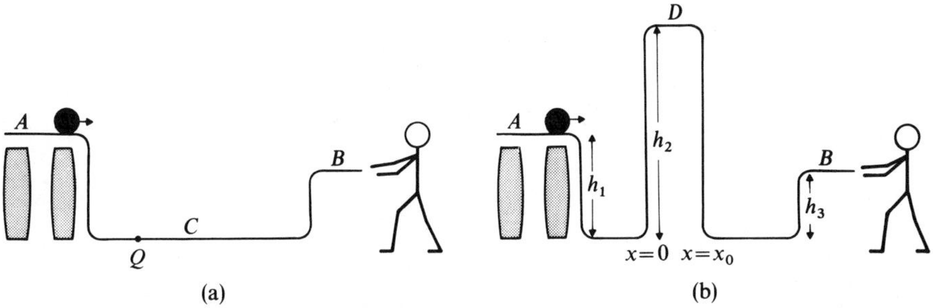

Fig. 1–8 Bowling alley problem.

to friction. When the ball is at point A, its energy is all in the form of potential energy. When it is rolling along towards B, say at Q, the ball has its entire energy in the form of kinetic energy. When the ball arrives at point B, the ball can have both potential and kinetic energy because B is lower in height than A.

Consider now what the prediction would be if the situation is as shown in Fig. 1–8b. Here a potential energy barrier is erected between points A and B at D. If point D is higher than A, as shown in Fig. 1–8b, and if the ball starts out from point A with zero initial velocity, most people would say that it is impossible for the bowler to collect the ball at point B. This conclusion follows from the fact that the ball does not have enough energy to classically overcome the gravitational potential barrier height h_2. The prediction of quantum mechanics to this situation is a surprising "yes, it is *possible* for the bowler to collect the ball!"

To see how the above answer is arrived at, let us consider the somewhat idealized potential energy diagram given in Fig. 1–9. The square potential barrier in Fig. 1–9a represents a barrier similar to that in Fig. 1–8b at point D. As stated earlier, we need only to solve for $\psi(x)$, because the time solution $T(t)$ has already been found, and if we are interested only in the probability of finding the particle beyond the barrier, then all we need to know is $\psi(x)$. The solution

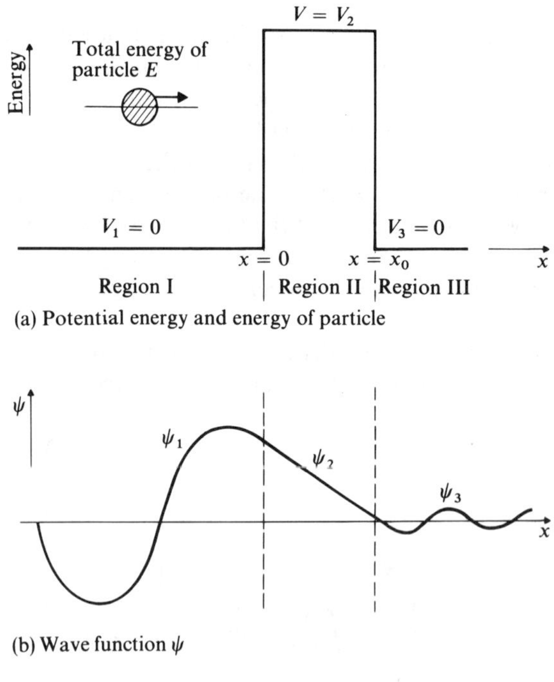

Fig. 1–9 Quantum-Mechanical tunneling.

given by Eq. (1–21) is valid in each of three regions I, II, and III. From this equation we can immediately write down the solutions valid in each region as follows. Note that, as indicated in Fig. 1–9a, E stands for the total energy of the particle as it approaches the potential barrier from the left in region I. Equation (1–21) takes the following forms in the three different regions.

Region I $\quad V = V_1 = 0,$ (1–39)

$$\psi_1(x) = F\sin\left(\frac{\sqrt{2mE}}{\hbar}x\right) + G\cos\left(\frac{\sqrt{2mE}}{\hbar}x\right). \tag{1–40}$$

Region II $\quad V = V_2 > E,$ (1–41)

$$\psi_2(x) = C\exp\left(-\frac{\sqrt{2m(V_2 - E)}\,x}{\hbar}\right) + D\exp\left(+\frac{\sqrt{2m(V_2 - E)}\,x}{\hbar}\right). \tag{1–42}$$

Region III $\quad V = V_3 = 0,$ (1–43)

$$\psi_3(x) = H\sin\left(\frac{\sqrt{2mE}}{\hbar}x\right) + L\cos\left(\frac{\sqrt{2mE}}{\hbar}x\right). \tag{1–44}$$

Coefficients F, G, C, D, H, and L are to be evaluated from boundary conditions. To begin with we note that for ψ_2, as we have not put any limits on the magnitude of x_0, the second term in Eq. (1–42) should be zero, otherwise ψ_2 could grow without bound and violate the reasonable condition given by Eq. (1–33). Thus

$$D = 0. \tag{1–45†}$$

Next we demand continuity of ψ_1 and ψ_2 at $x = 0$. Thus

$$\psi_1 = \psi_2, \text{ or } G = C. \tag{1–46}$$

For compactness we define

$$a \triangleq \frac{1}{\hbar}\sqrt{2mE} \tag{1–47}$$

$$b \triangleq \frac{1}{\hbar}\sqrt{2m(V_2 - E)} \tag{1–48}$$

We require that $d\psi/dx$ be continuous across both $x = 0$ and $x = x_0$. The former gives

$$Fa = -bC. \tag{1–49}$$

The requirements that $d\psi_2/dx = d\psi_3/dx$ at $x = x_0$ and $\psi_2 = \psi_3$, also at $x = x_0$, give, after algebraic manipulations, the following

†Strictly speaking this is a valid conclusion if $(x_0/\hbar)\sqrt{2m(V_2 - E)}\,x \gg 1$. We have for simplicity assumed this inequality to be valid here.

$$H = C\exp\left(-bx_0\right)\left[\sin ax_0 - \frac{b}{a}\cos ax_0\right] \qquad (1\text{–}50)$$

$$L = C\exp\left(-bx_0\right)\left[\frac{b}{a}\sin ax_0 + \cos ax_0\right]. \qquad (1\text{–}51)$$

Finally we can summarize the analytical results as follows,

$$\psi_1 = -\frac{b}{a}C\sin ax + C\cos ax \qquad (1\text{–}52)$$

$$\psi_2 = C\exp^{-bx} \qquad (1\text{–}53)$$

$$\psi_3 = C\exp^{-bx_0}\left[\left(\sin ax_0 - \frac{b}{a}\cos ax_0\right)\sin ax + \right.$$

$$\left. + \left(\frac{b}{a}\sin ax_0 + \cos ax_0\right)\cos ax\right]. \qquad (1\text{–}54)$$

It is important to note that ψ_3 is *not zero*, and neither is $\psi_3\,\psi_3^*$ zero. This tells us that the probability of finding the particle in region III is not zero. The probability of finding the particle in region III is significantly less than that in region I. The significant factor that makes ψ_3 smaller than ψ_1 is

$$\exp(-bx_0) = \exp(-\frac{1}{\hbar}\sqrt{2m(V_2 - E)}\,x_0). \qquad (1\text{–}55)$$

The relationship given by Eq. (1–55) tells us that the wider the potential barrier x_0, the lower the probability of finding the particle in region III. Also the more the energy difference $(V_2 - E)$, the lower is the probability of finding the particle in region III. Even though the probability of finding the particle in region III is quite small per particle, if a sufficient number of particles are moving towards the barrier, an appreciable number can, in fact, make it beyond the barrier. For example, if the probability per particle is only 0.01, a billion particles moving towards the barrier can result in 10 million particles making it across the barrier.

The process by which the particle gets across is called *tunneling*. The terminology betrays the fact that classical thinking was very much at the back of the minds of the people who thought up the name. The particle was probably visualized going through an *imaginary* tunnel across the barrier. Of course, no such real tunnel exists!

The prediction of tunneling is a completely new result. No such prediction is possible from classical theories. Note that quantum mechanics does not attempt to answer "*how* the particle gets across the barrier." It merely asserts that the probability of finding it on the other side of the barrier is not zero. This transport process is real and is important. For example, the most important feature, namely the negative resistance region in the current–voltage characteristics of a tunnel diode which will be discussed in Chapter 10, is due to the tunneling process.

It is instructive to note that the total solution to the tunneling problem given in Eqs. (1–52), (1–53), and (1–54) contains an arbitrary coefficient C which remains unevaluated. In the more ordinary problem such a constant would be evaluated by the normalization condition of Eq. (1–33). In the case at hand, normalization is impossible because in regions I and III the $\psi(x)$ function is periodic and the improper integral of Eq. (1–33) does not converge! The important question which is significant here is "what is the relative probability of the particle being in region III as compared to it being in region I?" This question can be answered by merely calculating the ratio $(|\psi_3|^2/|\psi_1|^2)$ which is independent of C! It is useful, in this context, to think of the wave nature of the particle in question. We may then say that ψ_1 is the sum total of incident plus reflected wave, whereas ψ_3 is the transmitted wave.

The tunnel effect is important not only in explaining the characteristics of tunnel diodes, but also high field emission phenomenon and even the transport of protons in a DNA molecule (a very important constituent of every living cell nucleus).

1-10 THE CORRESPONDENCE PRINCIPLE

If quantum theory is more general than classical Newtonian theory, it seems reasonable to expect that in those cases where we know the latter theory gives correct results, the former should be able to give proper explanation for the observed results also.

Consider, for example, that before the acceptance of quantum theory nobody reported detecting quantum of energy, e.g. photons. The explanation lies in the fact that the quantum was too small to be detected. Practical frequencies for which classical theories were experimentally found to be valid were usually less than 10^{11} cycles per second. The most energetic quantum consistent with Eq. (1–10) is about 10^{-15} erg. This is too small to be detected by conventional means. Newton's laws, which are purely deterministic, seem to give varied results over a wide range of experimental conditions. How can we reconcile this with the fact that quantum theory gives only probabilistic predictions? To answer this question, consider the *law of large numbers.*

The law of large numbers says that, in a sequence of independent trials of a process governed by probability, the percentage fluctuations from exact correspondence with the probabilities of the process approach zero as the sequence of independent trials becomes sufficiently long. We know, for instance, that the probability of a tossed coin landing "heads" is 50 per cent. If we toss the coin only once, it is just as likely not to be heads, and therefore we do not have confidence in predicting "heads." If, however, we toss the coin 10^{10} times, we know that "heads" will appear on very nearly half the tosses and that the percentage fluctuation from one-half will be very small indeed. This is exactly the implication of the law of large numbers. It is thus not hard to see why the deterministic and probabilistic predictions approach each other as the number of quanta involved in a process becomes large.

The above discussion suggests the following more general formulation. Bohr stated the correspondence principle which says that the laws of quantum theory lead to the classical results as the average when large numbers of quanta are involved. Most results of classical theory which are valid can be explained through the correspondence principle.

1-11 PAULI EXCLUSION PRINCIPLE

When we considered the one-dimensional Schrödinger's equation, we discovered that the separation constant, which is involved in the solution of this equation, is associated with energy. If instead of the one-dimensional equation, we had attempted to solve a three-dimensional problem, we would have needed two additional separation constants. These separation constants turn out to have only certain discrete allowed values as applied to electrons in an atom. Corresponding to the three separation constants we have three quantum numbers n, ℓ, and m. Later a fourth degree of freedom for electrons in an atom was discovered. This leads to an additional quantum number *spin* designated by s. These four quantum numbers can have the following permissible values

$$\begin{aligned} & & n &= 1, 2, 3, \ldots \\ &\text{for each } n: & \ell &= 0, 1, 2, 3, \ldots (n-1) \\ &\text{for each } \ell: & m &= 0, \pm 1, \pm 2, \pm 3, \ldots \pm \ell \\ &\text{for each } n, \ell, m: & s &= \pm 1/2. \end{aligned} \tag{1–56}$$

The derivation of Eq. (1–56) is too involved to be presented here.

What should be noticed is that the specification of the four quantum numbers essentially specifies the solution to Schrödinger's equation, just as specifying n for the hydrogen atom specifies both the energy and the orbit. A very important result of modern quantum theory is that: "No two identical particles which are described by *anti-symmetric* wave functions ψ and which are sufficiently close to each other so that there is non-negligible interaction can be described by the same quantum state (i.e. designated by the same four quantum numbers)." This is a more generally stated version of the *Pauli exclusion principle* which was first stated for electrons as a separate law. Some interpretations of this result will be found useful here. *Identical particles are those which are exactly similar, so that no inherent property of theirs can be used to distinguish between them.* Consider the situation portrayed in Fig. 1–10. Here two identical balls are shown headed for collision. On the right the situation is portrayed after collision. In Fig. 1-10a the situation is portrayed as it would be viewed classically. Here, according to Newtonian physics, if we know the initial position and velocity of each particle, we can predict the entire trajectory before and after the collision with certainty. Hence, we can be sure that after the collision the top ball is number 2 and the bottom ball is number 1. Thus, here the laws of classical mechanics permit us to distinguish between the two balls. Figure 1–10b shows how the nondeterministic quantum mechanics views the same event after collision. Here the top ball could be number 1 *or* number 2; one cannot be certain. The same is true for the ball at the bottom. This is primarily due to the probabilistic nature of quantum theory. From this rather simple example it is not hard to see that special statistics must be used when identical particles are being considered. The statistics in this case is expected to be different from that which is necessary for nonidentical particles.

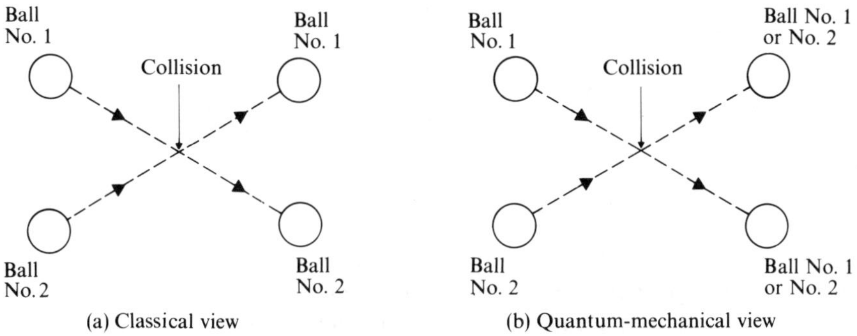

Fig. 1–10 Collision between two identical particles.

There are two different statistics for two generally different kinds of identical particles: those having symmetric wave function ψ, and those having anti-symmetric wave function ψ. The latter must obey the Pauli exclusion principle. *Fermi–Dirac statistics* apply to these latter class of particles called *Fermions*. Electrons, protons, neutrons, and mu-mesons are the only ones known so far that belong to this class. Particles with symmetric wavefunctions do not have to obey the Pauli exclusion principle. They follow the *Bose–Einstein statistics*, and are called *Bosons*. Examples of these are photons, pi-mesons.

As our attention will be focused mainly on electrons and holes which do obey the Pauli exclusion principle, it is worthwhile noting that electrons will not necessarily occupy the ground level energy, which is the lowest possible permissible energy, even at absolute zero degrees. To see this clearly let us see how many quantum states are possible for electrons corresponding to the ground state, that is $n = 1$. From the selection rules given in Eq. (1–56) we obtain

$$n = 1, \ell = 0, m = 0, s = +\tfrac{1}{2}$$

$$\text{and} \qquad n = 1, \ell = 0, m = 0, s = -\tfrac{1}{2} \qquad \text{(two states.)}$$

Thus we see that in the ground state in an atom there can be at most two states, i.e. at most two electrons can be simultaneously in the ground state. It is interesting to note that this conclusion is consistent with the fact that the innermost shell in an atom can at most have two electrons. We also see that the first two elements in the periodic table, hydrogen and helium, have one and two electrons respectively. Let us see how many quantum states are possible corresponding to $n = 2$. For $n = 2$, ℓ can be 0 or 1. For each of these ℓ values, m can have 0, or 0 and ± 1 respectively. Thus, we obtain the following eight combinations

	n	ℓ	m	s
1	2	0	0	$+\frac{1}{2}$
2	2	0	0	$-\frac{1}{2}$
3	2	1	0	$+\frac{1}{2}$
4	2	1	0	$-\frac{1}{2}$
5	2	1	1	$+\frac{1}{2}$
6	2	1	1	$-\frac{1}{2}$
7	2	1	-1	$+\frac{1}{2}$
8	2	1	-1	$-\frac{1}{2}$

Thus for $n = 2$ there can be at most eight quantum states. Thus in an atom there can be at most eight electrons in the second shell ($n = 2$). This conclusion, too, is correct. Indeed, Pauli was able to explain the Periodic table, then known, on the basis of the exclusion principle. This examination of the consequences of the Pauli exclusion principle tells us that *even at absolute zero temperature not all electrons can be in the ground state of an atom.* This is one of the explanations for why different elements have such drastically different properties.

1-12 SUMMARY

We have looked at some key historical experimental results which classical theories could not explain, and which led ultimately to the formulation of general quantum theory. This theory reconciles the wave-particle duality, and explains why one finds discreteness of energy levels in nature. The Heisenberg uncertainty principle and the probabilistic description of nature are both an integral part of this new theory.

Discreteness of energy levels in an atom and the Pauli exclusion principle enable us to look at a key concept in solid state physics, the energy band picture, which is presented in the next chapter. We shall refer to the present chapter in the rest of the book from time to time for some key ideas and a helping hand in obtaining new insights.

PROBLEMS

1–1 Monochromatic light of $\lambda = 300$ Å results in a stopping voltage of 6 V. Determine the work function of the emitter of the photocell.

1–2 Monochromatic light of fixed intensity and of $\lambda = 100$ Å is incident on the photosensitive cathode of a photocell. The resulting current is 50 μA. The stopping voltage is 4 V. The light intensity is altered to make the resulting current 400 μA. What is the stopping voltage under these circumstances?

1–3 Obtain an approximate expression for R_λ if the following magnitude relationship is valid

$$\nu \ll \frac{kT}{h}$$

1–4 What is the wavelength of a golf ball travelling in a straight line at 120 miles per hour? Assume the ball weighs 5 oz.

1–5 An electron gun delivers electrons of energy E from the right of the potential barrier of Fig. 1–9 having a height of 1 eV. The tunneling current collected to the left of the barrier is 10 mA. Estimate the resulting current if the barrier height is lowered by 0.2 eV.

1–6 Calculate the wavelength corresponding to the two lines having the longest wavelength in the Balmer series.

1–7 Calculate the wavelength corresponding to the two lines having the longest wavelength in the Lyman series.

1–8 Calculate the wavelength corresponding to the two lines having the longest wavelength in the Paschen series.

1–9 An electron of energy E_e is in a one-dimensional solid in which the potential energy is V_i. The potential energy outside the solid is V_2. Given that

$$V_2 > E_e > V_i$$

calculate the probability of finding the electron a distance x outside the body. The origin may be taken to be at the edge of the body. Interpret the result you get.

1–10 A light source emits photons of frequency f_1 whose value is known within $\pm 10^{-10}$ Hz. What is the most accurate statement that can be made about the time at which at given photon is emitted?

1–11 A photon is emitted by a cathode in a time interval known to within 10^{-10} sec. How accurately may we ideally hope to know the frequency ν of the emitted photon?

1–12 Describe the characteristics of solids in a world where the Pauli exclusion principle is not valid.

REFERENCES

Bohm, D., *Quantum Theory,* Chapters 1–10, Prentice-Hall, Englewood Cliffs, N.J. (1951).

Leighton, R. E., *Principles of Modern Physics,* Chapters 2, 4, 7, 8, McGraw-Hill Book Co., New York (1959).

McKelvey, John, P., *Solid State and Semiconductor Physics,* Chapters 1, 4, Harper & Row, New York (1966).

Richtmyer, F. H., Kennard, F. H., Lauritsen, T., *Introduction to Modern Physics,* 5th edition, Chapters 3–6, McGraw-Hill Book Co., New York (1955).

Sproull, R. L., *Modern Physics,* Chapters 3–6, 8, 12, John Wiley & Sons, New York (1956).

CHAPTER 2

ENERGY BANDS AND BASIC SEMICONDUCTOR PROCESSES

2-1 INTRODUCTION

In the preceding chapter we took a brief look at quantum theory as it is the basis of many important concepts and some processes of importance in semiconductor devices. In this chapter we shall study one of the key concepts important to the understanding of solid state and semiconductor phenomena known as *Energy bands*, which will allow us to correctly predict the behavior of solids under different circumstances. For example, band theory permits us to classify solids as conductors, semiconductors and insulators and to study the conduction process in solids in some detail. It also explains why there are some processes which are direction dependent in solids. The band theory is essential in understanding many light sensitive as well as light emission-absorption phenomena in solids and is important in explaining the different species of charge carriers in solids.

In this chapter we shall also consider many processes of vital importance to the understanding of semiconductor devices. Processes such as diffusion, recombination, generation, velocity saturation, and many others are considered in some detail. Chapters 1 and 2 are basic to the whole book and form the foundation on which the remaining chapters are based. We shall find many occasions in the rest of the book to refer back to these two chapters.

2-2 ONE-DIMENSIONAL ENERGY DIAGRAMS

We saw in Chapter 1 that electrons could occupy only discrete energy levels in an atom. A crystal is an orderly array of atoms, where a large number of atoms exist in a tightly packed arrangement. It turns out that in a crystal certain bands of allowed energies separated by energy gaps, where no electrons can exist, form the counterpart of the discrete energy levels in a single atom. Figure 2–1a shows four particular energy levels of a given atom. We now wish to consider a hypothetical experiment to see what happens to the energy diagram as we visualize assembling the crystal.

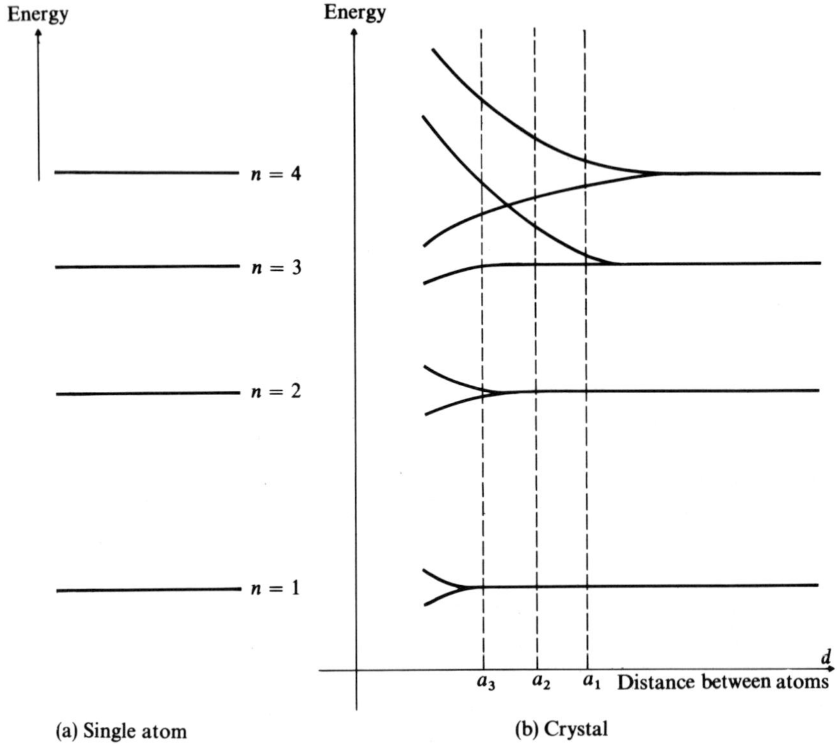

Fig. 2–1 Energy diagram for an atom and for a crystal.

Consider that first we have just two atoms located very far apart. The energy diagram for the electronic states, i.e. energy levels at which electrons can exist, looks like the extreme right-hand side of Fig. 2–1b. As the atoms are very far apart, the energy line for each atom is identical to the other as the Pauli exclusion principle is not involved. Thus we have two lines right on top of each other for large distances of separation d between the two. As the distance is diminished between the two, however, the Pauli exclusion principle will have to be satisfied, and the two lines now become distinct. This is sometimes called *energy splitting*. If instead of bringing two atoms close together, 10^{22} atoms are brought close together, each energy level splits into 10^{22} separate energy levels. It turns out that the separation between the outermost split levels remains about the same whether two atoms are involved or 10^{22}. For practical purposes these split energy levels become *bands of energies* separated by *band gaps* in which no energies exist which can be occupied by electrons. This means that the width of each band does not depend on the number of atoms.

The experiment of Fig. 2–1b is hypothetical. In an actual crystal, which is an orderly and compact arrangement of atoms, at a given temperature there is a fixed distance between adjoining atoms called the *lattice constant a.* Figure 2–1b shows the distances for three different crystals. For each such solid we can draw a one-dimensional *energy band* picture as in Fig. 2–2. Such a diagram gives a lot of information about the electrical properties of solids.

Let us see what we can learn about the conductivity of the solids whose energy band diagrams are shown in Fig. 2–2. Consider the solid that has a wide band gap energy E_g. *The valence band here is the highest completely full band at absolute zero degrees.* In this case and also in the narrow band gap solids, such band arises from the electronic states corresponding to the *valence electrons* of the host atoms. The conduction band, which is the next higher band to the valence band, is completely empty at absolute zero degrees temperature.

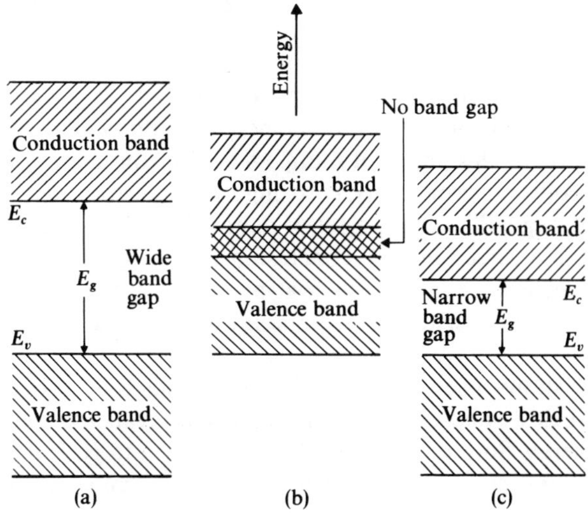

Fig. 2–2 One-dimensional energy band diagram. (Note there is no significance to abscissa.) (a) Insulator, lattice constant a_1 in Fig. 2–1b. (b) Conductor lattice constant a_3 in Fig. 2–1b. (c) Semiconductor lattice constant a_2 in Fig. 2–1b.

We observe that *neither a completely empty nor a completely full band can lead to electrical conduction.* It is easy to understand how an empty band cannot contribute because there are no charged particles to contribute to current flow. A completely full band also cannot lead to conduction because of the Pauli exclusion principle. To see why this is the case we note that we can talk about a *completely full* band due to the Pauli exclusion principle which essentially limits the electron density a band can have. This is limited by the number of *different* quantum states which can belong to the band. In a completely full band the electrons *cannot* respond to an externally applied electric field. If an electron were to change its state, it can do so only as another electron vacates its quantum state and there can perhaps be a simultaneous exchange of quantum states. If this happens there can be no net change in energy of any electron. Thus the external field cannot give up part of its energy to an electron in a completely full band.† *Thus a completely full band cannot contribute to conduction.* As the wide band gap solid has only completely full and completely empty bands, at absolute zero degrees, no conduction can occur. At higher temperatures not many electrons will make it to the conduction band because of the wide band gap. At temperatures at which significant numbers of electrons leave the valence band for the conduction band, conduction does not occur for the following reason. Electric conduction is caused by *the motion of charged particles along a preferred direction*, i.e. the direction of the electric field. In a crystal, the motion of electrons is a succession of collisions as shown in Fig. 2–3a. The collisions could be between electrons and electrons, or electrons and atomic cores, or electrons and vacancies in crystaline structure, etc. The most important of these is the collision between electrons and vibrating atomic cores.‡ The probability of collisions with these vibrating atomic cores goes up with temperature. Each time an electron suffers collision it loses its memory and starts all over again. Thus at temperatures at which significant numbers of electrons arrive in the conduction band, the collision probability is so high that electron motion along the field direction is negligible. Thus, a lot of agitated motion takes place but not along any preferred direction. The conclusion is that solids with a wide band gap are *insulators*, and behave as such at all temperatures. A typical though expensive example of an insulator is diamond which has a band gap of about 7 eV.

Figure 2–2b shows a solid with overlapping bands which are only *partially* full. Hence in such a solid electrons find plenty of available electronic states close to their energy. So it is quite easy for an electron to go from an occupied state to a nearby unoccupied state without violating the Pauli exclusion principle. We expect then that such a solid will easily conduct electricity. Such a solid is a *conductor*. As the temperature is raised the increasing collision probability will again decrease conductivity.

Figure 2–2c shows the energy band diagram of a solid which has a narrow band gap. Such a solid turns out to be a *semiconductor*. Here too, all bands up to and including the valence band are completely full at absolute zero degrees temperature. All bands including and above the conduction band are empty at absolute zero temperature. Thus, for exactly the same reason as the insulator, it is expected that semiconductors will not conduct at all at absolute zero temperature. As the temperature is raised above absolute zero, electrons will leave the valence band and go into the conduction band. At moderate temperatures at which this happens the collision probability is still not very high. So we expect

† We are here excluding a situation where the external electric field supplies so much energy to an electron that it makes a transition to the next band that has available electronic states.

‡ According to the quantum theory the vibrating atomic cores have a particle nature. They are called *phonons*. It is appropriate to refer to the collisions between electrons and phonons.

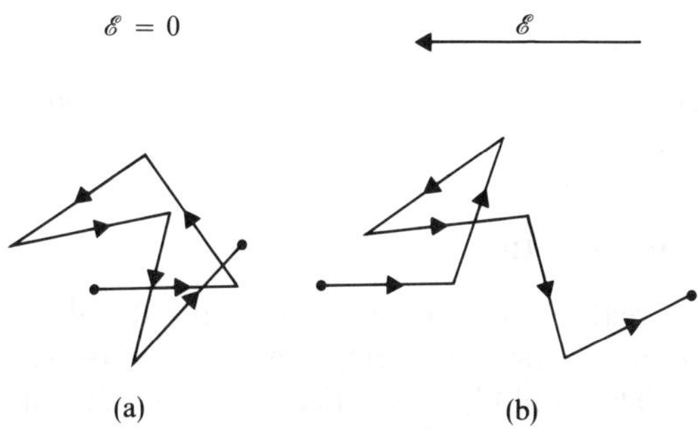

Fig. 2–3 Motion of electron in a crystal. Change of direction occurs after each collision.

semiconductors to conduct electricity at moderate temperatures, e.g. room temperature. Unlike the conductor, the conduction here involves two different independent kinds of motion; one in the conduction band, and the other in the valence band.

When an electron leaves the valence band and goes into the conduction band, the electron in the conduction band acts as a relatively "free" electron. By this we mean that the wave packet is not well localized in space. Such an electron, which is free to wander in the conduction band, contributes to electrical conduction. It turns out that in addition to the electron in the conduction band, the vacancy left behind in the valence band, called a *hole*, also can contribute to conduction. This is illustrated by Fig. 2–4a which shows schematically a parking garage with two floors of space. The upper floor is analogous to the conduction band, and the lower floor is analogous to valence band. Figure 2–4a corresponds to the situation in a semiconductor (and in an insulator) at absolute zero temperature where the valence band is completely full, and the conduction band is quite empty. The application of an electric field to a semiconductor is analogous to tilting the parking garage where gravity takes the role of the electric field. Figure 2–4b shows the same garage as Fig. 2–4a tilted. It is easy to see that no movement of cars can take place on both floors. If,

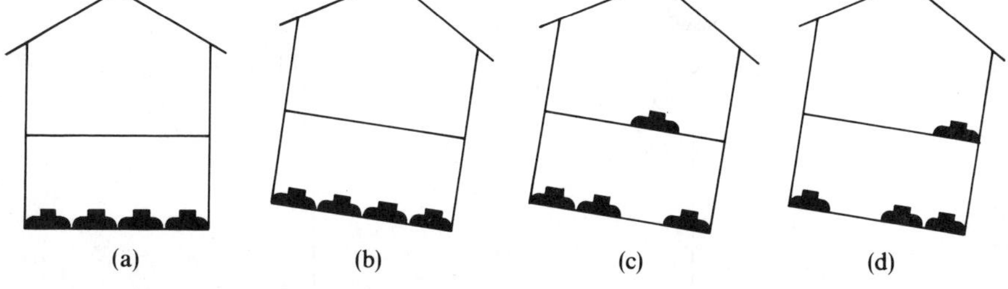

Fig. 2–4 Parking garage analogie to flow of electrons and holes.

however, one of the cars is moved from the first to the second floor and the garage is tilted as in Fig. 2–4c, and Fig. 2–4d, independent motion is possible for cars on both floors. Note that if the focus is shifted from the cars to the vacancy on the first floor the vacancy also seems to move!

2–3 THE CONCEPT OF A HOLE

It is important to realize that a hole is *not* simply the absence of an electron. To make this a little more concrete, consider the two-dimensional lattice shown in Fig. 2–5. This could be the lattice of any atom belonging to the fourth column of the periodic table which implies that each atom has four valence electrons. In Fig. 2–5 it can be clearly seen that each two neighboring atoms share their electrons mutually. For example, the central atom, which could be germanium or silicon (both of which belong to the fourth column of the periodic table), shares its four valence electrons with one from each of four nearest neighboring atoms. This makes a total of eight electrons associated with the central atom. In the preceding chapter we saw that eight electrons do correspond to a stable situation. In that situation the quantum number n was equal to two. Figure 2–5 shows the eight electrons consisting of four pairs. Each pair is held together by a *covalence bond.* This is a fairly strong bond and it takes energy ($\geqslant E_g$) to break it. When all covalence bonds are complete, the valence band is said to be completely full. If one of the electrons in this situation is missing from such a covalence bond, this particular vacancy is called a *hole.* It is important to note that a hole is not simply the absence of an electron. For example, the absence of an electron in free space is *not* a hole. Neither is the absence of an

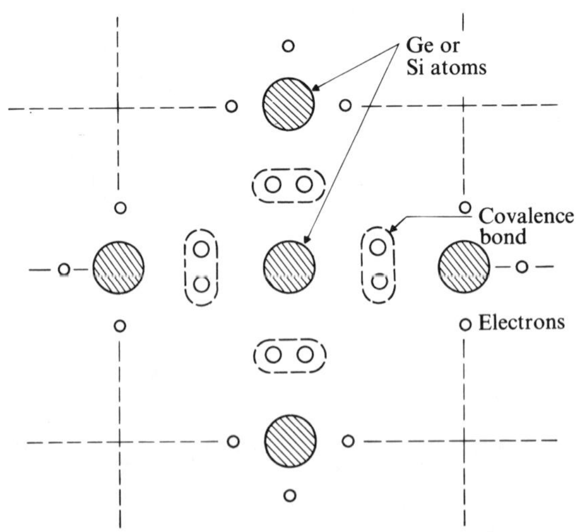

Fig. 2–5 Two-dimensional germanium or silicon lattice.

electron in the conduction band a hole. *It is the absence of an electron from valence band that is considered a hole.* The motion of a hole is a more convenient way to refer to the motion of electrons in the valence band. It turns out that it is actually more than convenient; it is indeed essential to postulate the existence of a particle called a *hole*. The postulation of the hole is important to the explanation of a whole range of experimental results, such as both positive and negative Hall coefficients†.

The hole can be shown to be a particle which has positive charge and positive *effective mass*. (We shall discuss the concept of effective mass in detail in Section 2–5.) First let us see how the hole can be shown to have a positive charge. Consider the valence band at absolute zero degrees. This band is completely full at this temperature. We have already seen that a completely full band cannot contribute to conduction.

In general we can write current density as the product of charge and its velocity. Thus for electrons in the valence band, we may write

$$J = -q \sum_{i=1}^{n} v_i, \tag{2–1}$$

where v_i = the velocity of the ith electron. In a completely full band current must be zero. Hence

$$J = 0 = -q \sum_{i=1}^{n} v_i = -q \sum_{\substack{i=1 \\ i \neq j}}^{n} v_i - qv_j. \tag{2–2}$$

In Eq. (2–2) we have written the jth term separately. If we designate by J^* the current that would result in the valence band *if only* j*th electron were* missing

$$J^* = -q \sum_{\substack{i=1 \\ i \neq j}}^{n} v_i. \tag{2–3}‡$$

By using Eq. (2–2) we have

$$J^* = +qv_j. \tag{2–4}$$

Equation (2–4) shows that *the hole can be characterized as having a positive charge.*

Now let us see how the hole can be thought to have also a positive effective mass. Again let us consider a completely full valence band. This time let us look at the time rate of change of J in this band.

$$\frac{dJ}{dt} = -q \sum_{i=1}^{n} \frac{dv_i}{dt} = -q \sum_{\substack{i=1 \\ i \neq j}}^{n} \frac{dv_i}{dt} - q\frac{dv_j}{dt}. \tag{2–5}$$

† See Section 2–17 for a detailed discussion of the Hall effect and the Hall coefficients.
‡ The designation $i \neq j$ on the summation means the jth term is left out.

But

$$\frac{dJ}{dt} = 0. \tag{2–6}$$

Equation (2–6) is true for the same reason as Eq. (2–2). Thus, if we designate by J^* the current which results when the jth electron is missing from valence band, we can say

$$\frac{dJ^*}{dt} = -q \sum_{\substack{i=1 \\ i \neq j}}^{n} \frac{dv_i}{dt} = q \frac{dv_j}{dt}. \tag{2–7}$$

We can use Newton's second law to evaluate dv/dt.

$$m_j^* \frac{dv_j}{dt} = F = -q\mathscr{E}, \tag{2–8}†$$

where F = force on electron,

$\mathscr{E}$ = electric field,

or

$$\frac{dv_j}{dt} = -\frac{q\mathscr{E}}{m_j^*}. \tag{2–9}$$

Substituting Eq. (2–9) into Eq. (2–7) gives

$$\frac{dJ^*}{dt} = -\frac{q^2\mathscr{E}}{m_j^*}. \tag{2–10}$$

In Eq. (2–10) m_j^* stands for the effective mass of the jth electron in the valence band. In the next section we shall show that the effective mass of electrons *in at least the top half of the valence band* is itself negative. This fact enables us to rewrite Eq. (2–10) as

$$\frac{dJ^*}{dt} = \frac{q^2\mathscr{E}}{|m_j^*|}. \tag{2–11}$$

Equation (2–11) shows that the *effective mass corresponding to the hole can be considered to have a positive effective mass.*

2–4 ENERGY MOMENTUM RELATIONSHIP

We have so far considered only the one-dimensional energy band picture. In this section we shall explore the relationship between energy and momentum in a crystal. This relationship gives much valuable information about the properties of the crystal.

We shall consider the *intrinsic* crystal (i.e. a crystal which is free of mechanical,

---

† Superscript star on m shall designate *effective mass.*

chemical, or other defects). The model for obtaining energy bands used in Section 2–2 is called the *tight-binding model.* We shall now start from the *free electron model.* We consider the crystal to be already assembled. Because of the periodic arrangement of atoms, the potential energy is also expected to be periodic. If the lattice spacing is a, the potential energy V can be written in the one-dimensional form as follows:

$$V(x) = V(x + a). \tag{2–12}$$

Bloch's theorem (Bloch, 1928) says that solutions to Schrödinger's equation for periodic potential energy such as Eq. (2–12) is

$$\psi(x) = U_{k_c}(x)\, e^{\pm j\mathbf{k}_c \cdot \mathbf{x}} \tag{2–13}$$

where $\mathbf{k}_c$ is a wave vector and $j = \sqrt{-1}$.
The function $U_{k_c}(x)$ is also periodic with the same period as the potential energy. Thus

$$U_{k_c}(x) = U_{k_c}(x + a). \tag{2–14}$$

Functions such as Eq. (2–13) are called *Bloch* functions. Three-dimensional version of Bloch functions can be written as follows:

$$\psi(\mathbf{r}) = U_{k_c}(\mathbf{r})\, e^{\pm j\mathbf{k}_c \cdot \mathbf{r}}, \tag{2–15}$$

where $\mathbf{r}$ = position coordinate vector.

It turns out that the quantity $\hbar\mathbf{k}_c$ behaves like momentum in that its time rate of change may be equated to force according to Newton's second law.

$$\mathbf{p} = \hbar\mathbf{k}_c. \tag{2–16}$$

The quantity on the right-hand side of this equation is called *crystal momentum,* which must be distinguished from *particle momentum.*

Kronig and Penney (1931) have solved the problem of determining, in one dimension, the permissible $E(\mathbf{k}_c)$ relationship. (The wave vector is $\mathbf{k}_c$, and k_c is the wave number.) The potential energy in a one-dimensional crystal is shown in Fig. 2–6a. The idealized potential energy used in the Kronig–Penney model is shown in Fig. 2–6b. Figure 2–6b is seen to be a rather crude approximation of Fig. 2–6a. The solution obtained according to this model is shown in Fig. 2–7a. It is readily seen that there appear discontinuities which occur at

$$k_c = \frac{n\pi}{a}, \tag{2–17}$$

where $n = 1, 2, \ldots$

† The Kronig–Penney model is an extremely simple one and is primarily for conceptual use only.

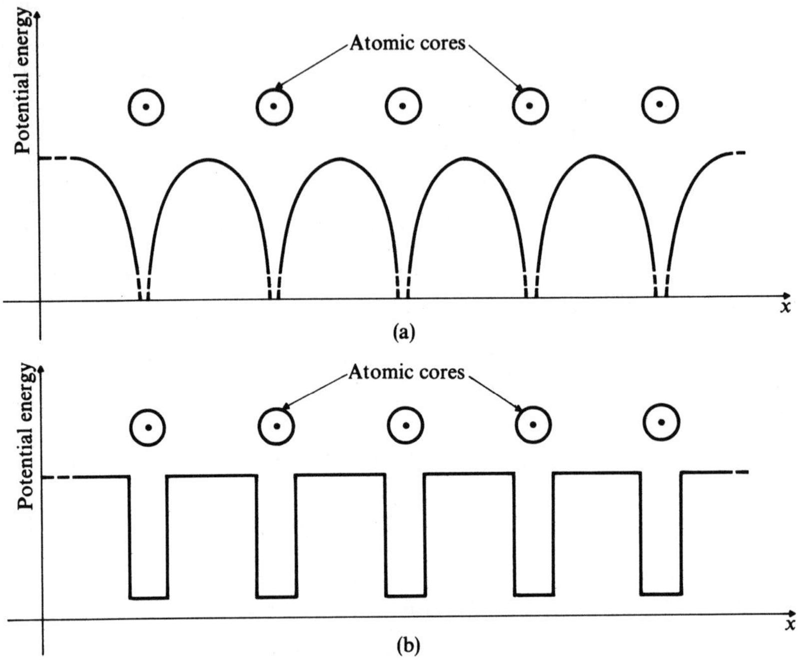

Fig. 2–6 Potential energy in a one-dimensional crystal. (a) Actual potential energy. (b) Idealized potential energy used in Kronig–Penney model.

At these discontinuities, gaps in permissible energies are seen to exist. These are indeed the band gaps we have seen before. Thus, the free electron model and the tight binding model of Section 2–2 *both predict the existence of band gaps.* No traveling wave solutions for electrons exist for energies corresponding to the gaps of Fig. 2–7a. Values given by Eq. (2–17) mark off the boundaries of first, second, third, . . ., *Brillouin zones* as shown in Fig. 2–7a. It can be seen that the $E(\mathbf{k}_c)$ is periodic with a period of $2\pi/a$, so that Bloch solutions repeat themselves when $\mathbf{k}_c$ is replaced by $\mathbf{k}_c + 2\pi n/a$. The periodicity here involves the reciprocal lattice constant $1/a$. Advantage can be taken of this periodicity to give the same information as Fig. 2–7a more compactly as shown in Fig. 2–7b. Note that curve 2 is obtained by moving the right-hand branch a distance $2\pi/a$ to the left, and the left-hand branch $2\pi/a$ to the right. Curves 3 and 4 are similarly obtained. This compact scheme is called the *reduced wave vector scheme.* In this connection it may be noticed that the valence bands of semiconductors look like curve 4, i.e. they are concave downwards. The conduction band of many, though not all, semiconductors look like curve 3, which looks like a valley. Real semiconductors frequently have multiple conduction band valleys. Indeed germanium has four valleys and silicon has six valleys in the conduction

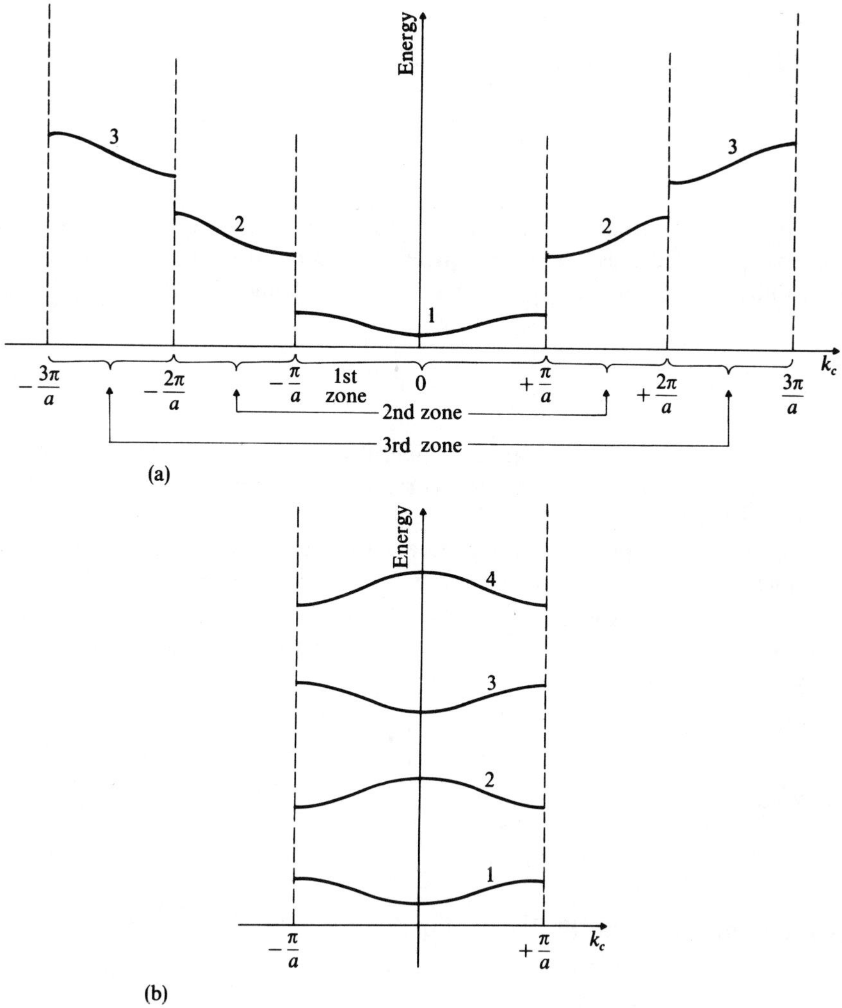

Fig. 2–7 $E(k_c)$ curves and Brillouin zones.

band. Note that the free electron model (Kronig–Penney model) does not predict overlapping bands which we know exist. Thus it is not accurate in this respect.

2-5 EFFECTIVE MASS

Let us consider an electron moving in a crystal under the influence of an external electric field. If Newton's second law is applied to this situation

$$F_{\text{total}} = F_{\text{int}} + F_{\text{ext}} = m\frac{dv}{dt}, \tag{2–18}$$

where F_{int} = forces internal to the crystal,
F_{ext} = forces external to the crystal,
m = electron mass,
v = velocity,
t = time.

Now the forces internal to the crystal consist of forces exerted by free electrons and holes, by fixed ions, by atomic nuclei, and so on. It is easy to see that it is quite difficult to compute F_{int}. For this reason we define *effective mass* m^* such that

$$F_{\text{ext}} = m^* \frac{dv}{dt}. \tag{2–19}$$

In Eqs. (2–18) and (2–19) dv/dt are the same. The reason for defining the effective mass through Eq. (2–19) is a desire to relate it to forces which are relatively easily measured.

To obtain an expression for effective mass of current carriers (i.e. electrons and holes), referred to simply as *carriers*, let us recognize that the particles are guided by their wavepackets. The wave packet moves with a velocity called *group velocity* which is then the velocity of the particle in question. Group velocity v can be related to k_c as follows:

$$v = \frac{d\omega}{dk_c}. \tag{2–20}$$

Now

$$E = h\nu = \left(\frac{h}{2\pi}\right) 2\pi\nu = \hbar\omega. \tag{2–21}$$

Substituting ω from Eq. (2–21) into Eq. (2–20) we obtain

$$v = \frac{1}{\hbar}\left(\frac{dE}{dk_c}\right). \tag{2–22}$$

If the only external force is due to the electric field $\mathscr{E}$, we may rewrite Eq. (2–19) as

$$F_{\text{ext}} = q\mathscr{E} = m^* \frac{dv}{dt}$$

or

$$q\mathscr{E} = \frac{m^*}{\hbar} \frac{d^2E}{dk_c^2} \frac{dk_c}{dt}. \tag{2–23}$$

In obtaining Eq. (2–23) we have used Eq. (2–22) and assumed that we are following the motion of a positive charge q. It will turn out that the end result is independent of q; so that it is equally valid also for a negative charge. The result which we are after is an expression for effective mass m^* as a function of E and $\mathbf{k}_c$ independent of time t. From Eq. (2–23) we must find a way of eliminating $d\mathbf{k}_c/dt$. To this end we observe that power W is the rate of change of energy E. Thus

$$W = \frac{dE}{dt}$$

or

$$dE = Wdt. \tag{2–24}$$

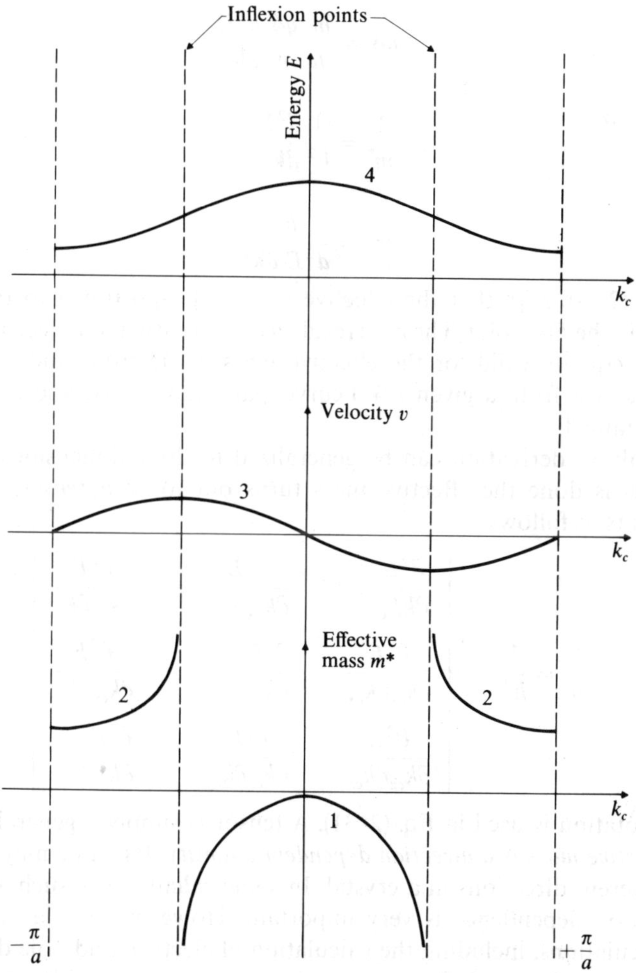

Fig. 2–8 Plot of energy E, velocity v, and effective mass m^* as a function of k_c.

But
$$dE = F_{\text{ext}}\, v dt \tag{2–25}$$
$$dE = q\mathscr{E}\left(\frac{1}{\hbar}\frac{dE}{dk_c}\right)dt. \tag{2–26}$$

In obtaining Eq. (2–26) use was made of Eq. (2–22). Equation (2–26) can be solved for $d\mathbf{k}_c/dt$,

$$\frac{dk_c}{dt} = \frac{q\mathscr{E}}{\hbar}. \tag{2–27}$$

Equation (2–27) can be substituted into Eq. (2–23) to obtain

$$q\mathscr{E} = \frac{m^*}{\hbar}\frac{q\mathscr{E}}{\hbar}\frac{d^2E}{d\mathbf{k}_c^2} \tag{2–28}$$

or

$$\frac{1}{m^*} = \frac{1}{\hbar^2}\frac{d^2E}{d\mathbf{k}_c^2}. \tag{2–29}$$

Thus

$$m^* = \frac{\hbar^2}{d^2E/d\mathbf{k}_c^2}. \tag{2–30}$$

Equation (2–30) says that the effective mass is proportional to the *radius of curvature* in the $E(\mathbf{k}_c)$ plot. The final result, as noted earlier, is indeed independent of q. It is equally valid for the effective mass of electrons and holes. Figure 2–8 shows how from a given $E(\mathbf{k}_c)$ curve, particle velocity, and effective mass can be obtained.

The above derivation can be generalized to three dimensions in **k**-space. When this is done the effective mass turns out to be a *tensor*, having nine components as follows

$$\frac{1}{m^*} = \frac{1}{\hbar^2}\begin{bmatrix} \dfrac{\partial^2 E}{\partial k_{cx}^2} & \dfrac{\partial^2 E}{\partial k_{cx}\partial k_{cy}} & \dfrac{\partial^2 E}{\partial k_{cx}\partial k_{cz}} \\ \dfrac{\partial^2 E}{\partial k_{cy}\partial k_{cx}} & \dfrac{\partial^2 E}{\partial k_{cy}^2} & \dfrac{\partial^2 E}{\partial k_{cy}\partial k_{cz}} \\ \dfrac{\partial^2 E}{\partial k_{cz}\partial k_{cx}} & \dfrac{\partial^2 E}{\partial k_{cz}\partial k_{cy}} & \dfrac{\partial^2 E}{\partial k_{cz}^2} \end{bmatrix}. \tag{2–31}$$

Matrix notation is used in Eq. (2–31). A tensor is simply a generalized vector. Thus *effective mass is a direction dependent quantity.* Its value may be different along different directions in a crystal. In some calculations, such as tunneling, the direction dependence is very important. However, in a large number of other calculations, including the calculation of electron and hole densities, the direction dependence of effective mass is not important, and it may be given a constant value.

Figure 2–8 shows that curve 4 is similar to the valence band $E(\mathbf{k}_c)$ curve for semiconductors. For this shape it is clear that $d^2E/d\mathbf{k}_c^2$ is negative. This indicates that the *effective mass for electrons in the valence band is negative.* Recall that in Section 2–3 this fact was utilized in coming to the conclusion that the *effective mass of a hole was positive.*

2–6 CONDUCTIVITY AND MOBILITY

As electrons and holes move through the crystal lattice, they encounter a series of collisions. It is thought that even when an electric field is applied to the crystal, the carrier loses all previous memory and starts all over again at each collision. Because of these collisions, it is experimentally observed that *a constant velocity results in the presence of a steady electric field.* This is to be contrasted with the steady acceleration which a charged particle would be expected to experience in free space in the presence of an electric field. Figure 2–9 symbolically shows electron density and hole density in the presence of an electric field. Electron and hole velocities in a crystal are proportional to velocity with the proportionality constant being *mobility.* Thus

$$\mathbf{v}_p = \mu_p \mathscr{E} \tag{2–32}$$

and

$$\mathbf{v}_n = -\mu_n \mathscr{E}, \tag{2–33}$$

where $\mathbf{v}_p, \mathbf{v}_n$ = hole and electron velocities (cm/sec.),

μ_p, μ_n = hole and electron mobilities (cm^2/Volt-sec.).

Note that Eqs. (2–32) and (2–33) are written such that *mobilities are positive constants.*

From electromagnetic field theory we may write current density as the product of charge density and velocity. Thus hole and electron current densities $\mathbf{j}_p, \mathbf{j}_n$ may be written as

$$\mathbf{j}_p = (qp)\mathbf{v}_p, \tag{2–34}$$

$$\mathbf{j}_n = (qn)\mathbf{v}_n, \tag{2–35}$$

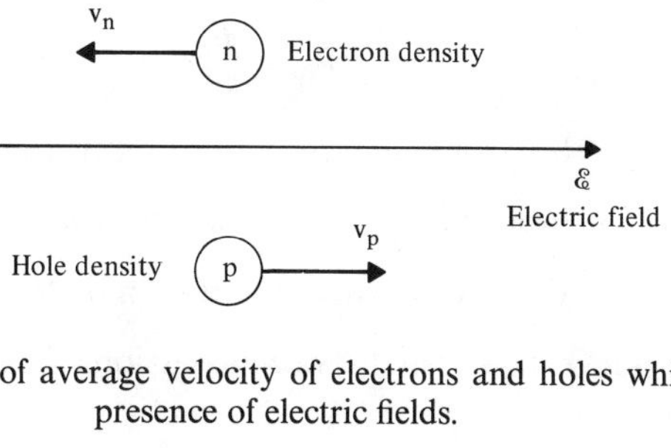

Fig. 2–9 The direction of average velocity of electrons and holes which results in the presence of electric fields.

where p, n = hole and electron densities (cm^{-3}).

Thus, total current may be written as

$$\mathbf{j} = \mathbf{j}_p + \mathbf{j}_n = (qp)\mathbf{v}_p + (-qn)\mathbf{v}_n. \tag{2–36}$$

Equation (2–32) and (2–33) may be substituted into Eq. (2–36) to obtain

$$\mathbf{j} = q[\mu_p p + \mu_n n]\mathscr{E}. \tag{2–37}$$

But

$$\mathbf{j} = \sigma\mathscr{E}, \tag{2–38}$$

where σ = conductivity.

So finally conductivity of a material which has both electrons and holes in it, can be written as

$$\sigma = q(\mu_p p + \mu_n n). \tag{2–39}$$

Mobility itself can be expressed in terms of the fundamental parameters of the crystal. If the average time between collisions (also called *mean free time*) is $\bar{t}$, then acceleration can be written

$$\mathbf{a} = \frac{\mathbf{v}}{\bar{t}}, \tag{2–40}$$

however,

$$\mathbf{a} = \frac{\mathbf{F}}{m^*} = \frac{q\mathscr{E}}{m^*}. \tag{2–41}$$

Substituting Eq. (2–40) into Eq. (2–41) we obtain

$$\mathbf{v} = \mathbf{a}\bar{t} = \frac{q\bar{t}}{m^*}\ \mathscr{E} = \mu\mathscr{E}. \tag{2–42}$$

Thus from Eq. (2–42) we can write

$$\mu_p = \frac{q\bar{t}_p}{m_h^*} \tag{2–43}$$

$$\mu_n = \frac{q\bar{t}_n}{m_e^*}, \tag{2–44}$$

where μ_p, μ_n = hole and electron mobilities respectively,

$\bar{t}_p, \bar{t}_n$ = hole and electron mean free times respectively,

m_h^*, m_e^* = effective masses for holes and electrons respectively.

The average distance travelled between collisions $\bar{\ell}$ is called *mean free path length.*

Mobility is expected to diminish with increasing temperature because the collision probability will increase and hence mean free time diminishes. *High densities of chemical or mechanical defects in an otherwise orderly arrangement of atoms also diminishes mobility.*

It can be seen from Eqs. (2–43) and (2–44) that the mobility depends importantly on the value of effective mass. This in turn depends on the $E(\mathbf{k}_c)$ curves according to Eq. (2–30). We need therefore to look at these curves in detail to determine what value effective mass should have. Figure 2–10a shows the $E(\mathbf{k}_c)$ curves for a very idealized simple case. Note that the conduction band valley and the valence band peak both occur precisely at $\mathbf{k}_c = 0$. Furthermore conduction and valence bands are *parabolic*, that is $E(\mathbf{k}_c)$ relationship is parabolic. Figure 2–10b shows the so called Fermi surface which links all $\mathbf{k}_c$'s that correspond to the same energy. In this simple case the energy on the Fermi surface and $\mathbf{k}_c$ are related as follows,

$$E = \frac{\hbar^2 k_c^2}{2m}. \qquad (2\text{–}45)$$

Several semiconductors in most common usage do not, unfortunately, have such simple bands as those of Fig. 2–10a and Fig. 2–10b. Let us consider the energy band diagram for both silicon and germanium which are two of the most widely used semiconductors. Figures 2–11a and 2–11b show the conduction and valence bands of germanium and silicon respectively. Several important features of this figure are worthy of note. First the conduction band has more than one valley. In germanium and silicon the valley corresponding to the lowest energy is *not* located at $\mathbf{k}_c = 0$. This seems to present a dilemma. Accord-

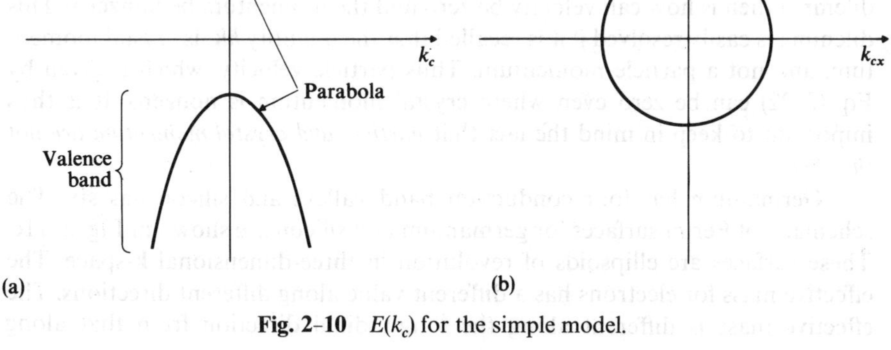

Fig. 2–10 $E(k_c)$ for the simple model.

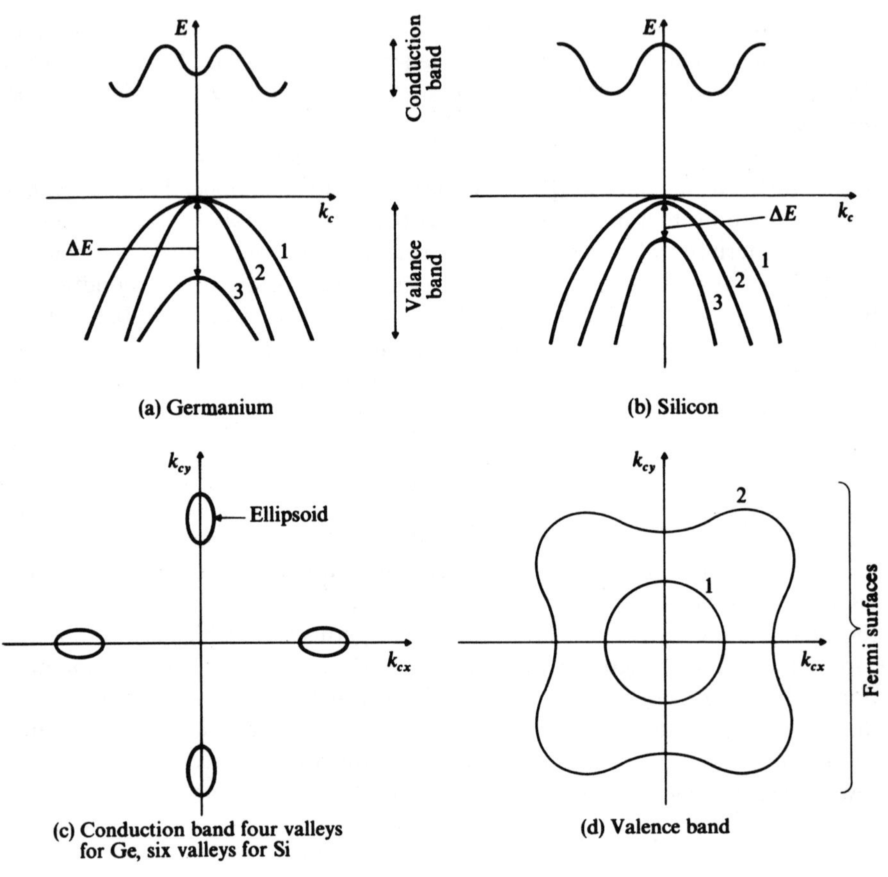

Fig. 2–11 $E(k_c)$ for germanium and silicon.

ing to Eq. (2–22), the velocity at the conduction band valley at $\mathbf{k}_c \neq 0$ is zero. From Eq. (2–16) the momentum is *nonzero* at such valleys because $\mathbf{k}_c \neq 0$. The dilemma then is how can velocity be zero and the momentum be nonzero? This dilemma is easily resolved if it is recalled that the quantity $\hbar\mathbf{k}_c$ is crystal momentum, and not a particle momentum. Thus particle velocity, which is given by Eq. (2–22) can be zero even where crystal momentum is nonzero. It is thus important to keep in mind the fact that *particle and crystal momentum are not the same.*

Germanium has four conduction band valleys and silicon has six. The schematic of Fermi surfaces for germanium and silicon are shown in Fig. 2–11c. These surfaces are ellipsoids of revolution in three-dimensional **k**-space. The effective mass for electrons has a different value along different directions. The effective mass is different along the longitudinal direction from that along transverse direction as can be seen in Table 2–1. For most of the calculations in *p-n* junction devices† the direction dependence of effective mass can be ignored. An outstanding example of an exception to this rule is the tunneling calculation which is very sensitive to the direction of the value of effective mass. In closing the discussion of conduction band it should be pointed out that the curves of Fig. 2–11a and Fig. 2–11b are themselves direction dependent. *That is, the shapes may look different along different directions.*

Next let us consider the valence band which is seen to consist of three different bands: bands 1, 2, and 3. It should be emphasized that these are *not* three different valence bands of the kind shown in Fig. 2–2c; rather the three bands are *within* the one valence band of Fig. 2–2c. All the holes in valence band of Fig. 2–2c will have to belong to band 1, 2, or 3 of Fig. 2–11a and Fig. 2–11b. Depending on which band a hole belongs to will determine the corresponding value of effective mass. This implies that there are in fact three species of holes in valence band having three different values of effective masses. As band 3, also called *split-off band*, is separated from bands 1 and 2 by an energy ΔE it is less likely to participate in the conduction mechanism unless the temperature is sufficiently high to populate band 3 with holes.

Table 2–1 Effective Mass of Electrons and Holes in Germanium and Silicon

	Conduction band		Valence band	
	Longitudinal effective mass m_ℓ^*	Transverse effective mass m_t^*	Band 1	Band 2
Germanium†	1.6 m	0.082 m	0.3 m	0.04 m
Silicon†	0.97 m	0.19 m	0.5 m	0.16 m

From the interpretation given to Eq. (2–30), namely that effective mass is proportional to the radius of curvature of the $E(\mathbf{k}_c)$ band, it is clear that the effective mass in band 1 will be greater than that in band 2. This is clearly the case as seen from Table 2–1.

From cyclotron resonance experiments (at 4.2°K) the effective mass and the population densities of holes in the top two bands can be determined. These are tabulated in Table 2–2. The holes of band 1 are called "heavy" holes or "slow" holes because their mass is higher and the mobility is lower. Holes of band 2 are called "light" holes or "fast" holes because the mass is lower, and the mobility is higher.

A number of experiments involving especially *p*-type germanium and silicon gave results which were baffling until the detailed band structure of these

---

† m is the mass of an electron in free space.

Table 2–2 Valence Band Holes in Germanium and Silicon

	Percentage of holes in band 1	Percentage of holes in band 2	ΔE
Ge	95	5	0.28 eV
Si	80	20	<0.05 eV

materials was known. The presence of at least two sets of holes must be taken into account when, for example, conductivity is calculated. For this case the conductivity of a pure *p*-type semiconductor may be written in general as

$$\sigma_p = q\left[\mu_{p1}p_1 + \mu_{p2}p_2 + \cdots \mu_{pm}p_m\right] \tag{2–46}$$

where p_m = population density of holes in band *m*,
μ_{pm} = mobility of holes in the *m*th valence band.

Also

$$\mu_{pm} = \frac{q\bar{t}_{pm}}{m^*_{pm}}, \tag{2–47}$$

where $\bar{t}_{pm}$ = mean free time for holes in the *m*th valence band,
m^*_{pm} = effective mass of holes in the *m*th valence band.

It is also permissible to define an *effective mobility* for holes in the valence band μ_p^* which can be analytically expressed as

$$\mu_p^* = \frac{\mu_{p1}p_1 + \mu_{p2}p_2 + \cdots \mu_{pm}p_m}{p_1 + p_2 \cdots p_m}. \tag{2–48}$$

Note that μ_p^* is the weighted average which takes account of the relative population densities of different hole densities in the different bands within the valence band. With the effective mobility defined by Eq. (2–48) we can use the simple expression of Eq. (2–39) with μ replaced by μ^*.

Example 2–1 Calculate, at room temperature, conductivity of *p*-type silicon which is doped to $N_a = 10^{15}$ per c.c. of acceptor atoms. $E_a = 0.01$ eV above the valence band. You may use the fact, which will be proved later, that the $np = n_i^2$ for a semiconductor held at constant temperature.

Solution. At room temperature $kT = 0.026$ eV. It is therefore reasonable to conclude that for the acceptor levels given, *all acceptor atoms are ionized.* Thus the total hole density is

$$p = N_a = 10^{15} \text{ per c.c.} \tag{1}$$

In silicon

$$np = n_i^2 = 2.25 \times 10^{20} \text{ per cm}^6 \tag{2}$$

$$n = \frac{n_i^2}{p} = \frac{2.25 \times 10^{20}}{10^{15}} = 2.25 \times 10^5 \text{ per c.c.} \tag{3}$$

So $n \ll p$, and electron density may be ignored compared with hole density. We may consider just the two bands in the valence band of silicon. In this case

$$\sigma_p = q\mu_p^* p. \tag{4}$$

The *measured* value of hole mobility in silicon is 480 cm²/Volt-sec. *This is necessarily the effective* μ_p^*. Thus

$$\sigma_p = 1.6 \times 10^{-19}(480)(10^{15}) = 7.68 \times 10^{-2}\,(\Omega\text{-}cm)^{-1}. \longleftarrow \tag{5}$$

Example 2–2 If in Example 2–1 it had been assumed that all holes are heavy holes, what would be the conductivity of this material?

Solution. Before we can proceed with the solution we need to know what is the value of μ_{p1}, i.e. mobility for heavy holes which reside in band 1. As this is not given we shall make the simplifying assumption that the mean free time between collisions is the same for all holes regardless of which band they belong to. This may or may not be a good assumption, but we need to start somewhere. In this case we note that

$$\frac{\mu_{p1}}{\mu_{p2}} = \frac{\dfrac{q\bar{t}_{p1}}{m_{p1}^*}}{\dfrac{q\bar{t}_{p2}}{m_{p2}^*}} = \frac{m_{p2}^*}{m_{p1}^*} = \frac{0.16m}{0.5m} = 0.32 \tag{1}$$

or

$$\mu_{p2} = \frac{\mu_{p1}}{0.32}. \tag{2}$$

Table 2–1 was used in arriving at the last result. Now the definition of effective mobility leads to

$$\mu_p^* = \frac{\mu_{p1}p_1 + \mu_{p2}p_2}{p_1 + p_2} = \mu_{p1}\left(\frac{p_1}{p_1 + p_2}\right) + \mu_{p2}\left(\frac{p_2}{p_1 + p_2}\right). \tag{3}$$

From Table 2–2 we have

$$\frac{p_1}{p_1 + p_2} = 0.80 \text{ and } \frac{p_2}{p_1 + p_2} = 0.2 \tag{4}$$

$$\mu_p^* = \mu_{p1}(0.8) + \mu_{p2}(0.2) = \mu_p(0.8) + \mu_{p1}\frac{0.2}{0.32}$$

or

$$\mu_p^* = \mu_{p1}[1.425] \text{ or } \mu_{p1} = \mu_p^*(1.425)^{-1} = (480)(1.425)^{-1} \tag{6}$$

$$\mu_{p1} = 336 \text{ cm}^2/\text{Volt-Sec.} \tag{7}$$

Thus if all holes are assumed to be heavy holes the conductivity would be

$$\sigma = q\mu_{p1}p = 1.6 \times 10^{-19}(336)(10^{15}) \tag{8}$$

$$\sigma = 0.0537\ (\Omega\text{-}cm)^{-1}. \longleftarrow \tag{9}$$

Thus we see from the above that the difference in the calculated conductivity is fairly sizeable when the assumption is made that all holes are "heavy."

2–7 SEMICONDUCTORS WITH CHEMICAL IMPURITIES

So far we have considered only the *intrinsic* semiconductor, i.e. a semiconductor which is free of impurities of all kinds including mechanical and chemical impurities. The energy band diagrams of the last section and of Section 2–2 correspond to intrinsic semiconductors. In this section we consider first the influence of chemical impurities on the band structure, and second predict the variation of conductivity with temperature. A semiconductor that has chemical impurities is called an *extrinsic semiconductor.*

The major kinds of impurities in a crystal are mechanical or chemical in

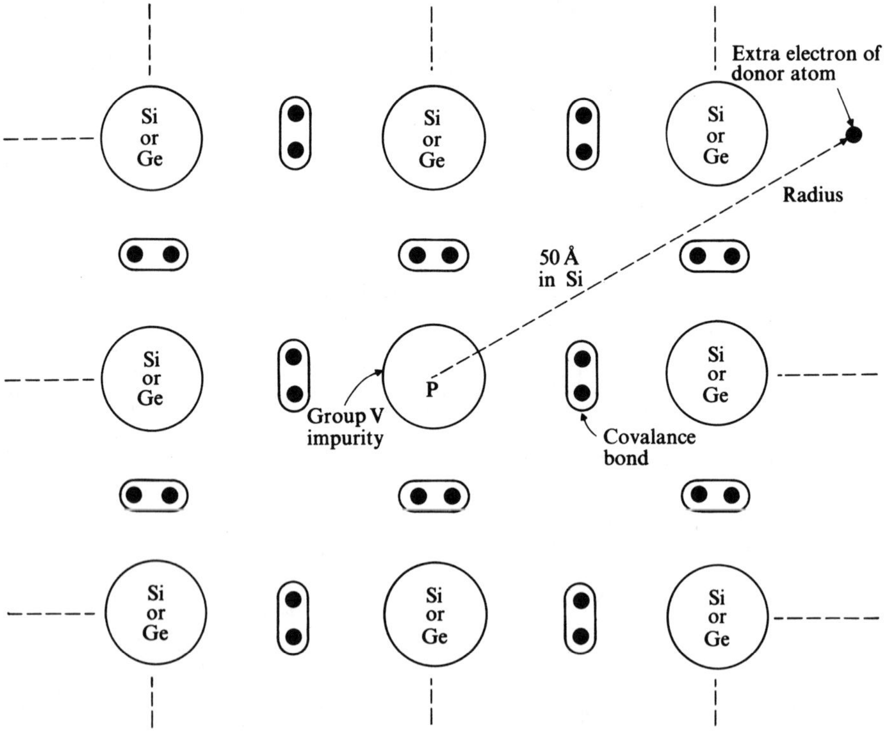

Fig. 2–12 Donor impurity in Si or Ge two-dimensional lattice.

nature. Mechanical defects can be a missing atom, an extra atom, an extra plane of atoms, a half plane of missing atoms, clustering of atoms, and so on. Chemical impurities constitute usually the presence of chemical atoms *other than the host atoms of the crystals.* These may be located at substitutional lattice sites or may exist interstitially, i.e. at sites in between lattice sites.

All the above kinds of impurities disturb the perfectly periodic nature of potential energy. The major consequence of this disturbance is the introduction of additional allowed electronic states. These could be within the main bands. The states which have the greatest influence, however, are located in the band gap. Technologically useful impurities are chemical. *Small controlled amounts of chemical impurities are found to have a rather profound influence on the conductivity of semiconductors.*

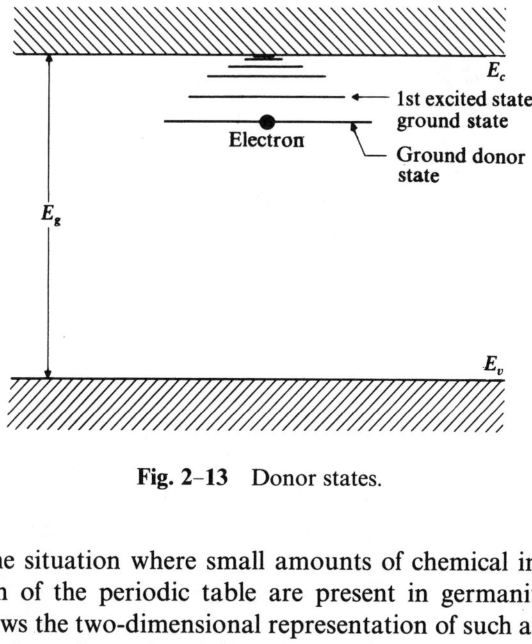

Fig. 2–13 Donor states.

Consider the situation where small amounts of chemical impurities from the fifth column of the periodic table are present in germanium or silicon. Figure 2–12 shows the two-dimensional representation of such a situation. The actual lattice is three-dimensional, but the two-dimensional representation is merely a flattened out version of the former. Typical examples of such chemical impurities are phosphorous (*P*), arsenic (As), and antimony (Sb). These impurities go into the crystal *substitutionally* as shown in Fig. 2–12. Four of the five valence electrons of such an impurity participate in the covalence bond with host atoms of silicon or germanium. The fifth electron is left out of the covalence bond. The relationship of such an electron with respect to an impurity, such as those mentioned above, is quite analogous to the electron and the hydrogen atom to which it belongs. That is in both cases there is a net positive charge at the center and the balancing negative charge is provided by the orbiting electron. From this observation we would expect hydrogen like bound electronic states for say the phosphorous atom. The most important of these is the *ground state.* Figure 2–13 shows a schematic of such states which start somewhat below the conduction band edge and merge with the bottom of the conduction band. We can utilize the above analogy and obtain the ground state energy. First we have seen that permissible energies in hydrogen atoms are given by

$$E_n = -\frac{mq^4}{2\hbar^2 n^2}. \tag{2–49}$$

Now if we set the energy corresponding to the bottom of the conduction band to zero, i.e. $E_c = 0$, then we expect that Eq. (2–49) can be directly converted to the energy states for the extra electron of impurities of the fifth column of the periodic table if we make two simple changes.

1. $m_e \longrightarrow m_e^*$.
2. $q^2 \longrightarrow \dfrac{q^2}{\varepsilon}$.

The first states the reasonable change of replacing the free electron mass by the effective mass m_e^*. The second takes account of the fact that the permittivity of germanium ($\varepsilon = 16$) and silicon ($\varepsilon = 12$) is significantly higher than that of air. Thus the energy states corresponding to these impurity atoms may be written:

$$E_{dn} = -\frac{m_e^* q^4}{2\varepsilon^2\hbar^2 n^2} \qquad n = 1, 2, \ldots \tag{2–50}$$

$$E_d = -\frac{m_e^* q^4}{2\varepsilon^2\hbar^2}. \tag{2–51}\dagger$$

The subscript d in Eqs. (2–50) and (2–51) refers to the fact that these energy levels are called *donor states.* Impurities from the fifth column of the periodic table are called *donor atoms* or *donor impurities* because the atom is capable of donating the fifth electron to the conduction process as we shall presently see. Figure 2–14a shows the one-dimensional energy diagram in which is also shown the donor ground state. At absolute zero degrees temperature the extra electron of the donor atom will be in the ground state as shown in Fig. 2–14a. A calculation of the radius of the electron in the ground state reveals that it is about 30 Å in silicon compared to lattice constant which is approximately 5.66 Å in Ge and 5.43Å in Si. Thus the electron in the donor ground state (also called *donor state*) spans quite a few host atoms ($\sim$1000). This implies that the energy necessary to free such an electron is expected to be small. The energy necessary to

†Here the left hand side of the equation is E_d rather than E_{d1} to conform to the notation in most published work for the ground state corresponding to donor atoms.

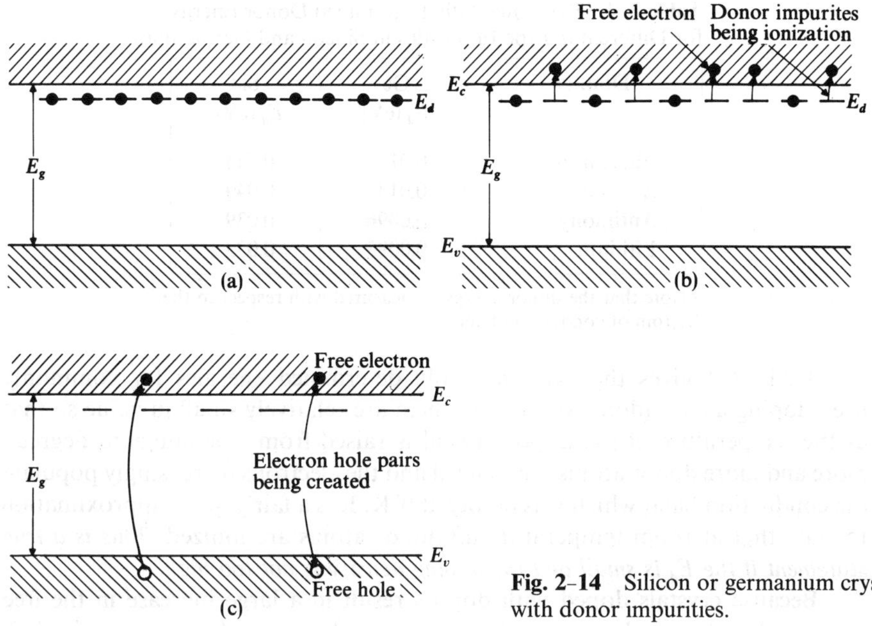

Fig. 2–14 Silicon or germanium crystal with donor impurities.

take an electron from the donor state into the conduction band is called the ionization energy and is equal to E_d. The computed value of E_d from Eq. (2–51) and observed values of ionization energy are shown in Table 2–3. Table 2–3 shows that Eq. (2–51) does predict the right order of magnitude of ionization energy. Equation (2–51) uses a single value of m_e^* and hence assumes that the effective mass is isotropic. If instead, the anisotropy of m_e^* is used taking into account the longitudinal and transverse m^* for germanium and silicon the observed and calculated values of ionization energies are in much better agreement than those shown in Table 2–3. Equation (2–51) does not give the correct prediction in one important respect. *It predicts that all donors will have the same ionization energy. This is not true because the potential energy is much more complicated in the close proximity of the donor atom than is assumed in Eq.* (2–51). This is expected to influence primarily the ground state. The *first excited* donor states (that is $n = 2$ in Eq. 2–50) are found to be almost the same for most donor impurities, thus confirming the prediction of Eq. (2–50).

Table 2–3 Donor Ionization Energies

	Permittivity ε	Observed ionization energy	Ionization energy computed using Eq. (2–51)
Si	12	0.04 eV	0.09 eV
Ge	16	0.01 eV	0.05 eV

Table 2–4 Experimentally Determined Donor Energy for Different *n*-Type Impurities in Silicon and Germanium

Donor†	Ge E_d (eV)	Si E_d (eV)
Phosphorous	0.012	0.044
Arsenic	0.013	0.049
Antimony	0.0096	0.039
Lithium	0.0093	0.033

† Note that the donor energy is measured with respect to the bottom of conduction band.

Table 2–4 gives the experimentally determined energies for commonly used doping agents (donors). Most of them are relatively small in value so that as the temperature of the doped crystal is raised from absolute zero degrees, more and more donor atoms are ionized and the electrons increasingly populate the conduction band which was empty at 0°K. It is a fairly good approximation to state that at room temperature *all* donor atoms are ionized. *This is a true statement if the* E_d *is small and the doping density is not too high.*

Because crystals doped with donors result in a large increase in the free electron density (an electron in the conduction band is a *free electron*, i.e. it is not bound to the donor atom), and because electrons carry negative charge *such crystals are called n-type* (*n* for negative charge; and donor impurities are called *n-type impurities*).

The introduction of low amounts of donor or *n*-type impurities has a profound effect on the conductivity of the material. The following illustrative example makes this point.

Example 2–3 Calculate the intrinsic conductivity of silicon at room temperature. Compare this to the conductivity of *n*-type silicon doped with phosphorous 10^{17} per cubic centimeter (c.c.) which is less than one impurity atom in 10^5 host atoms.

Solution: For intrinsic semiconductor the conductivity may be written from Eq. (2–39) as

$$\sigma = q(\mu_p p + \mu_n n) = qn_i(\mu_p + \mu_n).$$

The last equality utilized the fact that in an intrinsic semiconductor electron density is equal to hole density because all electrons in the conduction band must have come from valence band. The intrinsic carrier density is designated by n_i(1.50×10^{10} in silicon)

$$\begin{aligned}\sigma_i &= qn_i(\mu_p + \mu_n)\\ &= 1.6 \times 10^{-19}(1.50 \times 10^{10})(480 + 1350)\\ \sigma_i &= 4.39 \times 10^{-6}(\Omega\text{-cm})^{-1}.\end{aligned}$$

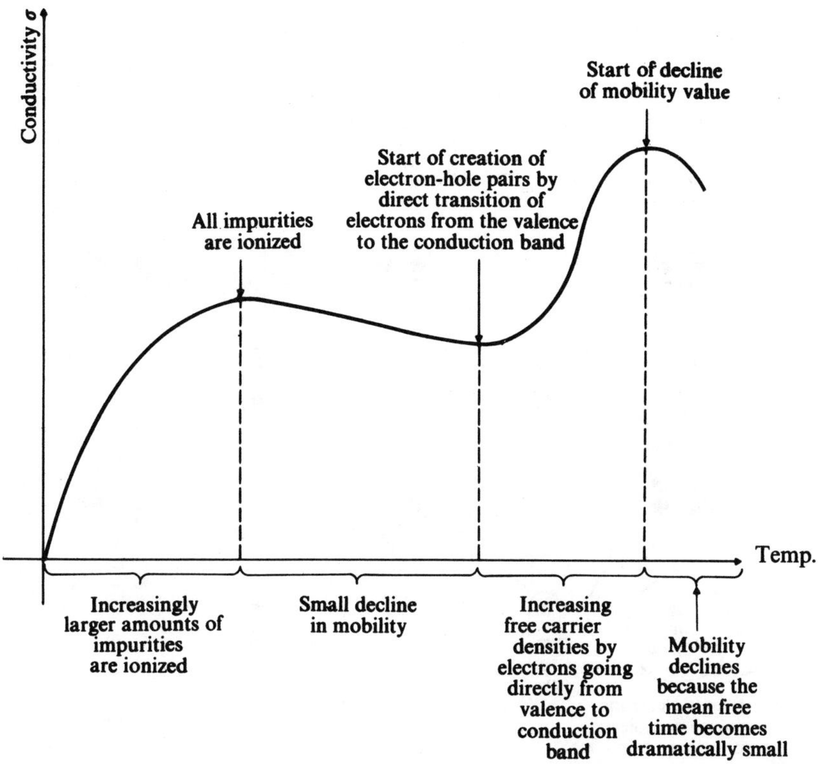

Fig. 2–15 Conductivity in an extrinsic semiconductor as a function of temperature.

Now consider the doped crystal. We shall assume that at room temperature all donor atoms are ionized. In this case

$$n \gg p$$
$$\sigma_n = q\,\mu_n n = 1.6 \times 10^{-19}\,(1350)(10^{17})$$
$$\sigma_n = 21.6\,(\Omega\text{-cm})^{-1}. \longleftarrow$$

Thus

$$\frac{\sigma_n}{\sigma_i} = \frac{21.6}{4.39 \times 10^{-6}} = 4.92 \times 10^{6}. \longleftarrow$$

Thus a doping level of less than one part in 10^6 host atoms produces a change in conductivity of over a factor of 10^5.

Let us now consider how we expect the conductivity of n-type semiconductor to change with temperature. At absolute zero degrees temperature all

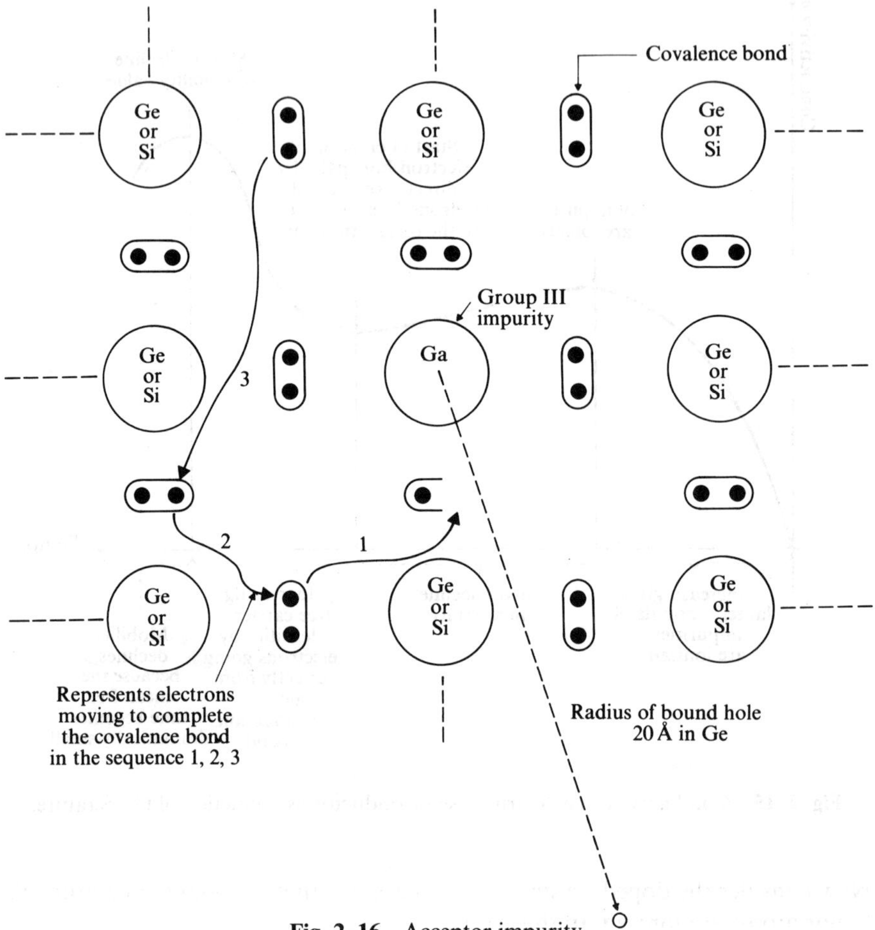

Fig. 2–16 Acceptor impurity.

bands including valence band and donor band are full, and all bands above donor band are empty. So as before conductivity is zero. As the temperature is raised, the donor states will start to give up electrons to the conduction band as shown in Fig. 2–14b. Electron density in the conduction band increases, so will conductivity. After all the donor states have been emptied out, no further increase in free carrier† density can occur until the temperature is high enough to have the valence band electrons jump directly to the conduction band. When this process becomes more probable, the electrons and holes begin to be created in pairs as shown in Fig. 2–14c. The free electron and hole densities now dramatically increase, and consequently the conductivity increases. Most useful

†Carrier refers to a charge particle that contributes to current flow.

electronic devices should be operated below the temperature where the electron density approaches hole density. When this occurs the material is said to have reached the *intrinsic stage* because in an intrinsic semiconductor $n = p$. The n-type material is no longer n-type. At sufficiently high temperature the mobility will decrease because of a decrease in mean free time (see Eq. 2–44), and the conductivity will drop. Figure 2–15 summarizes the variation in conductivity due to temperature.

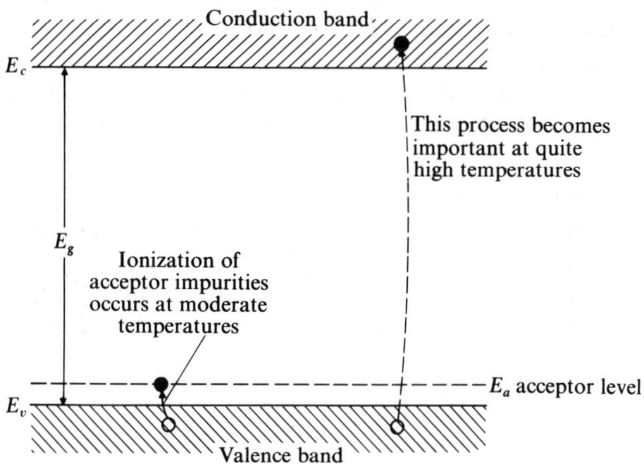

Fig. 2–17 Band picture for a semiconductor having acceptor impurities.

If in place of a chemical impurity from the fifth column of the periodic table we had selected an impurity from the third column of the periodic table, the two-dimensional lattice representation would look as shown in Fig. 2–16. These impurities go in substitutionally and have only three valence electrons. These form covalence bonds with three of their neighboring host atoms. The fourth nearest neighbor has no electron contributed by the impurity atom and hence has an incomplete covalence bond. It turns out that, as shown in Fig. 2–16, the neighboring electrons find it energetically conducive to complete the incomplete covalence bond, leaving an incomplete covalence bond of their own. Impurities from the third column of the periodic table are called *acceptor impurities* and the corresponding atoms as *acceptor atoms* because they accept electrons. Such impurities are also called *p-type* impurities because these lead to large numbers of free holes in the valence band. The positive charge of holes is responsible for the subscript p. Typical examples of acceptor impurities are gallium (Ga), boron (B), and indium(In).

When a crystal is doped with acceptor impurities the energy band picture looks as shown in Fig. 2–17. The presence of acceptor impurities do give rise to additional band gap energy states, *acceptor states*, analogous to the case of donor impurities. For commonly used acceptor impurities the acceptor energy band E_a ends up fairly close to the top of the valence band as can be seen from Table 2–5. An equation analogous to Eq. (2–51) should, in principle, be true for acceptor ionization energy; but because the valence band is degenerate (see Figs. 2–11a and 2–11b, the effective mass considerations are quite complicated.

The acceptor states are empty at absolute zero degrees. An alternative way of looking at this empty electronic state is to say that it is occupied by a *hole.* Such a hole is bound to the acceptor atom just as the extra electron associated with the donor atom is said to be bound to the donor atom. The radius of the orbit of the bound hole is about 20 Å in germanium and somewhat less than this in silicon. This radius also spans many host atom lattice sites. Thus we expect that it should take a relatively small amount of energy to emit this hole to the valence band, as is indeed confirmed by Table 2–5. When the hole has been emitted to the valence band where it is a *free hole,* we say that the corresponding acceptor atom has been *ionized.* Ionization energy is equal to E_a. An alternative description of this same phenomenon is to say that when an electron from the valence band goes into the acceptor state (which is the *ground state for hole* in the acceptor atom) a free hole is created in the valence band.

Table 2–5 Experimentally Determined Acceptor Energy for *p*-type Impurities In Silicon and Germanium

Acceptor	Ge E_a (eV)	Si E_a (eV)
Boron	0.010	0.045
Gallium	0.011	0.065
Indium	0.011	0.16
Alluminum	0.010	0.057

It is useful to note that when a donor atom is ionized it acquires a net positive charge. This can be understood by recalling that the donor atom is neutral when it has five electrons. Once it has lost its fifth electron, the atom necessarily acquires a net local positive charge. Unlike the electron, this positive charge is immobile and hence is called the *bound fixed charge.* Quite similarly an ionized acceptor acquires a net negative charge which is also immobile in contrast with the free hole. It is important to keep in mind that free holes reside only in valence band. Figure 2–15 also gives the variation of conductivity as a function of temperature for *p*-type semiconductor. The initial increase is due to the progressive ionization of acceptors.

Example 2–4 A silicon crystal 2 cm long and of 0.5 cm^2 cross-sectional area is known to have 10^{15} gallium atoms per cubic centimeter and 5×10^{15} lithium atoms per cubic centimeter. Calculate, at room temperature, the current that a 15 V battery will need to supply to the crystal when it is connected at its ends directly to the battery.

Solution. Ohm's law allows us to relate current density to conductivity σ

$$J = \sigma E = \frac{V\sigma}{\ell},$$

where ℓ = length of the bar along which current flows,
A = cross sectional area through which the current flows.

$$\text{Or current } I = JA = \frac{AV\sigma}{\ell} = \frac{AVq}{\ell}[\mu_p p + \mu_n n].$$

At room temperature we may assume that all acceptors and donors are ionized. *We must not, however, jump to the conclusion that electron density equals lithium atom density and hole density equals gallium atom density.* That is

$$n \neq N_d \quad \text{and} \quad p \neq N_a.$$

The reason for this is that the electrons in the donor states will have an appreciably higher probability of falling into the empty and available acceptor states rather than go into the conduction band. So of the 5×10^5 donor electrons, 10^5 will find themselves in the acceptor states thus occupying *all* acceptor states. Thus the electrons which will end up in the conduction band will be

$$n \doteq N_d - N_a = 5 \times 10^{15} - 1 \times 10^{15} = 4 \times 10^{15} \text{ per c.c.}$$

$p = 0$ because the electrons from valence band do not have any acceptor states into which to go.

$$I = \frac{AVq\mu_n n}{\ell} = \frac{(0.5)(15)(1.6 \times 10^{-19})(1350)(4 \times 10^{15})}{2}$$

$$I = 3.24 \text{ amp.}$$

Note that if $N_a = N_d$ the material will behave like an intrinsic crystal. Such a crystal is said to be *compensated.*

Example 2–5 A cylindrical p-type semiconductor crystal having a cross-sectional area of 0.2 cm^2 and a length of 0.8 cm is found to have 4×10^{17} electronic states per cubic centimeters in the middle of the band gap (1.0 eV). These states are occupied by electrons at 0°K. This crystal is now doped with 5×10^{17} acceptor atoms/c.c. whose energy is 0.005 eV above the top of the valence band. A 6 V battery is connected to the ends of the crystal. The valence band is populated by three types of holes having the *same mean free time* between collisions and having $m^*_{h1} = 0.8m$, $m^*_{h2} = 0.6m$, $m^*_{h3} = 0.4m$ (m is free electron mass). The mobility in band one is 1000 cm^2/Volt-Sec. Holes in the three bands consist of 30 per cent in band 1, 40 per cent in band 2, and 30 per cent in band 3. In this crystal $n_i = 10^{12}$/c.c.

a. How much current will the battery have to supply at room temperature?

b. What is the electron density at room temperature?

Solution.

a. $$I = \frac{V}{R} = \frac{VA\sigma}{\ell} \doteq \frac{VAq\left[\mu_{p1}\left(\frac{p_1}{p}\right) + \mu_{p2}\left(\frac{p_2}{p}\right) + \mu_{p3}\left(\frac{p_3}{p}\right)\right]p}{\ell} \tag{1}$$

$$\mu_{p2} = \frac{\left(\frac{q\bar{t}}{m^*_{h2}}\right)\mu_{p1}}{\left(\frac{q\bar{t}}{m^*_{h1}}\right)} = \frac{m^*_{h1}\mu_{p1}}{m^*_{h2}} = \left(\frac{0.8m}{0.6m}\right)\left(1000\right) = 1333. \tag{2}$$

Similarly:

$$\mu_{p3} = \left(\frac{m^*_{h1}}{m^*_{h3}}\right)\mu_{p1} = \frac{0.8m}{0.4m}(1000) = 2000. \tag{3}$$

Thus:

$$I = \frac{(6)(0.2)(1.6 \times 10^{-19})}{(0.8)}[(1000)(0.3) + (1333)(0.4) + (2000)(0.3)]10^{17}. \tag{4}$$

$$I = 34.4 \text{ amp.} \longleftarrow$$

Note that

$$p = 5 \times 10^{17} - 4 \times 10^{17} = 10^{17},$$

because the electrons from the states in the middle of the band gap will fall into 4×10^{17} of the empty acceptor states leaving 10^{17} states still vacant. Valence band electrons will, at room temperature, occupy these 10^{17} acceptor states creating that many holes in the valence band.

b. $n_p = n_i^2/p_p = (10^{12})^2/10^{17} = 10^7$ per c.c. $\longleftarrow$

2-8 DENSITY OF STATES FUNCTION

In order to be able to calculate free electron and free hole densities, two pieces of information are essential. The first of these is the information on how many electronic states are possible in a given energy interval. The second gives information on the probability that a given permissible energy state is occupied or not by an electron. In this section we consider the first of these called the *density of states function*, designated $N(E)$.

Formally $N(E)\,dE$ is the volume density of electronic states in the energy interval from E to $E + dE$. The calculation of $N(E)$ is quantum mechanical in nature, and will not be attempted here. In principle it counts the number of different electronic states permissible in a given energy interval. These calculations, based on certain simplifying assumptions, lead to the following expression for the density of states function for conduction and valence bands respectively

$$N_c(E) = \frac{4\pi}{h^3}(2m_e^*)^{3/2}(E - E_c)^{1/2}, \tag{2–52}$$

$$N_v(E) = \frac{4\pi}{h^3}(2m_h^*)^{3/2}(E_v - E)^{1/2}, \tag{2–53}$$

where m_e = effective mass for electrons in conduction band,
m_h = effective mass for holes in valence band,
E_c = energy at the bottom of conduction band,
E_v = energy at the top of valence band.

It is to be noted that Eqs. (2–52) and (2–53) take into account the Pauli exclusion principle.

Bands which follow the functional dependence on E as expressed by Eqs. (2–52) and (2–53) are called parabolic bands. They are derived from the simplifying assumption that the Fermi surface of each band is spherical in shape. Though this is not a good assumption for silicon and germanium, it nevertheless yields quite accurate results, because this assumption is fairly good in the vicinity of the bottom of the conduction band and the top of the valence band where almost all electrons and holes of importance reside. Figure 2–18 shows roughly how the actual $N(E)$ looks. We note that $N(E)$ goes to zero at the edge of each band. This is unfortunately *not* predicted by Eq. (2–52) at the top of the

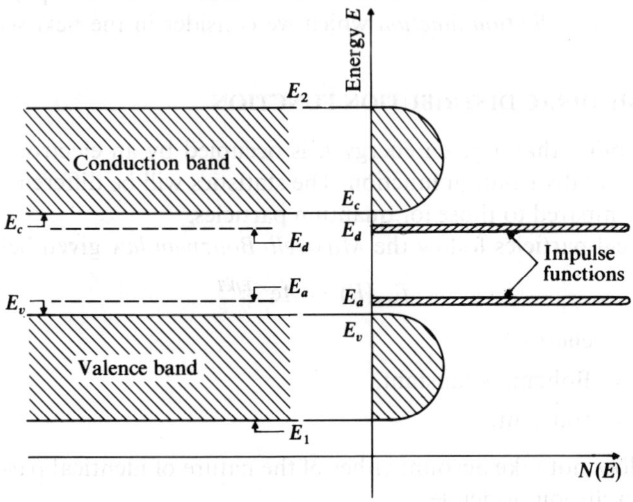

Fig. 2–18 Density of states function.

conduction band, nor by Eq. (2–53) at the bottom of the valence band. Again this will not lead to any serious errors because Eqs. (2–52) and (2–53) are both valid where electrons and holes, which are important to the conduction process, reside.

One more thing of interest to note in Fig. 2–18 *is that the area under each curve for conduction and valence bands must be equal because they have exactly the same number of states. Generally the higher the band is in energy the wider it is. The statement on equal areas then implies that the average height of* $N_c(E)$ *curve will be lower than that of* $N_v(E)$.

Table 2–6 gives the value of effective mass which should be used in the density of states functions Eqs. (2–52) and (2–53) in germanium and silicon. These take into account the rather detailed and complicated band structure of silicon and germanium to the extent that a single isotropic effective mass value can.

Table 2–6 Density of States Effective Mass†

	m_e^*	m_h^*
Ge	0.55 m	0.37 m
Si	1.1 m	0.59 m

† m^* is free electron mass.

It is important to emphasize that the density of states function gives merely the information on the *permissible* electron states that an electron can occupy. It *does not* say whether these states in a given case are occupied or not. To obtain the latter information the additional information is provided by the *Fermi–Dirac distribution function* which we consider in the next section.

2–9 FERMI–DIRAC DISTRIBUTION FUNCTION

The probability that a given energy E is occupied by an electron is described by a statistical distribution function. The statistics will be different for classical particles compared to those for quantum particles.

Classical particles follow the *Maxwell–Boltzman law* given below

$$f_{MB}(E) = A\mathrm{e}^{-E/kT}, \tag{2–54}$$

where E = energy,

k = Boltzman constant,

A = constant.

This law does not take account either of the nature of identical particles, nor of the Pauli exclusion principle.

Quantum particles with *symmetric wave functions* follow the *Bose–Einstein law* given below. The probability that an energy E is occupied by a particle subject to this law is given by

$$f_B(E) = \frac{1}{1 - e^{(E - E_b)/kT}} \tag{2–55}$$

where E_b = Bose level.

Particles which obey the distribution function given by Eq. (2–55) are called *Bosons.* They are particles characterized by symmetric wave functions. Examples of such particles are photons and pi-mesons.

By far the most useful distribution function for semiconductors is the *Fermi–Dirac distribution function* which is valid for quantum particles characterized by antisymmetric wave functions, and which takes account of the Pauli exclusion principle. Particles which obey the Fermi–Dirac distribution function are called *fermions.* These are particles characterized by antisymmetric wave functions. Examples of particles which obey this distribution are electrons, protons, and mu-mesons. For electrons this distribution function gives the probability that an electron occupies *a permissible electronic state,* i.e.

$$f(E) = \frac{1}{1 + e^{[E - E_f(T)]/kT}}. \tag{2–56}$$

where E_f is the Fermi level.

The functional notation in Eq. (2–56) indicates that the Fermi level is a function of temperature. It is, however, a relatively slowly varying function of temperature. To gain an insight into E_f, consider Eq. (2–56) at absolute zero degrees.

When $E < E_f$, the exponent in the exponential is minus infinity, hence this term vanishes and

$$f(E) = 1 \qquad \text{when} \qquad E < E_f \qquad \text{and} \qquad T = 0°\text{K}.$$

When $E > E_f$, the exponent to the exponential is positive infinity hence

$$f(E) = 0 \qquad \text{when} \qquad E > E_f \qquad \text{and} \qquad T = 0°\text{K}.$$

The resulting $f(E)$ distribution is shown in Fig. 2–19. *We may hence say that at absolute zero degrees all permissible energy states up to the Fermi level E_f are occupied by electrons, and all energy states above it are unoccupied.* The square look of the Fermi–Dirac distribution function at 0°K rounds out as the temperature is raised. At higher temperatures it can be seen from the distribution function, as portrayed in Fig. 2–19, that more and more electrons are likely to occupy the energy states higher than the Fermi level and that more electrons will be missing from energies lower than the Fermi level. It may also be noticed that at temperatures $T \neq 0$, at

$$E = E_f, f(E) = \tfrac{1}{2}.$$

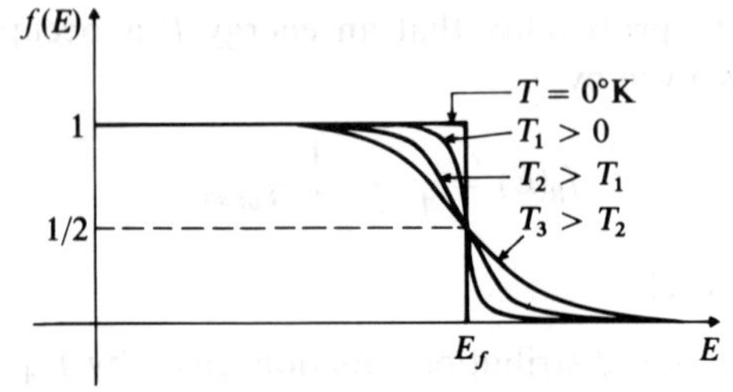

Fig. 2–19 Fermi–Dirac distribution function.

Figure 2–19 is slightly misleading in that *the Fermi level actually is a function of temperature unlike the portrayal of this figure.* This additional feature is left out of this figure to keep it simple. It should be noticed that the Fermi–Dirac distribution function is fairly sharp. That is, the probability changes from essentially one to essentially zero within a few kT's on either side of the Fermi level, and kT is 0.026 eV at room temperature.

2–10 CALCULATION OF ELECTRON AND HOLE DENSITIES, AND FERMI LEVEL

With the knowledge of the density of states function, and the Fermi–Dirac distribution function, we are ready to calculate the *free carrier* densities. Electrons in conduction band and holes in valence band are called *carriers* because they are carriers of current.

Electron density in the range of energies E to $E + dE$ is given by

$$dn = f(E)N_c(E)dE. \tag{2–57}$$

Thus electron density in an energy range E_1 to E_2 is obtained by integrating the Eq. (2–57).

$$n = \int_{E_1}^{E_2} f(E)N_c(E)dE. \tag{2–58}$$

To compute the free electron density we should let $E_1 = E_c$. Thus, using Eqs. (2–52) and (2–56) we have

$$n = \frac{4\pi}{h^3}(2m_e^*)^{3/2}\int_{E_c}^{E_2}\frac{(E - E_c)^{1/2}\, dE}{1 + e^{(E-E_f)/kT}}. \tag{2–59}$$

Unfortunately, the integral of Eq. (2–59) is neither standard nor tabulated. In order to proceed further, some simplifying assumptions are absolutely essential. First with very little error we can replace the upper limit of the integral by infinity. This may seem at first like a rather drastic change. The reason why this makes hardly any difference is that for the most reasonable doping densities the Fermi level ends up within the band gap, and the Fermi–Dirac distribution function, which appears as a factor in the integrand in Eq. (2–59), will force the integrand to have a negligible value for all energies higher than the vicinity of the bottom of conduction band. With this change we may rewrite Eq. (2–59) in terms of dimensionless parameters as

$$n = 4\pi \left(\frac{2mkT}{h^2}\right)^{3/2} \left(\frac{m_e^*}{m}\right)^{3/2} \int_0^\infty \frac{\eta^{1/2}\, d\eta}{1 + e^{\eta - \eta_f}} \tag{2–60}$$

or

$$n = N_0\, G_e\, F_{1/2}(\eta_f), \tag{2–61}$$

where

$$N_0 \triangleq 4\pi \left(\frac{2mkT}{h^2}\right)^{3/2} \tag{2–62}$$

$$G_e \triangleq \left(\frac{m_e^*}{m}\right)^{3/2} \tag{2–63}$$

$$F_{1/2}(\eta_f) \triangleq \int_0^\infty \frac{\eta^{1/2}\, d\eta}{1 + e^{\eta - \eta_f}} \tag{2–64}$$

$$\eta \triangleq \frac{E}{kT} \tag{2–65}$$

$$\eta_f \triangleq \frac{E_f}{kT}. \tag{2–66}$$

In Eq. (2–60) we have set $E_c = 0$ for reference. Equation (2–64) is called the *Fermi–Dirac integral* and is tabulated. This tabulation is graphically displayed in Fig. 2–20. As this integral does not have an analytic expression, a search for an analytic solution for electron density requires additional assumptions.

If

$$(E - E_f) \gg kT \tag{2–67}$$

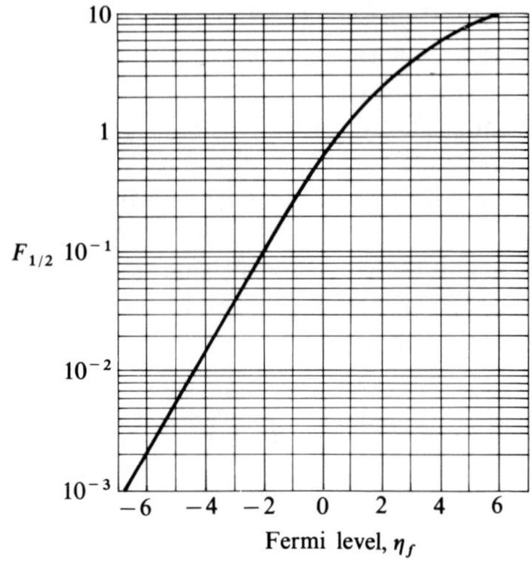

Fig. 2–20 Graphical plot of Eq. (2–64), the Fermi–Dirac integral.

then Eq. (2–59) can be written

$$n = 4\pi \left(\frac{2m_e^*}{h^2}\right)^{3/2} \int_0^\infty E^{1/2} \mathrm{e}^{-(E - E_f)/kT} \, dE. \tag{2–68}$$

In Eq. (2–68) the exponential in the denominator of the integrand of Eq. (2–59) has been considered much greater than 1.

To evaluate Eq. (2–68) let us define a new variable as follows:

$$x = \frac{E - E_c}{kT}, \qquad \text{thus} \qquad dE = kT\,dx. \tag{2–69}$$

So Eq. (2–68) can be rewritten as

$$n = 4\pi \left(\frac{2m_e^*}{h^2}\right)^{3/2} \mathrm{e}^{E_f/kT} kT \int_0^\infty (kTx)^{1/2} \, \mathrm{e}^{-(kTx + E_c)/kT} \, dx,$$

$$n = 4\pi \left(\frac{2m_e^*}{h^2}\right)^{3/2} \mathrm{e}^{-(E_c - E_f)/kT} \, (kT)^{3/2} \int_0^\infty x^{1/2} \, e^{-x} \, dx. \tag{2–70}$$

Now

$$\int_0^\infty x^{1/2} \, \mathrm{e}^{-x} \, dx = \frac{\sqrt{\pi}}{2}.$$

Thus

$$n = 4\pi \left(\frac{2m_e^*}{h^2}\right)^{3/2} \frac{\sqrt{\pi}}{2} (kT)^{3/2} e^{(E_f - E_c)/kT}. \tag{2–72}$$

More simply

$$n = N_c \, e^{(E_f - E_c)/kT}, \tag{2–73}$$

where

$$N_c \triangleq 2\left(\frac{2\pi m_e^* kT}{h^2}\right)^{3/2}. \tag{2–74}†$$

To calculate the free hole density in the valence band we may write

$$p = \int_{-\infty}^{E_v} f_p(E)\, N_v(E)\, dE. \tag{2–75}$$

In Eq. (2–75) $f_p(E)$ is the probability that an electronic state E, *in the valence band*, is occupied by a hole. In the valence band occupancy of an energy E by a hole is the same thing as its lack of occupancy by an electron. Thus,

$$f_p(E) = 1 - f(E) = \frac{1}{1 + e^{-(E - E_f)/kT}}. \tag{2–76}$$

If Eqs. (2–76) and (2–53) are substituted into Eq. (2–75) we shall, under conditions similar to those which led to Eq. (2–73), obtain the following expression for free hole density.

$$p = N_v \, e^{(E_v - E_f)/kT}, \tag{2–77}$$

where

$$N_v \triangleq 2\left(\frac{2\pi m_h^* kT}{h^2}\right)^{3/2}. \tag{2–78}‡$$

An interesting and very useful result is obtained by multiplying n by p from Eqs. (2–73) and (2–77).

$$np = N_c N_v \, e^{(E_v - E_f + E_f - E_c)/kT},$$

$$np = N_c N_v \, e^{(E_v - E_c)/kT} = N_c N_v \, e^{-E_g/kT} \triangleq n_i^2. \tag{2–79}$$

Equation (2–79) says that for a given material (which fixes E_g, m_e^*, and m_h^*) and given temperature (which fixes N_c and N_v), the product of electron and hole densities is a constant at thermal equilibrium. So if for some reason electron density is increased, for example by doping, the hole density will have to decrease to keep the product np constant. This product is also designated by n_i^2 to

†Note N_c must not be confused with $N_c(E)$ which is given by Eq. (2–52).
‡Note that $N_v \neq N_v(E)$. The latter is given by Eq. (2–53).

indicate that in an intrinsic material $n = p = n_i$, hence $np = n_i^2$. This fact was used in Example 2–1.

It is worthwhile noting that because of Eq. (2–79) the minority carrier density is quite small compared to the majority carrier density in semiconductors which are relatively heavily doped, i.e. where

$$N_d \quad \text{or} \quad N_a \gg n_i. \tag{2–79a}$$

It is important to note that in arriving at Eq. (2–73), the approximation of Eq. (2–67) was necessary. Similarly, in arriving at Eq. (2–77), the following approximation was necessary

$$(E_f - E) \gg kT. \tag{2–79b}$$

Equations (2–73) and (2–77) are quasi classical in that they follow Maxwell–Boltzman-like statistics. These are valid if the Fermi level is several kT's away from the conduction band edge and valence band edge.

Table 2–7 gives the value of n_i^2 in the three most commonly used semiconductors.

Table 2–7

n_i^2	5.6×10^{26}	2.25×10^{20}	81×10^{12}	cm^{-6}
at 300°K	Ge	Si	GaAs	

A knowledge of Fermi level gives a good idea of the distribution of carriers. The Fermi–Dirac distribution function changes rather abruptly from a value of close to one to close to zero within a few kT's on either side of E_f. Thus, it is important to know where the Fermi level is located. A good way of doing this is to require that the Fermi level will have to be consistent with space charge neutrality.

A statement of space charge neutrality is to say that the total positive charge equals the total negative charge. The net positive charge consists of the holes and *ionized donors.* The net negative charge consists of electrons and *ionized acceptors.* The above can be mathematically stated as an equation as follows

$$p + (N_d - n_d) = n + n_a, \tag{2–80}$$

where N_d = density of donor atoms,

n_d = density of electrons in the donor states,

n_a = density of electrons in the acceptor states.

The term in parenthesis $(N_d - n_d)$ represents the density of ionized donors because only those donor states which have "donated" the electrons are ionized. The apparent asymmetry of Eq. (2–80) is because the different conditions under which the acceptors are neutral compared to those when the donors are considered neutral. Donors are neutral when the donor states *are occupied* by electrons, whereas acceptors are neutral when acceptor states *are not occupied by* electrons.

Let us use the requirement of space charge neutrality to compute the Fermi level in three different useful cases.

Case A. Intrinsic Semiconductor

Here $n_d = N_d = 0 = n_a$. Hence from Eq. (2–80)

$$n = p. \tag{2–81}$$

Eqs. (2–73) and (2–77) may be used to obtain

$$N_c \, e^{(E_f - E_c)/kT} = N_v \, e^{(E_v - E_f)/kT} \tag{2–82}$$

or

$$\frac{N_c}{N_v} = e^{(2E_f - E_c - E_v)/kT}.$$

Taking the logarithm of both sides we obtain

$$\ln\left(\frac{N_c}{N_v}\right) = (2E_f - E_c - E_v)/kT$$

or

$$kT \ln\left(\frac{N_c}{N_v}\right) = 2E_f - E_c - E_v. \tag{2–83}$$

Using Eqs. (2–74) and (2–78) we can write

$$\frac{3kT}{2} \ln\left(\frac{m_e^*}{m_h^*}\right) = 2E_f - E_c - E_v. \tag{2–84}$$

The term on the left hand side of Eq. (2–84) is seen to be rather small compared to the right hand side and may hence be set equal to zero. Thus

$$2E_f - E_c - E_v = 0$$

or

$$E_f = \frac{E_c + E_v}{2}. \tag{2–85}$$

Equation (2–85) indicates that in intrinsic semiconductors, the Fermi level is located in the middle of the band gap. It actually departs from the middle of the band gap by the small logarithmic term in Eq. (2–84).

Case B. Pure n-Type Semiconductor at Room Temperature

Let us refer to Eq. (2–80), in this case $n_a = 0$ because of the absence of acceptors. Furthermore, at room temperature we may reasonably assume that all donors are ionized (this is true for most practical donors). This implies that there are no electrons in donor states, that is $n_d = 0$. If $N_d \gg n_i$ we may assume from Eq. (2–79) that $p \ll n$. Utilizing these observations we have

$$\not{p} + (N_d - \not{n}_d) = n + \not{n}_a \tag{2–86†}$$

or

$$N_d = n = N_c \exp\left(\frac{E_f - E_c}{kT}\right). \tag{2–87}$$

or

$$\frac{N_d}{N_c} = \exp\left(\frac{E_f - E_c}{kT}\right). \tag{2–88}$$

Taking the logarithm of both sides we obtain

$$\ell n\left(\frac{N_d}{N_c}\right) = \frac{E_f - E_c}{kT}. \tag{2–89}$$

Finally

$$E_f = E_c + kT\,\ell n\left(\frac{N_d}{N_c}\right) \tag{2–90}$$

For most *practical* doping densities

$$N_d < N_c, \tag{2–91}$$

For example at room temperature $N_c = 2.8 \times 10^{19}$ per c.c. in silicon. The normal doping densities are in the range 10^{14} to 10^{17} per c.c. Thus the second term on the right hand side of Eq. (2–90) is negative. Thus the Fermi level in a pure *n*-type semiconductor is somewhat below the bottom of the conduction band.

Case C. Pure p-Type Semiconductor at Room Temperature

In this case

$$N_d = n_d = 0 \qquad \text{and} \qquad p \gg n.$$

Using the above, Eq. (2–80) becomes

$$p = n_a. \tag{2–92}$$

---

†A line through a quantity indicates that it can be neglected.

If we assume that the doping density of acceptor atoms is N_a and assume that all acceptors are ionized, the following is obtained

$$n_a = N_a = N_v \, e^{(E_v - E_f)/kT}, \tag{2–93}$$

where we have used Eq. (2–77). Equation (2–93) can be rewritten as follows

$$\frac{N_a}{N_v} = e^{(E_v - E_f)/kT}. \tag{2–94}$$

If we take the logarithm of both sides of the above equation we obtain

$$\ell n\left(\frac{N_a}{N_v}\right) = (E_v - E_f)/kT. \tag{2–95}$$

Finally

$$E_f = E_v - kT\,\ell n\left(\frac{N_a}{N_v}\right). \tag{2–96}$$

Again for most practical doping densities

$$N_a \ll N_v.$$

So the second term on the right side of Eq. (2–96) is actually positive. Thus in a *p*-type semiconductor the Fermi level is somewhat above the top of the valence band.

If the approximation of Eq. (2–67) is not valid, then Eqs. (2–73) and (2–77) are also not valid. The calculation in such a case involves the solution of transcendental equations. Consider the case of pure *n*-type semi-conductor where Eq. (2–67) is not valid as indeed would be the case if E_f is say kT below the conduction band. In this case $n_a = 0$, $p \approx 0$ hence

$$n = N_d - n_d. \tag{2–97}$$

But

$$n_d = N_d f(\eta_d) \tag{2–98}$$

Equation (2–98) is obtained by applying Eq. (2–58) and integrating over the range of energies spanning the donor density of states function which is ideally assumed to be an impulse function enclosing an area of N_d in infinitesimal energy interval centered at E_d. Thus

$$n_d = \frac{N_d}{1 + e^{-\eta_d + \eta_f}}. \tag{2–99}$$

Substituting Eq. (2–99) into Eq. (2–97)

$$n = \frac{N_d}{1 + e^{-\eta_d + \eta_f}}. \tag{2–100}$$

Now we may use Eq. (2–61) for electron density to obtain

$$N_0 G_e F_{1/2}(\eta_f) = \frac{N_d}{1 + e^{-\eta_d + \eta_f}}. \tag{2–101}$$

Equation (2–101) can be solved graphically by plotting the left and right hand sides on the common abscissa of η_f. Intersection determines η_f. From η_f a value of E_f can be calculated by using Eq. (2–66). An alternative way of solving Eq. (2–101) is to use numerical trial and error procedure.

Lastly, in some calculations it is important to keep in mind that the band gap E_g is a function of temperature. In the restricted range of temperatures $150°\text{K} < T < 300°\text{K}$, a linear relationship is obtained in Ge and Si.

$$E_g(T) = E_g(0) - \beta_g T. \tag{2–102}$$

Table 2–8 gives values of β_g and $E_g(0)$ in Ge and Si.

Table 2–8

	$E_g(O)$ (eV)	β_g (eV/°K)
Ge	0.75	4.0×10^{-4}
Si	1.15	4.4×10^{-4}

Let us see if we can predict a general trend of how the Fermi level changes with temperature. We have just seen that the Fermi level is near the bottom of the conduction band in *n*-type material and near the top of the valence band in *p*-type material. As the temperature is raised sufficiently, we expect that electron-hole pairs will be created by electrons going directly from the valence to the conduction band. If this becomes the major process by which additional carriers are produced,

$$n = p.$$

This is precisely the case where we expect the Fermi level to be approximately in the middle of the band gap (see Eq. 2–85). Thus Fig. 2–21 shows the change of E_f with temperature. These curves need not be monotonic.†

2–11 RECOMBINATION‡

The process by which a free electron and a free hole come together to produce their mutual demise is called *recombination.*

In the early days the model for recombination was that of a wondering

†The interested reader should refer to Boltaks (1963).

‡A much more detailed discussion of the recombination–generation process will be found in Chapter 12.

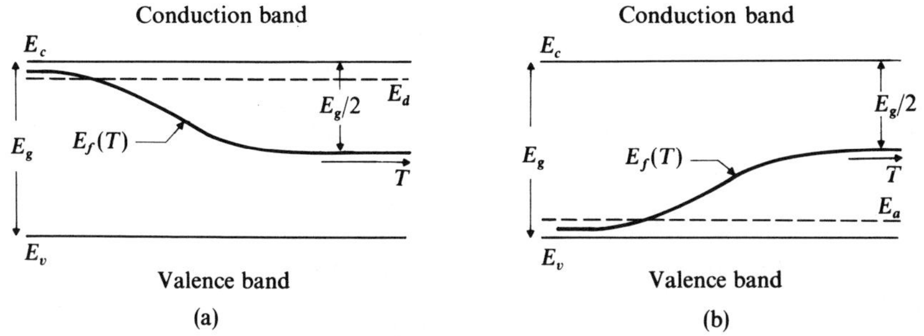

Fig. 2–21 Fermi level as a function of temperature. (a) $E_f(T)$ in pure n-type semiconductor. (b) $E_f(T)$ in pure p-type semiconductor.

free electron coming, by chance, by a wondering free hole to produce recombination. This model predicted a much longer average time before recombination than was experimentally observed in silicon and germanium. A successful model proposed by Shockley and Read (1952) is shown in Fig. 2–22. Note here that a band gap electronic energy level E_r is shown. An electron from the conduction band is initially trapped by this *empty* energy level E_r. Subsequently this electron falls into a hole in the valence band thus producing *recombination.* This last step is sometimes called the capture of the hole by the *electron occupied* energy state E_r. Thus the energy state is thought to alternate as an electron and hole trap. If the probability of capturing an electron, when the energy state at E_r is empty, is roughly comparable to the probability of capturing a hole when the energy state is occupied by an electron, then the energy state E_r is called a *recombination center.* The Shockley-Read theory *assumes that such recombination centers are uniformly distributed in space.* This model predicted an average lifetime of electrons or holes before recombination much closer to experimentally observed lifetime than the earlier model.

Another experimental finding of great importance is that any departure of carrier density from its thermal equilibrium value diminishes exponentially

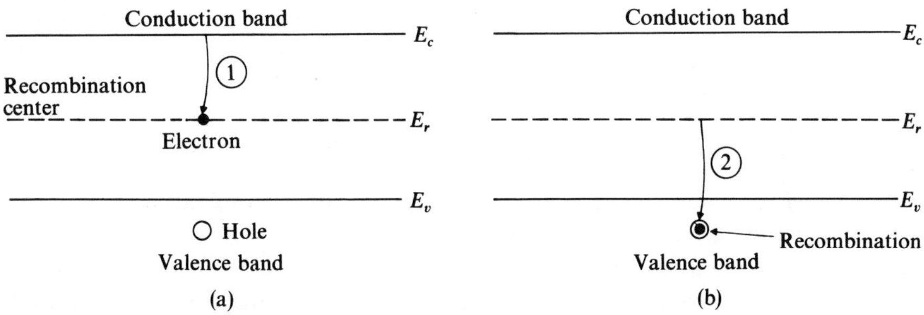

Fig. 2–22 Recombination through recombination center.

with time to its equilibrium value. Mathematically this can be stated as follows:

$$p(t) = p_0 + (p_i - p_0)\,e^{-t/\tau_p}, \tag{2–103}$$

where p_0 = thermal equilibrium value of hole density,
p_i = initial hole density,
τ_p = lifetime of holes.

The rate at which the density of holes changes with time is

$$\frac{\partial p}{\partial t} = -\frac{(p_i - p_0)\,e^{-t/\tau_p}}{\tau_p}. \tag{2–104}$$

From Eq. (2–103),

$$(p_i - p_0)\,e^{-t/\tau_p} = p - p_0. \tag{2–105}†$$

Equation (2–105) can be substituted into Eq. (2–104) to obtain

$$\frac{\partial p}{\partial t} = -\frac{p - p_0}{\tau_p}. \tag{2–106}$$

We shall return to this equation when we discuss the continuity and diffusion equations.

Quite analogously for electron density

$$\frac{\partial n}{\partial t} = -\frac{n - n_0}{\tau_n}, \tag{2–107}$$

where n_0 = thermal equilibrium value of electrons,
n = electron density at time t,
τ_n = lifetime of electrons.

The lifetimes τ_p and τ_n discussed so far are really effective life times which take account of both *bulk recombination* and *surface recombination* in the following way

$$\frac{1}{\tau} = \frac{1}{\tau_b} + \frac{1}{\tau_s}, \tag{2–108}$$

where τ = hole or electron lifetime,
τ_b = bulk lifetime,
τ_s = surface lifetime.

Bulk recombination takes place in the interior, i.e. away from surfaces, whereas the surface recombination takes place only at surfaces. The former does not depend on geometry of the crystal. The latter, unfortunately, is a function of the physical dimensions of the crystal. This is because the carrier *has to first*

†For simplicity here p is written to mean $p(t)$.

reach the surface before it recombines there. A quantity associated with τ_s which is *not* a function of the physical dimensions is called *surface recombination velocity* and is designated by *s*. It is defined as

$$s = \frac{\text{Recombination rate per unit area}}{\text{Excess concentration (i.e. above the thermal equilibrium value) just below surface}}.$$

Thus surfaces are characterized by stating their *s* value. Generally speaking

$$\tau_s = f(s). \tag{2–109}$$

In most of what follows we shall assume τ is a constant. In general it is a function of doping density, carrier densities, temperature, recombination center density, and crystaline defects density.

Shockley has calculated that in long rectangular rods of semiconductor where surface recombination is important,

$$\frac{1}{\tau_s} = D\left(\frac{a}{A^2} + \frac{b}{B^2}\right), \tag{2–109a}$$

where $a \tan a = s\dfrac{A}{D}$,

$b \tan b = s\dfrac{B}{D}$, and

$D =$ diffusion constant which will be discussed in Section 2–12.

2–12 DIFFUSION

One of the most important processes by which carriers are transported is *diffusion.* The other process is *drift* under the influence of electric field which was considered in Section 2–6. Diffusion is a very general physical process which applies to all groups of particles. It says that because of the random motion of particles at a given temperature their net motion will be in the direction of decreasing concentration of these particles. If the particles are charged, their motion constitutes current density. The resulting current density is proportional to the density gradient. The proportionality constant is called diffusivity or *diffusion constant.* Mathematically the following relationships may be written for hole and electron current densities

$$\mathbf{j}_p = -q\, D_p \nabla p \tag{2–110}†$$

$$\mathbf{j}_n = +q\, D_n \nabla n, \tag{2–111}†$$

---

†$\nabla\phi$ is called the *gradient* of ϕ. In Cartesian coordinates it is $\nabla\phi = \mathbf{i}\dfrac{\delta\phi}{\delta x} + \mathbf{j}\dfrac{\delta\phi}{\delta y} + \mathbf{k}\dfrac{\delta\phi}{\delta z}$ where **i**, **j**, and **k** are unit vectors along the *x*, *y*, and *z* axis.

where D_p, D_n = diffusion constants for holes and electrons respectively. Note that the negative sign in Eq. (2–110) indicates that holes go towards regions of *decreasing* density of holes. The latter part of this statement is also true for electrons. In connection with electrons it should be remembered that *conventional current is defined to be opposite the direction of electron movement.* Thus diffusion *current* due to electrons goes towards regions of increasing electron density and hence the positive sign to the right hand side of Eq. (2–111). Diffusion flow will stop when the concentration density is uniform and hence the gradient of density is zero. Figure 2–23 shows the change of hole density with time due to diffusion. This figure also shows on $t_0 = 0$ curve the relative magnitude of expected current density at different values of x.

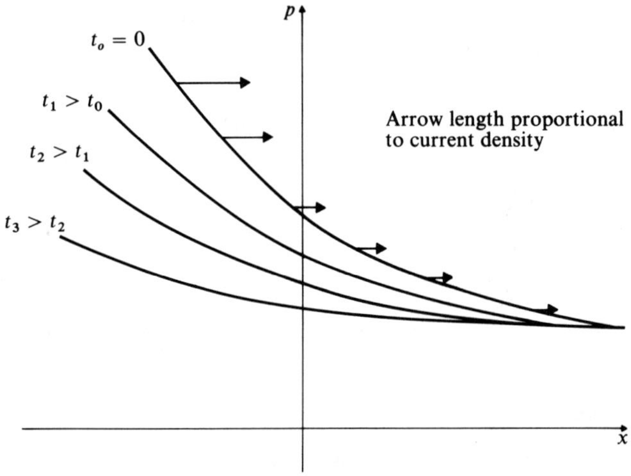

Fig. 2–23 Change of hole distribution due to diffusion.

When the density of carriers varies with position, and when there is an electric field present, the hole and electron current densities expected in general are

$$\mathbf{j}_p = q\,\mu_p p\mathscr{E} - qD_p\,\nabla p \qquad (2\text{–}112)$$

$$\mathbf{j}_n = q\,\mu_n n\mathscr{E} + qD_n\,\nabla n, \qquad (2\text{–}113)$$

where $\mathscr{E}$ = electric field.

The first terms on the right hand sides of Eqs. (2–112) and (2–113) are $\sigma\mathscr{E}$ terms due to the drift field. Equations (2–112) and (2–113) are called the *general transport equations*, and are quite important in semiconductor devices as they link the carrier densities to current densities.

2-13 CONTINUITY EQUATION

Consider a differential volume element as pictured in Fig. 2–24 within p-type semiconductor. We wish to write the equation of continuity. We shall first write the equation in words and then write the appropriate mathematical form for each term in words.

$$\overset{(1)}{\left[\text{Net current out of the volume}\right]} + q\overset{(2)}{\left[\begin{array}{c}\text{Rate at which holes are lost}\\ \text{due to recombination in the}\\ \text{volume}\end{array}\right]}$$

$$-q\overset{(3)}{\left[\begin{array}{c}\text{Rate at which holes are being}\\ \text{generated in the volume}\end{array}\right]} = q\overset{(4)}{\left[\begin{array}{c}\text{Net rate at which holes are}\\ \text{lost from the volume}\end{array}\right]} \tag{2–114}$$

$$(1) = \left[\begin{array}{c}\text{Net current out}\\ \text{of the volume.}\end{array}\right] = \left[\left(j_p + \frac{\partial j_p}{\partial x}dx\right) - j_p\right]dydx = \frac{\partial j_p}{\partial x}dxdydz. \tag{2–115}$$

The middle term in Eq. (2–115) $dydz$ is present to convert current density to current.

$$(2) = q\left[\begin{array}{c}\text{Rate at which holes}\\ \text{are lost due to re-}\\ \text{combination in the}\\ \text{volume.}\end{array}\right] = q\left(-\frac{\partial p}{\partial t}\right)dxdydz = q\left(\frac{p - p_n}{\tau_p}\right)dxdydz$$

$$= \frac{q(p - p_n)}{\tau_p}dxdydz. \tag{2–116}$$

In writing Eq. (2–116) we have used Eq. (2–106). Furthermore, we have substituted p_n for p_0 which was the steady-state value of holes. As the region of greatest interest for evaluating hole density will be in n-type region, where they are *minority carriers*, the thermal equilibrium value of holes in n-type regions shall be designated as p_n as done† in Eq. (2–116).

$$(3) = q\left[\begin{array}{c}\text{Rate at which holes}\\ \text{are being generated}\\ \text{in the volume.}\end{array}\right] = qG_p\,dxdydz, \tag{2–117}$$

where G_p = rate of generating holes.

$$(4) = q\left[\begin{array}{c}\text{Rate at which holes}\\ \text{are being lost in the}\\ \text{volume.}\end{array}\right] = q\left(-\frac{\partial p}{\partial t}\right) = -q\frac{\partial p}{\partial t}dxdydz. \tag{2–118}$$

†This is a general scheme of notation for carrier densities. The subscript designates the region in which the carrier density is evaluated.

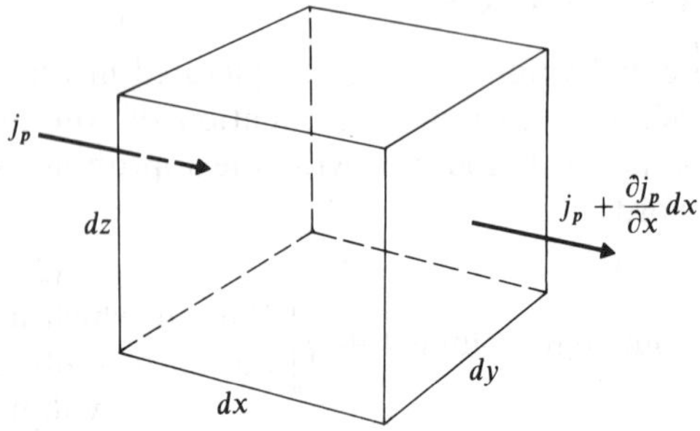

Fig. 2–24 Differential volume element for continuity equation.

In Eqs. (2–116), (2–117) and (2–118) the volume element multiplies each of the terms so that the volume *density* of holes is converted to the *number* of holes. Note further that Eq. (2–115) and Fig. 2–24 show that here we are considering a one-dimensional situation. Substituting Eqs. (2–115) through (2–118) into Eq. (2–114) we obtain

$$\frac{\partial j_p}{dx} dxdydz + q\frac{(p - p_n)}{\tau_p} dxdydz - q\, G_p dxdydz = -q\frac{\partial p}{\partial t} dxdydz \quad (2\text{–}119)$$

or

$$\frac{\partial j_p}{dx} + q\left[\frac{(p - p_n)}{\tau_p} - G_p\right] = -q\frac{\partial p}{\partial t}. \quad (2\text{–}120)$$

If we had considered a three-dimensional problem, instead of the one-dimensional problem, the following general continuity equation would have resulted

$$\nabla \cdot \mathbf{j}_p + q\left[\frac{p - p_n}{\tau_p} - G_p\right] = -q\frac{\partial p}{\partial t}. \quad (2\text{–}121)$$

It should be pointed out that the term involving generation of holes G_p is due to *causes other than thermal.* Thermal generation of carriers is taken into account by the recombination term of Eq. (2–116) which turns into generation if p is less than the thermal equilibrium value of hole density in n-type region p_n.

If we had considered the continuity of electrons in p-type region, the following equation analogous to Eq. (2–121) would have been obtained

$$-\nabla \cdot \mathbf{j}_n + q\left[\frac{n - n_p}{\tau_n} - G_n\right] = -q\frac{\partial n}{\partial t}, \quad (2\text{–}122)$$

where $\mathbf{j}_n$ = electron current density,

n_p = thermal equilibrium value of electrons in the p-type region,

G_n = generation rate of electrons.

If the transport Eq. (2–112) for holes is substituted into the continuity equation the following is obtained.

$$\nabla \cdot [q\mu_p p\,\mathscr{E} - qD_p \nabla p] + q\left[\frac{p - p_n}{\tau_p} - G_p\right] = -q\frac{\partial p}{\partial t} \tag{2–123}$$

or

$$q\mu_p \nabla \cdot (p\mathscr{E}) - qD_p \nabla^2 p + q\left[\frac{p - p_n}{\tau_p} - G_p\right] = -q\frac{\partial p}{\partial t}. \tag{2–124}$$

Following the rules for differentiation of the product in the first term of the left hand side of Eq. (2–124), the above equation can be written as

$$q\mu_p (p\nabla \cdot \mathscr{E} + \mathscr{E} \cdot \nabla p) - qD_p \nabla^2 p + q\frac{(p - p_n)}{\tau_p} - qG_p = -q\frac{\partial p}{\partial t}. \tag{2–125}$$

Eq. (2–125) is valid in a region which has both electric field and carrier gradient present. If we agree never to use Eq. (2–125) in regions where there is a net charge density present (hence an electric field-free region) the first two terms on the left hand side can be set equal to zero. The first term is zero because, from field theory,

$$\nabla \cdot \mathscr{E} = \frac{\sigma}{\varepsilon}, \tag{2–126}$$

where σ = net charge density

ε = permittivity

and $\sigma = 0$ by assumption. The second term which is proportional to $\mathscr{E}$ is zero because the electric field is zero.

Cancelling the q in the remaining terms, dividing through by D_p and multiplying the equation by -1 we obtain

$$\nabla^2 p - \frac{p - p_n}{\tau_p D_p} + \frac{G_p}{D_p} = \frac{1}{D_p}\frac{\partial p}{\partial t}. \tag{2–127}$$

In *Cartesian coordinates* the operator ∇^2 is given by

$$\nabla^2 = \frac{\partial^2}{\partial x^2} + \frac{\partial^2}{\partial y^2} + \frac{\partial^2}{\partial z^2}. \tag{2–128}$$

A new parameter, diffusion length, can now be defined

$$L_p^2 \triangleq D_p \tau_p, \tag{2–129}$$

where L_p = diffusion length for holes.

So Eq. (2–127) may be written somewhat more compactly as

$$\nabla^2 p - \frac{p - p_n}{L_p^2} + \frac{G_p}{D_p} = \frac{1}{D_p}\frac{\partial p}{\partial t}. \tag{2–130}$$

An exactly analogous equation for electrons obtained in a similar manner is

$$\nabla^2 n - \frac{n - n_p}{L_n^2} + \frac{G_n}{D_n} = \frac{1}{D_n}\frac{\partial n}{\partial t} \tag{2–131}$$

and

$$L_n^2 \triangleq \tau_n D_n, \tag{2–132}$$

where n_p = thermal equilibrium value of electron density in the p-region,
L_n = diffusion length for electrons.

2–14 CURRENT IN p-TYPE SEMICONDUCTOR

If a piece of semiconductor is connected at its ends through *ohmic contacts*† to a battery, the resulting current is due entirely to drift electric field. As there is no build-up of carriers in this case, there is no gradient of these carriers, and hence no diffusion currents. The transport Eqs. (2–112) and (2–113) then state that the drift currents are proportional to the carrier densities. In this section, we examine the relationship between the current in the external circuit and current internal to the semiconductor.

If the piece of semiconductor connected to the battery is n-type, no conceptual problems present themselves as the current in the n-type semiconductor is primarily carried by electrons as is the current in the external metal wires. The story about a p-type semiconductor crystal being connected to external metallic wires is quite different. In this case the transport equations say that the current is carried primarily by holes in the p-type semiconductor and by electrons in the external wire. The question therefore presents itself, "How does the electron current in the wires transform itself to hole current in the p-type semiconductor?"

Figure 2–25 shows a composite diagram of energy band and spatial arrangement. In Fig. 2–25a is shown the conditions prevailing with no bias applied. Electrons in the acceptor states and corresponding holes in the valence band are clearly seen. When the terminals of this crystal are connected to the battery as shown, an electric field is found to exist as shown. Under its influence

†Ohmic contacts are usually contacts between semiconductors and metals. These are non-*rectifying* contacts. There can be metal-semiconductor junctions which are rectifying, but we are not considering them here. Metal-semiconductor rectifying contacts, Shottky barriers are discussed in Section 10–6. Ohmic contacts can be represented by a simple resistance usually small in value.

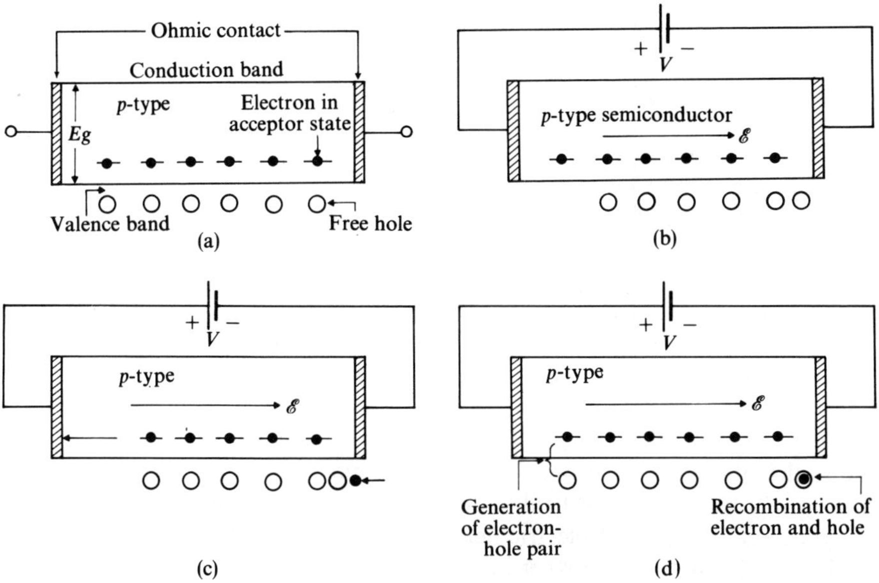

Fig. 2–25 Transport of carriers in *p*-type semiconductor.

the free holes will move to the right leaving a lack of holes on the left and an excess of holes on the right. This movement creates local electric fields at both ends because of an absence of space charge neutrality. The field on the right hand end entices an electron from the metal contact into the semiconductor; whereas the field on the left entices the electron from the acceptor state into the metal as shown in Fig. 2–25c. Figure 2–25d shows that finally an electron from the valence band goes into the available acceptor state on the left creating a free hole, whereas recombination accounts for the simultaneous disappearance of the free hole and the newly arrived electron. This completes the cycle which repeats again and again, and explains how hole flow in the *p*-type semiconductor transforms into electron flow in the metal wires. Two important conclusions follow from the above discussion. First, current flows with equal ease in either direction at metal–semiconductor junctions; hence these are nonrectifying junctions. Second, that recombination and generation of free holes takes place very close to the metal–semiconductor ohmic junctions.

2–15 SATURATION VELOCITY

Equations (2–32) and (2–33) assume that the velocity of a carrier is proportional to the electric field. This linear relationship is not unfortunately true at all field strengths. The velocity in fact follows a trend portrayed in Fig. 2–26. This

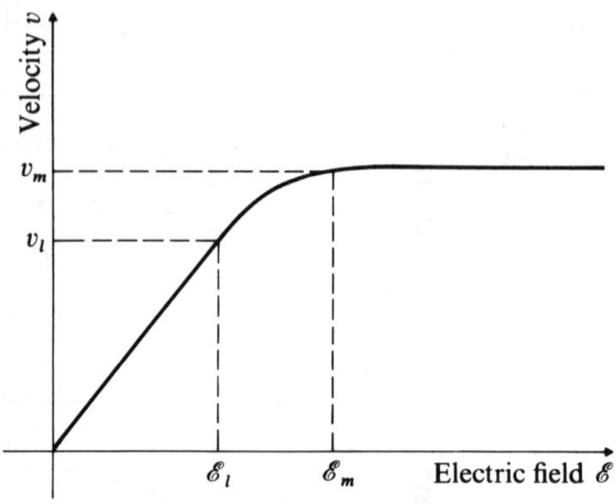

Fig. 2–26 $v(\mathscr{E})$.

shows that the linear relationship, which assumes that mobility is constant, is true up to a certain electric field strength, $\mathscr{E}_\ell$. The corresponding velocity is v_ℓ.

At a sufficiently high electric field $\mathscr{E}_m$ the velocity saturates to a value v_m, see Fig. 2–26. That is, no further increase in the velocity occurs no matter how high the electric field strength.† Table 2–9 gives a tabulation of $\mathscr{E}_\ell$, v_ℓ, $\mathscr{E}_m$, v_m. In general the saturation velocity occurs at a lower value of electric field as the temperature is lowered.

Table 2–9 Experimentally† obtained electric fields and drift velocities at 298°K

	Germanium		Silicon	
	Electrons	Holes	Electrons	Holes
Limiting linear electric field‡ $\mathscr{E}_\ell$ Volt/cm	900	1,400	2,500	7,500
Limiting linear velocity‡ v_ℓ cm/sec	3.2×10^6	2.4×10^6	3.3×10^6	2.8×10^6
Maximum limiting field‡ $\mathscr{E}_m$ Volt/cm	3,300	8,000	17,000	20,000
Maximum limiting velocity‡ v_m cm/sec	6.2×10^6	5.7×10^6	9×10^6	4×10^6

We can reasonably deduce that when the saturation velocity is reached the mobility varies inversely with the electric field. Note that Eqs. (2–32) and (2–33) assume that the mobility is independent of electric field. The fact that a limiting or saturation velocity exists is of great significance in many semiconductor devices.

†This statement is true for $\mathscr{E} < 10^5$ Volt/cm in Ge and Si so that the breakdown phenomena, explained in Section 3–6, is not approached.

2–16 NONUNIFORMLY DOPED SEMICONDUCTORS AND ASSOCIATED BUILT-IN FIELD

When semiconductor crystals are doped such that the doping level is a function of position along the crystal, they are called *graded semiconductors.* It is important to study them because they are in widespread use in the high frequency transistors. In this section we consider some general properties of graded semiconductors.

The following will be assumed in the analysis.

1. One-dimensional situation is being considered.
2. In every region space charge neutrality is satisfied (as a good approximation).
3. All donor (or acceptor) atoms are ionized.

Consider an n-type semiconductor which is doped nonuniformly. That is the donor concentration is a function of position x. For this reason the thermal equilibrium density of *both* electrons and holes will be functions of position and cannot, therefore, be considered to be constant. Thus,

$$n_n(x) = N_d(x). \tag{2–133}$$

The hole density may be deduced from the following:

$$n_n(x)\,p_n(x) = n_i^2. \tag{2–134}$$

Hence

$$p_n(x) = \frac{n_i^2}{n_n(x)} = \frac{n_i^2}{N_d(x)}. \tag{2–135}$$

Consider the situation prevailing at thermal equilibrium. We begin by writing the transport equations applicable to the present case.

$$j_p = q\mu_p\,p_n(x)\,\mathscr{E}_x - qD_p\frac{\partial p_n(x)}{\partial x}, \tag{2–136}$$

$$j_n = q\mu_n\,n_n(x)\,\mathscr{E}_x + qD_n\frac{\partial n_n(x)}{\partial x}. \tag{2–137}$$

In thermal equilibrium the *currents of both carriers must be individually zero.* Setting the left hand sides of Eqs. (2–136) and (2–137) to zero, and solving for the electric field we obtain respectively

$$\mathscr{E}_x = \frac{qD_p}{q\mu_p p_n(x)}\frac{\partial p_n(x)}{\partial x} = \frac{kT}{qp_n(x)}\frac{\partial p_n(x)}{\partial x} \tag{2–138}$$

and

$$\mathscr{E}_x = -\frac{qD_n}{q\mu_n n_n(x)}\frac{\partial n_n(x)}{\partial x} = -\frac{kT}{qn_n(x)}\frac{\partial n_n(x)}{\partial x}. \tag{2–139}$$

The extreme right hand sides of Eqs. (2–138) and (2–139) are obtained by the use of the Einstein relationships given below.

$$\frac{D_p}{\mu_p} = \frac{D_n}{\mu_n} = \frac{kT}{q}. \tag{2–140}$$

Equation (2–140) will be proved later. With the use of Eq. (2–133) we may rewrite Eq. (2–139)

$$\mathscr{E}_x = -\frac{kT}{qN_d(x)}\frac{\partial N_d(x)}{\partial x}. \tag{2–141}$$

The electric field $\mathscr{E}_x$ from Eq. (2–139) may be related to a potential function $v(x)$ according to electromagnetic field theory as follows:

$$\mathscr{E}_x = -\frac{\partial v(x)}{\partial x} = -\frac{kT}{qn_n(x)}\frac{\partial n_n(x)}{\partial x}. \tag{2–142}$$

From Eq. (2–142) the following may be written:

$$dv(x) = \frac{kT}{q}\frac{dn_n(x)}{n_n(x)}. \tag{2–143}$$

Equation (2–143) can be integrated from $x = a$ to $x = b$ as follows:

$$\int_a^b dv(x) = \frac{kT}{q}\int_{n_n(a)}^{n_n(b)} \frac{dn_n(x)}{n_n(x)} \tag{2–144}$$

$$v(b) - v(a) = \frac{kT}{q}\Big[\ln n_n(x)\Big]_{n_n(a)}^{n_n(b)} = \frac{kT}{q}\ln\left[\frac{n_n(b)}{n_n(a)}\right]. \tag{2–145}$$

Equation (2–145) is quite general. An important, special case is the one given below. If we let the value of carrier density at $x = a$ be the intrinsic carrier density and the potential at a be zero, then the following relationship is obtained.

$$v(b) - 0 = v(b) = \frac{kT}{q}\ln\left[\frac{n_n(b)}{n_i}\right]. \tag{2–146}$$

If we had started with Eq. (2–138) instead of Eq. (2–139), the following equation corresponding to Eq. (2–144) would have resulted.

$$v(b) - v(a) = -\frac{kT}{q}\ln\left[\frac{p_n(b)}{p_n(a)}\right]. \tag{2–147}$$

The special case corresponding to Eq. (2–146) is the following:

$$v(b) = -\frac{kT}{q}\ln\left[\frac{p_n(b)}{n_i}\right]. \tag{2–148}$$

Example 2–6 You are given a n-type material in which the density of holes varies from 10^{14} per c.c. at $x = c$ to 10^{17} per c.c. at $x = d$. Calculate the voltage at d with respect to voltage at c.

Solution. Here direct use of Eq. (2–147) may be made if it is recognized that points c and d replace points a and b respectively. Thus,

$$v(d) - v(c) = v_{dc} = -\frac{kT}{q}\ln\left[\frac{p_n(d)}{p_n(c)}\right] = -\frac{kT}{q}\ln\frac{10^{17}}{10^{14}} \tag{1}$$

$$v_{dc} = -(kT/q)(6.9). \tag{2}$$

At *room temperature*

$$kT/q = 26 \text{ mV}. \tag{2–149}$$

Hence

$$v_{dc} = -0.179 \text{ V}. \tag{2–150}$$

This voltage is due to impurity grading. Note that the voltage difference between two planes, defined by constant values of x, is a function only of the carrier density at these planes *and not on the detailed distribution between the two planes.*

If instead of an n-type crystal we had a p-type crystal at thermal equilibrium, the field $\mathscr{E}_x$ might be expressed as:

$$\mathscr{E}_x = -\frac{\partial v(x)}{\partial x} = \frac{kT}{qp_p(x)}\frac{\partial p_p(x)}{\partial x}. \tag{2–151}$$

If Eq. (2–151) is integrated from $x = a$ to $x = b$, the following is obtained:

$$v(b) - v(a) = -\frac{kT}{q}\ln\left[\frac{p_p(b)}{p_p(a)}\right]. \tag{2–152}$$

While the potential difference between point b and point a is independent of the detailed variation of carrier density between these points, the electric field is very much a function of this detailed distribution as may be seen from Eqs. (2–151) and (2–142). One special case worth noting is the field which results when the impurity grading is exponential.

Consider that the acceptor density varies as follows:

$$N_a(x) = N_a(0)\, e^{-x/L}, \tag{2–153}$$

where L = a characteristic length, a distance in which $N_a(x)$ diminishes $(1/e)N_a(0)$. If we assume that, say at room temperature, all acceptor atoms are ionized, we may write:

$$p_p(x) = p_p(0)\, e^{-x/L}. \tag{2–154}$$

Equation (2–151) allows us to evaluate the field

$$\mathscr{E}_x = \frac{kT}{q}\left(-\frac{1}{L}\right)\frac{p_p(0)\,e^{-x/L}}{p_p(0)\,e^{-x/L}} = -\frac{kT}{qL}. \tag{2–155}$$

Equation (2–155) reveals that *the electric field is constant when the impurity grading is exponential.* We note that if there are any minority carriers in a region where the field due to impurity grading exists, these carriers will be influenced by this electric field. If the minority carriers are being transported due to the diffusion process and this flow is aided by the electric field, such a field is called an *aiding field.* If the field opposes the transport due to diffusion, it is called an *opposing field.*

2-17 HALL EFFECT

An experimental procedure which allows the determination of the sign of charge of the dominant carriers, dominant carrier density, and mobility of the dominant carriers uses the *Hall effect.* The Hall effect consists of the following phenomenon: When a magnetic field **B** is applied to a rectangular-shaped *p*- or *n*-type semiconductor in which current density **J** is flowing, a voltage develops across the faces parallel to the plane defined by the vectors **B** and **J** as shown in Figs. 2–27a and 2–27b. This transverse potential together with the rotation of the equipotential lines in the semiconductor bar is called the Hall effect. Figure 2–27a shows an *n*-type semiconductor bar and Fig. 2–27b shows a *p*-type semiconductor bar. In both these figures the current density vector **J** and the magnetic field vector **B** are oriented in the same direction with respect to the semiconductor crystal. Note that while the current density vector is along the positive x direction in both figures, the carrier velocity is directed along the negative x axis in Fig. 2–27a while that in Fig. 2–27b is directed along the x axis. A charged particle of charge Q moving with a velocity **v** in a magnetic field **B** experiences a force **F** called the *Lorentz force*, according to the following equation:

$$\mathbf{F} = Q(\mathbf{v} \times \mathbf{B}). \tag{2–156}$$

The right hand side of Eq. (2–156) involves a vector cross product. The direction of **F** is determined by the direction along which a right hand screw would advance if **v** was rotated into **B**. Equation (2–156) specialized to Figs. 2–27a and 2–27b respectively gives

$$\mathbf{F} = -q(\mathbf{v}_n \times \mathbf{B}) \tag{2–157a}$$

$$\mathbf{F} = q(\mathbf{v}_p \times \mathbf{B}). \tag{2–157b}$$

The negative sign in Eq. (2–157a) is due to the negative charge of the electron. In *both* figures it is seen that the mobile carriers will accumulate toward the

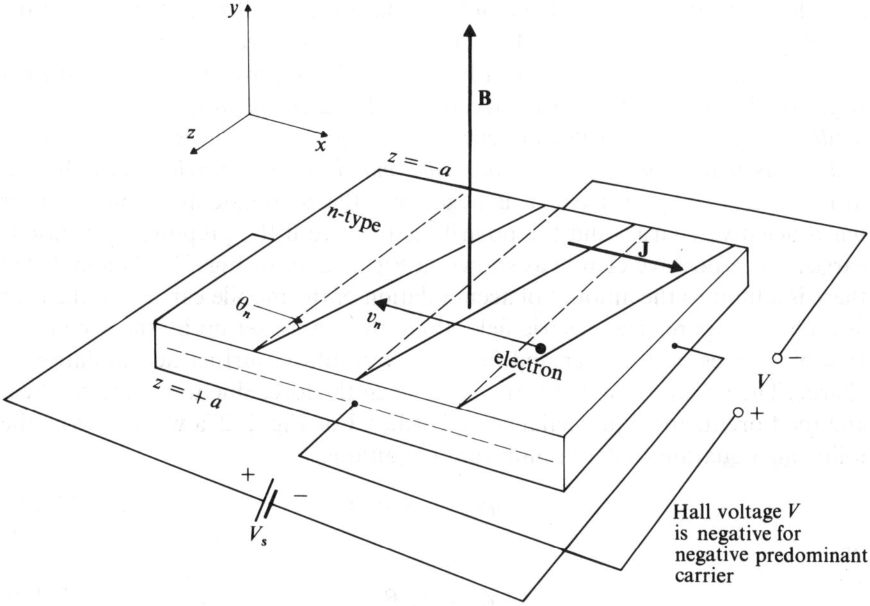

(a)

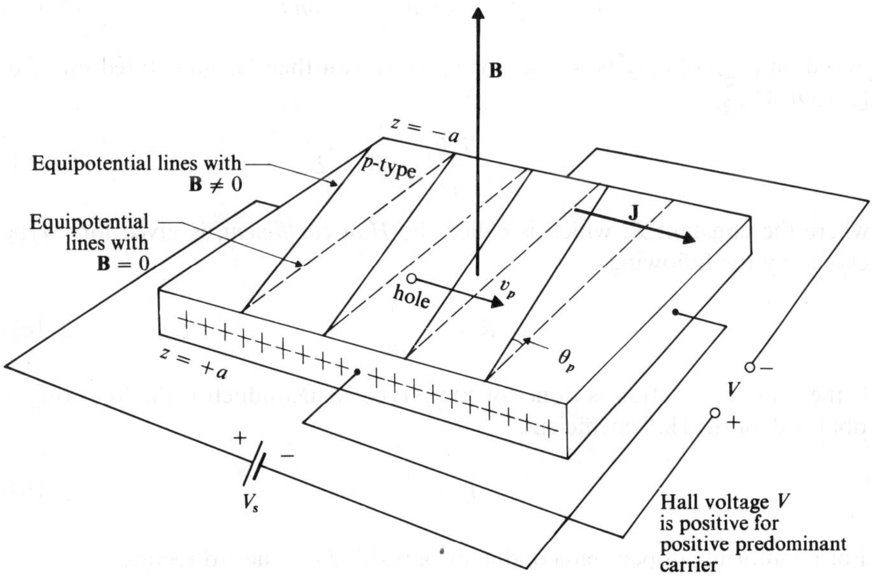

(b)

Fig. 2–27 Hall effect. (a) *n*-Type semiconductor. (b) *p*-Type semiconductor.

face closest to the viewer. At the opposite face there will be a depletion of the mobile carriers. In Fig. 2–27a the opposite face will have a net positive charge (due to ionized donors) whereas in Fig. 2–27b the opposite face will have a net negative charge (due to ionized acceptors). Thus, *the polarity of the voltage V (called simply the Hall voltage) between the near and far faces provides an unambiguous indication of the preponderant mobile charge carrier.* As indicated in Fig. 2–27a the polarity of V is negative if the preponderant mobile carrier has a negative charge, and the polarity is positive if the preponderant mobile carrier has a positive charge as shown in Fig. 2–27b. In Figs. 2–27a and 2–27b there is a limit to the amount of accumulation of the mobile carriers in the near face of each figure. The electric field along the z-axis, set up by the accumulation of charges on the near and far faces, will inhibit further accumulation of charge. The equilibrium will be reached when the force due to the electric field and the Lorentz force just balance each other. For Fig. 2–27a we may write the following equation under equilibrium conditions.

$$-q\mathscr{E}_z = -qv_nB_y \tag{2–158}$$

or

$$\mathscr{E}_z = v_n B_y. \tag{2–159}$$

In the n-type semiconductor the current density J_x may be expressed as follows:

$$J_x = \sigma_n\mathscr{E}_x = qn\,\mu_{dn}\mathscr{E}_x = qn\,v_n. \tag{2–160}$$

Equation (2–160) may be solved for v_n which can then be substituted into Eq. (2–159). Thus,

$$\mathscr{E}_z = -\frac{J_xB_y}{qn} = R\,J_x\,B_y, \tag{2–161}$$

where the constant R, which is called the *Hall coefficient*, is given for n-type crystal by the following:

$$R = -\frac{1}{qn}. \tag{2–162}$$

If the same calculation is done for the p-type semiconductor, the following is obtained for the Hall coefficient.

$$R = \frac{1}{qp}. \tag{2–163}$$

For a uniformly doped semiconductor of width $2a$ in the z-direction,

$$\mathscr{E}_z = \frac{V}{2a}. \tag{2–164}$$

Thus from Eqs. (2–161) and (2–164) we obtain

$$R = \frac{V}{(2a)\, J_x\, B_y}. \tag{2–165}$$

As all quantities on the right hand side, of Eq. (2–165) can be relatively easy to measure, R can be experimentally determined. Once R is experimentally determined, the carrier density is easily calculable through Eqs. (2–162) and (2–163).

In obtaining Eqs. (2–162) and (2–163), the average drift velocities v_n and v_p were used. A more careful averaging process leads to more accurate results as follows for n-type crystals,

$$R = -\frac{3\pi}{8}\left(\frac{1}{qn}\right) \quad (n\text{-type}) \tag{2–166}$$

and for p-type crystal

$$R = \frac{3\pi}{8}\left(\frac{1}{qp}\right) \quad (p\text{-type}). \tag{2–167}$$

If a crystal has both positive and negative charge carriers, the Hall coefficient may be shown to be the following:

$$R = \frac{3\pi}{8}\left[\frac{n\mu_{dn}^2 - p\mu_{dp}^2}{(n\mu_{dn} + p\mu_{dp})^2}\right], \tag{2–168}$$

where μ_{dn}, μ_{dp} = drift mobility for electrons and holes respectively, these coefficients are determined by the direct use of Eq. (2–39) for conductivity.

To understand the presence of the factor $3\pi/8$ in Eqs. (2–166) through (2–168) the main scattering mechanism needs to be determined for carriers. In Section 2–6 we saw that the carrier motion is a series of collisions and at the end of each, the carrier starts out anew with no memory of previous history. Electrons and holes may suffer collisions with one of several normal modes of crystal vibrations, for example transverse or longitudinal acoustic and/or "optical" mode. It is customary to refer to the particle description of these normal modes, i.e. acoustic phonons and/or optical phonons. The latter are called "optical" because in a lattice they can produce polarization which can interact with light. The numerical factor $3\pi/8$ assumed (a) lattice scattering due only to acoustic phonons, (b) conduction band and valence band Fermi surfaces are spherical. This numerical factor varies in Eqs. (2–166) through (2–168) at low magnetic fields between 1.0 and 1.7 if scattering due to impurities is considered together with lattice scattering. If all the scattering is due to the impurities alone, the numerical factor in R takes the value of 1.7.

In the absence of the transverse electric field $\mathscr{E}_z$, i.e. with $\mathbf{B}_y = 0$, the equipotential lines which must be perpendicular to $\mathscr{E}_x$, are along the z direction as shown by the dotted lines in Fig. 2–27. When the magnetic field is turned "on," the total electric field is the vector sum of $\mathscr{E}_x$ and $\mathscr{E}_z$. The equipotential lines are now as shown by the solid lines in Fig. 2–27. The angle θ_h called the *Hall angle* between the dotted and solid equipotential lines is equal to the angle between the total electric field and $\mathscr{E}_x$, and may be evaluated as follows for a small angle,

$$\tan\theta_h \doteq \theta_h = \frac{\mathscr{E}_z}{\mathscr{E}_x}. \tag{2–169}$$

From Eqs. (2–161) and (2–162) $\mathscr{E}_z$ may be obtained so that the Hall angle may be evaluated as follows:

$$\theta_n = \frac{\left(\frac{3\pi}{8}\right)\left(\frac{1}{qn} J_x B_y\right)}{\mathscr{E}_x} = \frac{\frac{3\pi\, q\mu_{dn} n\, \mathscr{E}_x B_y}{8qn}}{\mathscr{E}_x} \tag{2–170}$$

or

$$\theta_n = \frac{3\pi}{8}\mu_{dn} B_y. \tag{2–171}†$$

The proportionality constant between the Hall angle and the magnetic field $\mathbf{B}_y$ is called the *Hall mobility* identified by μ_H. Thus

$$\theta_n = \mu_{Hn} B_y = \frac{3\pi}{8}\mu_{dn} B_y \qquad (n\text{-type}). \tag{2–172}‡$$

Similarly

$$\theta_p = \mu_{Hp} B_y = \frac{3\pi}{8}\mu_{dp} B_y \qquad (p\text{-type}). \tag{2–173}‡$$

The field $\mathbf{B}_y$ may be measured and the Hall angle θ_h can also be measured by plotting the equipotential lines with $\mathbf{B}_y$ zero and then with $\mathbf{B}_y$ non-zero, and by measuring the angle between these sets of equipotential lines. The Hall mobility and drift mobility are related to each other by

$$\frac{\mu_{Hn}}{\mu_{dn}} = \frac{3\pi}{8}. \tag{2–174}$$

Similarly we expect the mobility ratio for holes to be

$$\frac{\mu_{Hp}}{\mu_{dp}} = \frac{3\pi}{8}. \tag{2–175}$$

---

† While θ_h stands for Hall angle in general, θ_n and θ_p stand for the Hall angle for *n*-type and *p*-type crystals respectively.
‡ While μ_H is Hall mobility, μ_{Hn} and μ_{Hp} stand for the Hall mobility for *n*-type and *p*-type crystals respectively.

The Hall coefficient may be expressed in yet another form. From Eq. (2–160) we may solve for $1/qn$

$$\frac{1}{qn} = \frac{\mu_{dn}}{\sigma_n}. \tag{2–176}$$

From Eq. (2–166) and (2–176) we have

$$R = -\frac{3\pi}{8}\left(\frac{1}{qn}\right) = -\frac{3\pi}{8}\left(\frac{\mu_{dn}}{\sigma_n}\right) = -\frac{\mu_{Hn}}{\sigma_n} \qquad (n\text{-type}) \tag{2–177}$$

$$R = +\frac{\mu_{Hp}}{\sigma_p} \qquad (p\text{-type}). \tag{2–178}$$

While Eq. (2–166) agrees reasonably well for n-type semiconductors, Eq. (2–167) does agree well with p-type semiconductors *only* if the different species of holes corresponding to the three bands within the valence band are recognized and their relative populations taken account of. Equation (2–166) agrees well also with most metals.†

Besides the many uses to which the Hall effect can be and usually is put to, the analytic equations Eq. (2–161) shows that the Hall effect can be used to produce a *perfect multiplier* where one quantity is made proportional to $\mathbf{J}_x$ and the other is made proportional to $\mathbf{B}_y$. The field in the z-direction then gives the quantity which is proportional to the product of $\mathbf{J}_x$ and $\mathbf{B}_y$.

2–18 MILLER INDICES

There are several properties of a crystal which are different along different directions (i.e. they are anisotropic). It is therefore important to be able to refer to different orientations of planes within a crystal in an unambiguous way. *Miller indices* provide a way of doing this. Figure 2–28 shows a two-dimensional crystal lattice. Directions of lines are designated by the Miller indices for the two-dimensional lattice. The scheme is the following: (1) any atom may be chosen as the origin of the coordinate system; (2) the intersection of a given line with the lattice is to be measured in terms of *lattice spacings*. If these are I and J, consider I' and J' which are the reciprocals respectively of I and J. Then i and j are the Miller indices when i and j are the smallest integers which have the same ratio as do I' and J' to each other. Note that parallel lines turn out to have the same Miller indices because they are oriented along the same direction. Note that if the intersection with a coordinate axis is negative, the corresponding Miller index carries the negative sign on top of the corresponding index, as shown in Fig. 2–28. You can check for yourself the designation of the Miller indices for the lines shown in Fig. 2–28.

† For a discussion of the statistical factor $3\pi/8$ and other related matters, see Conwell (1958) and Herman (1955).

The above scheme may be generalized to the three-dimensional crystal shown in Fig. 2–29a as follows:

1. Choose any atom as the origin of the coordinate system.
2. The intersection of the given plane with the coordinate axes is noted as A, B, and C respectively.
3. The distances OA, OB, and OC are measured in terms of respective lattice spacings as I, J, and K.
4. The smallest integers i, j, and k are determined such that they have the same ratios as $1/I$, $1/J$, and $1/K$ to each other.
5. The Miller indices i, j, and k are usually designated as follows: (ijk).

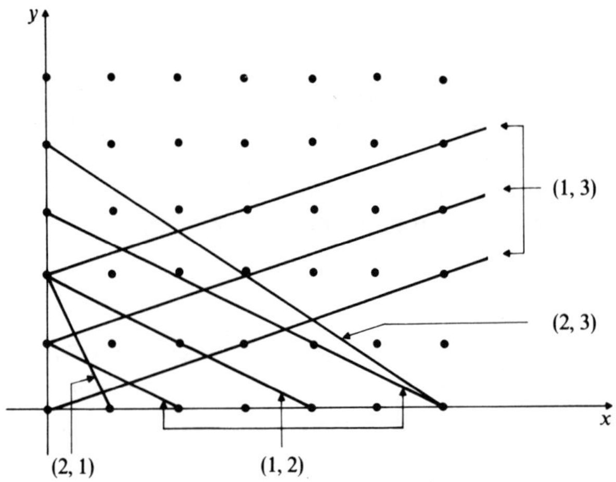

Fig. 2–28 Miller indices indentify different lines in a two-dimensional crystal lattice.

Note that if a given plane is parallel to a coordinate axis, its intersection with this axis is then at infinity whose reciprocal is zero. Figures 2 29b through 2–29d show some simple plane orientations and the corresponding Miller indices.

Example 2–7 Find the Miller indices of the plane shown in Fig. 2–29e.

Solution. The plane intersects the coordinate axes at $I = 4$, $J = 2$, $K = 3$ in terms of lattice spacings along the respective axes. Their reciprocals are 1/4, 1/2 and 1/3. These are in the same ratio as the integers $i = 3$, $j = 6$, and $k = 4$. Thus the Miller indices are: (3 6 4).

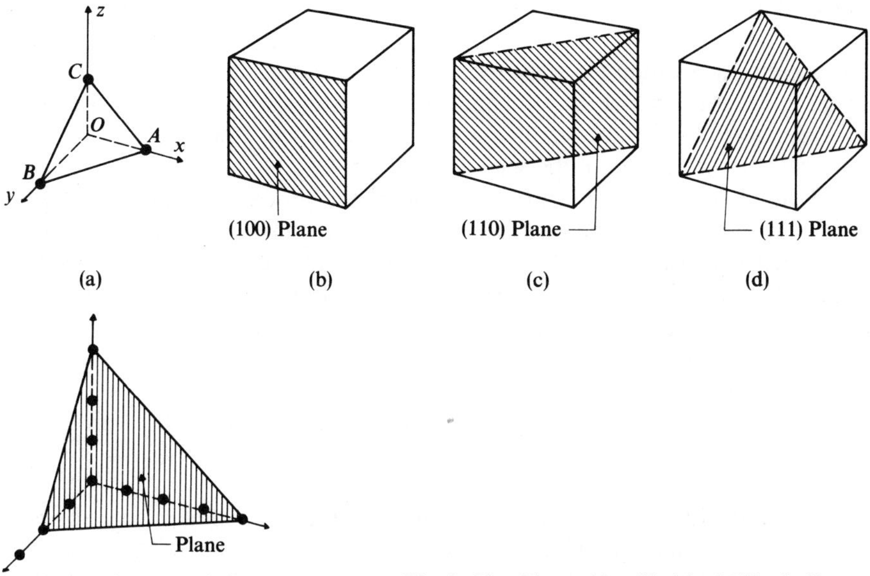

(e) Plane for Example 2–6

Fig. 2–29 Planes identified by Miller indices.

2–19 SUMMARY

In this chapter we considered a number of different concepts all of which will form the basis and foundation upon which we shall depend heavily in understanding how solid state and semiconductor devices work.

The band picture is fundamental to the understanding of solids. The energy band concept enabled us to classify solids into insulators, semiconductors, and conductors. The temperature variation of conductivity of these three kinds of solids could also be deduced from this band picture. We found in semiconductors that the current is carried by not only electrons but also by particles called holes. The hole may be characterized as having both a positive charge and a positive effective mass. The concept of effective mass allows us to simplify the mathematical treatment of the movement of carriers within a crystal. The $E(\mathbf{k}_c)$ curves were found to be very valuable in giving us a better insight into the details of the band picture than was possible to obtain from the simple one-dimensional energy band picture. We discovered, for example, that the conduction band of germanium and silicon contains several valleys, and that the valence band has three sub-bands within it. Corresponding to each is a different effective mass for holes belonging to each sub-band. The above complexity of the band picture is necessary to adequately explain the experimental findings about the commonly used semiconductors.

The carrier densities were calculated with the help of the density of states function and the relevant statistics of the Fermi–Dirac distribution function. At thermal equilibrium the product of free holes and free electrons in a semiconductor is a constant for a given material in both intrinsic as well as extrinsic crystals, i.e. those which have chemical impurities added to them. We found that the introduction of small amounts of chemical impurities profoundly altered the conductivity of the semiconductors.

Two of the most important semiconductor processes, diffusion and recombination, were considered in this chapter. The transport equations for holes and electrons, when concentration gradient and electric field were simultaneously present, were derived. The continuity equation taking account of the recombination process was obtained. Both of the above sets of equations will be needed to understand *p-n* junction devices.

We discovered that when a semiconductor is nonuniformly doped, electric fields develop which may aid the transport of minority carriers. For this reason, impurity grading is frequently employed in the manufacture of junction transistors.

We studied a number of different topics which will be found to be of considerable importance to the basic understanding of semiconductor devices. Hence, in the following several chapters we shall need to refer back to this chapter from time to time.

PROBLEMS

2–1 In a *p*-type semiconductor crystal the following electron and hole densities are found at room temperature: $p = 10^{14}$ per cubic centimeter; $n = 5 \times 10^{13}$ per cubic centimeter. If now this crystal is further doped with an additional 5×10^{14} acceptors/cm^3, what will be the new conductivity of this semiconductor? You may assume that the electron and hole mobilities are the same as those for germanium.

2–2 Consider a rectangular bar of *p*-type silicon which is 2 cm long. The cross section is a square 2 mm to a side. If the acceptor density is 10^{16} atoms/cm^3 and if the bar is connected to a battery of 3 Volts, what current will flow in this circuit at room temperature (a) if all the holes are considered to be "heavy holes", (b) if the results of the cyclotron experiment are considered to be valid at room temperature (as far as the relative population density of light and heavy holes is concerned)? The mobility of the heavy holes may be taken to be 350 cm^2/Volt-sec. For the purpose of this problem assume that the mean free time between collisions is the same for light and heavy holes.

2–3 Calculate the Fermi level at room temperature in an *n*-type germanium crystal doped with 10^{17} donor (As) atoms/cm^3. Use Eq. (2–90) for this calculation.

2–4 Use Eq. (2–101) to calculate the Fermi level at room temperature in Problem 2–3. What conclusions can you draw from the above calculation?

2–5 Calculate the Fermi level for the crystal of Problem 2–3 at the temperatures of liquid nitrogen and liquid helium and at 100°C by assuming that (a) the band gap has a constant value of 0.75 eV and (b) the band gap is a function of temperature, as given in Eq. (2–102). Also calculate the free-electron density.

2–6 Consider a rectangular bar of Ge of length 2 cm, cross-sectional area of 0.1 cm^2. This bar is doped with a *total* of 10^{15} donor atoms.
a) Calculate the majority carrier density.
b) Calculate the minority carrier density.

2–7 If the Ge bar of Problem 2–6 is connected at its ends to a 12Volts battery, how much current will the battery have to supply?

2–8 What percentage of the Ge atoms are replaced by the donor atoms in Problem 2–6? Ge has 72.6 g/(g-atom) and density of 5.33 g/cm^3. Use Avogadro's law that there are 6.023×10^{23} atoms/(g-atom).

2–9 A pure silicon cylindrical crystal having a cross-sectional area of 0.001 cm^2 and length of 1 mm is connected at its end to a 10 Volts battery. At room temperature it is desired that the current flowing through the semiconductor be 100 mA. Calculate,
a) The required total resistance R.
b) The required crystal conductivity.
c) N_d, the donor density which needs to be introduced into the crystal.

2–10 Consider a Si crystal which originally has 10^{16} donor atoms/c.c. Into this crystal are introduced 3×10^{16} acceptors atoms/c.c. Calculate the conductivity of this crystal. Are the approximations $n = N_d$, $p = N_a$ valid for this problem? Why?

2–11 The conductivity of a conductor at room temperature is 10^6 $(\Omega\text{-cm})^{-1}$. If the cross-sectional area of the conductor is 10^-2 cm^2 and a current of 1 Amp flows, calculate the average drift velocity of electrons. You may assume that for this conductor volume density is 10 g/cm^3, and 60 g/(g-atoms). You may assume that there is one free electron per atom.

REFERENCES

Blakemore, J. S., "Carrier Concentrations and Fermi Levels in Semiconductors," *Elec. Commun.*, **29,** 131–153, (1952).

Bloch, Z., *Physik*, **52,** 555 (1928).

Boltaks, B. I., *Diffusion in Semiconductors,* Academic Press (1963).

Bullis, W. M., "Properties of Gold in Silicon," *Solid State Electron.*, **9,** 143 (1966).

Carslaw, H. S. and Jaeger, J. C., *Conduction of Heat in Solids,* Clarendon Press, Oxford (1959).

Conwell, E., and Weisskopf, V. F., "Theory of Impurity Scattering in Semiconductors," *Phys. Rev.*, **77,** 388 (1950).

Conwell, E. M., "Properties of Silicon and Germanium: II," *Proc. IRE,* **46,** 1281–1300, (1958).

Dunlap, W. C., *An Introduction to Semiconductors,* John Wiley & Sons, New York (1957).

Ehrenreich, H., "Band Structure and Electron Transport in GaAs," *Phys. Rev.*, **120,** 1951 (1960).

Grove, A. S., *Physics and Technology of Semiconductor Devices,* John Wiley & Sons, New York (1967).

Hall, E. H., "On a New Action of the Magnet on Electric Currents," *Am. J. Math.,* **2,** 287 (1879).

Hall, R. N., "Electron-Hole Recombination in Germanium," *Phys. Rev.,* **87,** 387 (1952).

Haynes, J. R., and Shockley, W., "The Mobility and Life of Injected Holes and Electrons in Germanium," *Phys. Rev.,* **81,** 835 (1951).

Herman, F., "The Electronic Energy Band Structure of Silicon and Germanium," *Proc. IRE,* **43,** 1703 (1955).

Hilsum, C., and Rose-Innes, A. C., *Semiconducting III–V Compounds,* Pergamon Press, New York (1961).

Jost, W., *Diffusion in Solids, Liquids, Gases,* Academic Press (1952).

Kittel, C., *Quantum Theory of Solids,* John Wiley & Sons, New York (1963).

See for example, Kittel, C., *Introduction to Solid State Physics,* John Wiley & Sons., New York, 2nd ed. (1956), 3rd ed. (1966).

Kronig, R. de L., and Penney, W. G., *Proc. Royal Soc.,* **A130,** 499 (1931).

Lee, C. A., Logan, R. A., Batdorf, R. L., Kleimack, J. J., and Wiegmann, W., "Ionization Rates of Holes and Electrons in Silicon," *Phys. Rev.,* **134,** A761 (1964).

Moll, J. L., *Physics of Semiconductors,* McGraw-Hill, New York (1964).

Nanavati, R. P., *Introduction to Semiconductor Electronics,* Chapter 2, McGraw-Hill, New York (1963).

Norris, C. B., and Gibbons, J. F., "Measurement of High-Field Carrier Drift Velocities in Si by a Time-of-Flight Technique," *IEEE Trans. Electron Devices,* ED-14, 38 (1967).

Prince, M. B., "Drift Mobility in Semiconductors I, Germanium," *Phys. Rev.,* **92,** 681 (1953).

Pritchard, R. L., *Electrical Characteristics of Transistors,* McGraw-Hill, New York (1967).

Ruch, J. G., and Kino, G. S., "Measurement of the Velocity-Field Characteristics of Gallium Arsenide," *Appl. Phys. Letters,* **10,** 40 (1967).

Seidel, T. E., and Scharfetter, D. L., "Dependence of Hole Velocity upon Electric Field and Hole Density for *p*-Type Silicon," *J. Phys. Chem. Solids,* **28,** 2563 (1967).

Seitz, F., *The Modern Theory of Solids,* McGraw-Hill, New York (1940).

Shockley, W., *Electrons and Holes in Semiconductors,* D. Van Nostrand, Princeton (1950).

Shockley, W., and Read, W. T., "Statistics of the Recombination of Holes and Electrons," *Phys. Rev.,* **87,** 835 (1952).

Stevenson, D. T., and Keyes, R. J., "Measurement of Carrier Lifetime in Germanium and Silicon," *J. Appl. Phys.,* **26,** 190 (1955).

Sze, S. M., and Irvin, J. C., "Resistivity, Mobility, and Impurity Levels in GaAs, Ge, and Si at 300°K," *Solid State Electron,* **11,** 599 (1968).

Sze, S. M., *Physics of Semiconductor Devices,* Chapters 1, 2, John Wiley & Sons, New York (1969).

Trumbore, F. A., "Solid Solubilities of Impurity Elements in Germanium and Silicon," *Bell System Tech. J.,* **39,** 205 (1960).

Valdes, L. B., "Resistivity Measurement on Germanium for Transistors," *Proc. IRE,* **42,** 420 (1954).

Wolfstirn, K. B., "Holes and Electron Mobilities in Doped Silicon From Radio Chemical and Conductivity Measurements," *J. Phys. Chem. Solids,* **16,** 279 (1960).

van der Ziel, A., Solid State Physical Electronics, Chapters 3–5, Prentice-Hall, Englewood Cliffs, N.J. (1957).

CHAPTER 3

p-n JUNCTION DIODES

3–1 INTRODUCTION

The first two chapters are the foundation on which the understanding of semiconductor devices is to be based. In the present chapter we shall study the *p-n* junction diode. In Chapter 2 the characteristics of *p*-type and *n*-type semiconductors were examined separately. Junctions between *p*-type and *n*-type materials in single crystal semiconductors turn out to be rectifying junctions.

We wish to find a model which correctly predicts the terminal behavior of *p-n* junctions. The elements of the model will be expressed as a function of the fundamental device parameters. To this end we shall first need to solve the diffusion equation for the minority carriers in each of the regions. Boundary conditions for the minority carriers will first need to be derived. Terminal current–voltage relationship for the ideal *p-n* junction device will be obtained. The behavior of the ideal device will be related to that of an actual device. A more nearly complete low frequency model will be obtained for the actual *p-n* junction diode by the addition of extra elements to the ideal device model. The process of diffusion, recombination, and transport will all be needed for developing the above model.

In the following chapters multijunction devices such as the bipolar junction transistor will be studied. A thorough understanding of the single junction device considered in this chapter will be of a great help in analyzing the multijunction devices. Many results of the present chapter will be found to be of direct applicability to the more complex devices considered in the succeeding chapters.

3–2 UNBIASED *p-n* JUNCTION

When a *p*-type region occurs right next to an *n*-type region in such a way as not to destroy the crystaline integrity we have a *p-n* junction. (Note that it is not

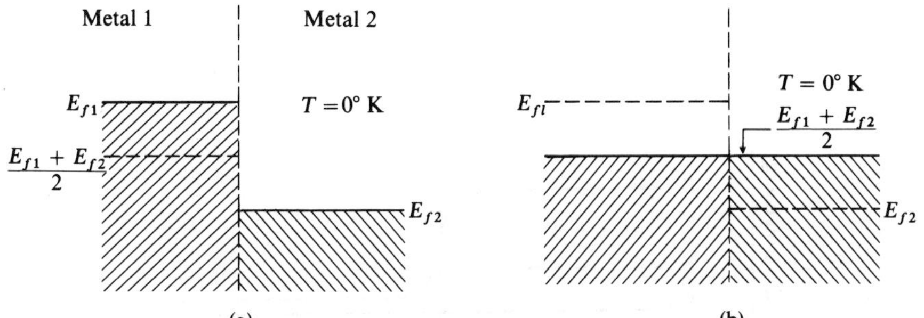

Fig. 3–1 Two different electron distributions at $T = 0°K$ for two different metals in contact. (b) Shows the rearrangement of electrons of Fig. 3–1a for lowest energy configuration.

simply a piece of *p*-type material and *n*-type material in physical contact.) Before we can see what the energy band picture looks like for the *p-n* junction we shall invoke an important result of statistical mechanics which requires that *the Fermi levels must be at the same energy in any number of materials in homogeneous contact and at thermal equilibrium.* To see the reasonableness of this requirement it will be easy to visualize the case of two different metals in homogeneous contact. We shall first assume that here the Fermi levels in these do not line up, and subsequently show that this is inconsistent with the requirement of thermal equilibrium. Consider the band picture corresponding to two different metals in homogeneous contact as shown in Fig. 3–1a. In metals the Fermi level is located in the overlapping bands as mentioned in Section 2–2. Note that E_{f1} is not E_{f2} by assumption. Let us consider the situation expected at $T = 0°K$. The permissible energy states up to E_{f1} must be occupied in metal 1 and all energy states up to E_{f2} must also be occupied in metal 2. If this situation obtains at thermal equilibrium it must correspond to lowest energy for the system. Thus if we can show a different distribution of electrons which corresponds to a lower energy for the system than that of Fig. 3–1a we shall have shown that the situation portrayed in Fig. 3–1a could not have been at thermal equilibrium. If we take the top layer of electrons in metal 1 corresponding to the energies from E_{f1} to $(E_{f1} + E_{f2})/2$ and relocate them on top of E_{f2} in metal 2 we have the situation as portrayed in Fig. 3–1b.

It is clear from Fig. 3–1b that this represents an overall lower energy system than that of Fig. 3–1a. Thus we have shown that the condition of the Fermi levels not lining up could not correspond to thermal equilibrium conditions. It is instructive to note that the above rearrangement of charge is not necessary when

$$E_{f1} = E_{f2}. \tag{3–1}$$

This is a general result though it was shown to be valid in metals.

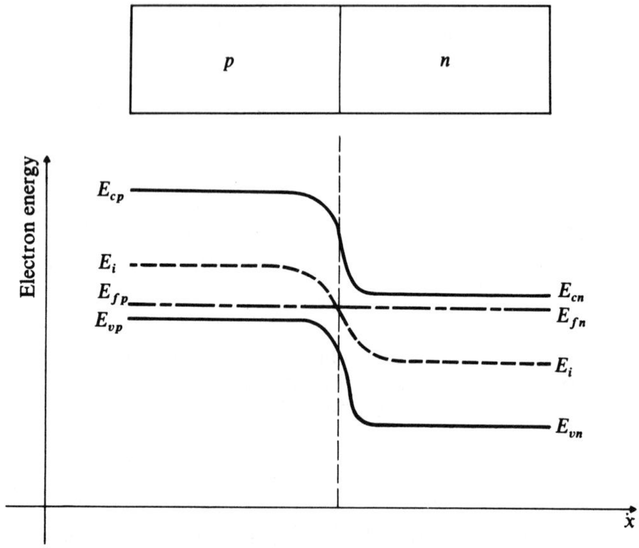

Fig. 3–2 Energy diagram for unbiased *p-n* junction.

Armed with the knowledge that the Fermi levels have to line up, the energy diagram can be constructed for a *p-n* junction as depicted in Fig. 3–2. An adjustment of bands occurs as a result of the above requirement on Fermi levels as seen in Fig. 3–2.

It is found useful to define an *intrinsic energy* E_i which is such that when it replaces E_f in Eqs. (2–73) and (2–77), it makes the electron density equal to hole density. Using this definition we can write:

$$n = p \tag{3–2}$$

or

$$N_c \exp[(E_i - E_c)/kT] = N_v \exp[-(E_i - E_v)/kT]. \tag{3–3}$$

Shifting the exponential terms to the right hand side of the equation we obtain

$$\frac{N_c}{N_v} = \exp[(-2E_i + E_v + E_c)/kT]. \tag{3–4}$$

Taking the logarithm of both sides we have

$$\begin{aligned} kT \ln\left(\frac{N_c}{N_v}\right) &= -2E_i + E_v + E_c \\ &= -2E_i + (E_v - E_c) + 2E_c \\ &= -2E_i - E_g + 2E_c \end{aligned}$$

or

$$E_c = \frac{kT}{2} \ln\left(\frac{N_c}{N_v}\right) + E_i + \frac{E_g}{2}. \tag{3–5}$$

Substituting this value of E_c into Eq. (2–73) we obtain

$$n = N_c \exp\left[\left(E_f - \frac{kT}{2} \ln(N_c/N_v) - E_i - E_g/2\right)/kT\right] \tag{3–6}$$

$$n = N_c \left(\frac{N_v}{N_c}\right)^{1/2} \left[\exp\left(-\frac{E_g}{2kT}\right)\right] \exp[(E_f - E_i)/kT]. \tag{3–7}$$

Now from Eq. (2–79) we have

$$N_c \left(\frac{N_v}{N_c}\right)^{1/2} e^{-E_g/2kT} = (N_c N_v)^{1/2} e^{-E_g/2kT} = n_i. \tag{3–8}$$

Thus Eq. (3–7) becomes

$$n = n_i \, e^{(E_f - E_i)/kT}. \tag{3–9}$$

Quite analogously the hole density can also be expressed in terms of E_i as

$$p = n_i \, e^{(E_i - E_f)/kT}. \tag{3–10}$$

As one of our aims is to obtain an equation relating current to voltage for the *p-n* junction device, it is essential to introduce voltage variables related to the energies. For this purpose we define the following two important potential functions

$$-q\psi_i \triangleq E_i, \tag{3–11}†$$

$$-q\varphi_f \triangleq E_f, \tag{3–12}$$

where ψ_i = intrinsic potential,

φ_f = Fermi potential.

The negative sign in Eqs. (3–11) and (3–12) is due to the fact that we are considering *electron* energies. It is the electron energy which has been plotted in the energy band diagrams so far.

In terms of the newly defined potentials given by Eqs. (3–11) and (3–12) we may rewrite Eqs. (3–9) and (3–10) as follows:

$$n = n_i \, e^{(q/kT)(\psi_i - \varphi_f)}, \tag{3–13}$$

$$p = n_i \, e^{(q/kT)(\varphi_f - \psi_i)}. \tag{3–14}$$

We have already noted that kT/q is 26 milliVolts at room temperature.

---

† ψ_i represents the general intrinsic potential. Intrinsic potentials in the p and n regions are designated ψ_p and ψ_n respectively.

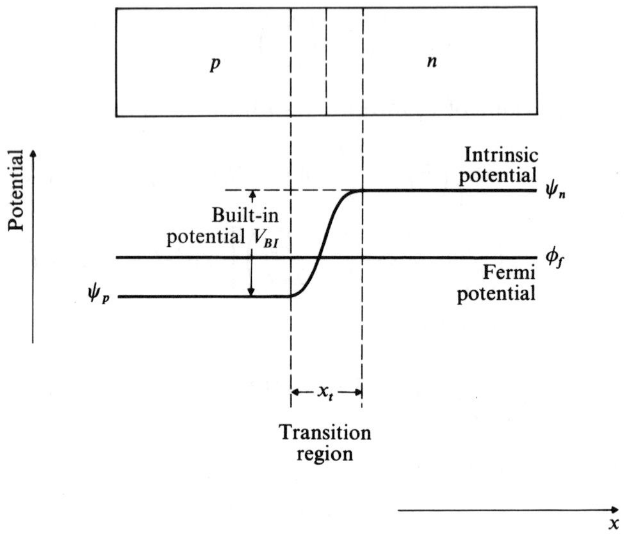

Fig. 3–3 Intrinsic and Fermi potentials in unbiased *p-n* junction.

The intrinsic energy E_i resides at about the middle of the band gap just as does the Fermi level of an intrinsic semiconductor. It may be recalled that in the intrinsic semiconductor $n = p$. This is precisely the condition required by the definition of E_i. In Fig. 3–2 is also shown E_i.

In Fig. 3–3 is shown the potential functions ψ_i and φ_f in the *p-n* junction. The fact that Fermi levels line up is shown by a perfectly horizontal line for φ_f. The curve for ψ_i in Fig. 3–3 seems to be "flipped" with respect to E_i in Fig. 3–2 because of the sign of Eq. (3–11). A very interesting thing seems to develop at the junction of p and n regions; namely, we find that there exists a change in potential ψ_i at the junction. As the electrostatic field is given by

$$\mathscr{E} = -\nabla\psi_i \tag{3–15}$$

an electric field is seen to exist in the *p-n* junction region as is also a *potential difference across the junction even though the device is unbiased.* This voltage difference and electric field are called *built-in voltage* V_{BI} and *built-in field* $\mathscr{E}_b$. It should be noticed that this voltage would not show itself at the terminals of this device. That is, a voltmeter across the external terminals of the device will not read any voltage. Figure 3–4 shows the complete potential distribution throughout the device including the ohmic contacts, and it is seen that the voltage difference between the external terminals is indeed zero.

The potential change at the ohmic junction results also from the requirement of Fermi levels having to line up at junctions of dissimilar substances at

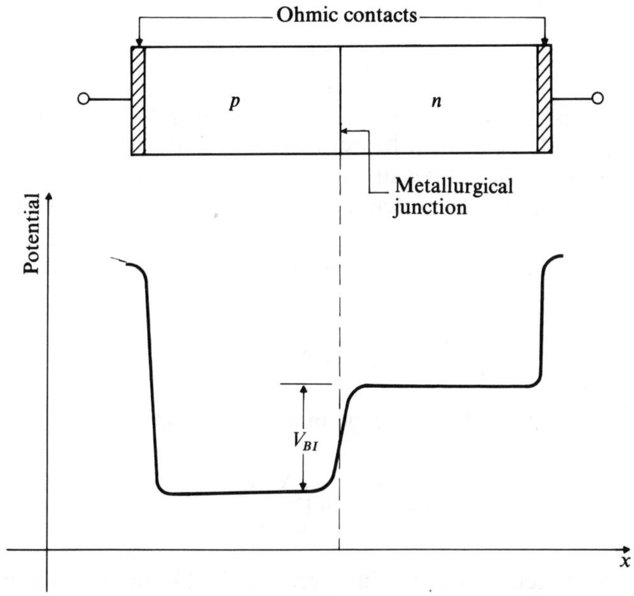

Fig. 3–4 Potential in *p-n* junction including ohmic contacts.

thermal equilibrium. When a voltage is applied at the external terminals, *the only potential difference which changes is the junction potential.*

Let us calculate the built-in voltage V_{BI} using Eqs (3–13) and (3–14) which are repeated for this case

$$n_n = n_i\, e^{(q/kT)(\psi_n - \varphi_f)}, \tag{3–16}$$

$$p_p = n_i\, e^{(q/kT)(\varphi_f - \psi_p)}, \tag{3–17}$$

where n_n = the thermal equilibrium density of electrons on the *n*-side,
p_p = the thermal equilibrium density of holes on the *p*-side.

Dividing Eqs. (3–16) and (3–17) by n_i and taking logarithms of both sides gives

$$\ln \frac{n_n}{n_i} = \frac{q}{kT}(\psi_n - \varphi_f), \tag{3–18}$$

$$\ln \frac{p_p}{n_i} = \frac{q}{kT}(\varphi_f - \psi_p). \tag{3–19}$$

Adding Eqs. (3–18) and (3–19) we obtain

$$\frac{kT}{q} \ln \frac{n_n p_p}{n_i^2} = \psi_n - \varphi_f + \varphi_f - \psi_p = \psi_n - \psi_p \triangleq V_{BI}. \tag{3–20}$$

It should be noted that here

$$n_n\, p_p \neq n_i^2$$

because the product np is equal to n_i^2 *only when n and p are evaluated in the same regions.* The quantity n_n and p_p, however, are evaluated in *different* regions. If acceptor states and donor states are sufficiently close to the band edges it is a good approximation to assume that all acceptors and donors are ionized. Then

$$n_n \doteq N_d \tag{3–21}$$

and

$$p_p \doteq N_a. \tag{3–22}$$

So we may express the built-in voltage in another form

$$V_{BI} \doteq \frac{kT}{q} \ln\left(\frac{N_d N_a}{n_i^2}\right). \tag{3–23}$$

So long as the terminals remain open-circuited (unbiased) the net current in the *p-n* junction has to be zero. This zero current can be thought to result from two sets of equal and opposite current components. Consider hole currents first. The diffusion component of hole current is expected from the *p*-side, where the holes are numerous, to the *n*-side where the holes are a small minority. At other than zero degrees, there will be some thermally generated holes on the *n*-side. The holes so generated, which find themselves close to the transition region edge, will find a favorable electric field and slide down the potential hill in the transition region and get into the *p* region. These two components exactly cancel each other.

Now consider the electron movements. We expect diffusion of electrons from high electron density *n*-region to low electron density *p*-region. This movement is balanced by the thermally generated electrons on the *p*-side, close to the junction, which find a favorable electric field and slide up the potential hill like helium filled balloons. The two corresponding electron current components exactly cancel each other to produce zero current.†

The above is a qualitative description of what happens. We can use the transport Eqs. (2–112) and (2–113) to obtain a very important relationship between mobility and diffusion constant. The transport equation for holes will be used first.

$$\mathbf{j}_p = 0 = q\, \mu_p\, p\, \mathscr{E} - qD_p\, \nabla p. \tag{3–24}$$

† Note that there is individual cancellation of diffusion and field induced drift for electrons and holes separately. This detailed balance is encountered in several situations in *p-n* junction theory.

Using Eq. (3–15) for $\mathscr{E}$ and Eq. (3–14) for p we get

$$0 = q\,\mu_p p(-\nabla\psi_i) - q\,D_p p\left(-\frac{q}{kT}\right)\nabla\psi_i \tag{3–25}$$

or

$$\mu_p = \frac{q\,D_p}{kT}. \tag{3–26}†$$

If the transport equation for electron current density had been used, the analogous equation for electron mobility and diffusion constant would be

$$\mu_n = \frac{q D_n}{kT}. \tag{3–27}†$$

While Eqs. (3–26) and (3–27) were derived for the no bias case, the results in the form of Eqs. (3–26) and (3–27) *are valid generally*. In particular, even when bias is applied, Eqs. (3–26) and (3–27) are still valid. This is analogous to calculating the value of the gravitational constant g from a particular experiment, for example of a block of wood sliding down an inclined plane. The value of g thus determined is valid generally even though it was computed from a particular situation.

3–3 CARRIER DENSITY AT THE EDGES OF THE TRANSITION REGION

In obtaining the solution to the diffusion equation which we shall need to solve for the case of the *p-n* junction, it will be necessary to know boundary conditions for minority carriers. For this purpose we shall now obtain these boundary conditions when a voltage appears across the transition region due to an *external* voltage source. This, however, presents a complication. The Fermi–Dirac statistic, which is appropriate to electrons and holes, is based on the important assumption that the carrier velocities are indeed *random*. When external bias is applied to the device the velocities *cannot be truly* random. For this reason the Fermi–Dirac statistics are not strictly valid. To take care of the most practical case of a biased *p-n* junction, we assert that the statistic which applies is of the same functional form as the Fermi–Dirac distribution function. *This is so because the thermal velocities, at most reasonable temperatures, are usually much higher than the field induced velocities.* Thus even though bias is applied, the velocities are still fairly random. Hence the statistical law appropriate to this situation is not altered radically from the truly random velocity case.

With the above observations in mind we *define* two functions called *quasi*

† Equations (3–26) and (3–27) are valid only for nondegenerate semiconductors in which the doping level is low enough so that the Fermi level is not in the conduction or valence bands.

Fermi levels for holes and electrons φ_p and φ_n, such that when these replace the Fermi level in Eqs. (3–13) and (3–14) respectively, the *actual* hole and electron densities are computed even when bias is applied to the *p-n* junction.

Figure 3–5 shows the *p-n* junction biased so that the *p*-side is made positive with respect to the *n*-side. We shall treat the *n*-side as a potential reference. This figure shows the changes in ψ_i, φ_p, and φ_n when the voltage is applied at the external terminals, and also when there is no bias applied. It is seen that the quasi Fermi levels are constant throughout the transition region. This is a result which can be analytically proved under certain conditions. We shall

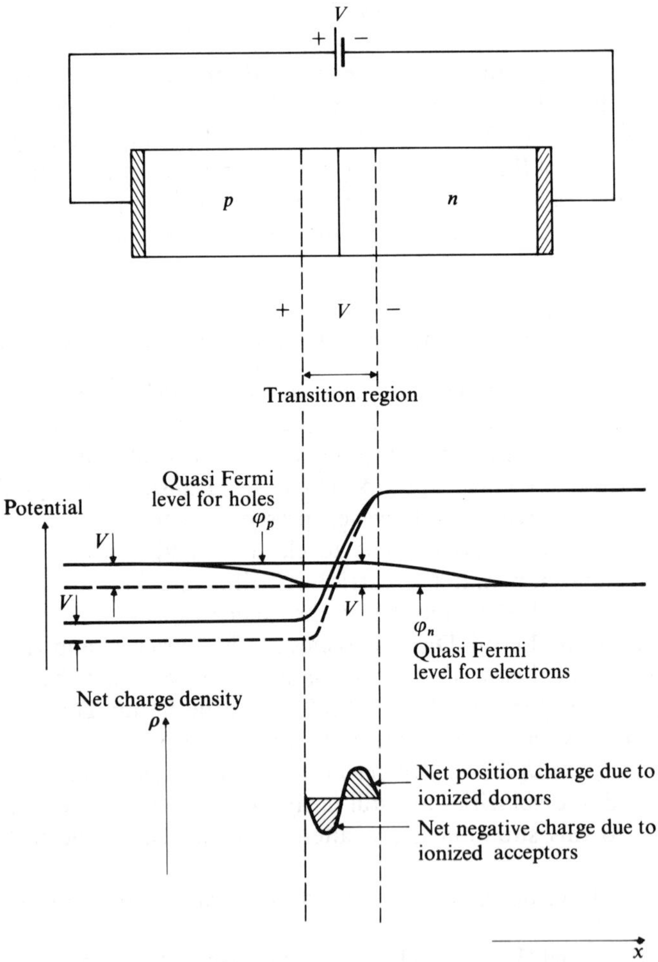

Fig. 3-5 Intrinsic and quasi Fermi potentials in a biased *p-n* junction.

not prove it here but shall take note of the conditions necessary for the statement to be true. These are:

1. The p- and n-sides are truly extrinsic that is

$$p_p \gg n_i, \text{ and } n_n \gg n_i. \tag{3–28}$$

2. $$L_p, L_n \gg x_t. \tag{3–29}$$

3. Injected minority carrier density is much less than the normally present majority carrier density in the region,

i.e. on the n-side $\quad p_{inj} \ll n_n \quad$ (3–30)

on the p-side $\quad n_{inj} \ll p_p. \quad$ (3–31)

All the above conditions are satisfied under most normal conditions. An important conclusion drawn from the above is that throughout the transition region width

$$\varphi_p - \varphi_n = V, \tag{3–32}$$

where V = the externally applied voltage which appears across the transition region.

To obtain the value of hole density at the transition region edge on the n-side, we begin by writing Eqs. (3–13) and (3–14) specialized to $x = x_{tn}$. Thus

$$p_{xtn} = n_i \, e^{(q/kT)(\varphi_p - \psi_i)}, \tag{3–33}$$

$$n_{xtn} = n_i \, e^{(q/kT)(\psi_i - \varphi_n)}. \tag{3–34}$$

By multiplying p_{xtn} with n_{xtn} we get the following:

$$p_{xtn} \, n_{xtn} = n_i^2 \, e^{(q/kT)(\varphi_p - \psi_i + \psi_i - \varphi_n)} = n_i^2 \, e^{(q/kT)(\varphi_p - \varphi_n)}. \tag{3–35}$$

By using Eq. (3–32) we have

$$p_{xtn} = \frac{n_i^2}{n_{xtn}} \, e^{qV/kT}. \tag{3–36}$$

Now

$$n_{xtn} = n_n + p_{inj} \doteq n_n, \tag{3–37}$$

where p_{inj} = holes injected into the n-region.

So we can write

$$p_{xtn} = \frac{n_i^2}{n_n} \, e^{qV/kT} = \frac{n_n p_n}{n_n} \, e^{qV/kT}$$

or

$$p_{xtn} = p_n \, e^{qV/kT}. \tag{3–38}$$

This is one boundary condition. The electron density at $-x_{tp}$ can be obtained in a precisely similar way. If this is done the result is

$$n_{-xtp} = n_p \, e^{qV/kT}. \tag{3–39}$$

3–4 IDEAL *p-n* JUNCTION *I-V* EQUATION

We shall solve the simple diffusion equation given by Eq. (2–130) for holes to the right of x_{tn} so that we are in fact in the field-free region. Before we do the analysis which will lead us to obtain current through the *p-n* junction as a function of the voltage across the *p-n* junction, it will be worthwhile to explicitly state the assumptions behind the analysis. These are:

1. Only a one-dimensional analysis will be attempted.
2. The transition region is considered to be sufficiently narrow so that no recombination or generation of carriers takes place within it.
3. The injected minority carrier density is much less than the thermal equilibrium value of the majority carrier density in the region. This will be ensured if

   $$p_n \, e^{qV/kT} \ll n_n \tag{3–40}$$

   and

   $$n_p \, e^{qV/kT} \ll p_p. \tag{3–41}$$

   This is called the *low injection assumption.*
4. Electric field outside the transition region will be considered zero. This implies that the resistance of the *n*- and *p*-regions outside the transition region is zero.
5. Leakage currents which go *around* the junction, rather than going *through* the junction, are assumed to be zero.
6. Only those surface effects which can be accounted for by a single effective lifetime Eq. (2–108) will be considered.
7. The electric field in the transition region is not so large as to produce breakdown.†
8. Causes of generating carriers other than thermal generation will not be considered. Thermal generation is already accounted for by the recombination term.
9. The width of the *p*- and *n*-region outside the transition region is much greater than L_n, and L_p respectively.

We shall begin by solving the diffusion equation for holes in the *n*-region

† The phenomenon of breakdown will be discussed in Section 3–5.

outside the transition region. As we are now considering the *d-c* case, $\partial p/\partial t$ may be set equal to zero. Under these conditions the simple diffusion equation is

$$\frac{\partial^2 p}{\partial x^2} - \frac{p - p_n}{L_p^2} = 0. \tag{3–42}$$

As the hole density in this case is only a function of position, Eq. (3–42) is an ordinary differential equation. The general form of the solution is for $x > x_{tn}$

$$p(x) = p_n + A e^{-(x - x_{tn})/L_p} + B e^{(x - x_{tn})/L_p}, \tag{3–43}$$

where A, B = arbitrary coefficients.

The first term on the right hand side is the *particular* solution of Eq. (3–42) whereas the next two terms constitute the complimentary solution.

The next task is to evaluate the coefficients A and B. We can easily convince ourselves that

$$B = 0. \tag{3–44}$$

If this were not the case the last term in Eq. (3–43) says that the hole density is a rapidly *increasing* function of distance as we go away from the junction. As the holes can come only from the p-side, the above statement is unreasonable. So B must be zero. To evaluate A we use the boundary condition Eq. (3–38)

$$p(x_{tn}) = p_n\, e^{qV/kT} = p_n + A$$

or

$$A = p_n(e^{qV/kT} - 1). \tag{3–45}$$

So

$$p(x) = p_n + p_n(e^{qV/kT} - 1)\, e^{-(x - x_{tn})/L_p}. \tag{3–46}$$

The hole density is seen to be an exponentially decreasing function of distance having p_n as its steady state value and L_p as the distance constant.

We may now utilize the transport equation to solve for current density as a function of x for $x \geqslant x_{tn}$. In this region there is no electric field so that for $x \geqslant x_{tn}$

$$j_p(x) = -qD_p \frac{\partial p}{\partial x} = \frac{qD_p p_n}{L_p} (e^{qV/kT} - 1)\, e^{-(x - x_{tn})/L_p}. \tag{3–47}$$

The hole current density is also an exponentially decreasing function of distance. The value of j_p at x_{tn} will be of particular interest to us.

$$j_p(x_{tn}) = \frac{qD_p p_n}{L_p} (e^{qV/kT} - 1). \tag{3–48}$$

If we had solved for the electron density and electron current density on the p-side we would expect to find the following results for $x \leqslant -x_{tp}$

$$n = n_p + n_p(e^{qV/kT} - 1)\, e^{(x + x_{tp})/L_n}. \tag{3-49}$$

for $x \leqslant -x_{tp}$

$$j_n(x) = q\frac{D_n n_p}{L_n}(e^{qV/kT} - 1)\, e^{(x + x_{tp})/L_n}. \tag{3-50}$$

In particular

$$j_n(-x_{tp}) = q\frac{D_n n_p}{L_n}(e^{qV/kT} - 1). \tag{3-51}$$

Now because of assumption 2

$$j_n(-x_{tp}) = j_n(+x_{tn}). \tag{3-52}$$

So that

$$j_{\text{total}}(x_{tn}) = j_p(x_{tn}) + j_n(x_{tn}) \tag{3-53}$$

$$j = q\left(\frac{D_p p_n}{L_p} + \frac{D_n n_p}{L_n}\right)(e^{qV/kT} - 1). \tag{3-54}$$

Writing in terms of current rather than current density we obtain

$$I = qA\left(\frac{D_p p_n}{L_p} + \frac{D_n n_p}{L_n}\right)(e^{qV/kT} - 1). \tag{3-55}$$

Alternatively

$$I = I_S(e^{qV/kT} - 1), \tag{3-56}$$

where

$$I_S \triangleq qA\left(\frac{D_p p_n}{L_p} + \frac{D_n n_p}{L_n}\right). \tag{3-56a}$$

Though Eq. (3–56) is derived at x_{tn}, it *is valid for all values of* x throughout transition region as well, because of Kirchhoff's current law which says that current is continuous.

Figure 3–6 shows a plot of Eq. (3–56). In the forward direction when $V \gg kT/q$ the exponential term is dominant, and the current is an exponentially rising function of voltage. When the voltage across the junction is negative and $|V| \gg kT/q$, the -1 term dominates the exponential term and the current is essentially $-I_S$. This current is often referred to as the *back saturation current* of the p-n junction diode and is quite small. Typical value of I_S in silicon is 10^{-10} Amp. The fact that extremely low and usually negligible values of currents flow in a p-n junction which is reverse biased is utilized in many integrated

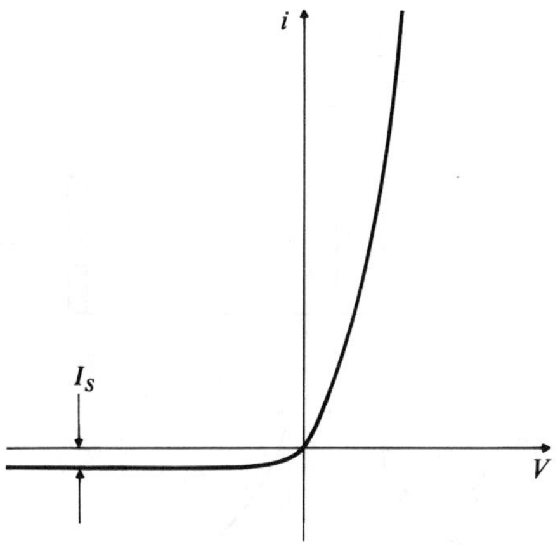

Fig. 3–6 *Ideal* $I(V)$ characteristics for the *p-n* junction diodes.

circuits where the reverse biased *p-n* junctions serve the function of isolating adjacent circuits on a single semiconductor crystal chip.

Figure 3–7 shows the hole and electron densities and hole and electron current densities plotted throughout the device in forward bias, as Fig. 3–8 shows these quantities for reverse bias.

The *p-n* junction passes current easily from *p-* to *n-*side, but allows only very small current from *n-* to *p-*side. So the *p-n* junction is an essentially unidirectional or a rectifying device. It is therefore called a *p-n junction diode.*

One interesting and informative result which can be calculated from Eq. (3–56) is the incremental resistance of the *p-n* junction diode in forward bias.

$$r_{\text{incremental}} \doteq \frac{dV}{dI}. \tag{3–57}$$

Taking the derivative of current from Eq. (3–56) with respect to the externally applied voltage V, we obtain

$$\frac{dI}{dV} = \frac{qI_S}{kT}e^{qV/kT}. \tag{3–58}$$

If

$$V \gg \frac{kT}{q}. \tag{3–59}$$

From Eq. (3–56)

$$I = I_S e^{qV/kT}, \tag{3–60}$$

where we have taken advantage of Eq. (3–59) and have dropped -1 in comparison with the exponential term in Eq. (3–56).

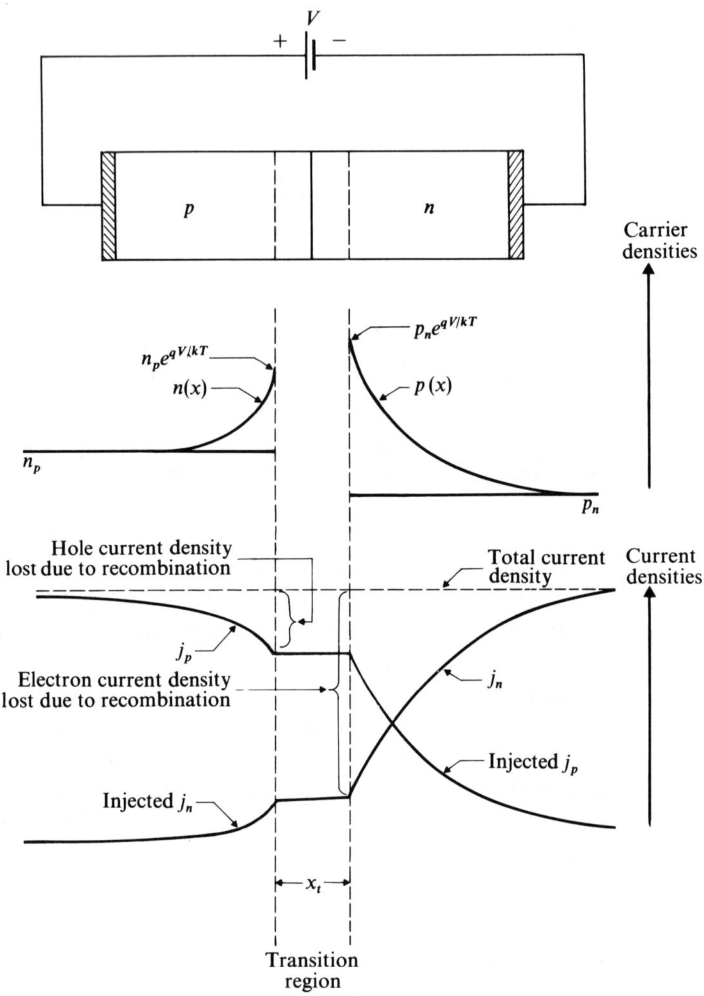

Fig. 3–7 Carrier densities and current densities in a forward biased *p-n* junction.

Thus the right hand side of Eq. (3–58) can be written in terms of current I if use is made of Eq. (12–60) to obtain the following:

$$\frac{dI}{dV} = \frac{qI}{kT} = g_d. \tag{3-61}$$

Finally

$$r_{\text{inc}} = r_d = \frac{dV}{dI} = \frac{kT}{qI}. \tag{3-62}$$

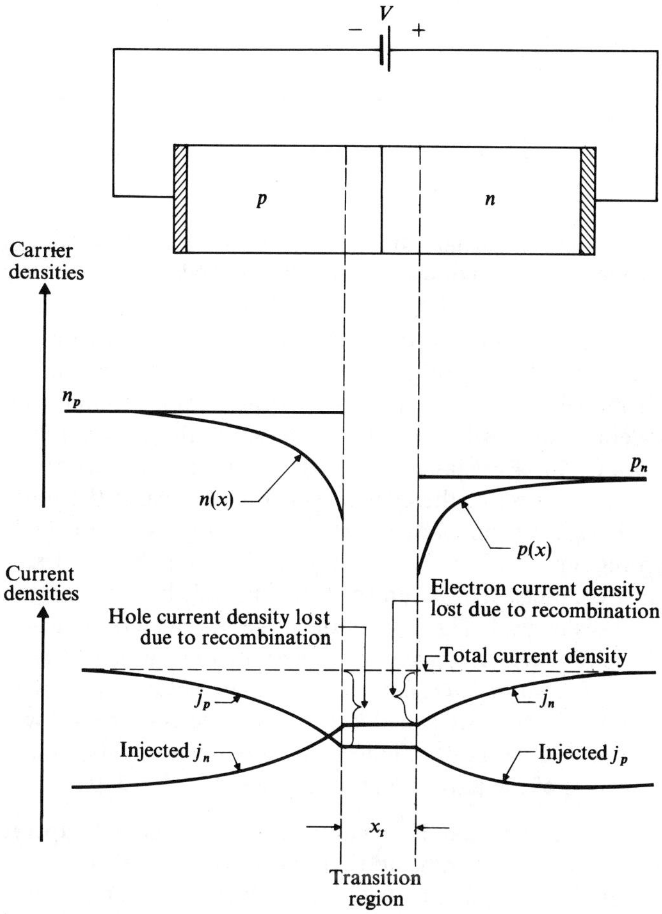

Fig. 3–8 Carrier densities and current densities in a reverse biased *p-n* junction.

Equations (3–61) and (3–62) are called *diffusion conductance* and *diffusion resistance* respectively because they arose out of the solution of the diffusion equation. This is therefore an *intrinsic* parameter. The result of Eqs. (3–61) and (3–62) is more general than the *p-n* junction. Diffusion resistances are generally found to be inversely proportional to current. Diffusion conductances are generally directly proportional to current. Note, however, the restriction of Eq. (3–59).

3–5 MODEL FOR PRACTICAL *p-n* JUNCTION

Equation (3–56) refers to an *ideal p-n* junction for which the nine assumptions listed in Section 3–4 are true. Not all of the assumptions are realistic for practical

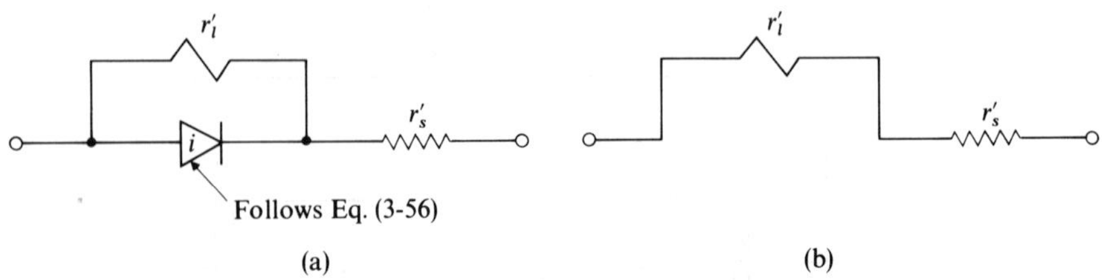

Fig. 3–9 Models of *p-n* junction. (a) Equivalent circuit of an actual *p-n* junction diode. (b) Incremental equivalent circuit when the diode is reverse biased.

p-n junctions. An attempt will be made in this section to find a model for practical *p-n* junctions. Assumptions 4 and 5 of Section 3–4 are not usually true. The simple model of Fig. 3–9a makes these assumptions unnecessary. In this simple model the diode with the letter *i* stands for the *ideal p-n* junction diode following Eq. (3–56). Resistance r'_ℓ represents the leakage resistance which accounts for that portion of the current which goes *around* the junction rather than going *through* the junction. The resistance r'_s represents the bulk resistance of *p*- and *n*-regions, outside the transition region, together with resistance due to the ohmic contacts. It is assumed in this model that the resistances r'_s and r'_ℓ are fixed resistances.† The primes on these parameters indicate that these are *extrinsic* parameters. *Extrinsic parameters are those which cannot be obtained from solutions of the diffusion equation.* Parameters obtainable from the solution of the diffusion equation are called *intrinsic* parameters. The following illustrative example will show how practical measurements can be made to deduce each of the values of the parameters of the model of Fig. 3–9a.

Example 3–1 The terminal *V-I* characteristics, at room temperature, of an actual *p-n* junction diode indicate that, when the forward current is 104 mA, the slope of the characteristics is 10.25 Ohms. When the reverse voltage is 2 Volts, the slope is 10^{-6} Amp/Volt. It is also observed that, when the terminal current is 10 mA, the voltage is 0.3 Volt. Calculate (a) r'_ℓ, (b) r'_s, (c) I_S.

Solution. At a reverse voltage of 2 Volts we note that $V = 2$ Volts $\gg kT/q = 0.026$ Volt at room temperature. This implies that at this reverse voltage

$$e^{qV/kT} \ll 1 \tag{1}$$

so that

$$I \doteq -I_S.$$

Thus

$$\frac{dI}{dV} \doteq 0 \tag{2}$$

† If the *n*- or the *p*-regions are very short compared with the corresponding diffusion lengths, r'_s may not be constant. This circumstance is ruled out because of assumption 9 in Section 3–4.

or for the *ideal* junction diode the incremental resistance at reverse voltage of 2 Volts is infinite. Thus for all practical purposes the ideal diode branch may be considered open circuited as shown in Fig. 3–9b. From the given information

$$10^6 \text{ Ohms} = r'_\ell + r'_s \doteq r'_\ell. \tag{3}$$

The approximation used above will be checked when we calculate r'_s next.

The forward current of 104 mA will go through the ideal diode and r'_s. The current through r'_ℓ may be neglected as it is such a high resistance compared to that offered by the diode. To calculate the resistance r'_s let us utilize the result given by Eq. (3–62) to obtain

$$r_d = \frac{kT}{qI} = \frac{26}{104} = 0.25\Omega. \tag{4}$$

The observed terminal slope at $I = 104$ mA that corresponded to 10.25 Ohm implies

$$r_{\text{observed}} = r_d + r'_s, \tag{3–63}$$

$$10.25 = 0.25 + r'_s, \tag{5}$$

$$r'_s = 10 \text{ Ohm}. \tag{6}$$

The value of r'_s obtained above justifies an assumption made earlier, that is

$$r'_s \ll r'_\ell. \tag{3–64}$$

In wiriting Eq. (3–63) we have neglected the current flowing through r'_ℓ as compared to the current through the ideal diode (i.e. through r_d). It is manifestly true from the above that

$$r_d \ll r'_\ell. \tag{3–65}$$

The third piece of data can now be used to calculate I_S. At a voltage of 0.3 V the current $I = 10$ mA. For this condition we can compute the voltage drop V' across r'_s due to this current

$$V' = Ir'_s = (10)(10) = 100 \text{ mV}.$$

Of the 300 mV applied externally only (300–100) or 200 mV appears across the ideal junction device. *It is important to note that it is only the voltage across the transition region of the* p-n *junction which is to be used for V in Eq. (3–56) and hence also in Eq. (3–60).* Equation (3–59) is here true, and thus Eq. (3–60) can be used

$$I = I_S\, e^{qV/kT} \tag{3–60}$$

or

$$I_S = \frac{I}{e^{qV/kT}} = \frac{10\,(\text{mA})}{e^{200/26}} = 4.53 \times 10^{-6} \text{ Amp}.$$

Thus the data given in the statement of the problem is indeed adequate to be able to compute the major parameters of the equivalent circuit of Fig. 3–9a.

Example 3–2 Consider a *p-n* junction diode with the following of its constants known:

$$\begin{aligned} \sigma_p \gg \sigma_n &= 1(\Omega\text{-cm})^{-1} \\ D_p &= 50 \text{ cm}^2/\text{sec} \\ \mu_n &= 1000 \text{ cm}^2/(\text{Volt-sec}) \\ L_p &= 1 \text{ cm} \\ I &= 100 \text{ mA} \\ V &\gg kT/q \\ A &= 1.25 \times 10^{-2} \text{ cm}^2 \end{aligned}$$

You may assume that L_n is the same order of magnitude as L_p.

Is the low injection assumption valid for this device? Explain in detail your answer including analytical and numerical calculations to justify your answer.

Solution. Because $\sigma_p \gg \sigma_n$ we may conclude that $p_n \gg n_p$ by reasoning as follows:

$$q\mu_p p_p \gg q\mu_n n_n \tag{1}$$

or

$$\mu_p \frac{n_i^2}{n_p} \gg \mu_n \frac{n_i^2}{p_n}. \tag{2}$$

As μ_p and μ_n are of the same order of magnitude numbers, it follows that

$$p_n \gg n_p. \tag{3}$$

Furthermore, L_p and L_n are the same order of magnitude numbers. Thus the ideal *p-n* junction equation may be approximated as follows:

$$I = qA\left[\frac{p_n D_p}{L_p} + \frac{n_p D_n}{L_n}\right](e^{qV/kT} - 1) \doteq \frac{qAp_n D_p}{L_p} e^{qV/kT}, \tag{4}$$

where the approximation

$$e^{qV/kT} \gg 1$$

is used, which is justified because

$$V \gg \frac{kT}{q}. \tag{5}$$

From Eq. (4) we see that

$$p_n e^{qV/kT} = \left(\frac{I}{A}\right)\frac{L_p}{D_p q} = \frac{0.1\,(\text{Amp}) \times 1\,(\text{cm})}{1.25 \times 10^{-2}\,(\text{cm}^2) \times 50\,(\text{cm}^2/\text{sec})\ 1.6 \times 10^{-19}\,(\text{coul.})} \tag{6}$$

$$p_n \, e^{qV/kT} = 10^{18}/\text{c.c.} \tag{7}$$

Now let us calculate the thermal equilibrium density of majority carriers on the n-side.

$$\sigma_n = q\mu_n n_n \qquad \text{or} \qquad n_n = \frac{\sigma_n}{q\mu_n} \tag{8}$$

$$n_n = \frac{1}{1.6 \times 10^{-19} \times 10^3} = 6.25 \times 10^{15}/\text{c.c.} \tag{9}$$

Note that Eq. (7) gives the *minority carrier density* at the edge of the transition region on the n-side and is seen to be greater than n_n. We must conclude that the low injection assumption

$$p_n \, e^{qV/kT} \ll n_n \tag{10}$$

is clearly *not* valid in this device, at this current level.

3–6 BREAKDOWN PHENOMENA

It is experimentally observed that when a p-n junction is sufficiently reverse biased, instead of the rather high incremental resistance, corresponding to r'_ℓ, which it normally presents in reverse bias, the device offers a very low incremental resistance as shown in Fig. 3–10. The point at which the current grows without bound is called the *breakdown* voltage. There are two different mechanisms which can explain this sudden increase in reverse current at breakdown voltage. These mechanisms are referred to as the *Zener effect* and the *avalanche effect*. The corresponding breakdowns are called *Zener breakdown* and *avalanche breakdown* respectively.

The Zener effect involves the tunneling mechanism. To see this consider the energy band picture in reverse biased junction as shown in Fig. 3–11b. Figure 3–11a shows the band picture at thermal equilibrium and hence at no

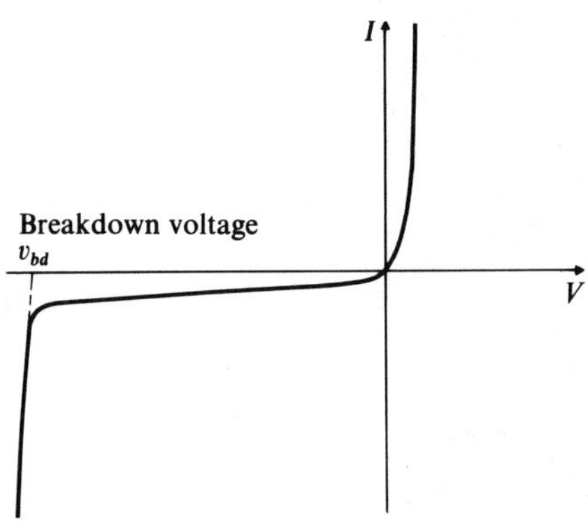

Fig. 3–10 Breakdown in reverse bias.

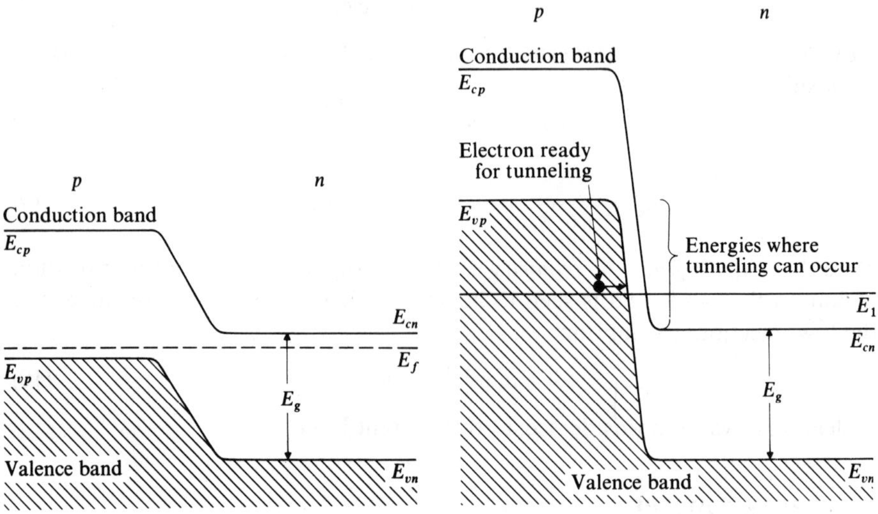

(a) Thermal equilibrium (b) Reverse bias

Fig. 3–11 Tunneling in Zener effect.

bias. Unlike the condition for forward bias or no bias, the reverse bias presents a unique situation. Here there exists a set of energies between the top of the valence band on the *p*-side and the bottom of conductance band on the *n*-side, such that there are filled states on the *p*-side opposite the empty states on the *n*-sides *at the same energy levels.*

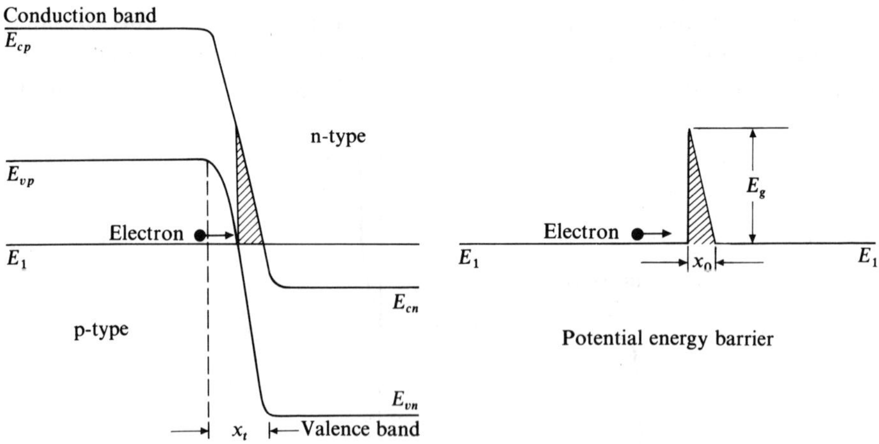

(a) (b)

Fig. 3–12 Band gap and potential barrier for Zener effect.

Consider an electron at the energy level E_1 which is marked on Fig. 3–12a. Such an electron will see a potential energy barrier shown on Fig. 3–12b. Electrons from the valence band on the *p*-side can indeed make a transition to the conduction band on the *n*-side provided that the following four conditions are fulfilled.

1. Electrons exist at permissible energy states on the *p*-side from which tunneling can occur.
2. Available, i.e. empty energy states, exist on the *n*-side into which the electrons from the *p*-side could go.
3. Overall energy and momentum conservation laws must be satisfied in the tunneling transition.
4. Tunneling probability for the carriers must not be too low.

It may be pointed out in connection with condition 4 that it is possible for appreciable tunneling current to flow even though the tunneling probability per electron is low, provided that a large number of electrons, that qualify for tunneling, exist.

Figure 3–12 clearly shows that the tunneling height is the band gap, i.e. E_g, and that the tunneling barrier width is less than the transition region width. That is

$$x_0 < x_t, \tag{3–66}$$

where x_0 = the tunneling barrier width,

x_t = transition region width.

Later it will be shown that x_t is directly proportional to some *fractional power* of *V*, the *total* junction voltage. The energy bands on the *p*-side shift with respect to *n*-side linearly with *V*. The net result of the above statements is for the tunneling barrier width x_0 to diminish with decreasing reverse voltage. From the results of Section 1–9 we know that when x_0 diminishes sufficiently, appreciable tunneling current will flow. From Eq. (1–45) we see that tunneling probability is exponentially dependent on the width of tunneling barrier;† so the onset of tunneling is expected to occur rather rapidly, and Zener breakdown will occur. It is expected that Zener effect will predominate in junctions having a rather narrow transition region width which in turn occurs in highly doped devices.

Historically the first mechanism offered to explain the sudden rise in reverse current at breakdown was the Zener effect. As the reverse voltage at breakdown does not change very much, these junctions could serve the function of

† Equation (1–45) was derived for a square potential barrier. The triangular barrier as considered here *does not* alter the essential nature of the solution in a major way. So the statement made here is correct.

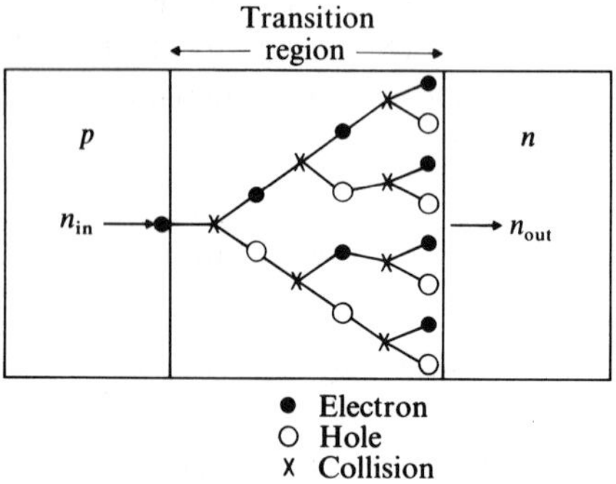

Fig. 3–13 Avalanche process in the transition region of a *p-n* junction.

reference fixed voltage useful in many different electronic circuits. These diodes were called Zener diodes. It was necessary to provide an external circuit element, usually a series resistance, to limit the current through the diode to prevent its destruction. It is now thought that many of these early diodes were experiencing the breakdown because of a process associated with the avalanche effect which is quite different from the Zener effect. The term Zener diode, however, still persists in literature.

When a *p-n* junction is reverse biased it is possible to have a very high electric field in the transition region. A carrier which enters such a transition region can acquire enough energy from the electric field which can be subsequently given up during collision with a lattice atom. An electron in the valence band can thus acquire enough energy to go directly into the conduction band. An electron-hole pair is thus created. These carriers can in turn acquire additional energy from the electric field to create additional electron hole pairs of their own. If the *mean free path length* $\bar{\ell}$, the average path length between collisions, is small compared with x_t, several collisions will be suffered by carriers before they leave the transition region. Thus carriers leaving the transition region, n_{out}, can be significantly higher than those which entered it, n_{in} (Fig. 3–13). The ratio (n_{out}/n_{in}) has been experimentally observed to follow the relationship

$$\frac{n_{out}}{n_{in}} \triangleq M = \frac{1}{1 - \left(\frac{V}{V_A}\right)^c}, \tag{3–67}$$

where V_A = avalanche breakdown voltage,

c = a numerical parameter which varies between 3 and 6 for germanium and between 2 and 4 for silicon,

M = multiplication factor.

In reverse bias the *p-n* junction $I(V)$ characteristics in those junctions in which the avalanche effect is dominant can be written

$$I = \frac{I_s(e^{qV/kT} - 1)}{1 - \left(\frac{V}{V_A}\right)^c}, \text{ for } V < 0. \tag{3–68}$$

Because the avalanche effect requires thick junctions it is expected to be dominant in less heavily doped junctions compared with those in which the Zener effect is important. It is observed that lightly doped junctions have high breakdown voltage, and heavily doped junctions have low breakdown voltage. As a rough guide, in *p-n* junctions whose breakdown voltage is $\geqslant -5$ Volts, the Zener effect is likely to be dominant, and in those whose breakdown voltage is $\leqslant -25$ Volts, the avalanche effect is dominant. Between -5 and -25 Volts it is possible to have both effects present simultaneously.

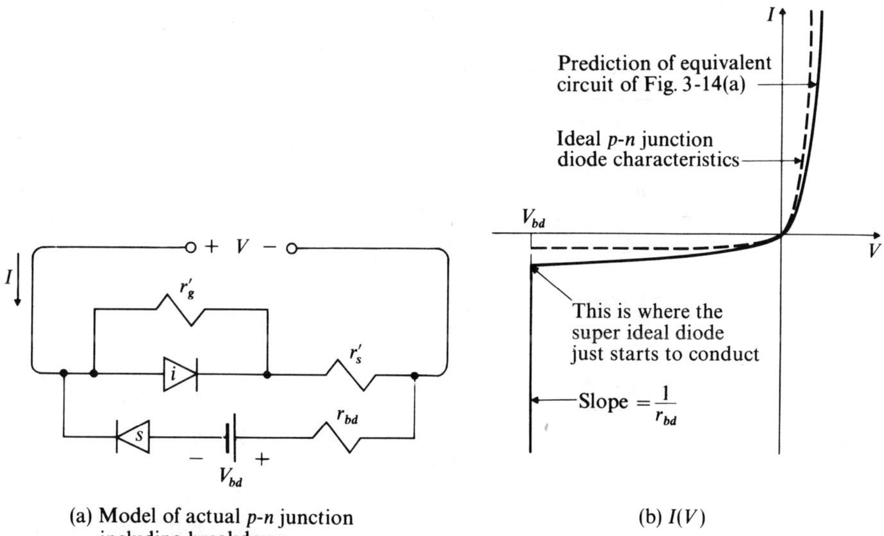

(a) Model of actual *p-n* junction including breakdown

(b) $I(V)$

Fig. 3–14 A *p-n* junction diode model including breakdown. (a) Model of actual *p-n* junction including breakdown. (b) $I(V)$.

The avalanche effect is used in very high cut-off frequency semiconductor *avalanche photo diodes*. The Zener effect is used in obtaining the useful negative resistance region of a *tunnel diode*.

An equivalent circuit for the *p-n* junction diode, which also takes the breakdown phenomenon into account, is shown in Fig. 3–14. The branch at the very

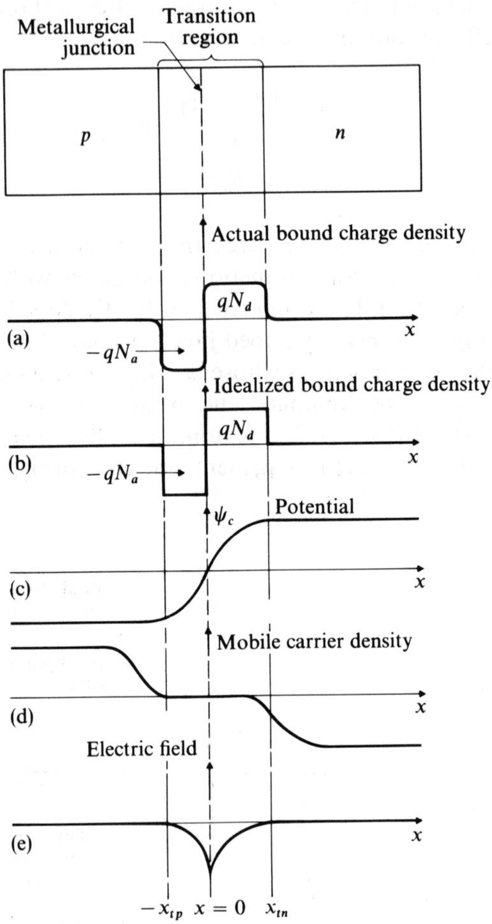

Fig. 3–15 Distribution of free and bound charge density, and potential in an abrupt or step *p-n* junction.

bottom consisting of a *super ideal*† diode (note *not* an ideal *p-n* junction diode which is designated‡ by *i* within the diode symbol) is the only addition to the equivalent circuit of Fig. 3–9. This extra branch conducts only when $V \leqslant$ breakdown voltage V_{bd}. When this voltage is reached the ideal diode *just begins* to conduct and the diode offers the rather low resistance of r_{bd} usually less than 1 Ohm. This is shown in Fig. 3–14b.

† A super ideal diode has zero voltage drop across it when the current is flowing in the forward direction, and has zero current flowing through it when the voltage is in the reverse direction.

‡ The letter *i* stands for "ideal" or "intrinsic."

3-7 JUNCTION CAPACITANCE

As shown in Fig. 3–5, there always exists net positive and negative charges due to ionized donors and acceptors in the transition region of *p-n* junctions. *Outside the transition region* too there exist ionized acceptors on the *p*-side, and ionized donors on the *n*-side. There is, however, no net charge here because the negative charge of ionized acceptors is neutralized by the positive charge of free holes on the *p*-side; just as the positive charge of ionized donors is neutralized by the negative charge of free electrons on the *n*-side. In the transition region the field which is always present tends to sweep out the mobile free carriers, and hence the net positive and negative charges remain unneutralized. The existence of net positive charge right next to net negative charge in the transition region gives rise to a capacitance which is called *junction capacitance.*

As we shall presently see, the junction capacitance, unlike the ordinary linear capacitance, is not only a function of voltage, but is also a *nonlinear* function of voltage. It is defined as

$$C_j \triangleq \frac{dQ}{dV_t}, \tag{3–69}$$

where C_j = junction capacitance,
Q = net charge,
V_t = total voltage across the transition region ($V_{BI} - V$).

The functional dependence of C_j on voltage is importantly determined by the detailed distribution of charge in the transition region. Consider the *abrupt* or *step*-like distribution of net charge shown in Fig. 3–15. The following steps will be taken to obtain an expression for the junction capacitance.

a) One dimensional Poisson's equation will be solved in the transition region.

$$\frac{\partial^2 \psi_c}{\partial x^2} = -\frac{\rho}{\varepsilon}, \tag{3–70}$$

where ψ_c = potential,
ε = permittivity,
ρ = charge density.

b) Equation (3–70) will be solved by integrating twice with respect to x.
c) Two constants of integration will be evaluated by requiring that (i) the potential ψ_c be zero at the metallurgical junction (this merely sets the reference for the potential), and (ii) the electric field be zero at the transition region edges.
d) The width of the transition region on the *p*- and *n*-regions are related to each other requiring that the net positive charge must equal the net negative charge.

e) The capacitance is found finally by using Eq. (3–69).

Let us integrate Eq. (3–70) from 0 to a general point x. Thus

$$\int_0^x \frac{\partial^2 \psi_c}{\partial x^2} dx = -\frac{\rho_1}{\varepsilon} \int_0^x dx, \tag{3–71}$$

$$\left.\frac{\partial \psi_c}{\partial x}\right|_{x=x} - \left.\frac{\partial \psi_c}{\partial x}\right|_{x=0} = \frac{\rho_1}{\varepsilon}(x - 0). \tag{3–72}$$

In particular let $x = x_{tn}$

$$\left.\frac{\partial \psi_c}{\partial x}\right|_{x=x_{tn}} - \left.\frac{\partial \psi_c}{\partial x}\right|_{x=0} = \frac{\rho_1}{\varepsilon} x_{tn}. \tag{3–73}$$

The first term on the left hand side of Eq. (3–73) must be zero because of the electric field being zero at the edge of the transition region

$$\text{Electric field} = -\left.\frac{\partial \psi_c}{\partial x}\right|_{x=x_{tn}} = 0, \tag{3–74}$$

therefore

$$\left.\frac{\partial \psi_c}{\partial x}\right|_{x=0} = \frac{\rho_1}{\varepsilon} x_{tn}. \tag{3–75}$$

Substituting this value into Eq. (3–72) we get

$$\left.\frac{\partial \psi_c}{\partial x}\right|_{x=x} = -\frac{\rho_1}{\varepsilon}(x - x_{tn}). \tag{3–76}$$

Integrating this equation once again between 0 and a general point x we obtain

$$\psi_c(x) - \psi_c(0) = -\frac{\rho_1}{\varepsilon}\left(\frac{x^2}{2} - x\, x_{tn}\right). \tag{3–77}$$

From condition (c) (i)

$$\psi_c(0) = 0.$$

So

$$\psi_c(x) = -\frac{\rho_1}{\varepsilon}\left(\frac{x^2}{2} - x\, x_{tn}\right). \tag{3–78}$$

For $x = x_{tn}$

$$\psi_c(x_{tn}) = -\frac{\rho_1}{\varepsilon}\left(\frac{x_{tn}^2}{2} - x_{tn}^2\right),$$

$$\psi_c(x_{tn}) = \frac{\rho_1 x_{tn}^2}{2\varepsilon} \tag{3–79}$$

If we had gone through an exactly similar development on the *p*-side, it is not hard to see that the result would have been

$$\psi_c(-x_{tp}) = \frac{\rho_2 x_{tp}^2}{2\varepsilon}. \tag{3–80}$$

The total voltage V_t across the junction is

$$V_t = \psi_c(x_{tn}) - \psi_c(-x_{tp}), \tag{3–81}$$

$$V_t = \frac{\rho_1}{2\varepsilon}x_{tn}^2 - \frac{\rho_2}{2\varepsilon}x_{tp}^2. \tag{3–82}$$

Overall space charge neutrality requires that

$$\rho_1 x_{tn} = -\rho_2 x_{tp} \tag{3–83}$$

or

$$qN_d x_{tn} = qN_a x_{tp}. \tag{3–84}$$

In writing Eq. (3–84) we have assumed that the *free carrier density in the transition region is negligible.* Using Eq. (3–83) we can rewrite Eq. (3–82):

$$V_t = \frac{\rho_1}{2\varepsilon}x_{tn}^2 + \frac{\rho_1}{2\varepsilon}x_{tn}x_{tp} \tag{3–85}$$

$$V_t = \frac{\rho_1 x_{tn}}{2\varepsilon}[x_{tn} + x_{tp}]. \tag{3–86}$$

Now Eq. (3–84) can be used to write:

$$V_t = \frac{\rho_1 x_{tn}}{2\varepsilon}\left(x_{tn} + \frac{N_d}{N_a}x_{tn}\right) = \frac{x_{tn}^2 qN_d}{2\varepsilon}\left(1 + \frac{N_d}{N_a}\right). \tag{3–87}$$

Equation (3–87) can now be solved for x_{tn}

$$x_{tn} = \left(\frac{2\varepsilon}{qN_d}\right)^{\frac{1}{2}}\left(\frac{N_a}{N_a + N_d}\right)^{\frac{1}{2}}(V_t)^{\frac{1}{2}}. \tag{3–88}$$

Using Eqs. (3–84) and (3–87) x_{tp} can be similarly evaluated to give

$$-x_{tp} = \left(\frac{2\varepsilon}{qN_a}\right)^{\frac{1}{2}}\left(\frac{N_d}{N_a + N_d}\right)^{\frac{1}{2}} V_t^{\frac{1}{2}}. \tag{3–89}$$

The charge Q on the n-side of the junction is given by

$$Q = qN_dAx_{tn} = qN_aAx_{tp}. \tag{3–90}$$

Substituting the right hand side of Eq. (3–88) for x_{tn} into Eq. (3–90) we obtain

$$Q = A\left[2\varepsilon qV_t\left(\frac{N_dN_a}{N_d + N_a}\right)\right]^{\frac{1}{2}} \tag{3–91}$$

We are now ready to calculate the junction capacitance by using Eq. (3–69)

$$C_j = \frac{dQ}{dV_t} = \frac{A}{2}\left(2\varepsilon q\frac{N_dN_a}{(N_d + N_a)V_t}\right)^{\frac{1}{2}}. \tag{3–92}$$

This may be written

$$C_j = \frac{k_a}{V_t^{\frac{1}{2}}} = \frac{k_a}{(V_{bi} - V)^{\frac{1}{2}}} \tag{3–93}$$

where

$$k_a \triangleq \frac{A}{2}\left[\frac{2\varepsilon qN_dN_a}{(N_d + N_a)}\right]^{\frac{1}{2}}. \tag{3–94}$$

Several comments are pertinent at this point. First, the junction capacitance of an abrupt *p-n* junction is inversely proportional to the square root of the *total* voltage across the junction. Note that the voltage V is an *algebraic* quantity. When the *p*-side is positive with respect to the *n*-side (forward bias), it is positive. When the *p*-side is negative with respect to the *n*-side, the voltage V is itself negative. Second, Eq. (3–84) can be written in a different form

$$\frac{x_{tn}}{(-x_{tp})} = \frac{N_a}{N_d}. \tag{3–95}$$

Equation (3–95) says that the penetration of the transition region into *n*- and *p*-regions is inversely proportional to doping in each region. Third, in the definition of junction capacitance which is really an *incremental capacitance* (analogous to the definition of incremental resistance) the incremental change of charge dQ at the *edges of the transition region* to a given incremental change in voltage is what is important. Figure 3–16 shows the relevant charge dQ at the edges of the transition region. It is evident from this picture that *this incremental capacitance looks exactly like a parallel plate capacitance* regardless of

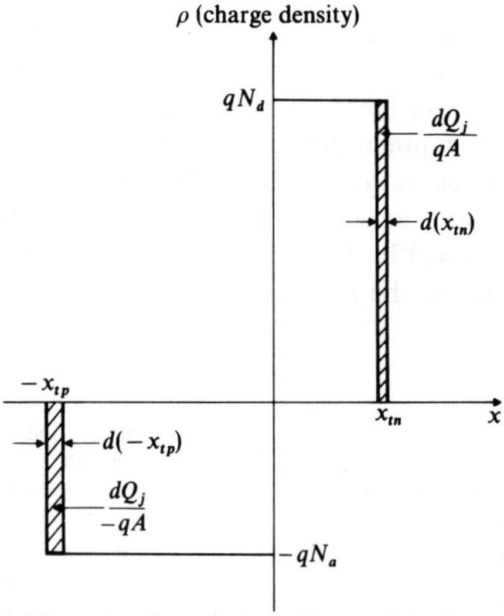

Fig. 3–16 Charge distribution for calculating junction capacitance in an abrupt *p-n* junction.

the detailed distribution of $\rho(x)$ having a separation of the width of the transition region x_t and area A. Thus the junction capacitance can be written as

$$C_j = \frac{A\varepsilon}{x_t}, \tag{3–96}$$

where

$$x_t = x_{tn} - (-x_{tp}). \tag{3–97}$$

If we substitute Eqs. (3–88) and (3–89) into Eq. (3–97) we obtain:

$$x_t = \left(\frac{2\varepsilon}{q}\right)^{\frac{1}{2}} \left\{\left[\frac{N_a/N_d}{N_a + N_d}\right]^{\frac{1}{2}} + \left[\frac{N_d/N_a}{N_a + N_d}\right]^{\frac{1}{2}}\right\} V_t^{\frac{1}{2}} \tag{3–98}$$

or

$$x_t = k_{at}(V_{BI} - V)^{\frac{1}{2}}, \tag{3–99}$$

where

$$k_{at} \triangleq \left(\frac{2\varepsilon}{q}\right)^{\frac{1}{2}} \left[\frac{(N_a/N_d)^{\frac{1}{2}} + (N_d/N_a)^{\frac{1}{2}}}{(N_a + N_d)^{\frac{1}{2}}}\right]. \tag{3–100}$$

We see that the transition region width of an abrupt *p-n* junction is directly proportional to the square root of the total voltage V_t.

If a different charge distribution in the transition region had been considered, a different functional dependence of C_j and x_t on V_t would have resulted than for an abrupt distribution. Consider a linear distribution of charge in the transition region as shown in Fig. 3–17a. Such junctions are called *linearly graded.* Two different device fabrication processes result in the charge distribution of Fig. 3–17b. Grown junction transistors of the 1950s and the more recent diffused base transistors sometimes lead to this charge distribution. The *gradient* of charge in the transition region is *a*. Note here that unlike the step junction, where transition region width of the *p*-side could be vastly different from the transition region width on the *n*-side, the linearly graded junction *must* have

$$x_{tn} = -x_{tp}. \tag{3–101}$$

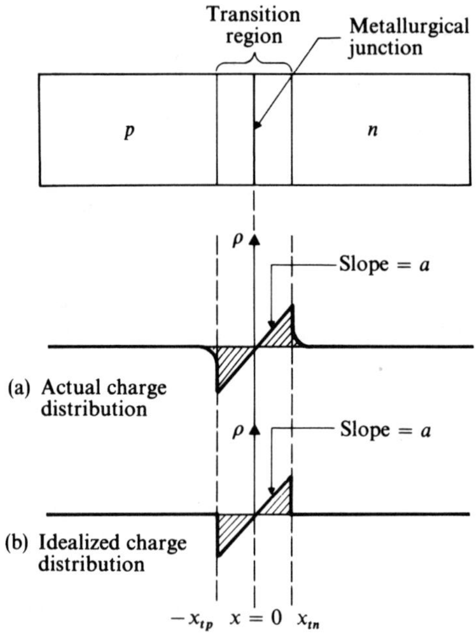

Fig. 3–17 Charge distribution in linearly graded *p-n* junctions.

Figure 3–17b shows the idealized distribution of charge in the transition region. Poisson's equation in this case is

$$\frac{\partial^2 \psi}{\partial x^2} = -q\frac{xa}{\varepsilon}. \tag{3–102}$$

The solution of Eq. (3–102) subject to the same conditions as for the abrupt junction case is the following:

$$x_t = \frac{\varepsilon 12}{qa}(V_{BI} - V)^{\frac{1}{3}} \tag{3–103}$$

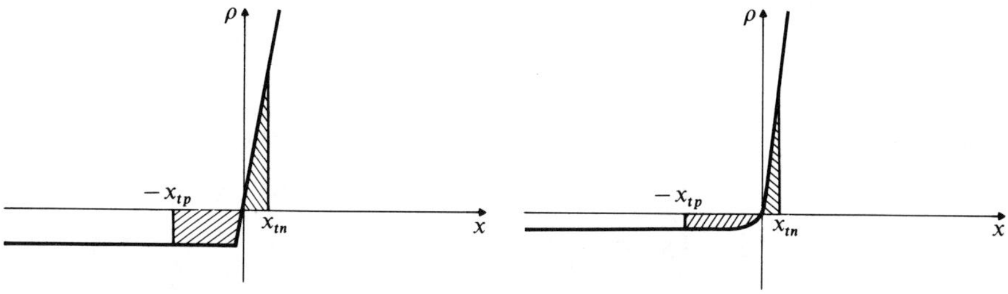

Fig. 3–18 Distribution of charge density in the transition region which give the same variation of x_t and C_j as that predicted by Eqs. (3–93) and (3–99) respectively provided that the transition region width is at least as wide as shown here.

and

$$C_j = \frac{k_\ell}{(V_{BI} - V)^{\frac{1}{3}}}, \tag{3–104}$$

where

$$k_\ell \triangleq A\left(\frac{qa\varepsilon^2}{12}\right)^{\frac{1}{3}}, \tag{3–105}$$

where A = cross-sectional area of the junction.

Equation (3–103) shows that the transition region width here too is directly proportional to a fractional power of total voltage across the transition region. The junction capacitance is *inversely* proportional to the fractional power of the total voltage across the transition region.

It is worthwhile noting that the abrupt distribution of net charge gives the results of Eqs. (3–99) and (3–93), as does the linear distribution of net charge give the results given by Eqs. (3–103) and (3–104). These, however, are not the only distributions which give these results. For example, Fig. 3–18a and b

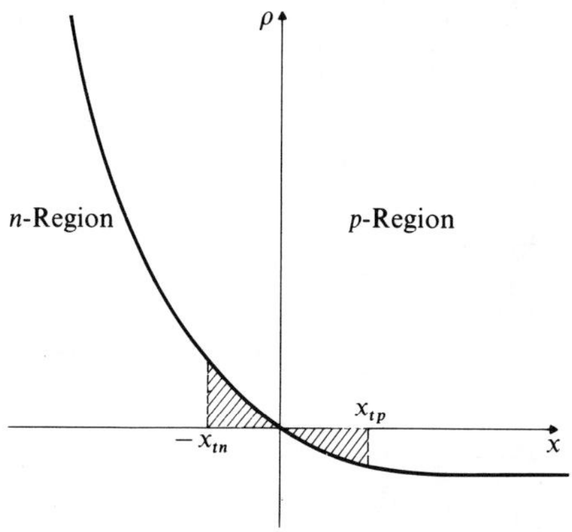

Fig. 3–19 Charge density in a diffused junction.

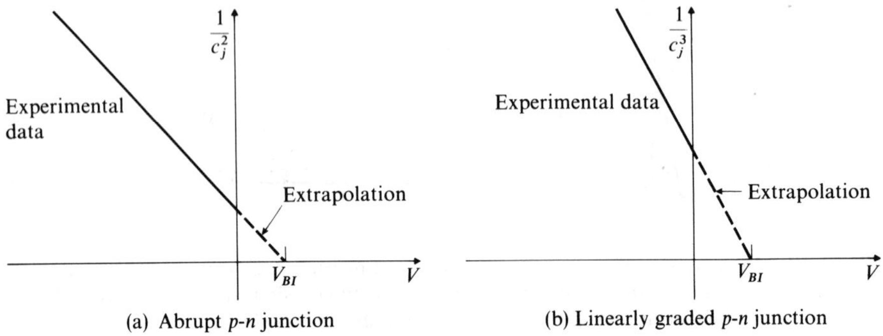

Fig. 3–20 Experimental determination of built-in voltage V_{BI} of abrupt and linearly graded *p-n* junctions.

show two of the charge distributions which give results corresponding to Eqs. (3–93) and (3–99) provided the transition region width is at least as wide as shown in this figure. Figure 3–19 shows a distribution which is encountered when doping is done by diffusion. Here if the bias on the junction is small so the transition region is quite narrow, the distribution looks linear and a cube root behavior is expected to result. When the reverse bias is large and negative, the transition region is broad mostly on the *p*-side in Fig. 3–19, and now a square root behavior is expected to be followed. Experimental results confirm these observations.

In the above derivation of analytic expressions for junction capacitance and the transition region width, a key assumption was that the density of free mobile carriers is negligible compared to the bound charge due to ionized

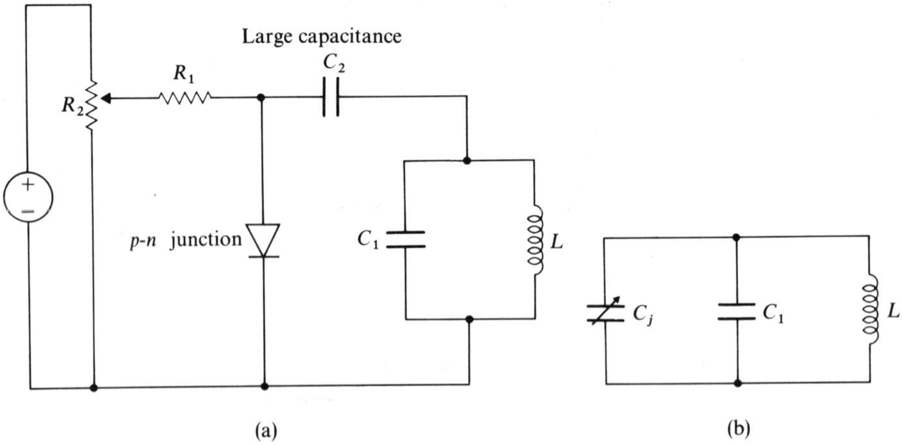

Fig. 3–21 Tuning frequency of an *L-C* circuit with a *p-n* junction diode.

acceptors and donors. This is called the *depletion assumption.* The free carrier distribution portrayed in Fig. 3–15 makes it clear that this assumption is not good at the edges of the transition region. Indeed in lightly doped junctions this approximation is poor. *The predicted transition region widths from the above equations are smaller than those which actually occur.* Nevertheless, because the junction capacitance is defined in terms of a derivative, the capacitance calculated is reasonably accurate. In general the above assumption gets better as the doping is increased and reverse bias in excess of a few Volts is considered.†

3–8 SUMMARY

In this chapter we considered the physics of *p-n* junctions. The phenomena of diffusion as an important transport process by which carriers are transported in *p-n* junctions was studied. The injection of minority carriers across *p-n* junctions and the recombination of electrons and holes was found to be of considerable importance in *p-n* junctions. The diffusion equation for minority carriers was solved subject to the boundary conditions which turned out to be exponentially dependent on the *externally* applied voltage across the *p-n* junction transition region. Finally through the use of transport equations the *I-V* equation which characterizes the intrinsic *p-n* junction diode was obtained. The extrinsic parameters were added to the intrinsic model to obtain the complete model for the *p-n* junction.

Two types of breakdown phenomena were considered for the *p-n* junction. The nonlinear junction capacitance was also derived.

The *p-n* junction theory forms the basis for understanding many other junction devices we shall consider in the chapters to follow.

PROBLEMS

3–1 In a *p-n* junction diode the following constants of the device are known

$$\begin{aligned} A &= 1\ \text{mm}^2 \\ \sigma_n &= 5(\Omega\text{-cm})^{-1} \ll \sigma_p \\ D_p &= 50\ \text{cm}^2/\text{sec.} \\ D_n &= 100\ \text{cm}^2/\text{sec.} \\ \tau_p &= 300\ \mu\text{sec.} \\ n_i &= 10^{14}\ \text{cm}^{-3} \end{aligned}$$

Calculate the current in the device at room temperature when $V = 0.3$ V. The bulk and leakage resistance may be neglected.

3–2 *A p-n* junction diode is made from a material which has $L_n/L_p = 10$ and $\sigma_p/\sigma_n = 12$. Find the ratio of electron to hole current crossing the junction when the diode is forward biased by 0.2 V. (Neglect bulk and leakage resistance.)

---

† At very high current densities the depletion assumption again becomes poor. The analysis needed to treat this case is far too involved to be presented here. The interested reader should look up Sah (1961).

3–3 A *p*-type semiconductor is found to have $p = 10^{14}$/c.c and $n = 5 \times 10^{13}$/c.c. Additional acceptors are now introduced so that the new $p = 10^{16}$/c.c. Now a *p-n* junction is formed with the resulting *p*-type material and an *n*-type material whose conductivity is 2 $(\Omega\text{-cm})^{-1}$. If this *p-n* junction is reverse biased by 26 mV, calculate the ratio of hole to electron current crossing the junction. In this device

$$\mu = 1{,}000 \text{ cm}^2/\text{Volt-sec.} \qquad L_p = 10^{-2} \text{ cm}$$
$$\mu_n = 2{,}000 \text{ cm}^2/\text{Volt-sec.} \qquad L_n = 10^{-1} \text{ cm}$$

Assume that this is an ideal *p-n* junction.

3–4 The actual current–voltage plot of a given *p-n* junction diode is given in Fig. P3–4. Calculate the value of all the elements in Fig. 3–14a.

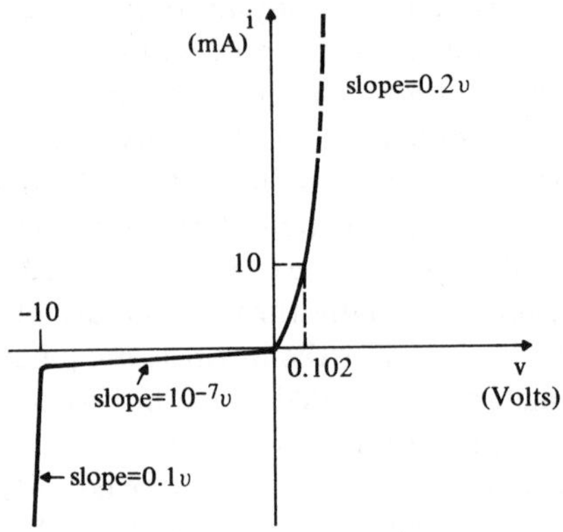

Fig. P3–4

3–5 Show that in a *p-n* junction in which $\sigma_p \gg \sigma_n$, the hole component of current crossing the junction is much bigger than the electron current. Assume that τ and L are the same order of magnitude for electrons as for holes.

3–6 Show that the back saturation current of an ideal *p-n* junction diode varies with temperature approximately as follows:

$$I_s = A(T)\, e^{m_s T_c^\circ},$$

where $m_s = E_g/k(273)^2$

$$A(T) = \frac{AqD_p 2}{L_p}\left[\frac{2\pi m_h^* k}{h^2}\right]^{3/2} T^{3/2}\, e^{-Eg/(273k)},$$

where T_c° means temperature in degrees centigrade. You may assume that $T_c^\circ \ll 273$ and $\sigma_p \gg \sigma_n$.

3–7 Calculate the maximum field strength in the transition region of an abrupt *p-n* junction.

3–8 You are given a reverse biased step *p-n* junction whose built-in voltage is 0.5 V. The junction capacitance is measured to be 10 pF at a reverse voltage of 1.5 V. Calculate the junction capacitance at a forward bias of 0.3 V.

3–9 An inductance of one millihenry is placed in parallel with the *p-n* junction of Problem 3–8. What frequency (Hz) will this circuit resonate at?

3–10 You are given a *p-n* junction which has a forward current of 10 mA going through it. For this device $r_s' = 10\ \Omega$. The voltage drop across the device is 0.230 V. What is I_S for this device? Neglect r_ℓ'.

3–11 We are comparing two *p-n* junctions having all physical constants (i.e. physical dimensions, life times, mobilities, etc.) the same except that $E_g = 0.5$ eV in *p-n* junction 1, and $E_g = 1.0$ eV in *p-n* junction 2. You may assume that $\sigma_p \gg \sigma_n$. If the forward current in device 1 is 10 mA when $V = +0.2$ V, what is the current in device 2 at this voltage?

REFERENCES

Crowell, C. R., and Sze, S. M., "Temperature Dependence of Avalanche Multiplication in Semiconductors," *Appl. Phys. Letters,* **9,** 242 (1966).

Gray, P. E., et al, *Physics Electronics and Circuit Models of Transistors,* Chapters 2–4, SEEC Vol. 2, John Wiley & Sons, New York (1964).

Grove, A. G., *Physics and Technology of Semiconductor Devices,* John Wiley & Sons, New York (1967).

Goetzberger, A., McDonald, B., Haitz, R. H., and Scarlet, R. M., "Avalanche Effects in Silicon *p-n* Junction. II. Structurally Perfect Junctions," *J. Appl. Phys.,* **34,** 1591 (1963).

Gummel, H. K., and Scharfetter, D. L., "Depletion-Layer Capacitance of *p-n* Step Junction," *J. Appl. Phys.,* **38,** 2148 (1967).

Hall, R. N., "Power Rectifiers and Transistors," *Proc. IRE,* **40,** 1512 (1952).

Kressel, H., "A Review of the Effect of Imperfections on the Electrical Breakdown of *p-n* Junctions," *RCA Review,* **28,** 175–207 (1967).

Lindmayer, J. and Wrigley, C. Y., Chapter 2, Van Nostrand, Princeton, N.J. (1964).

Moll, J. L., "The Evolution of the Theory of the Current–Voltage Characteristics of *p-n* Junctions," *Proc. IRE,* **46,** 1076 (1958).

Moll, J. L., *Physics of Semiconductors,* Chapter 7, McGraw-Hill, New York (1964).

Nanavati, R. P., *An Introduction to Semiconductor Electronics,* Chapter 3, McGraw-Hill, New York (1963).

Sah, C. T., Noyce, R. N., and Schockley, W., "Carrier Generation and Recombination in *p-n* Junction and *p-n* Junction Characteristics," *Proc. IRE,* **45,** 1228 (1957).

Sah, C. T., "Effect of Electrons and Holes on the Transition Layer Characteristics of Linearly Graded *p-n* Junctions," *Proc. IRE,* **49,** No. 2, pp. 603–618 (1961).

Schockley, W., "The Theory of *p-n* Junctions in Semiconductors and *p-n* Junction Transistors," *Bell Syst. Tech. J.,* **28,** 435 (1949).

Schockley, W., *Electrons and Holes in Semiconductors,* Van Nostrand (1950).

Sze, S. M., *Physics of Semiconductor Devices,* Chapter 3, John Wiley & Sons, New York (1969).

Sze, S. M., and Gibbons, G., "Avalanche Breakdown Voltages of Abrupt and Linearly Graded *p-n* Junctions in Ge, Si, GaAs, and GaP," *Appl. Phys. Letters* **8,** 111 (1966).

Veloric, H. S., and Prince, M. B., "High Voltage Conductivity-Modulated Silicon Rectifier," *Bell Syst. Tech. J.,* **36,** 975 (1957).

CHAPTER 4

D-C JUNCTION TRANSISTOR THEORY FOR DIFFUSION TRANSISTORS

4-1 INTRODUCTION

In this chapter we shall consider the first case of a multijunction device, namely the *junction transistor*. It is also called the *bipolar junction transistor*, thus indicating the use of both kinds of carriers: electrons and holes. Both minority and majority carriers play a role in the proper functioning of the device. The field effect transistors, which are considered in Chapters 8 and 9, are primarily majority carrier devices and hence are sometimes called *unipolar devices*. The bipolar junction transistor shall be called simply the transistor from now on. It has two *p-n* junctions in close proximity in a single crystal. We shall find that the information obtained from the previous chapter, which considered the single *p-n* junction, will be immensely valuable in this chapter.

The transistor was developed in 1951, two years after the first point-contact transistor was invented. Shockley, Bardeen, and Braittain were awarded the Nobel prize for this invention. There are two types of junction transistors, *p-n-p* and *n-p-n*. The device theory which will be developed in this chapter will be worked out for only the *p-n-p* transistor. In most cases the results for the *n-p-n* transistor will also be stated. The mathematical derivations for both devices are precisely identical in form, and hence needless duplication of effort will be avoided by the above procedure.

In this chapter we shall consider only the diffusion transistors, i.e. those transistors in which the diffusion process accounts for the transport of minority carriers across the base region. Transistors in which an electric field together with the diffusion process cooperate to transport the minority carriers will be considered in Chapter 7. In the present chapter the doping density is *not* a function of position within the emitter, base, and collector regions.

This chapter considers only the d-c or low frequency models for the transistor. Wide band models will be considered in the chapters which follow.

The transistor turns out to be a *minority carrier density controlled* device. If we know the distribution of minority carrier density in the device we can correctly predict the behavior of the device. The injection of minority carriers,

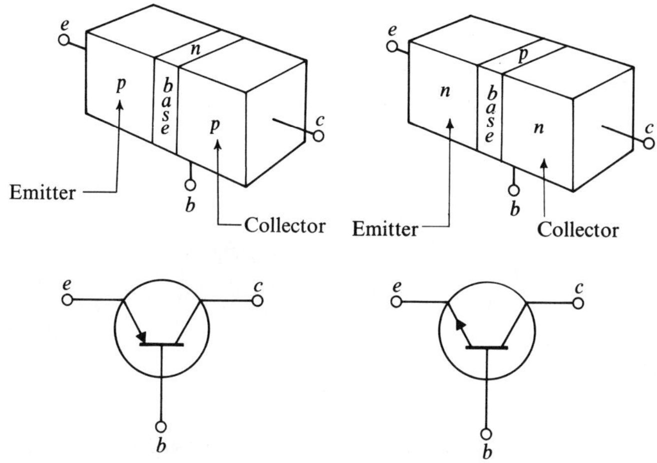

Fig. 4–1 *p-n-p* and *n-p-n* transistors.

transport across the base region and the unavoidable recombination of carriers will be all seen to be important to the understanding of the transistor action.

The transistor is the most widely used electronic amplifying device. It combines low weight and volume with high reliability and great economy. It can amplify over a very wide range of frequencies.

4–2 UNBIASED JUNCTION TRANSISTOR

A transistor has two junctions and three differently doped regions of usually a single crystal seminconductor. Figure 4–1 schematically shows a *p-n-p* and an *n-p-n* transistor. One of the three regions is called the *emitter*. The middle region is called the *base*. The remaining region is called the *collector*. The base region is usually extremely narrow in practical transistors for reasons which we shall presently see. Figure 4–1 also shows the symbols used for each transistor. Note that the arrow appears in the emitter leg only and allows the symbolic differentiation between *p-n-p* and *n-p-n* transistors.

If the three terminals of the transistor are left open-circuited the potential distribution in the device is as shown in Fig. 4–2. Again we see that there exist transition regions surrounding each junction. There is a built-in voltage for each junction. Note that the built-in voltage for the two junctions is *not necessarily the same.* These two built-in voltages would be equal to each other only if the thermal equilibrium value of majority carrier densities in the emitter and collector are the same, as can be seen from Eq. (3–20).

The condition of zero current can be visualized as the balance between diffusion components and the thermally generated components of current at each junction, in exactly the same way as in the case of the single junction given in Chapter 3.

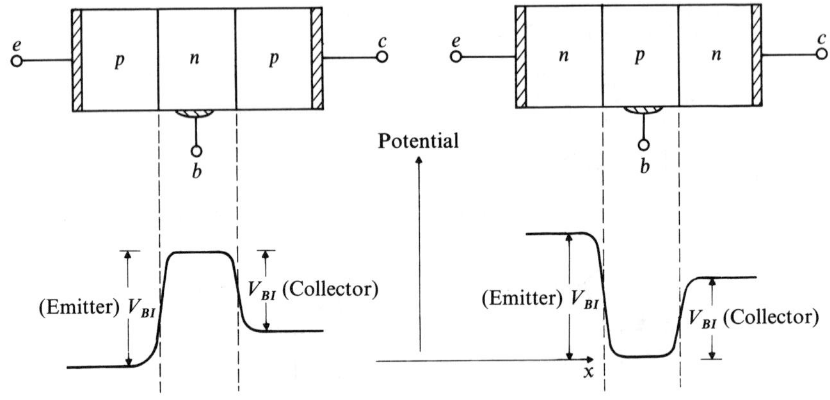

Fig. 4–2 Potential distribution in *p-n-p* and *n-p-n* transistors.

4–3 QUALITATIVE DESCRIPTION OF THE MAJOR PROCESSES IN THE NORMALLY BIASED JUNCTION TRANSISTORS

For amplifier application the transistor is biased so that the emitter junction is forward biased, and the collector junction is reverse biased. When the transistor is so biased, it is said to be normally biased. We shall see, in this chapter, that this choice of biasing leads to the highest gain as compared to other biasing choices. This is the reason for choosing this bias scheme for amplification applications. We shall consider primarily the *p-n-p* transistor for simplicity. Results for *n-p-n* transistors will, however, be stated.

The major processes of importance in the understanding of the transistor are the following. The forward biased emitter junction results in the injection of minority carriers.† (i.e., holes in *p-n-p* transistors, and electrons in *n-p-n* transistors) into the base region. Because of the existing gradient of minority carriers in the base region, where their density is high at the emitter and low at the collector, the minority carriers will be transported across the base region as a result of diffusion. Some holes may be lost in the base region by the process of recombination. These losses can be minimized by keeping the base width narrow. For this reason in practical transistors the base width is indeed extremely narrow. Finally those holes, which survive recombination in the base region, are collected by the collector. Figure 4–3 shows schematically how the minority carrier current distributes itself. It is clear from this figure that the base current consists of the electron current as well as the hole current crossing each junction.

In the transistor the major transport process consists of the holes going

† Unless otherwise designated minority carriers will mean minority carriers in the *base region.*

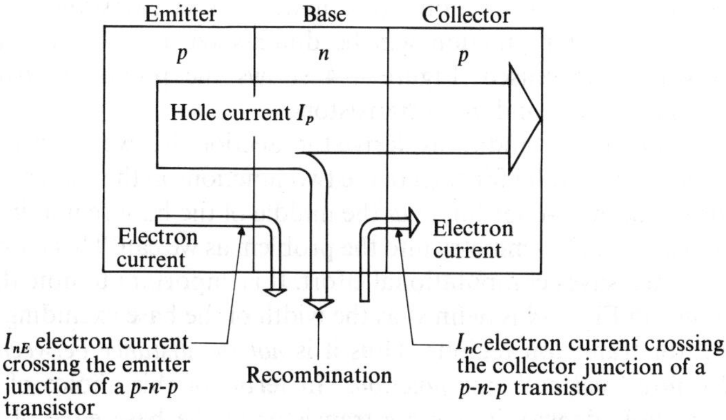

Fig. 4–3 Schematic of how the hole and electron currents are distributed in a normally biased *p-n-p* transistors.

from the emitter across the base to the collector. The magnitudes of the emitter current and the collector currents are roughly equal. It is easy to see that such a device is capable of power gain from the following argument. Note first that the emitter junction, being forward biased, will offer low input resistance. The collector junction, however, is reverse biased, and hence offers high resistance permitting the use of a load resistance of moderately high value in the collector circuit. Thus essentially the same current which flows at the input through low resistance also flows at the output through high resistance. This corresponds to power gain as can be seen from the following.

$$A_p = \frac{P_{out}}{P_{in}} = \frac{I_C^2 R_L}{I_E^2 r_{in}} = \frac{R_L}{r_{in}}, \tag{4–1}$$

where P_{out} = output power,

P_{in} = input power,

I_C = collector current,

I_E = emitter current,

R_L = load resistance in the collector circuit,

r_{in} = input resistance.

A_p = power gain.

The ratio on the right hand side of Eq. (4–1) can be made to be quite large; hence, power gain is possible in transistor amplifiers.

4–4 POTENTIAL DISTRIBUTION IN TRANSISTORS WITH BIAS APPLIED, AND BOUNDARY CONDITIONS FOR MINORITY CARRIERS

In normal bias the emitter junction is forward biased and the collector junction is reverse biased. We expect from the knowledge gained in the previous chapter in regard to single *p-n* junctions, that under this circumstance the potential barrier of the emitter junction will be diminished and that of the collector junction will be increased. Figure 4–4 shows the potential distribution in normally biased *p-n-p* and *n-p-n* transistors.

As the boundary conditions derived in Section 3–3 were for any *p-n* junction, they should be valid for each of the two junctions in the transistor. We shall set the origin of the *x*-coordinate in the middle of the base region in an attempt to introduce as much symmetry into the problem as we can. This is done because symmetry often saves computational effort. It is important to note that the base width shown in Fig. 4–4 is defined as the width of the base excluding the emitter and collector transition regions. Thus it is *not the distance between the metallurgical emitter and collector junctions.* In terms of this choice of coordinate system, the hole density in a *p-n-p* transistor at the base edge of the emitter junction is given by

$$p\left(-\frac{w_0}{2}\right) = p_n e^{qV_E/kT} \triangleq p_E, \tag{4–2}$$

where w_0 = base width,

p_n = thermal equilibrium value of hole density in the *n*-type base,

V_E = d-c voltage across the emitter junction due only to *external* sources.

The hole density at the base edge of the collector junction is given by

$$p\left(+\frac{w_0}{2}\right) = p_n e^{qV_C/kT} \triangleq p_C, \tag{4–3}$$

where V_C = d-c voltage across the collector junction due only to externally applied sources.

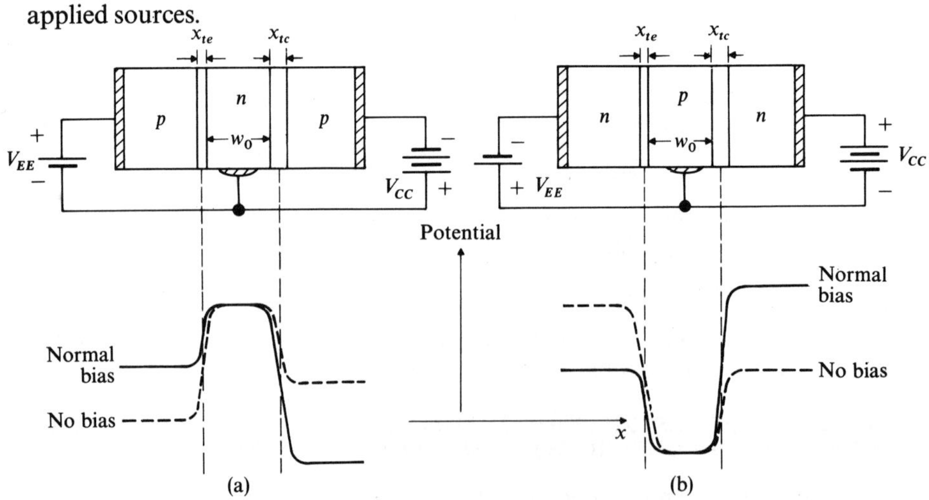

Fig. 4–4 Potential distribution under no bias and normal bias in *p-n-p* and *n-p-n* transistors.

4–5 D-C DIFFUSION THEORY FOR JUNCTION TRANSISTORS

The analysis which follows is based on exactly the same set of nine assumptions as were used in connection with the *p-n* junction as stated in Section 3–4. The only exception is that now the base width w_0 cannot be assumed to be much greater than the diffusion length L_p. In fact usually $w_0 \ll L_p$. Some statements which were made with respect to a single junction will need to be true for two junctions. For example, it will be assumed that there is no recombination or generation of carriers in the emitter and collector transition regions. The low injection assumption refers primarily to the base region, i.e. $p_{\text{inj}} \ll n_n$.

Having evaluated the boundary conditions for minority carriers in the last section, we are ready to solve the diffusion equation for these carriers in the base region. Before we proceed to do this it will be helpful to state the strategy for obtaining the terminal current-voltage relationships for the transistor. (1) First the diffusion equation for minority carriers in the base region will be solved. This yields the minority carrier density as a function of junction voltages. (2) Second, minority carrier current density will be obtained by using the transport equation with *zero field.* Thus current densities will be obtained as a function of voltage. This current density, however, will be a function of the position coordinate *x*. (3) The current density can be specialized to its value at the emitter and collector junctions where they may be multiplied by the respective junction areas to obtain the *minority carrier currents* crossing each junction. In *p-n-p* transistors when to the hole currents are added the electron currents crossing the emitter and collector junctions, the total current at each junction is obtained. Then the use of Kirchoff's law of continuity of currents enables us to relate terminal currents to the total currents crossing the junctions as a function of junction voltages.

We have purposely chosen the coordinate system in the base region which is symmetric with respect to the emitter and collector junctions because the origin of *x* coordinate is precisely in the middle of the base width. We shall take maximum advantage of this by also choosing the form of the solution for minority carriers which is likewise symmetric. To make these observations more concrete, we start by writing the d-c diffusion equation which is to be solved for minority carriers in the base region of a *p-n-p* transistor.

$$\frac{d^2p}{d^2x} - \frac{p - p_n}{L_p^2} = 0. \tag{4–4}$$

Equation (4–4) is the same equation as was used for the *p-n* junction diode.

The solution which we shall choose this time, is the following for

$$-\frac{w_0}{2} \leqslant x \leqslant \frac{w_0}{2},$$

$$p(x) = p_n + A \sinh \left[\frac{\left(\frac{w_0}{2} - x \right)}{L_p} \right] + B \sinh \left[\frac{\left(\frac{w_0}{2} + x \right)}{L_p} \right], \tag{4-5}$$

where L_p = diffusion length for minority carriers (holes) in the base region,
w_0 = base width.

Here too the term p_n is the particular solution, and the rest is the complimentary solution. This is a different form of the solution for Eq. (4–4) than was used for the single *p-n* junction in Eq. (3–43). These two forms are, however, equivalent. To see this let us use the defining equation for

$$\sinh y \triangleq \frac{e^y - e^{-y}}{2}. \tag{4-6}$$

Using Eq. (4–6) we may rewrite Eq. (4–5) as

$$p(x) = p_n + A\left[\frac{1}{2}\exp\left(\frac{w_0}{2L_p} - \frac{x}{L_p}\right) - \frac{1}{2}\exp\left(-\frac{w_0}{2L_p} + \frac{x}{L_p}\right)\right] + B\left[\frac{1}{2}\exp\left(\frac{w_0}{2L_p} + \frac{x}{L_p}\right) - \frac{1}{2}\exp\left(-\frac{w_0}{2L_p} - \frac{x}{L_p}\right)\right] \tag{4-7}$$

$$p(x) = p_n + \left[-\frac{A}{2}e^{-w_0/2L_p} + \frac{B}{2}e^{+w_0/2L_p}\right]e^{x/L_p} + \left[\frac{A}{2}e^{w_0/2L_p} - \frac{B}{2}e^{-w_0/2L_p}\right]e^{-x/L_p}. \tag{4-8}$$

In Eq. (4–8) if we realize that as A and B are arbitrary coefficients, and that w_0, L_p are but constants as far as functional dependence on x is concerned, this equation is of the same form as Eq. (3–43). The reason for choosing the solution as given in Eq. (4–5) is that it seems to display symmetry up to a point, and the evaluation of arbitrary coefficients is then quite simple.

Let us now evaluate the arbitrary coefficients by the use of the boundary conditions of Eq. (4–2) and (4–3). From Eqs. (4–5) and (4–2) we have

$$p(-w_0/2) \triangleq p_E = p_n + A \sinh(w_0/L_p) + 0$$

or

$$A = \frac{(p_E - p_n)}{\sinh(w_0/L_p)}. \tag{4-9}$$

From Eqs. (4–5) and (4–3) we have

$$p(+w_0/2) \triangleq p_C = p_n - B \sinh(w_0/L_p)$$

$$B = \frac{(p_C - p_n)}{\sinh(w_0/L_p)}. \tag{4-10}$$

Equations (4–9) and (4–10) allow us to write for the hole density in the base region

$$p(x) = p_n + \frac{(p_E - p_n)\sinh\left[\left(\frac{w_0}{2} - x\right)/L_p\right]}{\sinh(w_0/L_p)} + \frac{(p_C - p_n)\sinh\left[\left(\frac{w_0}{2} + x\right)/L_p\right]}{\sinh(w_0/L_p)} \tag{4–11}$$

As the base region has been defined *not* to include the transition regions of the emitter and collector junctions, the transport equation with the electric field zero can be used to express current as a function of p_E and p_C which are themselves functions of V_E and V_C. Thus

$$j_p(x) = -qD_p\frac{\partial p}{\partial x} = -qD_p\left\{-\left[\frac{(p_E - p_n)}{L_p\sinh(w_0/L_p)}\cosh\left(\frac{\frac{w_0}{2} - x}{L_p}\right) - \frac{(p_C - p_n)}{L_p\sinh(w_0/L_p)}\cosh\left(\frac{\frac{w_0}{2} + x}{L_p}\right)\right]\right\} \tag{4–12}$$

for

$$-\frac{w_0}{2} \leqslant x \leqslant \frac{w_0}{2}. \tag{4–13}$$

We may now convert the current density to current by multiplying Eq. (4–12) by the cross-sectional area A of the one-dimensional device. Then the current crossing the emitter and collector junctions will be obtained by simply specializing x to $-w_0/2$ and to $+w_0/2$, respectively. Thus

$$I_{pE} = \frac{qD_pA}{L_p}\left[(p_E - p_n)\coth\left(\frac{w_0}{L_p}\right) - (p_C - p_n)\operatorname{csch}\left(\frac{w_0}{L_p}\right)\right] \tag{4–14}$$

and

$$I_{pC} = \frac{qD_pA}{L_p}\left[(p_E - p_n)\operatorname{csch}\left(\frac{w_0}{L_p}\right) - (p_C - p_n)\coth\left(\frac{w_0}{L_p}\right)\right]. \tag{4–15}$$

We need now to calculate the electron current crossing each junction so that the total current can be computed. In obtaining the electron current crossing each junction, the following comments will be found instructive. We note, to begin with, that unlike the hole current which links both emitter and collector junctions, the electron currents link only one junction because the base terminal through which electron current can easily enter or leave the base region (owing to the ohmic contact to the external terminal). Thus the electron current crossing each junction is expected to be essentially the same as that for a single *p-n* junction. Note that such electron current crosses only one junction and not two. For example, electron current due to the motion of electrons in the emitter bulk region crosses only the emitter junction. This current goes to the base contact. Similarly, the electron current associated with the motion of electrons from the collector into the base region constitutes *current* going from the base to the collector thus linking only the collector junction (see Fig. 3–3). We can follow exactly the same procedure for electrons in the emitter and collector bulk regions as was just followed for the holes in the base region. We would expect a result in this case similar to that obtained for hole current crossing the emitter and collector junctions, except that the emitter and collector regions need not be as narrow as the base region. Because the electron current is crossing only one junction, we can obtain the electron current crossing the emitter junction by applying two simple rules to transform the hole currents of Eqs. (4–14) and (4–15) to electron currents.

1. Change all hole parameters to the corresponding electron parameters in the appropriate regions. This amounts to a changing of all *n*s to *p*s regardless of whether these appear in their own right or as subscripts. Thus the following transformations are needed

$$
\begin{aligned}
&D_p \to D_n \\
&D_n \to D_p \\
&p_E \to n_E \triangleq n_{pe}\, e^{qV_E/kT} \\
&p_C \to n_C \triangleq n_{pc}\, e^{qV_C/kT} \\
&p_n \to n_{pe} \text{ in emitter or } n_{pc} \text{ in collector} \\
&L_p \to L_{ne} \text{ in emitter or } L_{nc} \text{ in collector,}
\end{aligned}
\tag{4–16}
$$

where n_{pe} = thermal equilibrium values of electrons in the *p*-type emitter
n_{pc} = thermal equilibrium value of electrons in the *p*-type collector.

$$L_{ne}^2 \triangleq D_n\, \tau_{ne}; \tag{4–17}$$

$$L_{nc}^2 \triangleq D_n\, \tau_{nc}, \tag{4–18}$$

where τ_{ne} = life-time of electrons in the emitter,
τ_{nc} = life-time of electrons in the collector.

2. Convert the equations for hole current in the base region to those of electrons crossing each junction by taking the former applying the transformations indicated by Eq. (4–16) and taking the limit of resulting expressions in the limit as $w_0 \to \infty$. The last statement effectively puts the other junction at infinity, thus converting the transistor equations to those for a single junction as needed here for electron currents.

Following the above procedure we obtain

$$I_{nE} = \lim_{w_0 \to \infty} \left\{ \frac{qD_nA}{L_{ne}} \left[(n_E - n_{pe}) \coth\left(\frac{w_0}{L_n}\right) - (n_C - n_{pe}) \operatorname{csch}\left(\frac{w_0}{L_n}\right) \right] \right\} \tag{4–18a}$$

$$I_{nE} = \frac{qD_nA}{I_{ne}} (n_E - n_{pe}) \tag{4–19}$$

for

$$x < -(w_0/2 + x_{te}),$$

where x_{te} = emitter transition region width.

In obtaining Eq. (4–19) we have used the following results

$$w_0 \lim_{w_0 \to \infty} \coth\left(\frac{w_0}{L_n}\right) = 1 \tag{4–20}$$

$$w_0 \lim_{w_0 \to \infty} \operatorname{csch}\left(\frac{w_0}{L_n}\right) = 0. \tag{4–21}$$

Similarly,

$$I_{nC} = \lim_{w_0 \to \infty} \frac{qD_nA}{L_{nc}} \left[(n_E - n_{pe}) \operatorname{csch}\left(\frac{w_0}{L_n}\right) - (n_C - n_{pc}) \coth \frac{w_0}{L_n} \right]$$

or

$$I_{nC} = -\frac{qD_nA(n_C - n_{pc})}{L_{nc}} \tag{4–22}$$

for

$$x > -\left(\frac{w_0}{2} + x_{tc}\right),$$

where x_{tc} is the collector transition region width.

Because there is no recombination or generation of carriers in either transition region, we can simply add Eq. (4–14) to Eq. (4–19) to obtain total emitter current; and add Eq. (4–15) to Eq. (4–22) to obtain the total collector current. Thus

$$I_E = I_{pE} + I_{nE} = \frac{qD_pA}{L_p} [(p_E - p_n) \coth (w_0/L_p) - (p_C - p_n) \operatorname{csch} (w_0/L_p)] + \frac{qD_nA(n_E - n_{pe})}{L_{ne}}. \tag{4–23}$$

$$I_C = I_{pC} + I_{nC} = \frac{qD_pA}{L_p}\left[(p_E - p_n)\operatorname{csch}(w_0/L_p) - (p_C - p_n)\coth(w_0/L_p)\right] - \frac{qD_nA(n_C - n_{pc})}{L_{nc}}. \tag{4–24}$$

Now by virtue of Kirchhoff's current continuity law Eqs. (4–23) and (4–24) are the terminal emitter and collector currents for the intrinsic device. Here the "intrinsic" device means one for which the nine assumptions are valid. Equations (4–23) and (4–24) can be stated explicitly in terms of the voltage as follows

$$I_E = \left\{\left[\frac{qD_pAp_n\coth(w_0/L_p)}{L_p}\right] + \frac{qD_nAn_{pe}}{L_{ne}}\right\}(e^{qV_E/kT} - 1) - \left[\frac{qD_pAp_n\operatorname{csch}(w_0/L_p)}{L_p}\right](e^{qV_C/kT} - 1) \tag{4–25}$$

$$I_C = \frac{qD_pAp_n}{L_p}\operatorname{csch}(w_0/L_p)(e^{qV_E/kT} - 1) - \left[\frac{qD_pAp_n}{L_p}\coth(w_0/L_p) + \frac{qD_nAn_{pc}}{L_{nc}}\right](e^{qV_C/kT} - 1). \tag{4–26}$$

Equations (4–25) and (4–26) are the terminal current–voltage equations for the intrinsic transistor. It may be noted that these equations are quite general in that no commitments have been made as to which junction is forward or reverse biased. It is instructive to plot the minority carrier density in the base region for four different distinct biasing schemes. Figure 4–5 shows these distributions for each distinct case. Case (a) is where both junctions are forward biased. The transistor is said to be *current saturated* because the whole device offers rather a low resistance, and the currents are determined essentially by

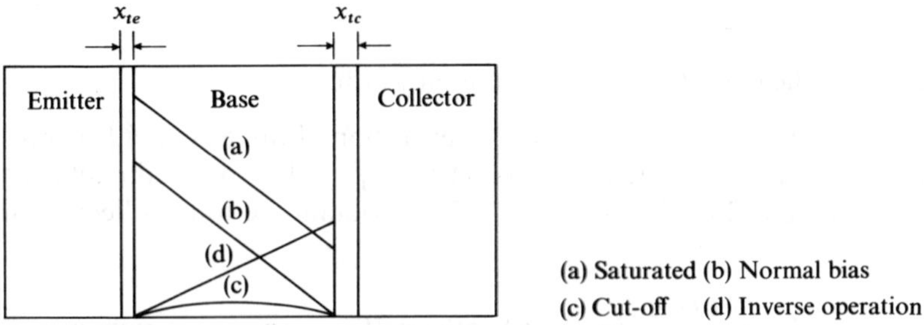

Fig. 4–5 Minority carrier density in the base region for different modes of operation. Distributions (a), (b), and (d) assume that they are linear which is an approximation which assumes w_0 much less than the minority carrier diffusion length. (a) Saturated. (b) Normal bias. (c) Cut-off. (d) Inverse operation.

external circuit elements. For example, currents *cannot exceed* the supply voltage at each terminal divided by the net series resistance. Boundary conditions Eq. (4–2) and (4–3) tell us that in this case if

$$V_E, V_C \gg \frac{kT}{q} = 0.026 \text{ Volt at room temperature} \tag{4–27}$$

then

$$p_E, p_C \gg p_n, \tag{4–28}$$

$$n_E \gg n_{pe}, \tag{4–29}$$

$$n_C \gg n_{pc}. \tag{4–30}$$

The distribution of case (a) assumed

$$V_E > V_C. \tag{4–31}$$

Case (b) corresponds to normally biased transistors, i.e.

$$V_E \gg \frac{kT}{q} \tag{4–32}$$

$$V_C \ll \frac{kT}{q}. \tag{4–33}$$

Equations (4–30) and (4–31) imply that

$$p_E \gg p_n, n_E \gg n_{pe}, \tag{4–34}$$

$$p_C, n_C \approx 0 \ll p_n, n_p. \tag{4–35}$$

Equation (4–35) follows from the fact that practical values of collector reverse bias are several Volts.

Case (c) corresponds to

$$V_E, V_C < 0. \tag{4–36}$$

Equation (4–36) implies

$$p_E, p_C \ll p_n, \tag{4–37}$$

$$n_E \ll n_{pe}, \tag{4–38}$$

$$n_C \ll n_{pc}. \tag{4–39}$$

In this case both junctions are *reverse biased*, and the whole device offers high resistance. The resulting currents are very small, and the device is considered to be *cut-off*.

Case (d) corresponds to *inverse bias* where

$$V_E < 0, \quad V_C > 0. \tag{4–40}$$

Thus

$$p_n \gg p_E, \tag{4-41}$$

$$n_{pe} \gg n_E, \tag{4-42}$$

$$p_C \gg p_n, \tag{4-43}$$

$$n_C \gg n_{pc}. \tag{4-44}$$

In cases (a), (b) and (d) the minority carrier density in the base region is shown to be essentially a straight line. This is only an approximation, because it is actually given by Eq. (4–11) which is not linear. The reason the straight line portrayal is not too wrong is that usually

$$L_p \gg w_0. \tag{4-45}$$

It is useful to note that the detailed slope of the minority carrier distribution in the base region determines how the transistor is functioning. It is noteworthy to observe that the transistor is a *minority carrier charge distribution controlled device.* Historically it has been variously suggested that the transistor is a current-controlled, and/or charge-controlled device. These are approximate statements. The statement in terms of a *charge distribution* controlled device is the *most accurate statement.*

4–6 SHORT-CIRCUIT CURRENT GAIN

A very important circuit parameter of the transistor is its short-circuit current gain from emitter to collector. This current gain can be written as the product of three parameters as follows

$$\alpha_0 = \frac{I_C}{I_E} = \left(\frac{I_{pE}}{I_E}\right)\left(\frac{I_{pC}}{I_{pE}}\right)\left(\frac{I_C}{I_{pC}}\right). \tag{4-46}$$

The first factor is called the *injection efficiency* γ_0 for it tells us what percentage of the total current crossing the emitter is made up of minority carrier current.

$$\gamma_0 \triangleq \frac{I_{pE}}{I_E}. \tag{4-47}$$

This is always less than 1, though ideally $\gamma_0 = 1$ is desirable. The second factor is called the *base transport factor* β_0^*

$$\beta_0^* = \frac{I_{pC}}{I_{pE}}. \tag{4-48}$$

This factor tells us how efficient is the transport of minority carriers across the base region, i.e. how many survive recombination. Ideally we would like $\beta_0^* = 1$, though it always is less than 1. Here β_0^* is to be distinguished from the common emitter current gain (I_C/I_B), i.e.

$$\beta_0^* \neq \beta_0. \tag{4-49}$$

The last factor in Eq. (4–46) is the *collector multiplication factor* α_0^*

$$\alpha_0^* = \frac{I_C}{I_{pC}}. \tag{4–50}$$

Unlike γ_0 and β_0^*, α_0^* *is always greater than one.* Again ideally a value of unity is desirable, indicating that the entire current crossing the collector junction is made up of holes (or minority carriers). Note that the common base† short circuit current gain should be carefully distinguished from the collector multiplication factor. Thus

$$\alpha_0^* \neq \alpha_0. \tag{4–51}$$

Neither is the collector multiplication factor to be confused with the collector avalanche multiplication factor associated with the collector breakdown. Usually of γ_0, β_0^* and α_0^*, β_0^* *is the most important,* since its departure from 1 is greater compared to the value of the other two parameters.

The subscript zero on all three of these parameters indicates that they are being considered here at *low frequencies only.*

4–7 BASE TRANSPORT FACTOR

With the device equations of Section 4–5 we can obtain β_0^* explicitly in terms of the fundamental device parameters. We shall assume that the transistor is normally biased and that Eqs. (4–34) and (4–35) are valid. Equations (4–14) and (4–15) may be utilized to obtain

$$\frac{I_{pC}}{I_{pE}} = \frac{\dfrac{qD_pA}{L_p}(p_E - p_n)\operatorname{csch}\left(\dfrac{w_0}{L_p}\right)}{\dfrac{qD_pA}{L_p}(p_E - p_n)\coth\left(\dfrac{w_0}{L_p}\right)}. \tag{4–52}$$

After cancellation of common factors, we obtain

$$\beta_0^* = \frac{I_{pC}}{I_{pE}} = \operatorname{sech}\left(\frac{w_0}{L_p}\right). \tag{4–53}$$

If

$$w_0 \ll L_p \tag{4–54}$$

we may expand sech (w_0/L_p) in the Taylor series and retain only the first two terms

$$\beta_0^* \doteq 1 - \tfrac{1}{2}\left(\frac{w_0}{L_p}\right)^2. \tag{4–55}$$

---

† A transistor amplifier configuration in which the base terminal is common to both the input and output pairs of terminals is called a *common base* circuit. Similarly if the emitter terminal is common to both input and output pairs of terminals it is called a *common emitter circuit.*

If instead an infinite product expansion is utilized a more accurate expression results even if only the first product is retained as given below

$$\beta_0^* = 1 - \frac{4}{\pi^2}\left(\frac{w_0}{L_p}\right)^2$$

$$\beta_0^* = 1 - (1/2.46)\left(\frac{w_0}{L_p}\right)^2. \tag{4–56}$$

The most accurate two term expression is

$$\beta_0^* = 1 - (1/2.43)\left(\frac{w_0}{L_p}\right)^2. \quad (p\text{-}n\text{-}p) \tag{4–57}$$

Equations (4–57) tells us that to maximize the base transport factor, it is most essential that

$$w_0 \ll L_p. \tag{4–58}$$

If we had considered an *n-p-n* transistor instead of a *p-n-p* transistor, the expression for β_0^* would have been

$$\beta_0^* = 1 - (1/2.43)\left(\frac{w_0}{L_n}\right)^2. \quad (n\text{-}p\text{-}n) \tag{4–59}$$

4–8 COLLECTOR MULTIPLICATION FACTOR

The derivation of the collector multiplication factor will not be given here because it is quite complicated by virtue of having to consider the *conductivity modulation* of the collector region. That is, the changes in the conductivity of the collector region when currents are flowing in the collector, should be taken into account.† Without deriving it, the result is stated here at low current density‡

$$\alpha^* = 1 + \frac{\sigma_{nc}}{2\sigma_{pc}}. \tag{4–60}$$

Where σ_{nc} = conductivity of *p*-type collector due to only electrons,
σ_{pc} = conductivity of *p*-type collector due to only holes

$$\sigma_{nc} = q\mu_n n_p, \tag{4–61}$$

where n_p = thermal equilibrium electron density in the *p*-type collector

$$\sigma_{pc} = q\mu_p p_{pc}, \tag{4–62}$$

† The interested reader is referred to Appendix B in Nanavati (1963).
‡ Low current density is defined as that which corresponds to $p_{ini} \ll n_n$ in *p-n-p* transistors.

where p_{pc} = thermal equilibrium density of holes in the p-type collector. At very high current densities

$$\alpha^* = 1 + \frac{\sigma_{nc}}{\sigma_{pc}}. \tag{4–63}$$

4–9 INJECTION EFFICIENCY

Injection efficiency γ_0 plays an important role in the design considerations particularly with regard to the choice of conductivities. In obtaining γ_0 as an explicit function of the fundamental device parameters, we shall begin by writing the defining equation in a slightly different form for p-n-p transistors

$$\gamma_0 \triangleq \frac{I_{pE}}{I_{pE} + I_{nE}} = \frac{1}{1 + I_{nE}/I_{pE}}. \tag{4–64}$$

Equations (4–14) and (4–19) can be substituted into Eq. (4–64) to obtain

$$\gamma_0 = \frac{1}{1 + \dfrac{\dfrac{qD_nA}{L_{ne}}(n_E - n_{pe})}{\dfrac{qD_pA}{L_p}[(p_E - p_n)\coth(w_0/L_p) - (p_C - p_n)\,\mathrm{csch}\,(w_0/L_p)}}. \tag{4–65}$$

For normal bias Eqs. (4–34) and (4–35) are valid, so that in the term in the denominator, i.e. $(p_C - p_n)\,\mathrm{csch}\,(w_0/L_p)$ may be neglected.

Thus γ_0 can be written as follows

$$\gamma_0 = \frac{1}{1 + \dfrac{\dfrac{qD_nAn_{pe}}{L_{ne}}(e^{qV_E/kT} - 1)}{\dfrac{qD_pAp_n}{L_p}(e^{qV_E/kT} - 1)\coth(w_0/L_p)}}$$

or

$$\gamma_0 = \frac{1}{1 + \left(\dfrac{D_n}{D_p}\right)\left(\dfrac{L_p}{L_{ne}}\right)\left(\dfrac{n_{pe}}{p_n}\right)\tanh\dfrac{w_0}{L_p}}. \tag{4–66}$$

Consider the ratio n_{pe}/p_n,

$$\frac{n_{pe}}{p_n} = \frac{n_i^2/p_{pe}}{n_i^2/n_n} = \frac{n_n}{p_{pe}}, \tag{4–67}$$

where n_n = thermal equilibrium electron density in n-type base region,
p_{pe} = thermal equilibrium hole density in p-type emitter.

The right hand side of Eq. (4–67) can be expressed in terms of the base and emitter conductivities.

$$\sigma_B = q\mu_n n_n \tag{4–68}$$

$$\sigma_E = q\mu_p p_{pe}. \tag{4–69}$$

Thus

$$\frac{n_n}{p_{pe}} = \frac{\sigma_B \mu_p}{\sigma_E \mu_n} = \frac{\sigma_B}{\sigma_E}\frac{D_p}{D_n}. \tag{4–70}$$

The last equation follows from the Einstein relations of Eqs. (3–26) and (3–27). Substituting Eq. (4–70) into Eq. (4–66), we have

$$\gamma_0 = \frac{1}{1 + \left(\frac{L_p}{L_{ne}}\right)\left(\frac{\sigma_B}{\sigma_E}\right)\tanh\frac{w_0}{L_p}}. \tag{4–71}$$

In those transistors where

$$w_0 \ll L_p \tag{4–72}$$

a one term Taylor series expansion is not a bad approximation. Thus

$$\tanh\frac{w_0}{L_p} \doteq \frac{w_0}{L_p}. \tag{4–73}$$

Use of Eq. (4–73) reduces γ_0 to a simple expression

$$\gamma_0 = \frac{1}{1 + \frac{\sigma_B}{\sigma_E}\frac{w_0}{L_{ne}}}. \tag{4–74}$$

It should be noted, however, that Eq. (4–74) is valid *only* if Eq. (4–72) is true.

Equation (4–71) indicates that the injection efficiency will be high if

$$\sigma_B \ll \sigma_E. \tag{4–75}$$

As this requirement does not conflict with anything else, *every transistor* manufactured *has the conductivity of the emitter and base chosen so that Eq. (4–75) is true:*

Example 4–1 An abrupt *n-p-n* transistor is known to have the following parameters:

V_{BI} (emitter) $= 1.0$ Volt
V_{BI} (collector) $= 0.5$ Volt
$w_0 = 2 \times 10^{-5}$ cm at $V_C = -5$ Volt and $V_E = 0.5$ Volt
$x_{tc} = 10^{-5}$ cm at $V_C = -5$ Volt

$$x_{te} = 10^{-6} \text{ cm at } V_E = 0.5 \text{ Volt}$$
$$\sigma_E = 10(\Omega\text{-cm})^{-1}$$
$$\sigma_C = 30(\Omega\text{-cm})^{-1}$$
$$\sigma_B = 1(\Omega\text{-cm})^{-1}$$
$$L_n = 1.5 \times 10^{-4} \text{ cm}$$
$$L_{pe} = 10^{-3} \text{ cm}$$
$$L_{pc} = 10^{-3} \text{ cm.}$$

If the transistor is operated so that $V_E = 0.5$ Volt and $V_C = -15.5$ Volt, calculate (a) β_0^*, (b) γ_0 and (c) α_0 if it is assumed that $\alpha_0^* = 1$.

Solution. (a) From Eq. (4–53) we see that for a *n-p-n* transistor the expression for the base transport factor, by analogy, is

$$\beta_0^* = \operatorname{sech}\left(\frac{w_0}{L_n}\right), \tag{4–75a}$$

where L_n = diffusion length for electrons in the base region.

In order to evaluate β_0^* we need to find what the base width w_0 is under the given condition. We should realize that w_0 is an indirect function of collector and emitter voltages. It is easy to understand this if we recall that the width of the emitter and collector transition regions is a function of emitter and collector voltages respectively. It is useful to remember that the base width was defined as the base region *excluding* the emitter and collector transition regions.

As the emitter voltage does not change in this problem, no change in the emitter transition region occurs. The collector voltage does change, and hence the collector transition region will also change. We may use Eq. (3–96) and the given data to calculate the constant k_{at}.

$$x_{tc} = k_{at}(V_{bi} - V)^{\frac{1}{2}} \tag{1}$$

$$10^{-5} = k_{at}[0.5 - (-5)]^{\frac{1}{2}} = k_{at}\sqrt{5.5} \tag{2}$$

or

$$k_{at} = \frac{10^{-5}}{2.35}. \tag{3}$$

We shall now compute the change in collector transition region when the collector voltage changes from -5 Volt to -15.5 Volt

$$\Delta x_{tc} = k_{at}(\sqrt{16} - \sqrt{5.5}) \tag{4}$$

$$\Delta x_{tc} = \frac{10^{-5}}{2.35}(4 - 2.35) = \frac{1.65}{2.35} \times 10^{-5} = 0.702 \times 10^{-5} \text{ cm.} \tag{5}$$

Because $\sigma_C \gg \sigma_B$, it follows that almost the entire collector transition region is on the *base side*. The change in this transition region can also be considered to be primarily on the base side. Thus, the *expansion* of the collector transition region must necessarily lead to the same *contraction* in the base width. Then

$$w_0^{\text{new}} = w_0^{\text{old}} - \Delta x_{tc} \tag{6}$$

$$w_0^{\text{new}} = 2 \times 10^{-5} = 0.702 \times 10^{-5} = 1.298 \times 10^{-5} \text{ cm.} \tag{7}$$

We note that $w_0^{\text{new}} \ll L_n$ so that the expression corresponding to Eq. (4–59) is valid. Thus finally

$$\beta_0^* = 1 - \frac{1}{2.43}\left(\frac{w_0}{L_n}\right)^2 = 1 - \frac{1}{2.43}\left(\frac{1.248}{1.5} \times 10^{-1}\right)^2 \tag{8}$$

$$\beta_0^* = 1 - 0.00306 = 0.997 \tag{9}$$

(b) To calculate γ_0 for the *n-p-n* transistor we write the analogous expression to Eq. (4–74) using the base width just found

$$\gamma_0 = \frac{1}{1 + \dfrac{\sigma_B}{\sigma_E}\dfrac{w_0}{L_{pe}}} \tag{4–75b}$$

$$= \frac{1}{1 + \dfrac{1}{10}\left(\dfrac{1.298 \times 10^{-5}}{10^{-3}}\right)} = \frac{1}{1.00130} \tag{10}$$

$$\gamma_0 = 0.999. \tag{11}$$

The expression used for γ_0 above is valid here because

$$w_0 \ll L_n \tag{12}$$

(c)

$$\alpha_0 = \gamma_0 \beta_0^* \alpha_0^* = (0.999)\,(0.997)\,(1) \tag{13}$$

$$\alpha_0 = 0.996. \tag{14}$$

Example 4–2 You are given a transistor *p-n-p* which has a base width of 1 μ, $L_p = 10\ \mu$, $L_{ne} = 5\ \mu$, $\sigma_B = 5 \times 10^{-3}(\Omega\text{-cm})^{-1}$, $\alpha^* = 1$. It is desired that α be 0.985. To achieve this, complete the transistor design by choosing the minimum value which σ_E must have to assure the required value of α.

Solution.

$$\alpha = \frac{\alpha^* \operatorname{sech}(w_0/L_p)}{I + \dfrac{\sigma_B L_p}{\sigma_E L_{ne}} \tanh(w_0/L_p)} \tag{1}$$

or

$$\sigma_E = \frac{\dfrac{\alpha\sigma_B L_p}{L_{ne}}\tanh(w_0/L_p)}{\alpha^* \operatorname{senh}(w_0/L_p) - \alpha} \tag{2}$$

$$\sigma_E = \frac{(0.985)(5 \times 10^{-3})(10^{-3})(0.09967)/(5 \times 10^{-4})}{(1)(0.995) - 0.985} = 0.0982$$

4–10 INCREMENTAL EQUIVALENT CIRCUIT FOR JUNCTION TRANSISTORS

We shall derive the incremental equivalent circuits for junction transistors in this section. The parameters of the equivalent circuit will be obtainable as a function of the fundamental device parameters. We shall use the device $I(V)$ equations for the transistor derived in Section 4–4. We rewrite these equations in the following simple form

$$I_E = a_{11}(e^{qV_E/kT} - 1) - a_{12}(e^{qV_C/kT} - 1) \tag{4–76}$$

$$I_C = a_{21}(e^{qV_E/kT} - 1) - a_{22}(e^{qV_C/kT} - 1), \tag{4–77}$$

where

$$a_{11} \triangleq \frac{qD_pAp_n\coth(w_0/L_p)}{L_p} + \frac{qD_nAn_{pe}}{L_{ne}} \tag{4–78}$$

$$a_{12} \triangleq \frac{qD_pAp_n\operatorname{csch}(w_0/L_p)}{L_p}. \tag{4–79}$$

$$a_{21} \triangleq \frac{qD_pAp_n\operatorname{csch}(w_0/L_p)}{L_p} \tag{4–80}$$

$$a_{22} \triangleq \frac{qD_pAp_n\coth(w_0/L_p)}{L_p} + \frac{qD_nAn_{pc}}{L_{nc}}. \tag{4–81}$$

We have thus lumped all the device parameters in the *a*s. Note that the set of *a*s is a remarkably symmetrical set. For example, from Eqs. (4–79) and (4–80)

$$a_{12} = a_{21} \tag{4–82}$$

and

$$a_{22} \doteq a_{11}. \tag{4–83}$$

The approximation of Eq. (4–83) would become exact in *p-n-p* transistors if the electron current components crossing the junctions are *zero.* It may be noted that *in practical p-n-p transistors the electron currents are indeed much less than the corresponding hole currents.*

As we wish to obtain *incremental* equivalent circuits for the transistor we must make sure that the reference directions of terminal voltages and currents

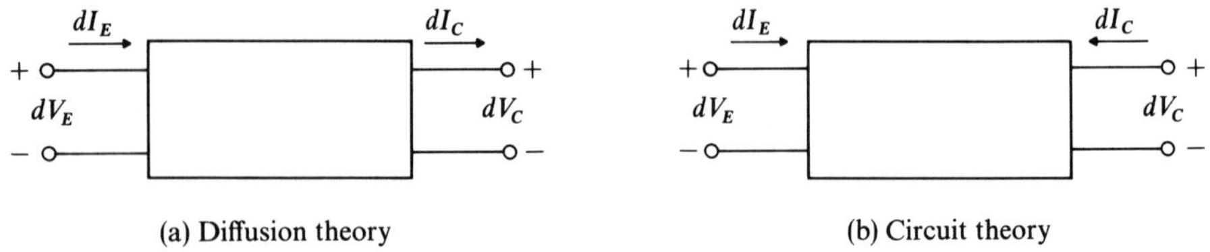

Fig. 4–6 Reference convention. (a) Diffusion theory. (b) Circuit theory.

are consistent with those which are associated with the usual circuit parameters, e.g. impedance, admittance, *h*-parameters, etc. Refer to Fig. 4–6a which shows the positive reference directions for *p-n-p* transistor device theory we have developed in this chapter. Figure 4–6b shows the reference conditions for circuit parameters defined by circuit theorists. It can be seen that these two sets of reference conditions are not mutually consistent. These can be *made* mutually consistent if the reference direction of collector current is turned around. This is easily accomplished by multiplying the right hand side of Eq. (4–77) by −1. The resulting equations now are given below

$$I_E = a_{11}(e^{qV_E/kT} - 1) - a_{12}(e^{qV_C/kT} - 1) \qquad (4\text{–}84)\dagger$$

$$I_C = -a_{21}(e^{qV_E/kT} - 1) + a_{22}(e^{qV_C/kT} - 1) \qquad (4\text{–}85)\dagger$$

We shall attempt to obtain the *incremental y-equivalent* circuit for the common base transistor. The incremental admittance parameters are defined in terms of the reference conditions in Fig. 4–6b by the following set of equations.

$$dI_E = y_{11}dV_E + y_{12}dV_C \qquad (4\text{–}86)$$

$$dI_C = y_{21}dV_E + y_{22}dV_C. \qquad (4\text{–}87)$$

Taking the differentials of Eqs. (4–84) and (4–85) we obtain

$$dI_E = \frac{a_{11}q}{kT}\, e^{qV_E/kT}\, dV_E - \frac{a_{12}q}{kT}\, e^{qV_C/kT}\, dV_C \qquad (4\text{–}88)$$

$$dI_C = -\frac{a_{21}q}{kT}\, e^{qV_E/kT}\, dV_E + \frac{a_{22}q}{kT}\, e^{qV_C/kT}\, dV_C. \qquad (4\text{–}89)$$

A comparison of Eqs. (4–88) and (4–89) with Eqs. (4–86) and (4–87) respectively yields the following:

$$y_{11} = \frac{a_{11}q}{kT}\, e^{qV_E/kT} \qquad (4\text{–}90)$$

† Equations (4–84) and (4–85) are also called the Ebers and Moll equations as they were first derived by J. J. Ebers and J. M. Moll (1954) in slightly different form.

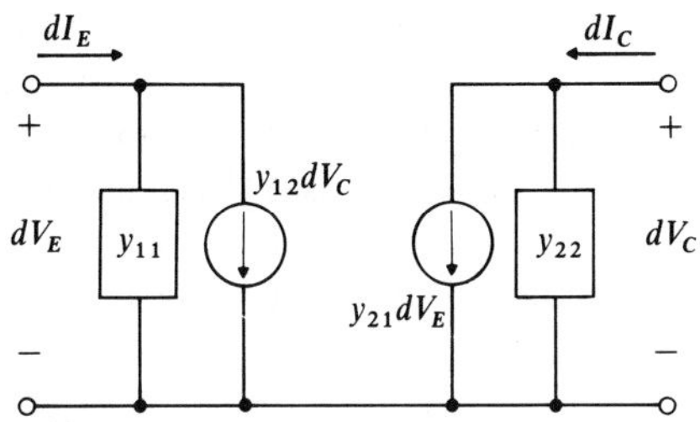

Fig. 4–7 Incremental admittance equivalent circuit.

$$y_{12} = -\frac{a_{12}q}{kT}\,e^{qV_C/kT} \tag{4–91}$$

$$y_{21} = -\frac{a_{21}q}{kT}\,e^{qV_E/kT} \tag{4–92}$$

$$y_{22} = \frac{a_{22}q}{kT}\,e^{qV_C/kT}. \tag{4–93}$$

Figure 4–7 shows the y-circuit which pictorially displays what Eqs. (4–86) and (4–87) mathematically state. Equation (4–86) says that the current dI_E consists of two components. One is the current flowing in the admittance y_{11} which has a voltage dV_E across it. As the other component of current $y_{12}dV_C$ is in terms of voltage dV_C in another part of the circuit (i.e. collector), it is represented by a *dependent current* source† as shown in Fig. 4–7. Equation (4–87) similarly says that the current dI_C consists of two components. One is the current flowing in admittance y_{22} which has a voltage dV_C across it. The other component depends on voltage in another part of the network (i.e. the emitter circuit), and is hence represented by a dependent current source† as shown in Fig. 4–7. The equivalent circuit of Fig. 4–7 is called an *incremental* circuit because its parameters are defined in terms of differentials of currents and voltages. Furthermore the circuit elements themselves are expressed in terms of derivatives. For example

$$y_{11} \triangleq \frac{\partial I_E}{\partial V_E} \tag{4–94}$$

$$y_{12} \triangleq \frac{\partial I_E}{\partial V_C} \tag{4–95}$$

$$y_{21} \triangleq \frac{\partial I_C}{\partial V_E} \tag{4–96}$$

$$y_{22} \triangleq \frac{\partial I_C}{\partial V_C}. \tag{4–97}$$

† Some authors refer to such sources as *controlled* sources.

It is useful to note that Eqs. (4–90) through (4–93) are valid for *any bias*. Equations (4–90) through (4–91) show that the admittances are functions of bias conditions. The dependence on voltage can be seen explicitly. There is also an implicit dependence of the admittance on current as well. To see this relationship explicitly we shall consider a specific bias condition, namely *normal bias*. Hence Eqs. (4–32) through (4–35) are valid. The use of these equations allows us to write Eq. (4–84) as

$$I_E \doteq a_{11}\, e^{qV_E/kT} \tag{4–98}$$

with the use of Eq. (4–98) we may rewrite Eq. (4–90)

$$y_{11} = \frac{q}{kT} a_{11}\, e^{qV_E/kT} \doteq \frac{qI_E}{kT}. \tag{4–99}$$

We see then that analogous to the diffusion conductance g_d of the single *p-n* junction, the input admittance of common base transistor is given by the same form of equations. Compare Eq. (3–61) and Eq. (4–99).

It is useful to note that in normal bias because of Eqs. (4–33) and (4–35) all admittances which depend on p_C such as y_{12} and y_{22} are approximately zero. That is

$$y_{12} \doteq 0 \tag{4–100}$$

$$y_{22} \doteq 0. \tag{4–101}$$

It should be remembered that Eqs. (4–99) through (4–101) are valid for *only normal bias*.

So far the equations and the equivalent circuits are for the *intrinsic* (*or ideal*) *device*. That is the extrinsic parameters must be added to obtain a fairly good model for a realistic device. Figure 4–8 shows such a complete circuit. The

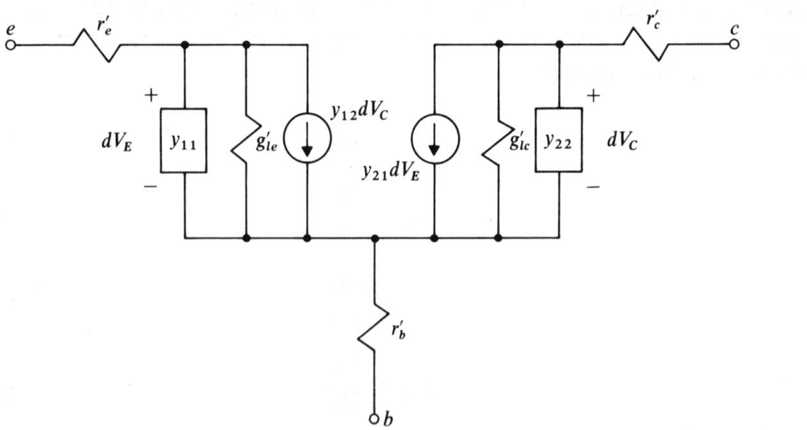

Fig. 4–8 Complete low frequency incremental equivalent circuit for junction transistor.

resistances r'_e, r'_b and r'_c represent the bulk resistances of the emitter, base, and collector bulk regions together with ohmic junction resistances. The leakage conductances $g'_{\ell e}$ and $g'_{\ell e}$ account for those currents which go around junctions rather than through them.

Example 4–3 Consider a *p-n-p* transistor which is biased in the inverse fashion (emitter junction reverse biased and collector junction forward biased). Assume that $e^{qV_C/kT} \gg 1$ and $e^{qV_E/kT} \ll 1$ and that electron currents crossing the junctions are negligibly small compared with hole currents. Express Y^i_{cc} in terms of I_C, temperature, and known constants. Calculate Y^i_{cc} at room temperature when $I_C = 2$ mA.

Solution:

$$I_C = -a_{21}(e^{qV_E/kT} - 1) + a_{22}(e^{qV_C/kT} - 1) \tag{1}$$

$$dI_C = -\frac{a_{21}q}{kT} e^{qV_E/kT} dV_E + \frac{a_{22}q}{kT} e^{qV_C/kT} dV_C \tag{2}$$

$$dI_C = Y^i_{ce} dV_E + Y^i_{cc} dV_C. \tag{3}$$

Hence:

$$Y^i_{cc} = \frac{qa_{22}}{kT} e^{qV_C/kT}. \tag{4}$$

Because:

$$e^{qV_E/kT} \ll 1, \text{ and } e^{qV_C/kT} \gg 1 \tag{5}$$

$$I_C \doteq a_{22}\, e^{qV_C/kT}. \tag{6}$$

Equation (4) may now be rewritten with the help of Eq. (6) to obtain finally

$$Y^i_{cc} = \frac{qI_C}{kT} \quad \longleftarrow \tag{7}$$

$$Y^i_{cc} = \frac{2 \text{ mA}}{26 \text{ mV}} = 0.077\,\mho \longleftarrow$$

Three comments are worthwhile here. First, r'_b is probably the most significant of the five extrinsic parameters. It is among the most important high frequency parameters. Second, Fig. 4–8 is an approximation in that the extrinsic parameters are shown as *lumped*, and the leakage conductances and r'_b are shown going to the common node. A more realistic portrayal would show $g'_{\ell e}$ and $g'_{\ell c}$ going to different intermediate points in the resistance r'_b. Third, the emitter and collector junction capacitances, while they are also extrinsic parameters, have been left out because we are here considering only the d-c or low frequency equivalent circuits. These will be included in the wide band equivalent circuits in the following chapters.

4–11 VOLTAGE LIMITATIONS OF JUNCTION TRANSISTORS

We expect that the breakdown voltage for each junction of the transistor is determined by the same considerations as for single *p-n* junctions. Since normally the emitter junction is more heavily doped than the collector junction, it is expected that the absolute value of breakdown voltage for the former is lower than that of the latter.

Besides the Zener and avalanche breakdown there is a third mechanism called *punch-through* which could set the limitation on voltage. In a *normally biased* transistor the collector junction is reverse biased. As the reverse bias is increased in absolute magnitude, the collector transition region width also is increased. It is possible for the base edge of the collector transition region to meet with the base edge of the emitter transition region. When this occurs the base width has been reduced to zero and very large currents can flow in the device. Punch-through is then said to have occurred. Of the three mechanisms Zener, avalanche, and punch-through, the one occuring at lowest absolute value of collector voltage sets the voltage limit.

4–12 CHOICE OF CONDUCTIVITIES

The device equations of the preceding section of this chapter can be utilized in making intelligent judgments as to the choice of conductivities of the emitter, base, and collector regions. An additional quantity, I_{CO} can also be calculated using the transistor device equations already obtained. As this is an important transistor parameter, it will be useful to calculate it. The calculation will reveal the fact that the results of this chapter are valuable in the calculation of a number of very useful quantities.

Recall that I_{CO} *is the collector current when the emitter current is zero* under the condition that the collector junction is reverse biased. Using this definition we may write from Eq. (4–77)

$$I_{CO} = a_{21}\,(e^{qV_E/kT} - 1) - a_{22}\,(e^{qV_C/kT} - 1) \tag{4–102}$$

The collector junction is reverse biased, so

$$I_{CO} = a_{21}\,(e^{qV_E/kT} - 1) + a_{22}. \tag{4–103}$$

We can solve for the factor $(e^{qV_E/kT} - 1)$ from Eq. (4–77) under the condition that $I_E = 0$,

$$0 = a_{11}\,(e^{qV_E/kT} - 1) + a_{12}$$

or

$$(e^{qV_E/kT} - 1) = -\frac{a_{12}}{a_{11}}. \tag{4–104}$$

Equation (4–104) can be substituted into Eq. (4–103) to obtain the following

$$I_{CO} = -\frac{a_{21}a_{12}}{a_{11}} + a_{22}. \tag{4–105}$$

We may substitute the values of a_{21}, a_{12}, a_{11}, and a_{22} previously obtained to get

$$I_{CO} = -\frac{qD_p^2\left(\frac{p_n}{L_p}\right)^2 \operatorname{csch}^2\left(\frac{w_0}{L_p}\right)}{D_p\left(\frac{p_n}{L_p}\right)\coth\left(\frac{w_0}{L_p}\right) + \frac{n_{pe}}{L_{ne}}D_n} + \frac{qD_p p_n A}{L_p}\coth\left(\frac{w_0}{L_p}\right) + \frac{AqD_n n_{pc}}{L_{nc}}. \tag{4–106}$$

But from Eq. (4–70) we know that

$$\frac{n_{pe}}{p_n} = \left(\frac{\sigma_B}{\sigma_E}\right)\left(\frac{\mu_p}{\mu_n}\right) \tag{4–107}$$

for reasons cited earlier $\sigma_B \ll \sigma_E$, and under these circumstances the second term in the denominator of the first term on the right hand side of Eq. (4–106) can be neglected. This allows us to simplify Eq. (4–106) to the following after some algebraic manipulations,

$$I_{CO} = \frac{qD_p A n_i^2 q\mu_n}{L_p\sigma_B}\tanh\left(\frac{w_0}{L_p}\right) + \frac{qD_n n_i^2 q\mu_p A}{L_{nc}\sigma_C}. \tag{4–108}$$

We are now prepared to consider how the conductivities ought to be chosen in a transistor. We may tabulate the reasons arguing for or against a particular choice.

1. To keep injection efficiency high

$$\sigma_E \gg \sigma_B. \tag{4–109}$$

   As this particular choice does not conflict with any other requirement, the emitter conductivity is always much greater than base conductivity.

2. Unfortunately the choice of collector and base conductivities is not quite so simple as the emitter and base conductivities. Equation (4–108) indicates that if σ_B and σ_C are high, then I_{CO} is low. This is desirable. However, a high choice for σ_B and σ_C leads to two unwelcome consequences: first, to a high value of collector junction capacitance; and second, the breakdown voltage is low. Both of these are not desirable.

3. If $\sigma_B \ll \sigma_C$ most of the collector transition region is on the base side. This results, for fixed bias, in a low value of w_0 which increases gain which is desirable. The disadvantages are that (a) the punch-through voltage is lower than if $\sigma_B \gg \sigma_C$, and (b) r_b' is high. Both of these are undesirable.

4. If $\sigma_B \gg \sigma_C$, I_{CO} is a function of fewer transistor parameters which is desirable. Extrinsic base resistance will be low which is also desirable. Extrinsic collector resistance r_c' is higher and hence, for a given collector current, a higher amount of voltage will be across the device when it is saturated. This is undesirable if the transistor is to be used as an electronic switch.

Also τ_{nc} will be high which is not good for storage time constant (to be discussed in Chapter 6).

We see then the choice of base and collector conductivities is subject to different requirements. Most instances may call for a compromise between conflicting requirements. If a specific application is in mind the conductivities can be chosen to accentuate the most important requirements.

An intriguing suggestion was made in the early 1950s by J. M. Early.† He proposed that if an intrinsic region can be interposed between the base and collector regions, the twin, apparently contradictory, requirements of high collector breakdown voltage and low r_c' can be fulfilled. To understand this, consider the following. First, the intrinsic region has low conductivity. *This is what allows high breakdown voltage.* This also allows the collector bulk region to be highly doped thus allowing r_c' to be low. A low r_c' makes the voltage across the saturated transistor low, a desirable feature. Furthermore this makes C_{tc} low which is also desirable.

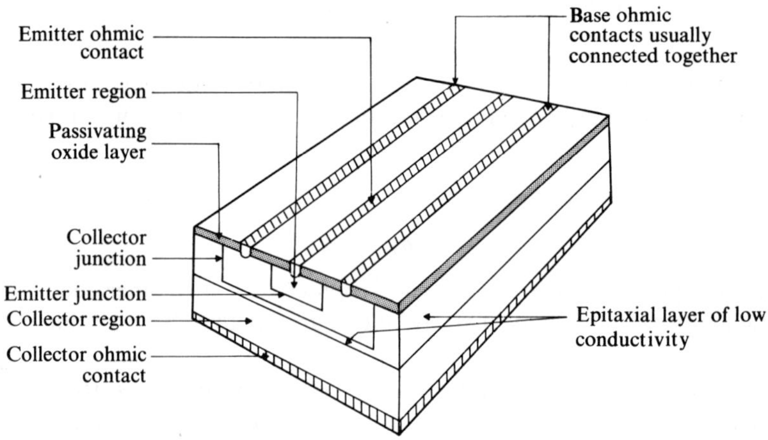

Fig. 4–9 Planar passivated epitaxial transistor.

Thus even though the advantages of an intrinsic material between the base and collector were recognized quite a while ago, the technology necessary to bring such a transistor into existence was not discovered until the early 1960s. The process which does the trick is called *epitaxy.* A *planar epitaxial* transistor is shown in Fig. 4–9. It may be noted that all junctions show up on a single plane (hence *planar*) which can be covered with a protective oxide which reduces leakage currents. Thus the advent of epitaxy relieves us from having

† J. M. Early, "*p-n-i-p* and *n-p-i-n* Junction Transistor Triodes", *The Bell System Technical Journal*, **33**, 517–533 (1954).

to make a compromise at least in the case of certain choices involving the collector.

4–13 SUMMARY

The d-c device equations relating currents and voltages in junction transistors were derived in this chapter. These analytic expressions were used to calculate a number of useful transistor parameters.

An incremental equivalent circuit was derived based on the nonlinear $I(V)$ relationships for the transistor. The elements of the circuit could be expressed as explicit analytic functions of fundamental device parameters and bias. Thus desirable equivalent circuit parameters may be related directly to the fundamental device parameters.

Finally the device equations were used to show how different considerations could be used to intelligently choose the relative values of the conductivities of the emitter, base, and collector.

PROBLEMS

4–1 Calculate the minority-carrier excess *charge* (i.e. in excess of the thermal-equilibrium value) in the base region as a function of the fundamental device parameters and emitter current. Assume that the transistor is normally biased and $p_E \gg p_n$, $p_C \ll p_n$, $w_0 \ll L_p$.

4–2 Calculate the excess minority-carrier *charge* in the base region when the transistor is saturated.

4–3 Calculate the *total charge* in the base region when the transistor is cut-off.

4–4 Calculate the minority-carrier charge in the collector when (a) the transistor is saturated, and (b) the transistor is normally biased.

4–5 How much minority-carrier charge needs to be removed from the base before the transistor can be brought from saturation to normal bias?

4–6 How much minority-carrier charge needs to be removed before the transistor can be brought from normal bias to cut-off?

4–7 When the transistor is saturated, the voltage drops between the terminals of the transistor are very small, and to a high degree of accuracy we may consider the currents as the independent variables and the voltages as the dependent variables. Treating the emitter and collector currents as independent variables, calculate the minority-carrier density in the base as a function of x.

4–8 Calculate the ratios (a) Q_b/I_E, (b) Q_b/I_C, (c) Q_b/I_B, where Q_b is the excess minority-carrier base charge. *Hint:* Use the results of Problems 4–1 as a starting point.

4–9 A step-junction transistor has $\beta^* = 0.99$ when it is normally biased and $V_C = -2$ Volts. Now the reverse collector voltage is changed so that $V_C = -5$ Volts. What is the new β^*? Express the answer in terms of w_0, L_p, k_{at}, $\sigma_C \gg \sigma_B$, built-in voltage of 0.5 Volt collector junction. Assume V_E = constant.

4–10 Obtain the incremental L equivalent circuit for the transistor. The L parameters are defined by the following two equations:

$$i_1 = L_{11}v_1 + L_{12}i_2 \qquad v_2 = L_{21}v_1 + L_{22}i_2.$$

Label the elements in terms of fundamental device parameters. Emitter and base are the input terminals (1–1′), and collector and base are the output terminals (2–2′).

4–11 In a silicon *p-n-p* transistor it is desired to have $\gamma_0 = 0.999$, $\beta_0^* = 0.98$. Choose N_a (emitter), N_d (base), w_0, L_p and L_{ne} so that the above requirements are met.

REFERENCES

Bardeen, J., and Brattain, W. H., "The Transistor, A Semiconductor Triode," *Phys. Rev.*, **74,** 230 (1948).

Ebers, J. J., and Moll, J. L., "Large-signal Behavior of Junction Transistors," *Proc. IRE,* **42,** 1761–1772 (1954).

Gray, P. E., Dewitt, D., Boothroyd, A. R., and Gibbons, J. F., *Physical Electronics and Circuit Models of Transistors*, SEEC, Vol. 2, John Wiley & Sons, New York (1966).

Middlebrook, R. D., *An Introduction to Junction Transistor Theory*, Chapter 8, John Wiley & Sons, New York (1957).

Moll, J. L., "The Evolution of the Theory of the Current-Voltage Characteristics of *p-n* Junctions," *Proc. IRE*, **46,** 1076 (1958).

Moll, J. L., *Physics of Semiconductors*, Chapter 8, McGraw-Hill Book Co., New York (1964).

Nanavati, R. P., *An Introduction to Semiconductor Electronics*, Chapter 4, McGraw-Hill Book Co., New York (1963).

Pritchard, R. L., "Advances in the Understanding of the *p-n* Junction Triode," *Proc. IRE*, **46,** 1130–1141 (1958).

Pritchard, R. L., *Electrical Characteristics of Transistors*, McGraw-Hill Book Co., New York (1967).

Shockley, W., "The Theory of *p-n* Junctions in Semiconductors and *p-n* Junction Transistors," *Bell System Tech. J.,* **28,** 435–489 (1949).

Shockley, W., Ceasson, G. L., and Haynes, J. R., "Hole Injection in Germanium-Quantitative Studies and Filamentary Transistors," *Bell System Tech. Journal*, **28,** 344 (1949).

Schockley, W., Sparks, M., and Teal, G. K., "*p-n* Junction Transistors," *Phys. Rev.*, **83,** 151–162 (1951).

Schockley, W., *Electrons and Holes in Semiconductors*, Chapters 3, 4, Van Nostrand, New York (1950).

Sze, S. M., *Physics of Semiconductor Devices*, Chapter 6, John Wiley & Sons, New York (1969).

CHAPTER 5

A WIDE BAND A-C MODEL FOR p-n JUNCTION

5-1 INTRODUCTION

In this chapter we consider for the first time the a-c equivalent circuit. It is fitting that we consider the *p-n* junction in detail so that the techniques developed here can also be utilized in developing models for the junction transistor. We shall consider only the sinusoidal excitation.

5-2 A-C DIFFUSION EQUATION

To consider the general a-c case we shall need to solve the general one-dimensional *time-dependent* diffusion equation which is repeated here

$$\frac{\partial^2 p}{\partial x^2} - \frac{p - p_n}{L_p^2} = \frac{1}{D_p}\frac{\partial p}{\partial t}. \tag{5–1}$$

It should be noted that, unlike the previous chapters where the minority carrier density was a function only of the position coordinate, e.g., $p(x)$, now the minority carrier density is a function of *both* position coordinate and time, that is $p(x, t)$.

The solution of Eq. (5–1) of the following form is expected to work

$$p(x, t) = p_0(x) + p_1(x)\, e^{j\omega t} \tag{5–2}$$

when the sources are sinusoidal of frequency ω. One assumption in writing Eq. (5–2) is that the minority carrier density varies sinusoidally with the same frequency as that of the driving sources. As the first term on the right hand side of Eq. (5–2) is a function only of x, it will correspond to the d-c solution. The second term is the a-c variation and is in the form of the product of two terms. One is a function *only* of position $p_1(x)$. The other is a function *only* of time. As we have seen in connection with Schrödinger's equation, a favorite method of solving partial differential equations is to assume a *product solution.*

Let us substitute the total solution of Eq. (5–2) into the general diffusion equation of Eq. (5–1) to see if the assumed solution works or not. For simplicity we shall simply designate $p_0(x)$ by p and $p_1(x)$ by p_1.

$$\frac{\partial^2 p_0}{\partial x^2} - \frac{p_0 - p_n}{L_p^2} + \frac{\partial^2(p_1 e^{j\omega t})}{\partial^2 x} - \frac{p_1 e^{j\omega t}}{L_p^2} = \frac{1}{D_p}\frac{\partial(p_1 e^{j\omega t})}{\partial t} = \frac{p_1 e^{j\omega t} j\omega}{D_p}. \tag{5–3}$$

Rearranging Eq. (5–3) we obtain

$$\frac{d^2 p_0}{dx^2} - \frac{p_0 - p_n}{L_p^2} = \frac{e^{j\omega t} d^2 p_1}{dx^2} - e^{j\omega t} p_1 \left[\frac{1}{L_p^2} + \frac{j\omega}{D_p}\right]. \tag{5–4}$$

Note first that the partial derivative of p_0 and p_1 with respect to x has been replaced by total derivatives because both p_0 and p_1 are functions only of x. Second, note that the left hand side of Eq. (5–4) may be recognized as the d-c diffusion equation which we have already solved in Chapter 4. The d-c solution can, therefore, be equated to zero as before. Thus, by doing this we obtain the following two equations.

$$\frac{d^2 p_0}{dx^2} - \frac{p_0 - p_n}{L_p^2} = 0, \tag{5–5}$$

$$\frac{d^2 p_1}{dx^2} - p_1\left(\frac{1 + j\omega\tau_p}{L_p^2}\right) = 0. \tag{5–6}$$

In obtaining Eq. (5–6) the common factor $e^{j\omega t}$ has been cancelled. It will be useful to define the following parameters:

$$C_p^2 \triangleq \frac{1 + j\omega\tau_p}{L_p^2}, \tag{5–7}$$

$$C_n^2 \triangleq \frac{1 + j\omega\tau_n}{L_n^2}. \tag{5–8}$$

With the use of Eq. (5–7) we may rewrite Eq. (5–6) as

$$\frac{d^2 p_1}{dx^2} - C_p^2 p_1 = 0. \tag{5–9}$$

It is useful to note that the *form* of the differential equation for the a-c hole density (minority carrier density) Eq. (5–9) is the same as that for the d-c hole density. This observation implies that the solutions to Eq. (5–9) should be similar in form to the solutions to Eq. (5–6). Thus, the *similarity of Eqs. (5-5) and (5–9) leads to the conclusion that the d-c solutions can be directly converted to the a-c solutions by the use of some simple transformations.* Before we proceed with this idea we must look at the boundary conditions.

The derivation of the boundary conditions of Eqs. (3–38) and (3–39) was quite general, and was not restricted to only d-c applied voltage. Thus, if d-c voltage V_0 and a-c voltage v_1 are simultaneously applied to a p-n junction, the value of hole density at the edge of the transition region on the n-side is given by

$$p_{xtn} = p_n\, \mathrm{e}^{q[V_0 + v_1(t)]/kT} = p_n\, \mathrm{e}^{qV_0/kT}\, \mathrm{e}^{qv_1(t)/kT} \tag{5–10}$$

If

$$|v_1(t)| \ll \frac{kT}{q} \tag{5–11}$$

the second exponential on the right hand side of Eq. (5–10) can be expanded in Taylor series as follows:

$$p_{xtn} = p_n\, \mathrm{e}^{qV_0/kT}\left[1 + \frac{qv_1(t)}{kT} + \text{higher order terms}\right]. \tag{5–12}$$

The higher order terms, which would be nonlinear, in Eq. (5–12) may be dropped because of Eq. (5–11). Equation (5–12) can be written

$$p_{xtn} = p_n\, \mathrm{e}^{qV_0/kT} + p_n\, \mathrm{e}^{qV_0/kT}\frac{qv_1(t)}{kT} \tag{5–13}$$

or

$$p_{xtn} = p_{\text{d-c}} + p_{\text{a-c}}, \tag{5–14}$$

where

$$p_{\text{d-c}} = p_n\, \mathrm{e}^{qV_0/kT} \tag{5–15}$$

and

$$p_{\text{a-c}} = p_n \mathrm{e}^{qV_0/kT}\frac{qv_1(t)}{kT} = p_{\text{d-c}}\frac{qv_1(t)}{kT}. \tag{5–16}$$

Equation (5–14) indicates that provided Eq. (5–11) is valid the general boundary condition separates into an a-c and a d-c part. Note that Eq. (5–11) represents the small signal assumption, not $|v_1| \ll V_0$.

Based on the above observations the following transformations will convert d-c solutions to the corresponding a-c solutions.

1. d-c boundary conditions → a-c boundary conditions.
2. $L_p \to \dfrac{1}{C_p}$, $L_n \to \dfrac{1}{C_n}$,

where

$$C_n^2 \triangleq \frac{1 + j\omega\tau_n}{L_n^2} \quad \text{and} \quad C_p^2 \triangleq \frac{1 + j\omega\tau_p}{L_p^2}. \tag{5–17}$$

3. p_n as a *particular* solution → 0
   and n_p as a particular solution → 0

Item 3 follows from the fact that the terms p_n/L_p^2 and n_p/L_n^2, which are present in the d-c diffusion equation, are absent in the a-c diffusion equation. They were these terms which gave the particular solutions $p = p_n$ and $n = n_p$ in the solution to the d-c diffusion equation. For example, the first term in Eq. (3–43), and in Eq. (4–11) is the particular solution. It is this solution which would be missing in the transformed a-c solution. Note that the transformation number 3 is *not* simply to replace p_n and n_p by zero everywhere. Only when p_n and n_p appear as the particular solutions are they to be replaced by zero.

Let us use the above transformation rules to convert the d-c *p-n* junction $I(V)$ equation to the corresponding a-c equation. The d-c equation is repeated for convenience.

$$I = qA\left[\frac{p_n D_p}{L_p} + \frac{n_p D_n}{L_n}\right](e^{qV_0/kT} - 1). \tag{5–18}$$

The (-1) term on the right hand side of Eq. (5–18) came from the particular solutions for electron and hole currents. Based on this observation we shall drop (-1) from Eq. (5–18) and slightly rearrange the terms to obtain

$$I = qA\left[\frac{D_p}{L_p}(p_n\, e^{qV_0/kT}) + \frac{D_n}{L_n}(n_p\, e^{qV_0/kT})\right]. \tag{5–19}$$

In Eq. (5–19) the terms within brackets ( ) are seen to be the d-c boundary conditions. We must replace these by a-c boundary conditions as given by Eq. (5–16) and replace L_p and L_n by $(1/C_p)$ and $(1/C_n)$ respectively. Using these ideas we may transform Eq. (5–19) to

$$\begin{aligned} I\, e^{j\omega t} = qA\Bigg[&\frac{D_p q p_n e^{qV_0/kT}}{kT}\cdot\frac{\sqrt{1 + j\omega\tau_p}}{L_p}\, V e^{j\omega\tau} \\ &+ \frac{D_n q n_p e^{qV_0/kT}}{kT}\cdot\frac{\sqrt{1 + j\omega\tau_n}}{L_n}\, V e^{j\omega\tau}\Bigg], \end{aligned} \tag{5–20}$$

where I = a-c amplitude of current,
V = a-c amplitude of voltage.

We have assumed that a-c current and voltage are both sinusoids of frequency ω. Equation (5–20) may be rewritten after cancelling the common factor $e^{j\omega\tau}$ as

$$I = \frac{VqA}{kT}\left[\frac{qD_p p_n e^{qV_0/kT}}{L_p}\sqrt{1 + j\omega\tau_p} + \frac{qD_n n_p e^{qV_0/kT}}{L_p}\sqrt{1 + j\omega\tau_n}\right]. \tag{5–21}$$

For a junction which is forward biased by a voltage

$$V_0 \gg \frac{kT}{q} \tag{5–22}$$

then

$$I_{p0} = \frac{qD_p A p_n e^{qV_0/kT}}{L_p} \tag{5–23}$$

$$I_{n0} = \frac{qD_n A n_p e^{qV_0/kT}}{L_n}. \tag{5–24}$$

Using Eqs. (5–23) and (5–24), Eq. (5–21) may be rewritten

$$I = \frac{qV}{kT}\left[I_{p0}\sqrt{1 + j\omega\tau_p} + I_{n0}\sqrt{1 + j\omega\tau_n}\right]. \tag{5–25}$$

We are now ready to write the *intrinsic admittance* of a forward biased *p-n* junction as follows:

$$Y^i = \frac{I}{V} = \left[\frac{I_{p0}q}{kT}\sqrt{1 + j\omega\tau_p} + \frac{qI_{n0}}{kT}\sqrt{1 + j\omega\tau_n}\right]. \tag{5–26}$$

Equation (5–26) is the intrinsic admittance because, so far, the extrinsic parameters have not been included. In an attempt to synthesize the intrinsic admittance Y^i we must regretably note that no linear, lumped, finite, passive, and bilateral (LLFPB) network can have an admittance such as that of Eq. (5–26). The functions which can be synthesized by LLFPB networks are called *positive real* (*PR*) functions which do not include those of Eq. (5–26).

In an attempt to obtain a workable equivalent circuit, we first note that no real measuring system can accurately detect frequencies from 0 to ∞. Thus we may be willing to accept a bandwidth limitation on the validity of the equivalent circuit representation of Eq. (5–26). With this limitation accepted, we may expand the square roots in Eq. (5–26) by the use of binomial expansion. We shall retain only two terms to obtain

$$Y^i = \frac{q}{kT} I_{p0}\left(1 + \frac{j\omega\tau_p}{2} + \cdots\right) + I_{n0}\left(1 + \frac{j\omega\tau_n}{2} + \cdots\right). \tag{5–27}$$

A rearrangement of terms gives

$$Y^i = \frac{q}{kT}(I_{p0} + I_{n0}) + j\omega\left(\frac{qI_{p0}}{2kT}\tau_p + \frac{qI_{n0}}{2kT}\tau_n\right) \tag{5–28}$$

$$y^i = g_d + j\omega C_d, \tag{5–29}$$

where

$$g_d \triangleq \frac{q}{kT}(I_{p0} + I_{n0}) = \frac{qI_0}{kT}, \tag{5–29}$$

where I_0 = total d-c current.

Here g_d is diffusion conductance which is seen to be identical to Eq. (3–61). Diffusion capacitance is given by

$$C_d \triangleq \frac{q}{2kT}(I_{p0}\tau_p + I_{n0}\tau_n). \tag{5–30}$$

Two limiting forms of the diffusion capacitance are of interest. If

$$\sigma_p \gg \sigma_n \tag{5–31}$$

it can be shown that

$$I_{p0} \gg I_{n0}. \tag{5–32}$$

Then

$$C_d = \frac{qI_{p0}\tau_p}{2kT} \doteq \frac{qI_0\tau_p}{2kT}. \tag{5–33}$$

If the junction is symmetric so that

$$\tau_p I_{p0} = \tau_n I_{n0} \tag{5–34}$$

then Eq. (5–30) becomes

$$C_d = \frac{q\tau_p I_{p0}}{kT} = \frac{q\tau_n I_{n0}}{kT} = \frac{q\tau_p I_0}{2kT}. \tag{5–35}$$

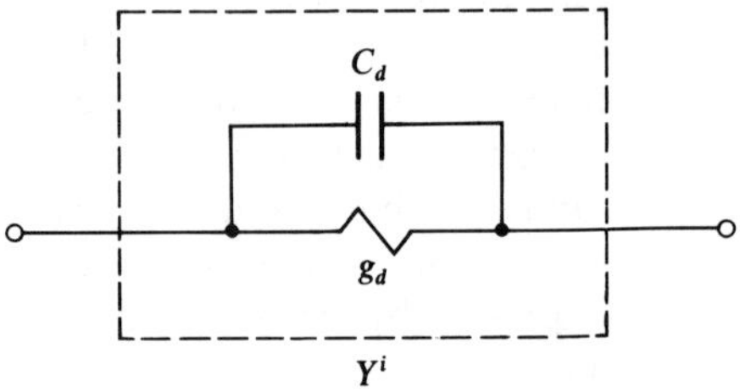

Fig. 5–1 Intrinsic incremental model for forward biased *p-n* junctions.

It is instructive to note first that diffusion conductance and diffusion capacitance are *directly* proportional to d-c bias current. Second, the time constant T for the intrinsic network shown in Fig. 5–1 is independent of bias if either of the two limiting expressions Eqs. (5–33) or (5–35) is used. For example

$$T = \frac{C_d}{g_d} = \frac{\dfrac{qI_0\tau_p}{2kT}}{\dfrac{qI_0}{kT}} = \frac{\tau_p}{2}. \tag{5–36}$$

5–3 COMPLETE EQUIVALENT CIRCUIT FOR THE *p-n* JUNCTION

In the previous section we obtained the intrinsic equivalent circuit of a *p-n* junction. To obtain the complete model for a forward biased *p-n* junction the *extrinsic* parameters must be added to the intrinsic model. These extrinsic parameters are the bulk resistance and ohmic contact resistance r_s', leakage conductance g_ℓ', the junction capacitance C_j, and linear stray capacitance C_ℓ. Figure 5–2 shows the complete equivalent circuit. It may be noted that unlike all other circuit elements whose value depends on the current and/or voltage

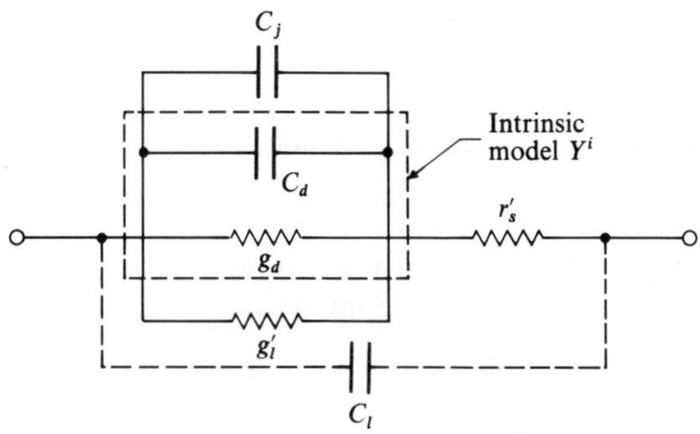

Fig. 5–2 Complete incremental equivalent circuit for forward biased *p-n* junctions.

due to only the *external* sources, the junction capacitance depends upon the total voltage, i.e. $(V_{BI} - V)$ across the junction.

Several features of the model obtained turn out to be rather general. They are similar to what is found in the models of wide band transistor equivalent circuits. The common elements are:

1. Diffusion capacitances and conductances are proportional to d-c bias currents.
2. Practical equivalent circuits are generally obtained by the use of expansion of the admittance functions and retraining a finite number of terms in the expansion. The throwing away of the higher order terms is what sets the upper bound on the frequency up to which the equivalent circuit is a good model for the device.
3. Generally speaking each element of the equivalent circuit is expressible as an explicit function of the d-c bias.

Example 5–1 The following experimental data are obtained on an abrupt *p-n* junction having $\sigma_p \gg \sigma_n$, $\tau_p = 10^{-9}$ sec., $\tau_n = 5 \times 10^{-9}$ sec. at room temperature.

(i) at high reverse voltage it is observed that the capacitance of the junction reaches a limiting value of 2 pF as a function of reverse voltage.

(ii) at a reverse voltage of 15 Volts the device capacitance is 18 pF.

(iii) when $1/C^2$ is plotted as a function of voltage the extrapolated data points intersect the voltage axis at 1.0 Volt.

(iv) At a reverse voltage of 1 Volt the slope of the $I(V)$ characteristics is 10^{-7}℧.

(v) at a forward current of 130 mA the slope of the $I(V)$ characteristics is 1/2.2 mho.

(vi) when $I = 10$ mA, $V = 0.2$ Volt.

From the above data, calculate, C_d, C_j, r_d, r'_ℓ, r'_s, when the forward current is 78 mA.

Solution: The data of (i) allows us to conclude

$$C_\ell = 2\,\text{pF} \tag{1}$$

because C_ℓ and C_j are effectively in parallel with each other and $C_j \to 0$ for sufficiently high reverse voltage. The data of (ii) leads us to write

$$18\,\text{pF} = C_\ell + C_j = 2 + C_j \tag{2}$$

or

$$C_j = 16\,\text{pF} = \frac{k_a}{(V_b' - V)^{\frac{1}{2}}}. \tag{3}$$

From (iii) we deduce that the voltage axis intercept is

$$V_{BI} = 1.0\ \text{Volt}. \tag{4}$$

Thus

$$C_j = 16 = \frac{k_a}{[1 - (-15)]^{\frac{1}{2}}} = \frac{k_a}{4} \tag{5}$$

or

$$k_a = 64. \tag{6}$$

From the data of (iv)

$$r_\ell' = 10\ \text{meg-}\Omega. \tag{7}$$

From the data of (v) the following may be written:

$$r_d + r_s = 1/(1/2.2) = 2.2\ \text{Ohm}. \tag{8}$$

But

$$r_d = \frac{kT}{qI} = \frac{26\,\text{mV}}{130\,\text{mA}} = 0.2\ \text{Ohm}. \tag{9}$$

Thus

$$r_s' = 2.2 - 0.2 = 2.0\ \text{Ohm}. \tag{10}$$

At 78 mA, the diffusion resistance r_d and diffusion capacitance C_d may be calculated as follows:

$$r_d = \frac{kT}{qI} = \frac{26}{78} = 0.33\ \text{Ohm}. \tag{11}$$

$$C_d = \frac{q\tau_p I_0}{2kT} = \frac{10^{-9} \times 78}{52} = 1.5 \times 10^{-9}\,\text{F}. \tag{12}$$

To calculate C_j it is necessary to know the voltage across the transition region at the given current of 78 mA. To this end we shall make use of the following.

$$\frac{I_2}{I_1} = \frac{I_S \exp\left[\frac{q}{kT}(V_2 - I_2 r_s')\right]}{I_S \exp\left[\frac{q}{kT}(V_1 - I_1 r_s')\right]}, \tag{5–37}$$

when

$$V_1, V_2 > \frac{kT}{q} \tag{5–38}$$

$$\frac{78}{10} = \frac{\exp\left[\frac{q}{kT}(V_2 - 78 \times 2)\right]}{\exp\left[\frac{q}{kT}(V_1 - 10 \times 2)\right]} = \frac{\exp\left[\frac{1}{26}(V_2 - 156)\right]}{\exp\left[\frac{1}{26}(200 - 20)\right]} \tag{13}$$

$$7.8 \exp\left(\frac{180}{25}\right) = \exp\left[\frac{1}{26}(V_2 - 156)\right] \tag{14}$$

$$7870 = \exp\left[\frac{1}{26}(V_2 - 156)\right] \tag{15}$$

$$26 \ell\text{n}\,(7870) = (V_2 - 156) \tag{16}$$

$$233 = (V_2 - 156) \tag{17}$$

or

$$V_2 = 389 \text{ mV} = 0.389 \text{ V}. \tag{18}$$

It should be noted that to obtain the voltage across the transition region due only to external sources, the voltage drop across r_s' must be subtracted from the terminal voltages V_1 and V_2. Thus, when the terminal current is 78 mA the voltage across the transition region is 233 mV.

We are now ready to calculate C_j at this voltage.

$$C_j = \frac{k_a}{\sqrt{V_{BI} - V}}, \tag{19}$$

$$C_j = \frac{64}{\sqrt{1 - 0.233}} = \frac{64}{\sqrt{0.767}} = \frac{64}{0.878} = 73 \text{ pF}. \tag{20}$$

5–4 EQUIVALENT CIRCUIT FOR REVERSE BIASED *p-n* JUNCTION

For reverse bias and

$$|V| \gg \frac{kT}{q} \tag{5–39}$$

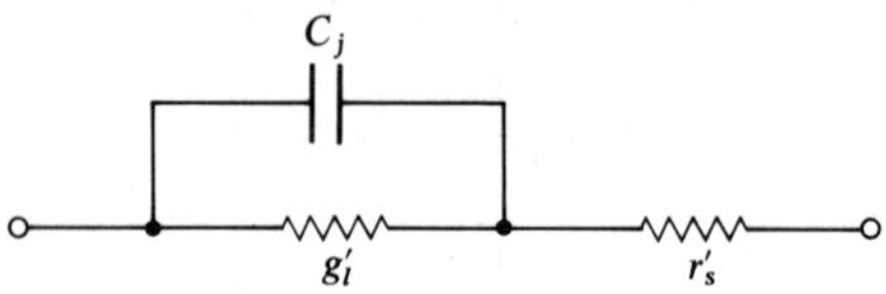

Fig. 5–3 Incremental equivalent circuit for reverse biased *p-n* junctions.

the intrinsic incremental model for *p-n* junction is essentially an open circuit.

$$g_d = 0$$

and C_d is nonexistent. Thus the equivalent circuit consists of only extrinsic parameters. Figure 5–3 shows this incremental equivalent circuit for a reverse biased *p-n* junction for which Eq. (5–39) is true.

5–5 SUMMARY

We obtained wide band equivalent circuits for a forward and reverse biased *p-n* junction. These were obtained by recognizing that the form of the d-c diffusion equation was the same as for the a-c diffusion equation. Transformations were deduced which would directly convert d-c solutions to the corresponding a-c solutions. An important limitation of the model is an upper bound in frequency up to which the equivalent circuits are valid. Another significant limitation of the forward biased *p-n* junction is the size of the a-c amplitude V of voltage which is limited by

$$V \ll \frac{kT}{q} = 0.026 \text{ Volt} \tag{5–40}$$

at *room temperature.* This is the real *small signal* limitation and not

$$|V_{\text{a-c}}| \ll |V_{\text{d-c}}| \tag{5–41}$$

as may be intuitively expected. This limitation *does not* apply to the equivalent circuit of reverse biased *p-n* junction.

PROBLEMS

5–1 Devise an experimental way by which the three capacitances C_ℓ, C_j, and C_d can be measured and distinguished from one another.

5–2 The following experimental data is obtained on a *p-n* junction.

Device capacitance (pF)	Voltage (Volts)
5	0
1	−10
1	−20

The device is known to be symmetric and to have $\tau_p = 10^{-11}$ second, $r'_s = 0.5\,\Omega$, $r'_\ell = 10$ meg-Ω, built-in voltage = 1.0 Volt, $\sigma_p \gg \sigma_n$. Calculate C_ℓ, C_j, and C_d at $V = 0.4$ Volt, current of 52 mA. Assume step junction.

5-3 For the device of Problem 5–2 calculate r_d at $V = 0.4$ Volt.

5-4 What is the time-constant T_i of the intrinsic part of the device specified in Problem 5–2.

5-5 Consider two step junction *p-n* junctions *A* and *B*. Device *B* is made from a semiconductor having twice the band gap as that of device *A*. Which parameters of the complete incremental wide band model of the *p-n* junction are expected to be different in the two devices? Give reasons and indicate which parameters of device *A* are expected to be bigger than those of device *B*. You may assume that k_a μ_p, μ_n, r'_s, r'_ℓ are identical for both devices and $\sigma_p \gg \sigma_n$.

REFERENCES

Gray, P. E. et al, *Physical Electronics and Circuit Models of Transistors,* Chapter 5, SEEC Vol. 2, John Wiley & Sons, New York (1964).

Moll, J. L., *Physics of Semiconductors,* Chapter 7, McGraw-Hill Book Co., New York (1964).

Nanavati, R. P., *An Introduction to Semiconductor Electronics,* Chapter 7, McGraw-Hill Book Co., New York.

Schockley, W., *Electrons and Holes in Semiconductors,* Chapter 12, Van Nostrand, New York (1950).

CHAPTER 6

WIDE BANDWIDTH MODELS FOR DIFFUSION TRANSISTORS

6-1 INTRODUCTION

In the previous chapter a wide bandwidth model for the *p-n* junction diode was developed. This served as a good simple illustration of what is necessary to obtain wide band† models for junction devices. We shall follow a similar procedure for the diffusion transistor in the present chapter. Many things which we found to be true for the *p-n* junction diode will turn out to be true also for the transistor. For example, the intrinsic equivalent circuit elements are functions of the d-c bias and the frequency range of validity of the model is *finite.*

We shall consider only the diffusion transistor in this chapter. By *diffusion transistor* is meant one in which the major transport mechanism in the base region is *diffusion.* This will be true for transistors which have a uniformly doped base region and which are operated under conditions of low injection, i.e. the injected minority carrier density is significantly less than the majority carrier density in the base region. High injection effects will be considered in Chapter 12.

The transistor device theory will be covered in two parts. Part A will present the *unperturbed case.* Despite the fact that time varying a-c voltages are being considered and the transition region widths of the emitter and collector junctions are therefore also changing with time, we shall consider the base width to be *constant* with time in Part A. Recall that the base width was defined as the region *between* the emitter and collector transition regions (see Fig. 4–4). Thus, strictly speaking when the emitter and collector transition regions change with time so does the base width. We are ignoring this fact in only Part A. The change in base width with time will be called the *perturbation.* Hence the solutions we obtain in Part A are the *unperturbed* solutions, and will serve as the *zero* order solutions to the more general case including the perturbation which will be considered in Part B. Thus, we shall build on to the unper-

† A *wide bandwidth model* shall be designated simply as the *wide band model* from now on. By this we mean that the model is valid for a relatively wide bandwidth of frequencies.

turbed solution. The first order solutions to be obtained in Part B will be added to the zero order solutions of Part A.

PART A | UNPERTURBED CASE

6–2 BOUNDARY CONDITIONS AND TRANSFORMATIONS

The time dependent diffusion equation is that of Eq. (5–1). The solution for the minority carrier density in the base region of *p-n-p* transistors is given by Eq. (5–2). The total solution is the sum total of a-c and d-c solutions as follows

$$p(x, t) = p_0(x) + p_1(x)\, e^{j\omega t}. \tag{6–1}$$

Equation (6–1) is the same as Eq. (5–2). As before the first and second terms on the right hand side of Eq. (6–1) are the d-c and the a-c solutions. The diffusion equation which applies to the a-c part of the minority carrier density in the base region is Eq. (5–6) which is repeated here for convenience.

$$\frac{d^2 p_1}{dx^2} - p_1\left(\frac{1 + j\omega\tau_p}{L_p^2}\right) = 0. \tag{6–2}$$

The *boundary condition* which is appropriate to the minority carrier density at the emitter edge of the base region is the following:

$$p(-w_0/2) = p_n\, e^{(q/kT)[V_E + v_e(t)]} \tag{6–3}$$

or

$$p(-w_0/2) = p_n\, e^{qV_E/kT} \cdot e^{qv_e(t)/kT}, \tag{6–4}$$

where V_E = emitter to base d-c voltage,

$v_e(t)$ = emitter to base a-c voltage.

The second exponential on the right hand side of Eq. (6–4) may be expanded in Taylor series and only two terms may be retained if

$$|v_e(t)| \ll kT/q \tag{6–5†}$$

then

$$p(-w_0/2) \doteq p_n\, e^{qV_E/kT}\,[1 + qv_e(t)/kT] = p_E + p_e, \tag{6–6}$$

where

$$p_e \triangleq \frac{qp_E v_e(t)}{kT} \tag{6–7}$$

and p_e = a-c boundary condition.

Similarly at the collector edge of the base region the boundary condition for

† This is the small signal assumption in the transistor device theory and not the usual one like $|v_e(t)| \ll V_E$. Equations (6–5) and (6–11) are a more stringent restriction than the one above. We follow the notation for current and voltage where capital letters with capital subscripts stand for d-c quantities, lower case letters with lower case subscripts designate signal or a-c quantities. Lower case letters with capital subscripts stand for total quantities.

Table 6–1 Transformation Laws Which Convert d-c Solutions Into a-c Solutions In *p-n-p* Transistors

	d-c solutions		a-c solutions	
1.	L_p	→	$\dfrac{1}{C_p} \triangleq \dfrac{L_p}{\sqrt{1 + j\omega\tau_p}}$	(6–12)
2.	L_{en}	→	$\dfrac{1}{C_{en}} \triangleq \dfrac{L_{en}}{\sqrt{1 + j\omega\tau_{en}}}$	(6–13)
3.	L_{cn}	→	$\dfrac{1}{C_{cn}} \triangleq \dfrac{L_{cn}}{\sqrt{1 + j\omega\tau_{cn}}}$	(6–14)
4.	p_n as a particular solution	→	0	
5.	n_{pe} as a particular solution	→	0	
6.	n_{pc} as a particular solution	→	0	
7.	d-c boundary conditions	→	a-c boundary conditions	

where n_{pe} = thermal equilibrium density of electrons in the *p*-type emitter
n_{pc} = thermal equilibrium density of electrons in the *p*-type collector
τ_{en} = minority carrier lifetime of electrons in the emitter
τ_{cn} = minority carrier lifetime of electrons in the collector

$$L_{en} \triangleq \sqrt{D_n\tau_{en}} \qquad (6\text{–}15)$$

$$L_{cn} \triangleq \sqrt{D_n\tau_{cn}} \qquad (6\text{–}16)$$

minority carriers may be written as

$$p(+w_0/2) = p_C + p_c, \qquad (6\text{–}8)$$

where

$$p_C \triangleq p_n e^{qV_C/kT} \qquad (6\text{–}9)$$

and

$$p_c \triangleq \frac{qp_C v_c(t)}{kT} \qquad (6\text{–}10)$$

provided that

$$|v_c(t)| \ll kT/q, \qquad (6\text{–}11)\dagger$$

where V_C = collector to base d-c voltage,

$v_c(t)$ = collector to base a-c voltage.

Recall that p_n is the thermal equilibrium value of holes in the *n*-type base region.

†See footnote on page 177.

Table 6–2 Transformation Laws Which Convert d-c Solutions Into a-c Solutions In *n-p-n* Transistors

	d-c solutions		a-c solutions	
1.	L_n	$\longrightarrow$	$\dfrac{1}{C_n} \triangleq \dfrac{L_n}{\sqrt{1 + j\omega\tau_n}}$	(6–17)
2.	L_{ep}	$\longrightarrow$	$\dfrac{1}{C_{ep}} \triangleq \dfrac{L_{ep}}{\sqrt{1 + j\omega\tau_{ep}}}$	(6–18)
3.	L_{cp}	$\longrightarrow$	$\dfrac{1}{C_{cp}} \triangleq \dfrac{L_{cp}}{\sqrt{1 + j\omega\tau_{cp}}}$	(6–19)
4.	n_p as a particular solution	$\longrightarrow$	0	
5.	p_{ne} as a particular solution	$\longrightarrow$	0	
6.	p_{nc} as a particular solution	$\longrightarrow$	0	
7.	d-c boundary condition	$\longrightarrow$	a-c boundary condition	

where p_{ne} = thermal equilibrium density of holes in the n-type emitter,
p_{nc} = thermal equilibrium density of holes in the n-type collector,
τ_{ep} = minority carrier lifetime of holes in the emitter,
τ_{cp} = minority carrier lifetime of holes in the collector

$$L_{ep} \triangleq \sqrt{D_p\tau_{ep}} \tag{6–20}$$

$$L_{cn} \triangleq \sqrt{D_p\tau_{cp}} \tag{6–21}$$

We saw that the general time dependent diffusion equation of Eq. (5–1) conveniently separates into the d-c diffusion equation of Eq. (5–5) and the a-c diffusion equation of Eq. (5–6). We have just seen that the total boundary condition separates into a-c and d-c parts provided the magnitude relationships (6–5) and (6–11) are valid. These facts allow us to treat the a-c diffusion equation and the corresponding a-c boundary conditions by themselves.

Because of the similarity of the mathematical forms of the d-c and a-c diffusion equations, d-c solutions may be converted directly into a-c solutions if certain transformation laws are observed. Such transformation laws Eq. (5–17) were discussed in connection with the *p-n* junction. The same kind of transformation laws also apply to the transistor because the reasoning that led to the relationship Eq. (5–17) is valid also for the transistor. In the case of the transistor, proper account must be taken of the existence of two junctions. These observations result in the transformation laws given in Table 6–1. Note that in Tables 6–1 and 6–2 some parameters have a double subscript because for example the *p-n-p* transistor has two *p*-type regions and hence the double subscript removes the ambiguity which would otherwise be present.

6–3 A-C SOLUTIONS

With the use of the results of the last section we shall be able to directly convert the d-c transistor solutions to the a-c unperturbed solution.

Equations (4–14) and (4–15), which are valid for *p-n-p* intrinsic transistor, may now be converted to the corresponding a-c equations. The d-c Eqs. (4–14) and (4–15) are first stated here for convenience and the transformed a-c unperturbed equations are stated directly after the d-c equations.

$$I_{pE} = \frac{qD_pA}{L_p}\left[(p_E - p_n)\coth(w_0/L_p) - (p_C - p_n)\operatorname{csch}(w_0/L_p)\right] \quad (4\text{–}14)$$

$$I_{pC} = \frac{qD_pA}{L_p}\left[(p_E - p_n)\operatorname{csch}(w_0/L_p) - (p_C - p_n)\coth(w_0/L_p)\right]. \quad (4\text{–}15)$$

With the help of Table 6–1 the above equation may be converted to the a-c equations as follows.

$$i_{pe}(t) = qD_pAC_p\frac{q}{kT}\left[v_e(t)p_E\coth(C_pw_0) - v_c(t)p_C\operatorname{csch}(C_pw_0)\right] \quad (6\text{–}22)$$

$$i_{pc}(t) = qD_pAC_p\frac{q}{kT}\left[v_e(t)p_E\operatorname{csch}(C_pw_0) - v_c(t)p_C\coth(C_pw_0)\right], \quad (6\text{–}23)$$

where $i_{pe}(t)$ = a-c minority carrier current crossing the emitter junction,

$i_{pc}(t)$ = a-c minority carrier current crossing the collector junction,

$v_e(t)$ = a-c voltage appearing across the emitter junction due to only external sources,

$v_c(t)$ = a-c voltage appearing across the collector junction due to only external sources.

The d-c electron current components crossing the emitter and collector junctions are given by Eqs. (4–19) and (4–22) and are repeated below.

$$I_{En} = \frac{qD_nA}{L_{en}}(n_E - n_{pe}) \quad (4\text{–}19)$$

$$I_{Cn} = -\frac{qD_nA}{L_{cn}}(n_C - n_{pc}). \quad (4\text{–}22)$$

The corresponding a-c equations obtained from the use of the transformation laws of Table 6–1 are the following:

$$i_{en}(t) = \frac{q^2D_nAC_{en}n_Ev_e(t)}{kT} \quad (6\text{–}24)$$

$$i_{cn}(t) = \frac{q^2D_nAC_{cn}n_Cv_c(t)}{kT}, \quad (6\text{–}25)$$

where

$$i_{en}(t) = \text{a-c electron current crossing the emitter junction,}$$

$$i_{cn}(t) = \text{a-c electron current crossing the collector junction.}$$

In obtaining Eqs. (6–22) through (6–25) we have used all seven items of Table 6–1 *simultaneously*. Consider for example the term

$$\frac{qD_pA}{L_p}(p_E - p_n) \coth (w_0/L_p).$$

First p_n is dropped in accordance with item 4 of Table 6–1 because this term arose from the particular solution to the diffusion equation. Thus the above term took the following form.

$$\frac{qD_pA}{L_p} p_E \coth (w_0/L_p).$$

Now p_E, which is the d-c boundary condition for the minority carrier density at the emitter edge of the base region, must be replaced by the a-c boundary condition p_e. Thus, the above term becomes

$$\frac{qD_pA}{L_p} p_e \coth (w_0/L_p).$$

Now p_e may be replaced by its evaluation as given by Eq. (6–7) to obtain the following.

$$\frac{qD_pAqp_E}{L_pkT} v_e(t) \coth (w_0/L_p).$$

Then finally L_p is replaced by $1/C_p$ according to item 1 to obtain

$$\frac{q^2D_pAp_EC_pv_e(t)}{kT} \coth (C_pw_0).$$

This is how the first term on the right hand side of Eq. (6–22) was obtained. Other terms in Eqs. (6–22) through (6–25) were obtained in a precisely similar way.

The total emitter current is obtained by adding the hole and electron currents Eqs. (6–22) and (6–24) crossing the emitter junction,

$$i_e(t) = \frac{q^2D_pAC_p}{kT}[v_e(t)p_E \coth (C_pw_0) - v_c(t)p_C \operatorname{csch} (C_pw_0)] + \frac{q^2D_nAC_{en}n_Ev_e(t)}{kT}. \tag{6–26}$$

The total collector current cannot be obtained simply by adding the hole and electron currents crossing the collector junction because Eq. (6–25) is not true! The reason for this is that the transformations (of Table 6–1) which were used to obtain Eq. (6–25) do not take account of a phenomenon which is important in the bulk of the collector. The *finite* conductivity of the collector permits the setting up of voltage gradients and drift fields in the bulk of the collector region. Hole flow will be varying with time and so will be the bulk voltage variations due to the hole flow. The current due to time varying hole flow sets up an a-c field in the bulk of the collector. This field modulates the flow of electrons in the collector bulk. This phenomenon is also called *conductivity modulation.* This effect was not considered before, and hence the transformation laws are not valid for this particular case. The calculations taking account of the conductivity modulation in the collector are fairly involved and only the results for collector multiplication factor α^* will be stated here for the low injection† or the very high injection‡ limits.

At low injection (low current densities) for *p-n-p* transistors

$$\alpha^* \triangleq \frac{i_c}{i_{cp}} = 1 + \frac{\sigma_{nc}}{2\sigma_{pc}}. \qquad (6\text{–}27)‡§$$

At high injection (high current densities)

$$\alpha^* = 1 + \frac{\sigma_{nc}}{\sigma_{pc}}, \qquad (6\text{–}28)†‡$$

where σ_{nc} = conductivity of the *p*-type collector due to only electrons,

σ_{pc} = conductivity of the *p*-type collector due to only holes.

With the collector multiplication factor in hand we are ready to write the expression for the a-c unperturbed collector current as follows:

$$i_c(t) = \alpha^* i_{cp}(t). \qquad (6\text{–}29)$$

With the help of Eq. (6–23), Eq. (6–29) may be written

$$i_c(t) = \frac{q^2 D_p C_p \alpha^* A}{kT}\left[v_e(t)p_E \operatorname{csch}(C_p w_0) - v_c(t)p_C \coth(C_p w_0)\right]. \qquad (6\text{–}30)$$

† High injection phenomena will be discussed in detail in Ch. 12.

‡ The more accurate expression for α^* shows its frequency dependence. Generally the collector multiplication factor does not set the limit on frequency cut-off of the current gain hence the simpler expressions given here are adequate.

§ The interested reader should read either or both of the following references: Nanavati (1963) and Early (1953). 1. R. P. Nanavati, *An Introduction to Semiconductor Electronics* Secs. 7-6, 7-7. McGraw-Hill Book Co. New York, (1963). 2. J. M. Early, "Design Theory of Junction Transistors", Bell System Tech. J. Vol. 32, pp. 1271–1312, Nov. 1953.

We are particularly interested in sinusoidal excitations. We shall therefore now make the assumption, which is valid for linear systems, that the current and voltage are characterized by the same frequency. Thus,

$$i_e(t) = I_e \, e^{j\omega t} \tag{6–31}$$

$$i_c(t) = I_c \, e^{j\omega t} \tag{6–32}$$

$$v_e(t) = V_e \, e^{j\omega t} \tag{6–33}$$

$$v_c(t) = V_c \, e^{j\omega t}. \tag{6–34}$$

With Eqs. (6–31) through (6–34) substituted into Eqs. (6–26) and (6–30) the following equations are obtained after the cancellation of the exponential factor $e^{j\omega t}$,

$$I_e = \frac{q^2 D_p A C_p}{kT} [V_e p_E \coth (C_p w_0) - V_c p_C \operatorname{csch} (C_p w_0)] + \frac{q^2 D_n A C_{en} n_E V_e}{kT} \tag{6–35}$$

$$I_c = \frac{q^2 D_p A C_p \alpha^*}{kT} [V_e p_E \operatorname{csch} (C_p w_0) - V_c p \;\coth (C_p w_0)]. \tag{6–36}$$

It should be recalled that Eqs. (6–35) and (6–36) are valid only if the magnitude relationships (6–5) and (6–11) are true. This means that the amplitude of the a-c emitter and collector voltages must be less than kT/q which is 26 mV at room temperature. While it is reasonable to restrict the a-c amplitude of emitter voltage to a few milliVolts, it is certainly unreasonable to place a similar restriction on collector voltage which could easily be a few Volts. Is this a contradiction? Not really! The only reason why the collector junction can sustain a few Volts is that this junction is reverse biased, usually by a few Volts. When this fact is true, as in normal bias case, the minority carrier density is essentially zero. Thus,

$$p_C \doteq 0 \tag{6–37}$$

and

$$p_c = \frac{qp}{kT} v_c(t) \doteq 0. \tag{6–38}$$

Thus, the very terms which would have violated Eq. (6–11) go to zero and no contradiction of the device small signal condition is involved. Equation (6–35) and (6–36) take on the following simpler form under *normal bias*

$$I_e = \left[\frac{q^2 D_p A C_p p_E}{kT} \coth (C_p w_0) + \frac{q^2 D_n A C_{en} n_E}{kT} \right] V_e \tag{6–39}$$

$$I_c = \left[\frac{q^2 D_p A C_p \alpha^*}{kT} p_E \operatorname{csch} (C_p w_0) \right] V_e. \tag{6–40}$$

6–4 SHORT-CIRCUIT CURRENT GAIN

Incremental a-c short-circuit current gain is expressed analogous to the d-c gain of Eq. (4–47)

$$\alpha = \left(\frac{i_c}{i_e}\right) = \left(\frac{i_{pe}}{i_e}\right)\left(\frac{i_{pc}}{i_{pe}}\right)\left(\frac{i_c}{i_{pc}}\right) \tag{6–41}$$

$$\alpha^* \triangleq \frac{i_c}{i_{pc}}. \tag{6–42}$$

Equation (6–42) is the collector multiplication factor α^* which has already been evaluated in Eqs. (6–27) and (6–28).

The second factor on the right hand side of Eq. (6–41) is the base transport factor β^*. This may be obtained by the use of the transformation laws of Table 6–1 applied to the d-c β_0^* of Eq. (4–54).

$$\beta^* \triangleq \frac{i_{pc}}{i_{pe}} = \operatorname{sech}(C_p w_0) = \operatorname{sech}\left[\frac{w_0}{L_p}\sqrt{1 + j\omega\tau_p}\right] \quad p\text{-}n\text{-}p \tag{6–43}$$

$$\beta^* \triangleq \frac{i_{nc}}{i_{ne}} = \operatorname{sech}(C_n w_0) = \operatorname{sech}\left[\frac{w_0}{L_n}\sqrt{1 + j\omega\tau_n}\right] \quad n\text{-}p\text{-}n. \tag{6–44}$$

The first factor on the right hand side of Eq. (6–41) is the incremental a-c injection efficiency γ which may be obtained also by the use of the transformation laws of Table 6–1 on Eq. (4–71).

$$\gamma \triangleq \frac{i_{pe}}{i_e} = \frac{1}{1 + \left(\frac{\sigma_B}{\sigma_E}\right)\left(\frac{C_{en}}{C_p}\right)\tanh(C_p w_0)}, \tag{6–45}$$

where C_{en} is given by Eq. (6–13).

Thus, with the use of Eqs. (6–42), (6–43) and (6–45) we may write

$$\alpha = \gamma\beta^*\alpha^*. \tag{6–46}$$

It is instructive to consider both the high frequency limit of γ and its upper cut-off frequency. First the high frequency limit is obtained as

$$\lim_{\omega\to\infty} \frac{1}{1 + \left(\frac{\sigma_B}{\sigma_E}\right)\left(\frac{C_{en}}{C_p}\right)\tanh(C_p w_0)}. \tag{6–47}$$

Note first that

$$\lim_{\omega\to\infty} \tanh(C_p w_0) = \lim_{\omega\to\infty} \tanh\left(\frac{w_0}{L_p}\sqrt{1 + j\omega\tau_p}\right) = 1 \tag{6–48}$$

and

$$\lim_{\omega \to \infty} \left(\frac{C_{en}}{C_p} \right) = \left(\frac{L_p}{L_{en}} \right) \left(\frac{\tau_{en}}{\tau_p} \right)^{\frac{1}{2}}. \tag{6–49}$$

Using Eq. (6–48) and (6–49) we may now evaluate Eq. (6–47) as follows:

$$\lim_{\omega \to \infty} [\gamma] = \frac{1}{1 + \left(\frac{\sigma_B}{\sigma_E} \right) \left(\frac{L_p}{L_{en}} \right) \left(\frac{\tau_{en}}{\tau_p} \right)^{\frac{1}{2}}}. \tag{6–50}$$

Another simpler form of Eq. (6–50) is the following:

$$\lim_{\omega \to \infty} [\gamma] = \frac{1}{1 + \left(\frac{\sigma_B}{\sigma_E} \right) \left(\frac{D_p}{D_n} \right)^{\frac{1}{2}}}. \tag{6–51}$$

The upper cut-off frequency ω_{c0} of a transfer function is defined as the frequency at which the magnitude of the transfer function is $(1/\sqrt{2})$ times its mid-band value. If the transfer function is a *single pole function* having no finite plane zeros, it can always be put in the following form

$$T(s) = \frac{T_0}{1 + (s/\omega_{coh})} \tag{6–52}$$

or

$$T(j\omega) = \frac{T_0}{1 + (j\omega/\omega_{coh})}, \tag{6–53}$$

where T_0 = low frequency (or mid-band) gain,

ω_{coh} = upper cut-off frequency,

s = the complex variable $\sigma + j\omega$.

The fact that ω_{coh} is the upper cut-off frequency is easily ascertained by the direct use of the definition and Eq. (6–53), when

$$\omega = \omega_{coh} \tag{6–54}$$

$$T(j\omega_{co}) = \frac{T_0}{1 + j1} \tag{6–55}$$

$$|T(j\omega_{co})| = T_0/\sqrt{2}. \tag{6–56}$$

Equation (6–56) shows that ω_{coh} is really the upper cut-off frequency because it satisfies the definition. Thus, any transfer function which can be put in the form of either of Eq. (6–52) or Eq. (6–53) allows us at a glance to recognize the low frequency (or mid-band) gain and the upper cut-off frequency.

As a direct application of the above remarks let us find the upper cut-off frequency of γ. We shall assume that

$$w_0 \ll L_p \tag{6–57}$$

which must be true in any realistic good transistor. The magnitude relationship (6–57) allows us to take the one term Taylor series expansion of Eq. (6–45) as follows

$$\gamma = \frac{1}{1 + (\sigma_B/\sigma_E) C_{en} w_0}. \tag{6–58}$$

By the use of Eq. (6–13) we obtain

$$\gamma = \frac{1}{1 + (\sigma_B/\sigma_E)(w_0/L_{en}) \sqrt{1 + j\omega\tau_{en}}}. \tag{6–59}$$

Using the binomial expansion of the square root and retaining two terms Eq. (6–59) becomes

$$\gamma = \frac{1}{1 + \dfrac{\sigma_B}{\sigma_E}\dfrac{w_0}{L_{en}}\left(1 + j\omega\dfrac{\tau_{en}}{2}\right)} \tag{6–60}$$

$$\gamma = \frac{1}{\left(1 + \dfrac{\sigma_B}{\sigma_E}\dfrac{w_0}{L_{en}}\right) + j\omega\dfrac{\sigma_B w_0 \tau_{en}}{\sigma_E 2L_{en}}} \tag{6–61}$$

or

$$\gamma = \frac{1\Big/\left(1 + \dfrac{\sigma_B}{\sigma_E}\dfrac{w_0}{L_{en}}\right)}{1 + j\omega\left(\dfrac{\sigma_B w_0 \tau_{en}}{\sigma_E 2L_{en}}\right)\Big/\left(1 + \dfrac{\sigma_B w_0}{\sigma_E L_{en}}\right)}. \tag{6–62}$$

Equation (6–62) is of the form of Eq. (6–53). Thus the cut-off frequency is

$$\omega_{c0} = \frac{1 + \dfrac{\sigma_B w_0}{\sigma_E L_{en}}}{\dfrac{\sigma_B w_0 \tau_{en}}{2\sigma_E L_{en}}} = \left(\frac{2\sigma_E L_{en}}{\sigma_B w_0 \tau_{en}} + \frac{2}{\tau_{en}}\right). \tag{6–63}$$

Let us now calculate the cut-off frequency for the base transport factor β^*. In preparation for doing this we list the Taylor series of the hyperbolic functions of the variable x.

$$\sinh x = x + \frac{x^3}{3!} + \frac{x^5}{5!} + \cdots \tag{6–64}$$

$$\cosh x = 1 + \frac{x^2}{2!} + \frac{x^4}{4!} + \cdots \tag{6–65}$$

$$\tanh x = x - \frac{x^3}{3} + \frac{2x^5}{15} + \cdots \tag{6–66}$$

$$\operatorname{csch} x = \frac{1}{x} + \frac{x}{3!} - \cdots \tag{6–67}$$

$$\operatorname{sech} x = 1 - \frac{x^2}{2!} \cdots \tag{6–68}$$

$$\coth x = \frac{1}{x} + \frac{x}{3} - \frac{x^3}{45} + \cdots \tag{6–69}$$

The right hand side of Eq. (6–43) may be expanded with the help of Eq. (6–65) to obtain the following

$$\beta^* = \frac{1}{\cosh(C_p w_0)} \doteq \frac{1}{1 + \frac{1}{2}\frac{w_0^2}{L_p^2}(1 + j\omega\tau_p)} \tag{6–70}$$

$$\beta^* = \frac{1\Big/\left[1 + \frac{1}{2}\left(\frac{w_0}{L_p}\right)^2\right]}{1 + j\omega\tau_p \frac{w_0^2}{2L_p^2}\left[1 + \frac{1}{2}\left(\frac{w_0}{L_p}\right)^2\right]^{-1}}. \tag{6–71}$$

Equation (6–71) is of the form of Eq. (6–53). Thus the low frequency β_0^* is

$$\beta_0^* = \frac{1}{1 + \frac{1}{2}\left(\frac{w_0}{L_p}\right)^2} \doteq 1 - \frac{1}{2}\left(\frac{w_0}{L_p}\right)^2 \tag{6–72}$$

$$\omega_\beta^* = \frac{1}{\tau_p} + \frac{2L_p^2}{\tau_p w_0^2} = \frac{1}{\tau_p} + \frac{2\tau_p D_p}{\tau_p w_0^2} \tag{6–73}$$

$$\omega_\beta^* = \frac{1}{\tau_p} + \frac{2D_p}{w_0^2}. \tag{6–74}$$

Generally

$$\frac{1}{\tau_p} \ll \frac{2D_p}{w_0^2}. \tag{6–75}$$

With the use of the magnitude relationship (6–75), Eq. (6–74) becomes

$$\omega_\beta^* \doteq \frac{2D_p}{w_0^2} \qquad (p\text{-}n\text{-}p). \tag{6–76}$$

If we use the infinite product expansion for $\cosh(C_p w_0)$ instead of the Taylor series expansion, a more accurate calculation results. Thus, we may use the following

$$\cosh x = \prod_{n=1}^{\infty}\left[1 + \frac{4x^2}{(2n-1)^2\pi^2}\right]. \tag{6–77}$$

Using only one product the following results:

$$\beta^* = \frac{1}{1 + \left(\frac{4}{\pi^2}\right)\left(\frac{w_0}{L_p}\right)^2(1 + j\omega\tau_p)} \tag{6–78}$$

or

$$\beta^* = \frac{1/[1 + (4/\pi^2)(w_0/L_p)^2]}{1 + j\omega\tau_p\left[\frac{(4/\pi^2)(w_0/L_p)^2}{1 + (4/\pi^2)(w_0/L_p)^2}\right]}. \tag{6–79}$$

Equation (6–79) is of the form

$$\beta^* = \frac{\beta_0^*}{1 + j\frac{\omega}{\omega_\beta^*}}, \tag{6–80}$$

where

$$\beta_0^* \doteq \frac{1}{1 + \left(\frac{4}{\pi^2}\right)\left(\frac{w_0}{L_p}\right)^2} \doteq 1 - \left(\frac{4}{\pi^2}\right)\left(\frac{w_0}{L_p}\right)^2 \tag{6–81}$$

and

$$\omega_\beta^* = \frac{1}{\tau_p} + \left(\frac{\pi^2}{4}\right)\frac{D_p}{w_0^2} \doteq \frac{2.46D_p}{w_0^2}. \tag{6–82}$$

where Eq. (6–75) has been used.

If the definition of the cut-off frequency is used directly on Eq. (6–43), the following is obtained for the cut-off frequency

$$\omega_\beta^* = \frac{2.43D_p}{w_0^2} \qquad (p\text{-}n\text{-}p). \tag{6–83}$$

We shall use Eq. (6–83) as it is the most accurate. *Note that the numerical factor 2.43 replaces $\pi^2/4$ in the infinite product expansion.* If

$$\gamma \doteq \gamma_0 \tag{6–84}$$

and

$$\alpha^* \doteq \alpha_0^*. \tag{6–85}$$

Then

$$\alpha_0 = \gamma_0 \alpha_0^* \left[1 - \frac{1}{2.43} \left(\frac{w_0}{L_p} \right)^2 \right] \tag{6–86}$$

and

$$\omega_\alpha = \omega_\beta^* = \frac{2.43 D_p}{w_0^2}. \tag{6–87}$$

Thus, the common base *intrinsic* short circuit current gain cut-off frequency is given by Eq. (6–87). This is the *intrinsic* cut-off frequency because only the intrinsic elements have figured in the calculation so far.

The *intrinsic* short circuit current gain for the *common emitter* circuit β may be easily calculated by using the following.

$$\beta = \frac{\alpha}{1 - \alpha} = \frac{1}{\dfrac{1}{\alpha} - 1}. \tag{6–88}$$

It is possible to express β as follows if we assume both γ_0 and α_0^* to be unity

$$\beta = \frac{1}{\cosh (C_p w_0) - 1} = \frac{\beta_0}{1 + j\dfrac{\omega}{\omega_\beta}}. \tag{6–89}$$

Using a precisely similar procedure as for β, (see Problem 6–26) the following is obtained

$$\beta_0 = 2.43 \left(\frac{L_p}{w_0} \right)^2 \tag{6–90}$$

$$\omega_\beta = \frac{1}{\tau_p}. \tag{6–91}$$

The above quantities which we have calculated are not affected by the perturbation to be discussed in Section B and hence these calculations, though calculated for the unperturbed case, are nevertheless valid generally.

Finally, from our knowledge of the *h*-parameters we may also note that

$$h_{21b} = \alpha \tag{6–92}$$

$$h_{21e} = \beta, \tag{6–93}$$

where the third subscript indicates the circuit configuration; *b* for common base circuit and *e* for common emitter circuit. Some people also use the following terminology

$$h_{fb} \triangleq h_{21b} \tag{6–94}$$

$$h_{fe} \triangleq h_{21e}, \tag{6–95}$$

where h_{fb} = forward short circuit current gain in the common base circuit,

h_{fe} = forward short circuit current gain in the common emitter circuit.

PART B | THE PERTURBED CASE

6-5 BOUNDARY CONDITIONS

In the previous section we had worked out the a-c solution to the a-c diffusion equation subject to the restriction that the emitter and collector edges of the base region remained *fixed.* In the a-c case the emitter and collector ends of the base region are expected to be functions of time because the emitter and collector transition region widths will vary with time as the respective voltages across these junctions vary with time. Recall that the base width has been defined as the region *between* the emitter and collector *transition regions.* With both the transition regions varying sinusoidally with time, the base width will vary with time as follows,

$$w(t) = w_0 + w_1 e^{j\omega t}, \tag{6–96}$$

where w_0 = average base width,

w_1 = amplitude of the sinusoidally varying base width.

We shall list the assumptions of the analysis which follows.

1. Only the one-dimensional diffusion equation will be solved.
2. The low injection condition is valid, i.e. in *p-n-p* transistors

$$p_{\text{injected}} \ll n_n \text{ (base).} \tag{6–97}$$

3. The voltage drops in the bulk emitter, base, and collector regions will be neglected when the diffusion equation solutions are obtained.
4. The distributed extrinsic parameters will be approximated by lumped time invariant circuit elements.
5. The emitter edge of the base region boundary position will be considered as *fixed.*
6. The perturbed solution $\tilde{p}(x, t)$ is small compared with the unperturbed solution $p_0(x, t)$.
7. Only *first order* terms in the perturbed solution will be retained.
8. The major influence of the collector end of the base region variation with time is on the d-c minority carrier density in the base region. We therefore assume

$$p_{\text{a-c}} \ll p_{\text{d-c}} \tag{6–98}$$

and
$$\frac{dp_{\text{a-c}}}{dx} \ll \frac{dp_{\text{d-c}}}{dx}. \tag{6–99}$$

9. There is negligible influence of perturbation on the extrinsic resistances r'_e, r'_c.
10. The d-c reverse biased collector voltage V_C is less than the break down voltage and the punch-through voltage.
11. The leakage resistance of both junctions will be neglected.
12. The total solution consists simply of the unperturbed solutions *plus* the perturbed solution.
13. The perturbed solution will be evaluated at the collector end of the base region at $x = +w_0/2$ and *not* $x = w_0/2 + w_1\, e^{j\omega t}$. The latter procedure will give correction terms of a higher than first order and hence will not be considered.

Assumptions 3 and 11 will be easily removed by a simple modification by the addition of simple circuit elements to the intrinsic equivalent circuit. Assumption 5 is justified because the emitter junction is forward biased normally for amplifier application. The a-c voltage across a forward biased junction must be significantly less than kT/q for linear amplification. Thus the a-c voltage amplitude across the forward biased emitter junction is indeed small and hence the variation of the transition region, due to the time varying emitter voltage, is entirely negligible. In contrast the collector transition region width can change significantly because the reverse biased collector junction can sustain several Volts compared with a few milliVolts across the emitter junction.

The total solution $p_t(x, t)$ including the d-c, the a-c unperturbed, and the a-c perturbed solutions may be expressed as follows

$$p_t(x, t) = p_0(x, t) + \tilde{p}(x, t), \tag{6–100}$$

where $p_0(x, t)$ = solution to the d-c and the a-c unperturbed case,

$\tilde{p}(x, t)$ = perturbed solution.

Note that the perturbed solution exists only when the perturbing influence, the time varying collector edge of the base region, is nonzero.

If the origin of the coordinate system is chosen to be the middle of the average base width as was done in Chapter 4, the following boundary conditions must be satisfied,

$$p_t(-w_0/2, t) = p_E + p_e\, e^{j\omega t} \tag{6–101}$$

and

$$p_t(w_0/2 + w_1\, e^{j\omega t}, t) = p_C + p_c\, e^{j\omega t}. \tag{6–102}$$

Note that in Eq. (6–102) the collector edge of the base width is located not at

$w_0/2$ but instead at $(w_0/2 + w_1\, e^{j\omega t})$. This is what makes the boundary condition different from standard. Proper cognizance has been taken of the perturbing influence. We have already noted that the base width is a function of time. The base width, however, is also a function of the total collector voltage v_C. We first express the total collector voltage as follows:

$$v_C = V_C + V_c\, e^{j\omega t}. \tag{6–103}$$

Now $w(v_C)$ is a well behaved function, and hence may be expanded in Taylor series about the point V_C as follows:

$$w(v_C) = w_0 + \left.\frac{\partial w}{\partial v_C}\right|_{V_C} (v_C - V_C) + \text{higher order terms.} \tag{6–104}$$

According to assumption 7 we are after the first order theory, *the higher order terms may be neglected.* Equation (6–103) permits us to write

$$v_C - V_C = V_c\, e^{j\omega\tau} \tag{6–105}$$

Furthermore, Eq. (6–96) permits us to write the following

$$w(t) - w_0 = w_1\, e^{j\omega t}. \tag{6–106}$$

We may also recognize the following

$$\left.\frac{\partial w}{\partial v_C}\right|_{V_C} = \frac{\partial w_0}{\partial V_C}. \tag{6–107}$$

With the use of Eqs. (6–104), (6–105) and (6–106), Eq. (6–104) may be written in the following form after the higher order terms have been dropped because of assumption number 7.

$$w_1\, e^{j\omega t} = \frac{\partial w_0}{\partial V_C} V_c\, e^{j\omega t}$$

or

$$w_1 = \frac{\partial w_0}{\partial V_C} V_c. \tag{6–108}$$

Thus, the amplitude of the base width variation is proportional to the amplitude of the collector voltage with the proportionality constant depending on the rate of change of the average base width with d-c collector to base voltage.

It can be shown that for a step junction transistor where the collector doping density is much higher than that of the base, thus most of the collector transition region is on the base side (see Problem 6–25)

$$\frac{\partial w_0}{\partial V_C} = \left(\frac{\varepsilon\mu_n}{2\sigma_B}\right)^{\frac{1}{2}} (v_t)^{-\frac{1}{2}}. \tag{6–109}$$

In a linearly graded junction it can be shown that

$$\frac{\partial w_0}{\partial V_C} = \frac{2\varepsilon}{aq}\left(\frac{aq}{12\varepsilon}\right)^{\frac{2}{3}} v_t^{-\frac{2}{3}}, \tag{6–110}$$

where ε = permittivity,

a = charge density gradient in the transition region.

Next we shall obtain the boundary conditions pertaining to the perturbed solution $\tilde{p}(x, t)$. Because the standard boundary conditions have already been incorporated into the unperturbed solution $p_0(x, t)$, the perturbed solution needs to account for the boundary conditions due only to the perturbation alone. Keeping the above observations in mind we note that because the emitter edge of the base region is considered fixed according to assumption 5, the perturbation must vanish at the emitter edge of the base region. Thus,

$$\tilde{p}(-w_0/2, t) = 0. \tag{6–111}$$

To evaluate the boundary condition for the perturbation term we shall need to expand the total solution in Taylor series about the point $x = w_0/2$. Thus,

$$p_t(x, t) = p_t(w_0/2, t) + \left.\frac{\partial p_t}{\partial x}\right|_{x=w_0/2} (x - w_0/2) + \text{higher order terms.} \tag{6–112}$$

Particularizing x to

$$x = \frac{w_0}{2} + w_1\, e^{j\omega t} \tag{6–113}$$

$$p_t(w_0/2 + w_1\, e^{j\omega t}, t) = p_t\left(\frac{w_0}{2}, t\right) + \left.\frac{\partial p_t}{\partial x}\right|_{x=w_0/2} (w_0/2 + w_1\, e^{j\omega t} - w_0/2) + \\ + \text{higher order terms.} \tag{6–114}$$

Some thought and a fair amount of analysis indicates that from Eq. (6–114) we may write

$$p_t(w_0/2 + w_1\, e^{j\omega\tau}, t) - p_t(w_0/2, t) = p(w_0/2, t) = \left.\frac{\partial p_t}{\partial x}\right|_{x=w_0/2} w_1\, e^{j\omega\tau} + \\ + \text{higher order terms.} \tag{6–115}$$

Now the higher order terms may be neglected. Because of assumptions 6, 7 and 8 we may rewrite Eq. (6–115) as follows

$$\tilde{p}(w_0/2, t) = \left.\frac{\partial p_t}{\partial x}\right|_{x=w_0 2} w_1\, e^{j\omega t} = -\left.\frac{\partial p_0(x)}{dx}\right|_{x=w_0/2} w_1\, e^{j\omega t}. \tag{6–116}$$

Where the following magnitude relationship is used

$$\frac{\partial p_0(x)}{\partial x} \gg \frac{\partial \tilde{p}}{\partial x}, \tag{6–117}$$

where $p_0(x)$ = d-c solution of Chapter 4.

Equation (4–10) may be substituted on the right hand side of Eq. (6–116) to obtain

$$\tilde{p}(w_0/2, t) = -\left[-(p_E - p_n)\operatorname{csch}(w_0/L_p) + (p_C - p_n)\coth(w_0/L_p)\right] \times \frac{w_1}{L_p} e^{j\omega t}. \tag{6–118}$$

Equations (6–111) and (6–118) constitute the boundary conditions for the perturbed solution. Note that we have evaluated the perturbed solution at the collector edge of the base region at $w_0/2$ rather than $(w_0/2 + w_1 e^{j\omega t})$; this is because we are retaining only the first order effects, and the addition of $w_1 e^{j\omega t}$ will lead to taking second order effects into account which we are not considering here.

6–6 SOLUTIONS TO THE PERTURBED CASE

Armed with the boundary conditions for $\tilde{p}(x, t)$ we may proceed to obtain the solution. Before we do this it will be worthwhile to outline the method of attack as well as to clarify the objective of this development. The following steps will be followed.

1. Obtain the perturbed solution $\tilde{p}(x, t)$ with the help of the knowledge that it must satisfy the time dependent diffusion equation subject to its boundary conditions Eqs. (6–111) and (6–118).
2. The perturbed solution $\tilde{p}(x, t)$ will be added to the unperturbed solution $p_0(x, t)$ already obtained to get the total minority carrier density in the base region as a function of x and t.
3. Minority carrier current density is obtained in the base region as a function of x, and t by the use of the diffusion equation subject to the condition that the electrostatic field is zero.
4. The minority carrier current density crossing the emitter and collector junctions is obtained from step 3 by specializing x to $-w_0/2$ for the emitter junction and to x to $+w_0/2$ for the collector junction.
5. Electron current crossing the emitter junction will next be evaluated by taking the following limit

$$i_{en} = \lim_{\omega \to \infty} (i_{ep}) \tag{6–119}$$

and by subsequently replacing all quantities pertaining to the transport of holes to the corresponding quantities pertaining to the transport of electrons in the emitter regions.

6. The addition of i_{en} to i_{ep} obtains the total emitter current i_e.
7. The total collector current i_c is obtained by the use of the following equation

$$i_c = \left(\frac{i_{cp} + i_{cn}}{i_{cp}} \right) i_{cp} = \alpha^* i_{cp}. \tag{6–120}$$

8. The resulting equations for the total emitter and collector currents will be functions of emitter and collector voltages. Thus, these equations are appropriate to obtain directly the *common base intrinsic* parameters. These are common base parameters because the voltages are collector-to-base and emitter-to-base voltages. The parameters are intrinsic because they are the direct result of the solution of the diffusion equation. The extrinsic parameters such as junction capacitances and bulk resistances have been left out.
9. The wide band intrinsic admittance parameters obtained in item 8 will turn out to be hyperbolic functions of irrational argument. Hence, the driving point admittances such as Y^i_{ee} or Y^i_{cc} cannot be synthesized by LLFPB networks just as was the case of the intrinsic admittance of the p-n junction. The intrinsic admittances will be expanded in Taylor series or infinite product expansion and *only a finite* number of terms will be retained. These trunketed expansions will enable us to synthesize the admittances in LLFPB networks for Y^i_{ee} and Y^i_{cc}.
10. The retention of a finite number of terms will put a limit on the frequency range over which the equivalent circuit is valid.
11. The *extrinsic* parameters will next be added to the *intrinsic* parameters to obtain the complete common base equivalent circuit.
    To obtain the common emitter equivalent circuit the following additional steps will be necessary.
12. The *complete common base* circuit should now be transformed to the common emitter configuration.

It is important to note that the *complete* common base circuit including the extrinsic elements should be transformed to the common emitter equivalent circuit. An alternate procedure *which is incorrect* that can be used is the following. First the common base intrinsic equivalent circuit is obtained. This circuit is transformed to the common emitter equivalent circuit. Then the extrinsic elements are added. This procedure is incorrect because the extrinsic elements which are added at the end *will not be influenced by the transformation laws* which transform the common base to common emitter equivalent circuits. The correct procedure will permit the transformation laws influencing *both intrinsic and extrinsic elements.*

We proceed to implement the above procedure. Because $\tilde{p}(x, t)$ is also an a-c solution and must satisfy the diffusion equation, it should have the same form as obtained for the unperturbed a-c solution

$$\tilde{p}(x, t) = [G \sinh C_p(w_0/2 - x) + H \sinh C_p(w_0/2 + x)]\, e^{j\omega t}. \quad (6\text{–}121)$$

The coefficients G and H now will have a different value than the coefficients found in the a-c unperturbed solutions because now the boundary conditions are different.

The coefficient may now be obtained readily as follows. For

$$x = -w_0/2,$$

$$\tilde{p}(-w_0/2, t) = 0 = e^{j\omega t} G \sinh(C_p w_0) + 0 \quad (6\text{–}122)$$

$$G = 0. \quad (6\text{–}123)$$

For

$$x = +w_0/2$$

$$\tilde{p}(w_0/2, t) = 0 + e^{j\omega t} H \sinh(C_p w_0) \quad (6\text{–}124)$$

Equation (6–118) may be substituted into Eq. (6–124) to evaluate H as follows

$$H = -\frac{1}{L_p \sinh(C_p w_0)}[-(p_E - p_n)\operatorname{csch}(w_0/L_p) + (p_C - p_n)\coth(w_0/L_p)]\, w_1. \quad (6\text{–}125)$$

The perturbed solution may hence be written

$$\tilde{p}(x, t) = -\frac{e^{j\omega t} w_1}{L_p}[-(p_E - p_n)\operatorname{csch}(w_0/L_p) + (p_C - p_n)\coth(w_0/L_p)]\frac{\sinh[C_p(w_0/2 + x)]}{\sinh(C_p w_0)}. \quad (6\text{–}126)$$

We can now calculate the hole current density due to the *perturbed* minority carrier density by the direct use of the transport equation to be used in the field-free base region as follows

$$j_p(x, t) = -qD_p \frac{\partial \tilde{p}(x, t)}{\partial x} = +\frac{qD_p w_1 e^{j\omega t}}{L_p}[-(p_E - p_n)\operatorname{csch}(w_0/L_p) + (p_C - p_n) \times \coth(w_0/L_p)]\left\{C_p \frac{\cosh C_p(w_0/2 + x)}{\sinh(C_p w_0)}\right\}. \quad (6\text{–}127)$$

The total hole current density $j_{pt}(x, t)$ may be written as the sum of three terms as given below:

$$j_{pt}(x, t) = -qD_p\left[\frac{\partial p_0(x)}{\partial x} + \frac{\partial p_1(x, t)}{\partial x} + \frac{\partial \tilde{p}(x, t)}{\partial x}\right] \quad (6\text{–}128)$$

$$j_{pt}(x, t) = j_{p0}(x) + j_{p1}(x, t) + \tilde{j}_p(x, t), \quad (6\text{–}129)$$

where $j_{p0}(x)$ = d-c minority carrier current density in the base due to d-c voltages across the emitter and collector junctions,

$j_{p1}(x, t)$ = unperturbed a-c minority carrier current density in the base due to a-c voltages across the emitter and collector junctions,

$\tilde{j}_p(x, t)$ = perturbed a-c minority carrier current density in the base due to a-c voltages across the emitter and collector junctions.

In Chapter 4 $j_{p0}(x)$ was obtained. In Part A of this chapter $j_{p1}(x, t)$ was obtained. Equation (6–127) gives $\tilde{j}_p(x, t)$. Our major interest is the a-c solutions. The *total* a-c minority carrier current density shall be designated by $j^*_{pt}(x, t)$. Thus,

$$j^*_{pt}(x, t) = j_{p1}(x, t) + \tilde{j}_p(x, t). \qquad (6\text{–}130)$$

Substituting the values of the two quantities on the right hand side of Eq. (6–130) we have

$$j^*_{pt}(x, t) = \frac{qC_pD_p}{\sinh(C_pw_0)}\left[p_e\cosh[C_p(w_0/2 - x)] - p_c\cosh[C_p(w_0/2 + x)]\right]e^{j\omega t}$$
$$-\frac{qC_pw_1D_pe^{j\omega t}}{L_p\sinh(C_pw_0)}\cdot\left[\cosh[C_p(w_0/2 + x)]\right]\left[(p_E - p_n)\operatorname{csch}(w_0/L_p)\right.$$
$$\left.-(p_C - p_n)\coth(w_0/L_p)\right]. \qquad (6\text{–}131)$$

For sinusoidally varying minority carrier current density the following may be used.

$$j^*_{pt}(x, t) = J^*_{pt}e^{j\omega t}. \qquad (6\text{–}132)$$

Furthermore,

$$v_e(t) = V_e e^{j\omega t} \qquad (6\text{–}133)$$

$$v_c(t) = V_c e^{j\omega t}. \qquad (6\text{–}134)$$

When Eqs. (6–132) through (6–134) are substituted into Eqs. (6–131) and x is set equal to $(-w_0/2)$ and $(+w_0/2)$ the minority carrier current density crossing the emitter and collector junctions respectively are obtained as follows

$$J^*_{pte} = J^*_{pt}(-w_0/2) = qC_pD_p\left\{\frac{qp_E}{kT}V_e\coth(C_pw_0) - \frac{p_Cq}{kT}V_c\operatorname{csch}(C_pw_0)\right.$$
$$\left.-\frac{w_1}{L_p}\operatorname{csch}(C_pw_0)[(p_E - p_n)\operatorname{csch}(w_0/L_p) - (p_C - p_n)\coth(w_0/L_p)]\right\}. \qquad (6\text{–}135)$$

$$J^*_{ptc} = J^*_{pt}(+w_0/2) = qC_pD_p\left\{\frac{qp_E}{kT}V_e\operatorname{csch}(C_pw_0) - \frac{qp_C}{kT}V_c\coth(C_pw_0)\right.$$
$$\left.-\frac{w_1\coth(C_pw_0)}{L_p}[(p_E - p_n)\operatorname{csch}(w_0/L_p) - (p_C - p_n)\coth(w_0/L_p)]\right\} \qquad (6\text{–}136)$$

Equations (6–135) and (6–136) are general for virtually *any* bias. As our purpose here is to obtain wide band models for the *normally biased* transistor, simplification in the above equations would result if the following approximations are used.

$$p_C \approx 0 \tag{6–137}$$

$$p_c \approx 0 \tag{6–138}$$

$$p_E \gg p_n. \tag{6–139}$$

We shall first obtain the d-c hole current density crossing the emitter and collector junctions subject to the approximations of Eq. (6–137) and (6–139) on Eqs. (4–14) and (4–15).

$$J_{pE} \doteq \frac{qD_p p_E}{L_p} \coth(w_0/L_p) \tag{6–140}$$

and

$$J_{pC} \doteq \frac{qD_p}{L_p}[(p_E - p_n)\operatorname{csch}(w_0/L_p) + p_n \coth(w_0/L_p)]. \tag{6–141}$$

Equations (6–135) and (6–136) may be rewritten with the help of Eqs. (6–137) through (6–141) and Eq. (6–108) as follows

$$J^*_{pte} = \left[\frac{qJ_{pE}}{kT} \cdot \frac{\sqrt{1 + j\omega\tau_p}\tanh(w_0/L_p)}{\tanh(C_p w_0)}\right] V_e - \left[J_{pC}\frac{\partial w_0}{\partial V_C} \cdot \frac{C_p}{\sinh(w_0 C_p)}\right] V_c \tag{6–142}$$

$$J^*_{ptc} = \left[\frac{qJ_{pE}}{kT} \cdot \frac{\sqrt{1 + j\omega\tau_p}\tanh(w_0/L_p)}{\sinh(w_0 C_p)}\right] V_e - \left[J_{pC}\frac{\partial w_0}{\partial V_C} \cdot \frac{C_p}{\tanh(C_p w_0)}\right] V_c. \tag{6–143}$$

The electron current density crossing the emitter junction may be obtained by taking the limit of J^*_{pte} when $w_0 \to \infty$ and then interchanging all hole transport related terms to the electron transport related terms. Thus,

$$J^*_{nte} = \frac{qJ_{nE}}{kT}\sqrt{1 + j\omega\tau_{en}}\, V_e. \tag{6–144}$$

The total emitter *current* may now be obtained by adding the hole current and electron currents crossing the emitter junction. Thus we obtain

$$I_{te} = \frac{q}{kT}\left[\frac{I_{pE}\sqrt{1 + j\omega\tau_p}\tanh(w_0/L_p)}{\tanh(C_p w_0)} + I_{nE}\sqrt{1 + j\omega\tau_{en}}\right] V_e - \left[J_{pC}A_e\frac{\partial w_0}{\partial V_C} \cdot \frac{C_p}{\sinh(w_0 C_p)}\right] V_c, \tag{6–145†}$$

where A_e = area of the emitter junction.

† The use of possibly different areas of the emitter and collector junctions A_e and A_c is only an approximate way of taking account of non-one-dimensionality of the problem. The more accurate way would be to solve the three-dimensional diffusion equation itself—quite a cumbersome problem.

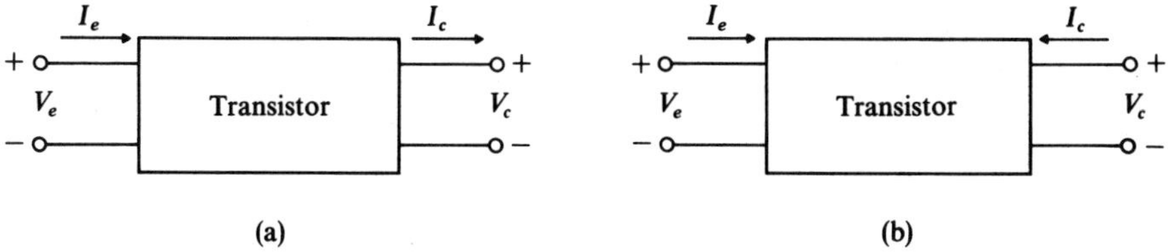

Fig. 6–1 Reference conditions on transistor currents and voltages. (a) Reference conditions for the diffusion equation solution. (b) Reference conditions appropriate to the circuit theory definitions for parameters such as the Y's, Z's, h's, etc.

To obtain the total collector current we may utilize the ideas of Eq. (6–120) as follows:

$$I_{tc} = \alpha^* I_{ptc}. \tag{6–146}$$

With the help of Eq. (6–146), Eq. (6–143) may be rewritten as follows:

$$I_{tc} = \left[\frac{qJ_{pE}A_c\alpha^*\sqrt{1 + j\omega\tau_p}\tanh(w_0/L_p)}{kT\sinh(C_p w_0)}\right]V_e - \left[I_{pC}\alpha^*\frac{\partial w_0}{\partial V_C}\cdot\frac{C_p}{\tanh(w_0 C_p)}\right]V_c, \tag{6–147†}$$

where A_c = area of the collector junction.

Equations (6–146) and (6–147) are *almost* in the appropriate form from which the intrinsic admittance parameters may be obtained except for reference direction of currents. The diffusion equation which we have solved to obtain Eqs. (6–146) and (6–147) assumes the reference direction of currents and voltages shown in Fig. 6–1a. The admittance parameters defined by circuit theorists assume the reference direction shown in Fig. 6–1b. It is clear that only the collector current reference direction of Fig. 6–1a needs to be mathematically turned around. To this purpose we shall multiply the right hand side of Eq. (6–147) by (-1) and obtain

$$I_{tc} = -\left[\frac{qJ_{pE}A_c\alpha^*\sqrt{1 + j\omega\tau_p}\tanh(w_0/L_p)}{kT\sinh(C_p w_0)}\right]V_e + \left[I_{pC}\alpha^*\frac{\partial w_0}{\partial V_C}\cdot\frac{C_p}{\tanh(w_0 C_p)}\right]V_c. \tag{6–148}$$

The defining equations for the intrinsic admittance parameters Y_s^i are

$$I_{te} = Y_{ee}^i V_e + Y_{ec}^i V_c \tag{6–149}$$

$$I_{tc} = Y_{ce}^i V_e + Y_{cc}^i V_c. \tag{6–150}$$

†See footnote on page 198.

Comparing Eq. (6–146) and Eq. (6–148) with Eqs. (6–149) and (6–150) we directly obtain the *intrinsic* admittance parameters for the *p-n-p* bipolar junction transistor

$$Y^i_{ee} = \frac{q}{kT} \cdot \frac{I_{pE}\sqrt{1 + j\omega\tau_p}\tanh(w_0/L_p)}{\tanh(C_p w_0)} + \frac{q}{kT} I_{nE}\sqrt{1 + j\omega\tau_{en}} \quad (6\text{–}151)$$

$$Y^i_{ec} = -J_{pC} A_e \frac{\partial w_0}{\partial V_C} \cdot \frac{C_p}{\sinh(C_p w_0)} \quad (6\text{–}152)$$

$$Y^i_{ce} = -\left[\frac{qJ_{pE}A_c}{kT}\left(1 + \frac{\sigma_{nc}}{2\sigma_{pc}}\right)\frac{\sqrt{1 + j\omega\tau_p}\tanh(w_0/L_p)}{\sinh(C_p w_0)}\right] \quad (6\text{–}153)$$

$$Y^i_{cc} = I_{pC}\left(1 + \frac{\sigma_{nc}}{2\sigma_{pc}}\right)\frac{\partial w_0}{\partial V_C} \cdot \frac{C_p}{\tanh(C_p w_0)}. \quad (6\text{–}154)$$

Several comments are worth making at this point. First, Eqs. (6–151) through (6–154) *are the key results* necessary to obtain wide band equivalent circuits for the bipolar diffusion transistors. Second, the frequency dependence is both explicit and implicit. The implicit frequency dependence is through the C_p and C_{en} which were defined in Eqs. (6–12) and (6–13). Third, the admittances Y^i_{ee} and Y^i_{cc}, which are the so-called *driving point admittances*, are not *positive real* functions and hence cannot be synthesized by LLFPB (linear, lumped, finite, passive, and bilateral) networks. Fourth, it will be necessary to use some kind of expansions of the expressions of the above equations for admittances and retain a finite number of terms in order to obtain an LLFPB circuit realization for them. This procedure puts a limit on the range of frequencies for which the obtained circuit is valid. Fifth, the dependence of the intrinsic admittances on the bias current and voltage is again explicitly seen through Eqs. (6–151) through (6–154). There is also an implicit dependence through the base width w_0 which we have already seen in Chapter 4 to be a function of the emitter and collector bias. Sixth, we shall need to add the extrinsic elements to the intrinsic circuits to obtain the complete model. Seventh, the intrinsic admittances given in Eqs. (6–151) through (6–154) are for a *p-n-p* transistor. The analogous set of equations for the *n-p-n* transistor are given below.

$$Y^i_{ee} = \frac{q}{kT}\left[I_{nE}\frac{\sqrt{1 + j\omega\tau_n}\tanh(w_0/L_n)}{\tanh(C_n w_0)} + I_{pE}\sqrt{1 + j\omega\tau_{ep}}\right] \quad (6\text{–}155)$$

$$Y^i_{ec} = -J_{nC} A_e \frac{\partial w_0}{\partial V_C} \cdot \frac{C_n}{\sinh(C_n w_0)} \quad (6\text{–}156)$$

$$Y^i_{ce} = -\frac{qJ_{nE}A_c\alpha^*}{kT} \cdot \frac{\sqrt{1 + j\omega\tau_n}\tanh(w_0/L_n)}{\sinh(C_n w_0)} \quad (6\text{–}157)$$

$$Y^i_{cc} = J_{nC} A_c \alpha^* \frac{\partial w_0}{\partial V_C} \cdot \frac{C_n}{\tanh(w_0 C_n)} \quad (6\text{–}158)$$

where τ_n = minority carrier lifetime in the p-type base region,

J_{nE} = d-c electron current density components crossing the emitter junction,

L_n = minority carrier (electron) diffusion length in the base region.

$$C_n \triangleq \frac{\sqrt{1 + j\omega\tau_n}}{L_n}, \tag{6–159}$$

I_{nE} = d-c electron current component crossing the emitter junction,

I_{pE} = d-c hole current component crossing the emitter junction.

6–7 WIDE BAND MODELS AND EQUIVALENT CIRCUITS

In the preceding section we obtained the analytic expression for the *intrinsic* admittance parameters *convenient* for the common base configuration. This is not a complete model for the transistor as the extrinsic elements are yet to be added. We define the intrinsic parameters as those obtainable through the solution of the current continuity law and the diffusion equation. Parameters which cannot be so obtained shall be designated as extrinsic parameters. Included in the extrinsic elements are the bulk and leakage resistances, junction, and stray capacitances.

Figure 6–2 shows a model which has both intrinsic and extrinsic elements. A new element not yet discussed is shown in the base leg of the circuit, namely $\mu_{bc}V_c$ dependent voltage source. This source shows the influence of the perturbation (collector boundary position changing with time) on the extrinsic base resistance. The perturbation changes the effective base width. The base resistance is inversely proportional to the effective base width as a very simple one-dimensional analysis of the majority carrier flow through the base contact indicates

$$r_b' \doteq \frac{\rho_B}{w_0}. \tag{6–160}$$

Thus, the perturbation causes the extrinsic resistance to vary with time due to the a-c variation in the collector boundary of the base region. This variational term is represented by the dependent source which is evaluated below:

$$\mu_{bc}V_c = I_B \frac{\partial r_b'}{\partial V_C} V_c = I_B \frac{\partial r_b'}{\partial w_0}\frac{\partial w_0}{\partial V_C} V_c \tag{6–161}†$$

so that

$$\mu_{bc} = I_B \frac{\partial r_b'}{\partial w_0}\frac{\partial w_0}{\partial V_C}. \tag{6–162}$$

† Do not confuse r_c' here with r_c. Here r_c' is the resistance of the *bulk* collector – typically a few Ohms – whereas r_c is the leakage resistance of the collector junction typically hundreds or thousands of megaOhms. Also distinguish r_b' and r_e' from r_b and r_e.

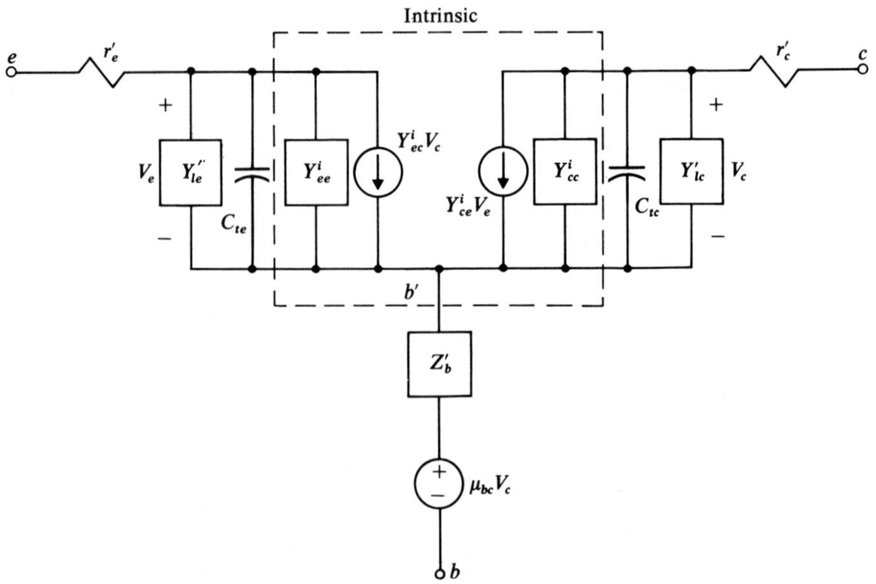

Fig. 6–2 A complete equivalent circuit including extrinsic parameters.

C_{te} = emitter transition (or junction) capacitance.
C_{tc} = collector transition (or junction) capacitance.
$Y'_{\ell e}$ = emitter-junction leakage admittance.
$Y'_{\ell c}$ = collector-junction leakage admittance.
r'_e, r'_c = extrinsic emitter and collector bulk resistances
Z'_b = extrinsic base impedance.
μ_{bc} = feedback factor.

Note that the intrinsic part of the model is enclosed in a dotted box to clearly separate the intrinsic from the extrinsic parts of the model. The resistances due to the finite conductivities of the emitter base and collector are represented by† r'_e, r'_b, and r'_c. The extrinsic base impedance Z'_b is simply the generalization of r'_b which is necessary in some transistors, e.g. linearly graded devices. The leakage paths for current flow *around* the emitter and collector junctions rather than *through* them are represented by $Y'_{\ell c}$ and $Y'_{\ell c}$ respectively. Generally these are ohmic and are frequency independent. Thus, usually they may also be alternately designated by $g'_{\ell e}$ and $g'_{\ell c}$. One major simplifying assumption of the model of Fig. 6–1 is that all the extrinsic elements C_{te}, C_{tc}, emitter and collector junction capacitances, $Y'_{\ell e}$, $Y'_{\ell c}$, and Z'_b share a common node. This may or may not be a good approximation. Probably in the very high frequency transistors it is not a good approximation. In truth, the extrinsic elements are distributed in nature. Showing them as a finite number of lumped elements is a major approximation, and showing several of them sharing a common node is a further approximation. A somewhat more accurate representation would show the C_{te} and $Y'_{\ell e}$ going from some interior point of r'_e to some interior point of Z'_b. Similarly C_{tc} and $Y'_{\ell e}$ would be better shown going from some interior point of r'_c to some interior point of Z'_b. We have not done this in an effort to keep the model simple. It may be necessary to re-examine this assumption in the case of some transistors. This point will be discussed further in Chapter 7.

A fairly detailed calculation (Middlebrook, 1957) shows that the influence of $\mu_{bc}V_c$ is negligible compared to that of r'_b (or Z'_b) in most devices. Hence, we shall ignore the presence of $\mu_{bc}V_c$ in any further considerations of the junction transistor.

6–8 SHORT-CIRCUIT CURRENT GAIN

With the help of the model of Fig. 6–2 we shall obtain an expression for the short-circuit current gain, i.e. the ratio of the current in the short-circuit (put across the output terminals) to the input current. We shall first obtain an expression for the common base circuit. The short-circuit current gain is one of the most important parameters of the transistor.

The short-circuit current gain which will be calculated first will assume the following.

1. $r'_e, r'_c, Z'_b = 0$
2. $C_{te} = 0$
3. $Y'_{\ell e} = 0$
4. $\mu_{bc} = 0$

Under the assumptions listed above the equivalent circuit which results is shown within the dotted line in Fig. 6–3 which is set up to enable us to calculate

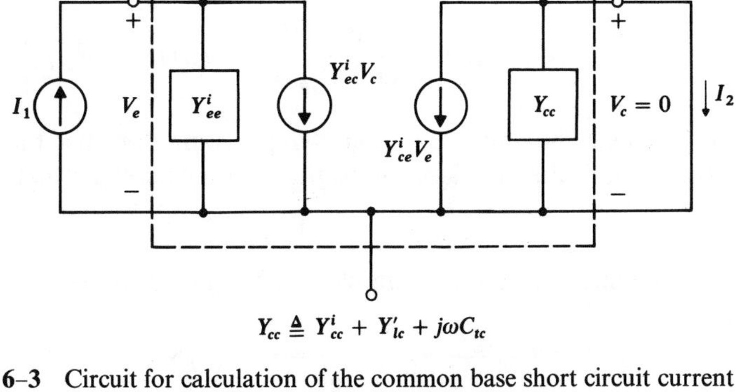

$$Y_{cc} \triangleq Y^i_{cc} + Y'_{lc} + j\omega C_{tc}$$

Fig. 6–3 Circuit for calculation of the common base short circuit current gain.

the short-circuit current gain. Note that we have utilized the *complete* Y_{cc} admittance *including the extrinsic elements* such as $Y'_{\ell c}$ and C_{tc}. Thus,

$$Y_{cc} \triangleq Y^i_{cc} + Y'_{\ell c} + j\omega C_{tc}. \tag{6–163}$$

The current in the output short-circuit is clearly the current from the output dependent current source $Y^i_{ce} V_e$. Thus

$$I_2 = -Y^i_{ce} V_e. \tag{6–164}$$

Note here that the value of I_2 is independent of the value of Y_{cc}. Hence for greater generality we included the extrinsic elements in the collector admittance Eq. (6–164) as they are seen not to influence the short-circuit current in the output circuit.

Because the output is short-circuited

$$V_c = 0. \tag{6–165}$$

Hence, the dependent current source $Y^i_{ec} V_c$ in the input circuit has a zero value and hence may be open-circuited. It is now easy to calculate V_e as follows:

$$V_e = \frac{I_1}{Y^i_{ee}}. \tag{6–166}$$

Substituting the value of V_e into Eq. (6–164) we obtain

$$\left.\frac{I_2}{I_1}\right|_{V_C=0} = \alpha = h_{21b} = -\frac{Y^i_{ce}}{Y^i_{ee}}. \tag{6–167}$$

Note that the short-circuit current gain for common base is designated as α or as h_{21b}. Recall that h_{21} is a short-circuit current gain. The additional subscript of b is to remind us that the calculation is being made in the *common base circuit* configuration. Equations (6–155) and (6–157) may be substituted into Eq. (6–167)

$$\alpha = h_{21b} = \frac{\dfrac{q}{kT}\left[I_{pE}(1 + j\omega\tau_p)^{\frac{1}{2}} \dfrac{\tanh(w_0/L_p)}{\sinh(C_p w_0)}\right]\alpha^*}{\dfrac{q}{kT}\left[I_{pE}(1 + j\omega\tau_p)^{\frac{1}{2}} \dfrac{\tanh(w_0/L_p)}{\tanh(C_p w_0)} + I_{nE}(1 + j\omega\tau_{en})^{\frac{1}{2}}\right]}. \tag{6–168}$$

Now if we make the additional simplifying assumption that the electron currents crossing both the junctions are negligible thus implying that

$$\gamma \approx 1, \alpha^* \approx 1 \tag{6–169}$$

then Eq. (6–168) takes on the rather simple and compact expression

$$\alpha = h_{21b} = \frac{1}{\cosh(C_p w_0)} = \operatorname{sech}(C_p w_0). \tag{6–170}$$

There is much published work which considers Eq. (6–170) as "exact". From the derivation given above it is quite clear that a number of assumptions need to be made before the so-called "exact expression" for the short-circuit current gain is obtained. It is *not* an exact expression at all! It is simply more accurate than the more approximate and simple expressions which we shall presently proceed to obtain.

Equation (6–52) is seen to have a single pole and no zeros. We have already seen that T_0 is the low frequency gain and that ω_{coh} is the upper cut-off frequency. A transfer function such as Eq. (6–53) is known as a *low-pass* transfer function. The term is descriptive in that it says that low frequencies are amplified and the very high frequencies are *not* amplified. At sufficiently high frequencies the magnitude of $T(j\omega)$ approaches zero. As the frequency approaches zero,

$$\lim_{\omega \to 0} [T(j\omega)] = T_0. \tag{6–171}$$

We have established that T_0 is the low frequency value of $T(j\omega)$. For

$$\omega \ll \omega_{coh} \tag{6–172}$$

$$|T(j\omega)| \doteq T_0. \tag{6–173}$$

Thus, if a single pole transfer function can be put in the form of Eq. (6–52) or (6–53), T_0 is the low frequency or also, in this case, the midband gain, and ω_{c0} is the upper cut-off frequency in radians per seconds.

Let us use the expansion of Eq. (6–77) to obtain α retaining only the first product

$$h_{21b} = \alpha \doteq \frac{1}{1 + \left(\dfrac{4}{\pi^2}\right)\left(\dfrac{w_0}{L_p}\right)^2 (1 + j\omega\tau_p)} \tag{6–174}$$

$$h_{21b} = \alpha = \frac{1\bigg/\left[1 + \left(\dfrac{4}{\pi^2}\right)\left(\dfrac{w_0}{L_p}\right)^2\right]}{1 + j\omega\tau_p\left[\dfrac{(4/\pi^2)(w_0/L_p)^2}{1 + (4/\pi^2)(w_0/L_p)^2}\right]}. \tag{6–175}$$

We shall write Eq. (6–175) in the same form as Eq. (6–52), i.e.

$$\alpha = \frac{\alpha_0}{1 + j\dfrac{\omega}{\omega_\alpha}}. \tag{6–176}$$

Here ω_α is called the *alpha cut-off frequency.*

$$\omega_\alpha = \frac{1 + (4/\pi^2)(w_0/L_p)^2}{\tau_p(4/\pi^2)(w_0/L_p)^2} = \frac{1}{\tau_p} + \frac{\pi^2}{4}\left(\frac{D_p}{w_0^2}\right). \tag{6–177}$$

Usually,

$$\frac{1}{\tau_p} \ll \frac{\pi^2}{4} \frac{D_p}{w_0^2} \tag{6–178}$$

then

$$\omega_\alpha = \frac{\pi^2}{4} \cdot \frac{D_p}{w_0^2} = 2.46 \frac{D_p}{w_0^2}. \tag{6–179}$$

A more accurate expression for ω_α by the direct use of Eq. (6–170) rather than the infinite product expansion yields the following:

$$\omega_\alpha = 2.43 \frac{D_p}{w_0^2} \qquad (p\text{-}n\text{-}p) \tag{6–180}$$

and

$$\alpha_0 = \frac{1}{1 + \left(\dfrac{1}{2.43}\right)\left(\dfrac{w_0}{L_p}\right)^2} \doteq 1 - \left(\frac{1}{2.43}\right)\left(\frac{w_0}{L_p}\right)^2 \quad (p\text{-}n\text{-}p). \tag{6–181}$$

We see that a replacement of $(\pi^2/4)$ by 2.43 gives a better approximation. The magnitude of the approximate expression for α, Eq. (6–176), agrees quite well with Eq. (6–170) over a wide range of frequencies. The phase angle of Eq. (6–176), however, is poor. For example, at

$$\omega = \omega_\alpha \tag{6–182}$$

the phase angle obtained from Eq. (6–177) is in error by about 12 degrees. Note that the right hand side of Eqs. (6–180) and (6–181) are the same as the right hand side of Eq. (6–83) and (6–81) respectively. We see that the short-circuit current gain for the common base circuit turns out to be the same as the base transport factor β^*. From the derivation here though we see in detail under what circumstances the short circuit common base current gain may be approximated by the base transport factor.

The short-circuit current gain in the common base and the common emitter circuit are related to each other by Eq. (6–88). Thus, the use of Eqs. (6–89) and (6–170) give

$$h_{21e} = \beta = \frac{\alpha}{1-\alpha} = \frac{1/\cosh(C_p w_0)}{1 - \dfrac{1}{\cosh(C_p w_0)}} = \frac{1}{\cosh(C_p w_0) - 1}. \tag{6–183}$$

The use of the infinite product expansion and retaining only one term of Eq. (6–77) yields the following:

$$h_{21e} = h_{fe} = \beta = \frac{\dfrac{\pi^2}{4}\left(\dfrac{L_p}{w_0}\right)^2}{1 + j\omega\tau_p}. \tag{6–184}$$

Equation (6–184) may be put in the form

$$h_{21e} = \beta = \frac{\beta_0}{1 + j\omega/\omega_\beta}. \tag{6–185}$$

If we again use the numerical transformation

$$\frac{\pi^2}{4} \rightarrow 2.43 \tag{6–186}$$

for greater accuracy, the low frequency β beta or h_{21e} is given by

$$\beta_0 = h_{FE} = 2.43\left(\frac{L_p}{w_0}\right)^2 \qquad (p\text{-}n\text{-}p) \tag{6–187}$$

and the upper cut-off frequency for β and h_{fe} is

$$\omega_\beta = \frac{1}{\tau_p} \qquad (p\text{-}n\text{-}p). \tag{6–188}$$

The expressions for *n-p-n* transistors for the quantities corresponding to Eqs. (6–180), (6–181), (6–187), and (6–188) are

$$\omega_\alpha = 2.43\frac{D_n}{w_0^2} \qquad (n\text{-}p\text{-}n) \tag{6–189}$$

$$\alpha_0 \doteq 1 - \left(\frac{1}{2.43}\right)\left(\frac{w_0}{L_n}\right)^2 \qquad (n\text{-}p\text{-}n) \tag{6–190}$$

$$\beta_0 = 2.43\left(\frac{L_n}{w_0}\right)^2 \qquad (n\text{-}p\text{-}n) \tag{6–191}$$

$$\omega_\beta = \frac{1}{\tau_n} \qquad (n\text{-}p\text{-}n). \tag{6–192}$$

We see by comparing Eqs. (6–181), (6–180), (6–187), and (6–188) with Eqs. (6–86), (6–87), (6–90), and (6–91) respectively, that the perturbation theory and the unperturbed calculations give the same results as far as the intrinsic short-circuit current gains are concerned. It was already noted that γ and α^* and β^* are not expected to be affected by the perturbation considerations. In these cases the calculations of Part A should agree with those of Part B. We have just now confirmed this fact by doing the actual calculations.

The common base and common emitter short-circuit current gain parameters are *not* independent of each other. It is useful to obtain the dependence of one set of parameters on the other set. Thus, Eq. (6–181) may be rewritten in terms of Eq. (6–187).

$$\alpha_0 = \frac{1}{1 + \left(\frac{1}{2.43}\right)\left(\frac{w_0}{L_p}\right)^2} = \frac{1}{1 + \frac{1}{\beta_0}} = \frac{\beta_0}{\beta_0 + 1} \tag{6–193}$$

or

$$\beta_0 = \frac{\alpha_0}{1 - \alpha_0}. \tag{6–194}$$

Equation (6–181) may be rewritten as follows:

$$\omega_\alpha = 2.43 \frac{D_p}{w_0^2} = 2.43 \frac{D_p \tau_p}{w_0^2 \tau_p} = \left[2.43 \left(\frac{L_p}{w_0}\right)^2\right] \left(\frac{1}{\tau_p}\right), \tag{6–195}$$

where use has been made of

$$L_p^2 \triangleq D_p \tau_p. \tag{6–196}$$

Equations (6–187) and (6–188) allow us to write Eq. (6–195) as follows:

$$\omega_\alpha = \beta_0 \omega_\beta. \tag{6–197}$$

The cut-off frequencies of Eqs. (6–180) and (6–188) are called the *intrinsic cut-off frequencies* because their relationship to the transistor fundamental parameters involve only the *intrinsic* parameters. When extrinsic parameters are important the short circuit cut-off frequency may depend importantly on them as we shall see in Section 6–13.

6–9 COMMON BASE WIDE BAND EQUIVALENT CIRCUIT

We seek a practical wide band equivalent circuit for the common base configuration. This will be obtained under the following conditions for *p-n-p* transistors.

1. The electron component of current crossing each junction shall be neglected.
2. The transistor is *normally* biased.
3. The extrinsic parameters shall be taken account of as shown in Fig. 6–2.
4. Collector multiplication factor is frequency independent. In Part A we did see that at low current density as well as at very high current densities α^* is truly frequency independent.

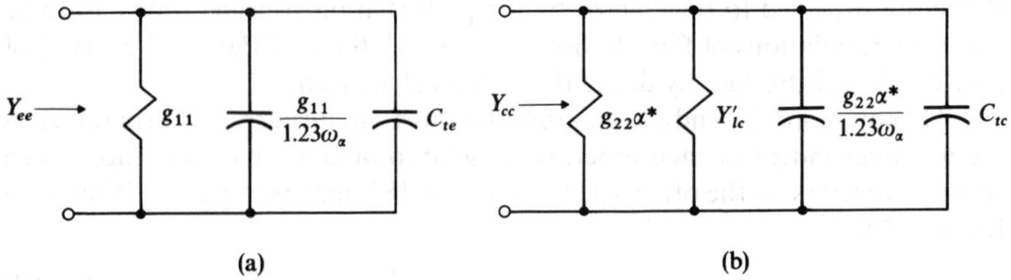

Fig. 6–4 Synthesized circuit representations of Y_{ee} and Y_{cc}. (a) Lumped circuit representation of Y_{ee}. (b) Lumped circuit representation of Y_{cc}.

We shall find it convenient to define the following two conductances.

$$g_{ee} \triangleq qI_{pE}/kT \tag{6–198}$$

$$g_{cc} \triangleq \frac{\partial w_0}{\partial V_C} \cdot \frac{I_{pC}}{L_p \tanh(w_0/L_p)}. \tag{6–199}$$

With the help of the above conductances and the four conditions stated at the beginning of this section the common base admittances assume the following form:

$$Y_{ee} = \frac{g_{ee}(1 + j\omega\tau_p)^{\frac{1}{2}} \tanh(w_0/L_p)}{\tanh(C_p w_0)} + j\omega C_{te} \tag{6–200}$$

$$Y_{ec}^i = -\frac{g_{cc}(A_e/A_c)(1 + j\omega\tau_p)^{\frac{1}{2}} \tanh(w_0/L_p)}{\sinh(C_p w_0)} \tag{6–201}$$

$$Y_{ce}^i = -\frac{g_{ee}\alpha^*(A_c/A_e)(1 + j\omega\tau_p)^{\frac{1}{2}} \tanh(w_0/L_p)}{\sinh(C_p w_0)} \tag{6–202}$$

$$Y_{cc} = \frac{g_{cc}\alpha^*(1 + j\omega\tau_p)^{\frac{1}{2}} \tanh(w_0/L_p)}{\tanh(C_p w_0)} + j\omega C_{tc} + Y'_{\ell c}. \tag{6–203}$$

It may be noted here that because of condition 2, $Y'_{\ell e}$ is neglected in writing Y_{ee}. Note that Y_{ee} and Y_{cc} include both intrinsic as well as extrinsic parameters.

When the hyperbolic functions are expanded as in Eqs. (6–64) through (6–69) and only the terms up to and including the second power in x are retained, the admittances of Eqs. (6–200) through (6–203) take relatively simple forms. First Y_{ee} will be obtained to illustrate the method; and then all the admittances will be written in their usable form. In this case,

$$Y_{ee} = g_{ee}(1 + j\omega\tau_p)^{\frac{1}{2}}(w_0/L_p)\left[\frac{(L_p/w_0)}{(1 + j\omega\tau_p)^{\frac{1}{2}}} + \frac{w_0}{3L_p}(1 + j\omega\tau_p)^{\frac{1}{2}}\right] + j\omega C_{te}. \tag{6–204}$$

The term in the square brackets is the two-term expansion of

$$\coth[(w_0/L_p)(1 + j\omega\tau_p)^{\frac{1}{2}}].$$

Equation (6–204) now simplifies to

$$Y_{ee} \doteq g_{ee}\left[1 + \frac{j\omega\tau_p w_0^2}{3L_p^2}\right] + j\omega C_{te}. \tag{6–205}$$

With the help of Eq. (6–180), we may rewrite Eq. (6–205) as follows:

$$Y_{ee} = g_{ee} + \frac{g_{ee} j\omega}{1.23\omega_\alpha} + j\omega C_{te}. \tag{6–206}$$

The emitter admittance is seen to be the parallel combination of a conductance g_{ee}, the capacitance $g_{ee}/(1.23\omega_\alpha)$, and the capacitance C_{te}. Figure 6–4a shows the network of Y_{ee} synthesized from Eq. (6–206). Figure 6–4b shows a similar network synthesized for Y_{cc} which is obtained from the following equation for Y_{cc}.

$$Y_{cc} = g_{cc}\alpha^* + \frac{g_{cc}\alpha^* j\omega}{1.23\omega_\alpha} + j\omega C_{tc} + Y'_{\ell c}. \tag{6–207}$$

Comparing Eq. (6–200) with Eq. (6–203) the only difference is that $g_{cc}\alpha^*$ replaces g_{ee}, and C_{tc} replaces C_{te}, and now $Y'_{\ell c}$ is present because the collector junction is reverse biased and the leakage admittance cannot be neglected. The leakage admittance of the emitter junction was neglected because the junction was *forward biased*.

The remaining admittances are obtained by the above method. The results are given below:

$$Y^i_{ec} = -\frac{g_{cc}(A_e/A_c)[1 - \frac{1}{3}(w_0/L_p)^2]}{1 + j\omega/(2.45\omega_\alpha)} \tag{6–208}$$

$$Y^i_{ce} = -\frac{g_{ee}\alpha^*(A_c/A_e)[1 - (\frac{1}{3})(w_0/L_p)^2]}{1 + j\omega/(2.45\omega_\alpha)}. \tag{6–209}$$

With the above results obtained, we can finally obtain the essentially complete common base equivalent circuit as shown in Fig. 6–5.

Some observations on the above results are worthwhile. First, the capacitances

$$C_{de} = \frac{g_{ee}}{1.23\omega_\alpha} \tag{6–210}$$

$$C_{dc} = \frac{g_{cc}\alpha^*}{1.23\omega_\alpha} \tag{6–211}$$

in the emitter and collector circuits are called *diffusion capacitances* because they come from the solution of the diffusion equation. Second, these diffusion capacitances are directly proportional to the d-c minority carrier currents crossing the junction in question. Thus, C_{de} is proportional to I_{Ep} and C_{dc} is proportional to I_{Cp}. Equations (6–198) and (6–199) were used to draw these conclusions. Conductances g_{ee} and g_{cc} are called diffusion conductances. Third, each of the *diffusion* R-C circuits has a time constant related simply to ω_α as follows:

$$T_{RC} = \left(\frac{1}{g_{11}}\right)\left(\frac{g_{11}}{1.23\omega_\alpha}\right) = \left(\frac{1}{g_{22}\alpha^*}\right)\left(\frac{g_{22}\alpha^*}{1.23\omega_\alpha}\right) = \frac{1}{1.23\omega_\alpha}. \tag{6–212}$$

Fourth, because the original Y^i_{ee} and Y^i_{cc} were not *positive real PR* functions, the truncation of the series expansion of the hyperbolic functions was necessary so that LLFPB networks could represent these functions. In the process, however, an upper bound is introduced to the frequency for which the equivalent

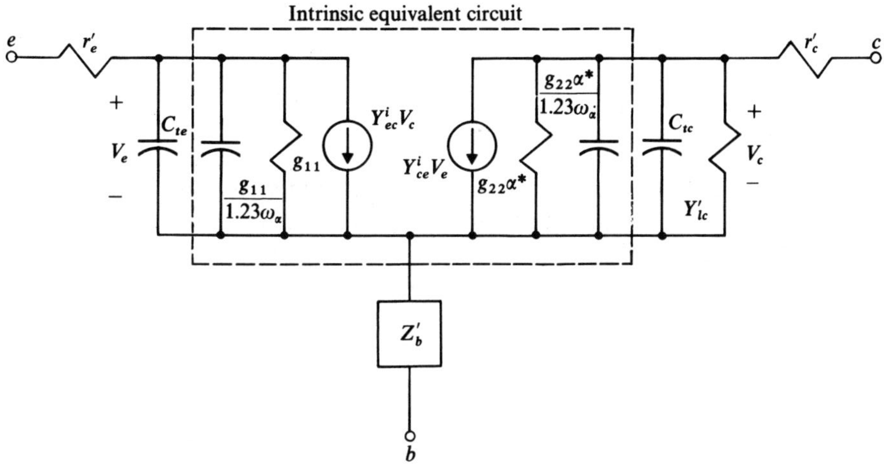

Fig. 6–5 The "complete" equivalent circuit for the common base circuit.

circuit is valid. If a wider frequency range is desired over which the equivalent circuit is to be valid, more terms may be utilized in the expansions for hyperbolic functions.

6–10 COMMON-EMITTER WIDE BAND EQUIVALENT CIRCUIT

Because the common-emitter circuit gives the maximum power gain (current and voltage gain significantly greater than one) it is the most widely used configuration. It is therefore worthwhile to obtain a wide band equivalent circuit for this circuit configuration. Because h-parameters of the transistor are easiest to measure, we shall hence attempt to obtain the h-equivalent circuit for the common emitter configuration.

Our starting point is the common base Y-parameters. We shall now make the further simplifying assumption that the electron current components crossing both junctions are zero. Alternately,

$$\gamma \approx 0 \tag{6–213}$$

$$\alpha^* \approx 0. \tag{6–214}$$

The admittance parameters of Eqs. (6–151) through (6–154) take the following simpler form for p-n-p transistors.

$$Y^i_{ee} \doteq \frac{g_{ee}(1 + j\omega\tau_p)^{\frac{1}{2}} \tanh (w_0/L_p)}{\tanh (C_p w_0)} \tag{6–215}$$

$$Y^i_{ec} \doteq -\frac{g_{cc}(1 + j\omega\tau_p)^{\frac{1}{2}} \tanh (w_0/L_p)}{\sinh (C_p w_0)} \tag{6–216}$$

$$Y^i_{ce} \doteq -\frac{g_{ee}(1 + j\omega\tau_p)^{\frac{1}{2}} \tanh(w_0/L_p)}{\sinh(C_p w_0)} \tag{6–217}$$

$$Y^i_{cc} \doteq \frac{g_{cc}(1 + j\omega\tau_p)^{\frac{1}{2}} \tanh(w_0/L_p)}{\tanh(C_p w_0)}. \tag{6–218}$$

Our strategy is to express the h-parameters as a function of the Y-parameters. In many transistors further approximations are valid for transistor operation in the active region,

$$r'_c \approx 0 \tag{6–219}$$

$$r'_e \approx 0 \tag{6–220}$$

$$C_{te} \ll \frac{g_{11}}{1.23\omega_\alpha} \tag{6–221}$$

$$C_{tc} \gg \frac{g_{22}\alpha^*}{1.23\omega_\alpha}. \tag{6–222}$$

The approximations of (6–219) through (6–222) allow us to write

$$Y_{ee} = Y^i_{ee}. \tag{6–223}$$

With the above approximations taken account of we may draw the common base equivalent circuit convenient for the *common emitter configuration* as shown in Fig. 6–6. Note that the extrinsic base impedance as well as collector

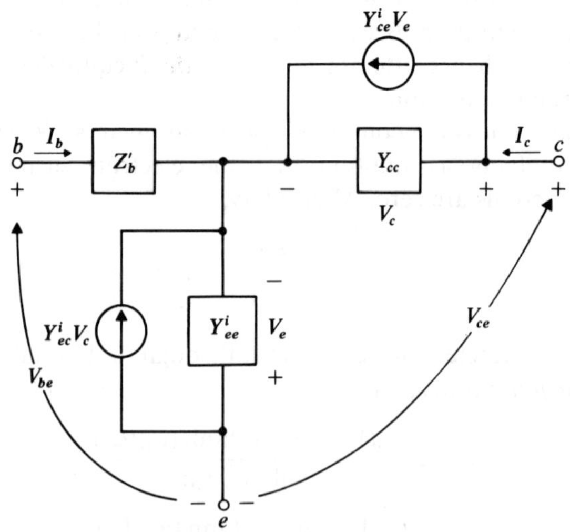

Fig. 6–6 Y-Equivalent circuit redrawn to allow us to calculate the h-parameters.

extrinsic parameters have been retained. The h-parameters for the common emitter circuit are defined as follows:

$$V_{be} = h_{11e}I_b + h_{12e}V_{ce} \tag{6–224}$$

$$I_c = h_{21e}I_b = h_{22e}V_{ce}. \tag{6–225}$$

The h-parameters may be obtained in operational terms as follows:

$$h_{11e} = \left.\frac{V_{be}}{I_b}\right|_{V_{ce}=0} \tag{6–226}$$

$$h_{12e} = \left.\frac{V_{be}}{V_{ce}}\right|_{I_b=0} \tag{6–227}$$

$$h_{21e} = \left.\frac{I_c}{I_b}\right|_{V_{ce}=0} \tag{6–228}$$

$$h_{22e} = \left.\frac{I_c}{V_{ce}}\right|_{I_b=0}. \tag{6–229}$$

Equations (6–226) through (6–229) applied to the circuit of Fig. 6–6 will allow us to calculate the h-parameters in terms of the common base Y-parameters. Note that the important extrinsic parameters have been included in the common base Y-parameters before the common emitter equivalent circuit is obtained, as indeed we should do. An incorrect procedure would be to take the intrinsic common base Y-circuit, transform it to the common emitter h-circuit, and then add the extrinsic parameters.

We shall illustrate the procedure in detail by calculating h_{11e}. The result of precisely similar calculations on other h-parameters will then be stated. We shall use Eq. (6–226) which along with Eqs. (6–227) through (6–229) are *ratios of response to excitation* under specified conditions. Thus, to use Eq. (6–226) we must short circuit the collector and emitter terminals to each other, then drive the input terminals with a current source I_b and calculate the response which is the voltage V_{be}. Figure 6–7 shows the circuit ready to calculate h_{11e}. We start by writing Kirchhoff's voltage law (KVL) on the circuit of Fig. 6–7 as follows,

$$V_{be} = I_b Z_b' - V_c. \tag{6–230}$$

The use of Kirchhoff's current law (KCL) on the circuit of Fig. 6–7 leads to

$$I_b = -V_c(Y_{ee}^i + Y_{cc}) - (Y_{ce}^i + Y_{ec}^i)V_c. \tag{6–231}$$

Equation (6–231) allows us to solve for V_c. Thus,

$$V_c = -\frac{I_b}{Y_{ee}^i + Y_{cc} + Y_{ec}^i + Y_{ce}^i} = -\frac{I_b}{\Sigma Y}, \tag{6–232}$$

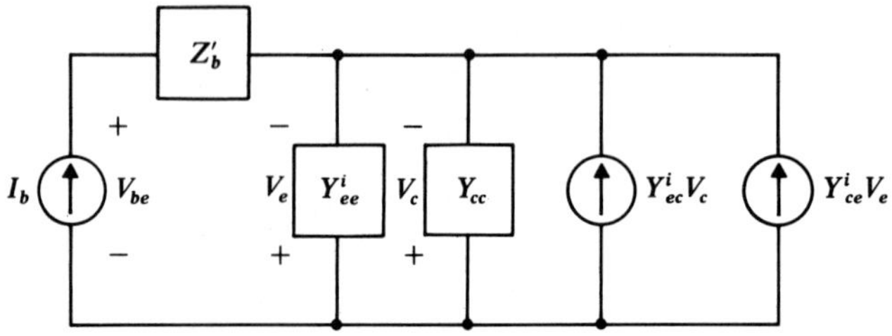

Fig. 6–7 Circuit for calculating h_{11e}.

where for compactness we have used

$$\Sigma Y \triangleq Y^i_{ee} + Y_{cc} + Y^i_{ec} + Y^i_{ce}. \tag{6–233}$$

Equation (6–232) may be substituted into Eq. (6–230) to obtain

$$V_{be} = I_b Z_b' + \frac{I_b}{\Sigma Y}. \tag{6–234}$$

Hence, finally h_{11e} may be calculated as follows:

$$h_{11e} = \left.\frac{V_{be}}{I_b}\right|_{V_{ce}=0} = Z_b' + \frac{1}{\Sigma Y}. \tag{6–235}$$

Quite similarly the other h-parameters are determined to be

$$h_{12e} = \frac{Y^i_{ec} + Y_{cc}}{\Sigma Y}, \tag{6–236}$$

$$h_{21e} = \frac{-(Y^i_{ce} + Y_{cc})}{\Sigma Y}, \tag{6–237}$$

$$h_{22e} = \frac{Y^i_{ee}Y_{cc} - Y^i_{ec}Y^i_{ce}}{\Sigma Y}. \tag{6–238}$$

When the Ys are substituted into Eqs. (6–235) through (6–238) we obtain

$$h_{11e} = Z_b' + \frac{\sinh(C_p w_0)}{(1 + j\omega\tau_p)^{\frac{1}{2}}(g_{ee} + g_{cc})[\tanh(w_0/L_p)][\cosh(C_p w_0) - 1] + (j\omega C_{tc} + Y_{\ell c}')\sinh(C_p w_0)} \tag{6–239}$$

$$h_{12e} = \frac{g_{cc}(1 + j\omega\tau_p)^{\frac{1}{2}}[\tanh(w_0/L_p)][\cosh(C_p w_0) - 1] + (j\omega C_{tc} + Y_{\ell c}')\sinh(C_p w_0)}{(g_{ee} + g_{cc})[\tanh(w_0/L_p)][\cosh(C_p w_0) - 1](1 + j\omega\tau_p)^{\frac{1}{2}} + (j\omega C_{tc} + Y_{\ell c}')\sinh(C_p w_0)}, \tag{6–240}$$

$$h_{21e} = \frac{(1 + j\omega\tau_p)^{\frac{1}{2}}\,[\tanh(w_0/L_p)]\,[g_{ee} - g_{cc}\,\mathrm{csch}\,(C_p w_0)] - (j\omega C_{tc} + Y'_{\ell c})\sinh(C_p w_0)}{(g_{ee} + g_{cc})\,[\tanh(w_0/L_p)]\,[\cosh(C_p w_0) - 1]\,(1 + j\omega\tau_p)^{\frac{1}{2}} + (j\omega C_{tc} + Y'_{\ell c}\sinh(C_p w_0)}, \tag{6–241}$$

$$h_{22e} = \frac{g_{ee}g_{cc}(1 + j\omega\tau_p)\,[\tanh^2(w_0/L_p)]\,[\sinh^2(C_p w_0)] + (j\omega C_{tc} + Y'_{\ell c})g_{ee}(1 + j\omega\tau_p)^{\frac{1}{2}}\cdot\tanh(w_0/L_p)\cosh(C_p w_0)}{(g_{ee} + g_{cc})\,[\tanh(w_0/L_p)]\,[\cosh(C_p w_0) - 1]\,(1 + j\omega\tau_p)^{\frac{1}{2}} + (j\omega C_{tc} + Y'_{\ell c})\sinh(C_p w_0)}. \tag{6–242}$$

As predicted earlier, the above expressions of the admittances involve hyperbolic functions of *irrational* arguments. As the driving point h-parameters, h_{11e}, h_{22e} are *not positive-real functions* they cannot be synthesized by LLFPB networks. In obtaining Eqs. (6–245) through (6–248) below, the following magnitude relationships were assumed to be valid which are usually satisfied in most transistors.

$$|Y'_{\ell c} + j\omega C_{tc}| \ll g_{11}, \tag{6–243}$$

$$g_{11} \gg g_{22}. \tag{6–244}$$

As the next step we obtain the Taylor series expansions of the h-parameters and retain only terms up to second order,

$$h_{11e} = Z'_b + \frac{\beta_0/g_{ee}}{1 + j\omega/\omega_\beta}, \tag{6–245}$$

$$h_{12e} = \frac{g_{cc} + g'_{\ell c}}{g_{ee}} + \frac{j\omega C_{tc}\beta_0/g_{ee}}{1 + j\omega/\omega_\beta}, \tag{6–246}$$

$$h_{21e} = \frac{\beta_0}{1 + j\omega/\omega_\beta}, \tag{6–247}$$

$$h_{22e} = 2g_{cc} + \frac{j\omega C_{tc}\beta_0}{1 + j\omega/\omega_\beta} + g'_{\ell c} + j\omega C_{tc}. \tag{6–248}$$

The frequency limitation imposed on the expressions of Eqs. (6–245) through (6–248) is that the magnitude of the argument of the frequency dependent hyperbolic terms be less than one. Thus,

$$|C_p w_0|^2 = \left(\frac{w_0}{L_p}\right)^2 \left|1 + j\omega\tau_p\right| < 1$$

or

$$\left(\frac{w_0}{L_p}\right)^2 \sqrt{1 + \omega^2\tau_p^2} < 1. \tag{6–249}$$

Usually for the upper limits on frequency

$$\omega\tau_p \gg 1. \tag{6–250}$$

Then the magnitude relationship (Eq. 6–249) becomes

$$\frac{w_0^2}{D_p\tau_p}\omega\tau_p < 1 \tag{6–251}$$

or

$$\omega < \frac{D_p}{w_0^2} \sim \omega_\alpha/2. \tag{6–252}$$

The last term on the right hand side of Eq. (6–245) represents a parallel *R-C* network and the second term on the right hand side of Eq. (6–248) represents a parallel *R-C* network as can be seen from the following:

$$Z_{R\|C}(s) = \frac{(R)\left(\frac{1}{Cs}\right)}{R + \frac{1}{Cs}} = \frac{\frac{R}{Cs}}{\frac{RCs + 1}{Cs}} = \frac{R}{RCs + 1} \tag{6–253}$$

for $s = j\omega$,

$$Z_{R\|C}(j\omega) = \frac{R}{j\omega RC + 1} \tag{6–254}$$

$$Y_{R\text{ series }C}(s) = \frac{(Cs)\left(\frac{1}{R}\right)}{\frac{1}{R} + Cs} = \frac{\frac{Cs}{R}}{\frac{RCs + 1}{R}} = \frac{Cs}{RCs + 1}. \tag{6–255}$$

For $s = j\omega$

$$Y(j\omega) = \frac{j\omega C}{j\omega C + 1}. \tag{6–256}$$

With the above observations kept in mind, the *h*-equivalent circuit of Fig. 6–8 is obtained.

Another common emitter equivalent circuit in widespread use is the Giacoletto and Johnson (1959) circuit. This circuit is based on the circuit of Fig. 6–9. Though Giacoletto and Johnson had derived this circuit for the alloy junction transistors of the 1950s, the circuit configuration seems to work well for many of the present day transistors. One of the advantages of this circuit is that the dependent source is essentially independent of frequency, unlike, for example, h_{21e} which is frequency dependent. The actual frequency dependence of the transconductance g_m is

$$g_m = \frac{qI_C/kT}{1 + j\omega/2.4\omega_\alpha}. \tag{6–257}$$

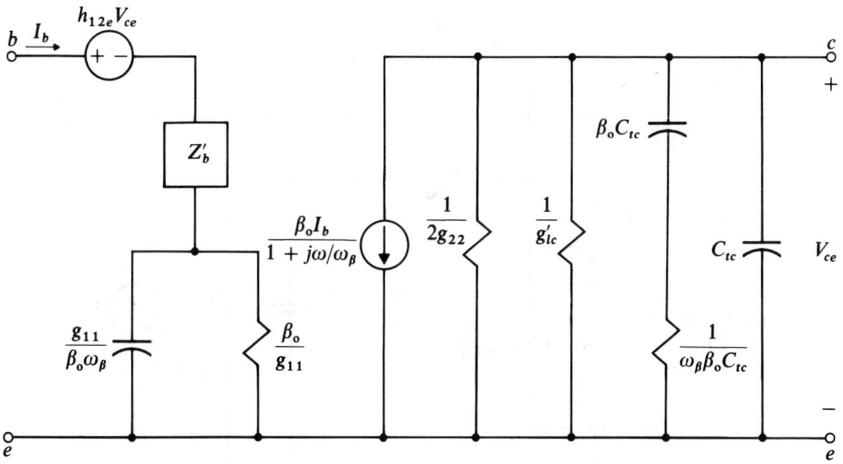

Fig. 6–8 Common-emitter wideband h-equivalent circuit.

Because the equivalent circuit of Fig. 6–10 has the identical frequency limitation as that of the h-equivalent circuit, the magnitude relationship (6–252), within the range of validity of the circuit of Fig. 6–10, g_m is given by

$$g_m \doteq qI_C/kT. \tag{6–258}$$

Many more essentially equally valid equivalent circuits for the transistor can be obtained as well as the ones already obtained. Suffice it to say that the procedure for obtaining these equivalent circuits is similar to the derivation of the circuits so far derived in this chapter. Which particular equivalent circuit out of many possible ones should be utilized by a given person depends upon the particular use intended for them and the personal inclination of the user of the equivalent circuit.

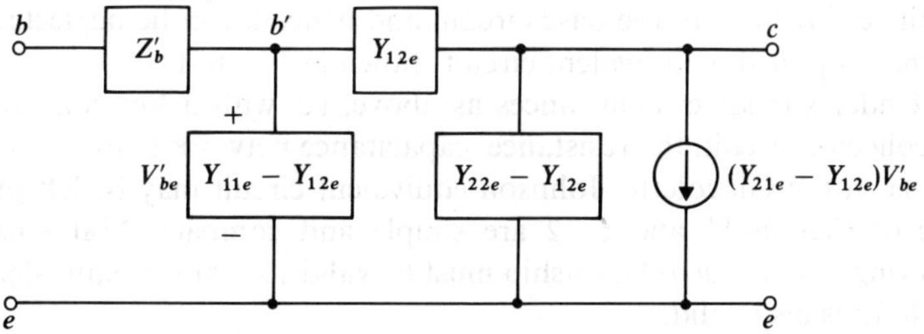

Fig. 6–9 One dependent source admittance equivalent circuit for the common emitter circuit configuration.

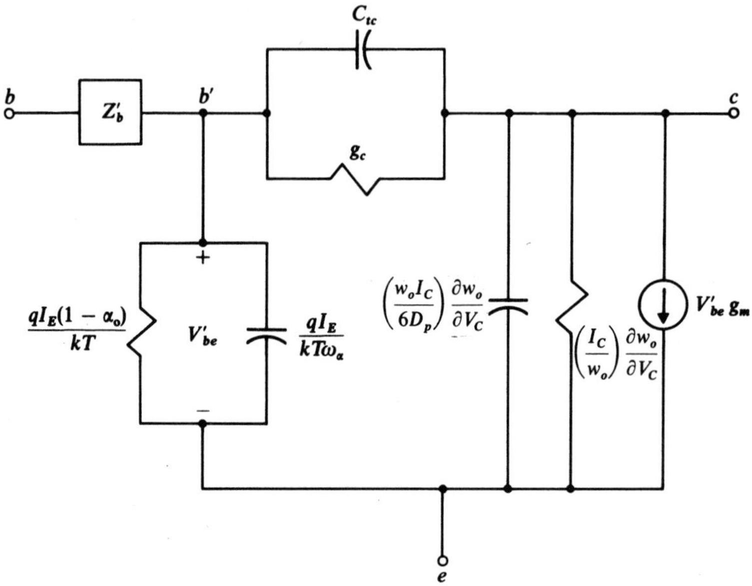

Fig. 6–10 Hybrid-π equivalent circuit of Giacoletto and Johnson.

6–11 SOME SIMPLE EQUIVALENT CIRCUITS

It turns out that under certain circumstances the equivalent circuits of the previous section may be further simplified without loss of accuracy. If the load resistance R_L in the collector circuit is a few hundred Ohms, its admittance will in general be much greater than that of h_{22e}. Thus, if

$$\frac{1}{R_L} \gg |h_{22e}| \tag{6–259}$$

both h_{22e} and h_{12e} may be neglected. Usually the magnitude of h_{12e} is quite small. When the load resistance is of a low value as indicated above, the product $I_c R_L$ will give a negligible voltage for the dependent source (that is V_{ce} when multiplied by h_{12e}) in the base circuit, and hence it can be neglected. Thus, we get the simplified *h*-equivalent circuit shown in Fig. 6–11.

Under similar circumstances as above, i.e. with a low load resistance in the collector circuit, the resistance-capacitance network between collector and emitter in the Giacoletto–Johnson equivalent circuit may be left out. The circuits of Figs. 6–11 and 6–12 are simple and compact. Mathematically the following magnitude relationship must be valid so that the equivalent circuit of Fig. 6–12 is also valid.

$$\frac{1}{R_L} \gg \left(\frac{I_C}{w_0}\right)\frac{\partial w_0}{\partial V_C} \gg g_c \triangleq g'_{\ell c} - \frac{I_C w_0 \partial w_0}{2L_p^2 \partial V_C}. \tag{6–260}$$

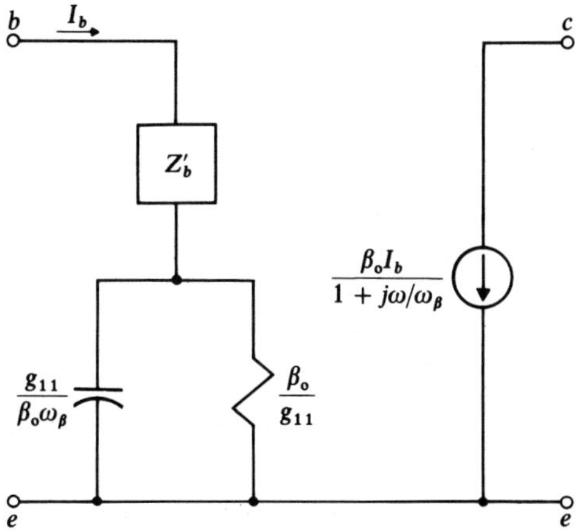

Fig. 6–11 Simplified *h*-equivalent circuit valid when Eq. (6–259) is valid.

It is not hard to see from the simple circuit of Fig. 6–11 that when the common emitter circuit has a load resistance which satisfies the magnitude relationship (6–259) the *circuit* current gain is given as follows

$$A_i = \frac{I_c}{I_b} = h_{21e} = \frac{\beta_0}{1 + j\dfrac{\omega}{\omega_\beta}}, \tag{6–261}$$

whose low frequency current gain is β_0 and whose upper cut-off frequency ω_{coi}, is given by

$$\omega_{coi} = \omega_\beta. \tag{6–262}$$

Because the input circuit of both Figs. 6–11 and 6–12 involve a parallel R–C circuit, the cut-off frequency ω_{cov} is expected to be higher than the current gain cut-off frequency ω_β. The reason for the higher cut-off frequency for A_v is the decrease of input impedance with increasing frequency which results in an increase in I_b when the input is driven by a *voltage* source. The increase in I_b with increasing frequency partially compensates for the decrease of β with increasing frequency. The circuit analysis of this case, for a load resistance satisfying the magnitude relationship (6–259), and for the extrinsic base impedance being r_b', yields the low frequency voltage gain A_{vo} and cut-off frequency ω_{cov} given below:

$$A_{vo} = -\frac{\beta_0 R_L}{\beta_0/g_{11} + r_b'}, \tag{6–263}$$

$$\omega_{cov} = \omega_\beta\left(1 + \frac{\beta_0}{g_{11} r_b'}\right). \tag{6–264}$$

Equation (6–264) makes clear that ω_{cov} can be significantly greater than ω_β.

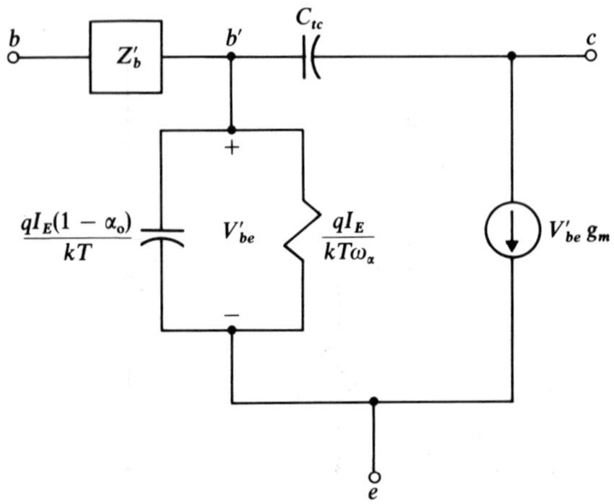

Fig. 6–12 Simplified hybrid-π equivalent circuit valid when Eq. (6–260) is valid.

6–12 SOME IMPORTANT OBSERVATIONS ON DEPENDENT SOURCES IN EQUIVALENT CIRCUITS

The equivalent circuit of Fig. 6–8 demonstrates that if we had added the extrinsic elements *after* the transformation from common base to common emitter, h_{22e} would not have had the series *R-C* branch. Note that each of the elements of this branch is a function of C_{tc}. Note also that the elements of this branch are expressed as a mixture of *both extrinsic* C_{tc} and intrinsic ω_β, β_0, parameters. Because the calculations in obtaining Fig. 6–8 were done correctly, the transformation laws which convert the common base to the common emitter circuit have influenced *both* the extrinsic as well as intrinsic circuit elements. Thus, it is most important that the common base circuit must be complete, i.e. must have *both* extrinsic as well as intrinsic elements, *before* it is transformed to obtain a common emitter or common collector equivalent circuit.

Another important point to be noted is that *the dependent sources of the equivalent circuit are always functions of currents and voltages in only the intrinsic part of the equivalent circuit.* The reason for this restriction is that the dependent sources are obtained from the *intrinsic part* of the transistor model as seen in the previous sections. In the common base equivalent circuit of Fig. 6–2 the values of the dependent current sources are proportional to the voltages across the intrinsic admittances Y^i_{ec} and Y^i_{cc} and not the terminal voltages which would then necessarily include the extrinsic parameters such as r'_e, r'_c, Z'_b as well. Note that the voltage V_e in Fig. 6–2 across C_{te} and $Y^i_{\ell e}$ is *not* a contradiction of the general rule stated above. This voltage V_e is nevertheless the voltage across the intrinsic admittance Y^i_{ee}. Thus, the dependent current source $Y^i_{ce}V_e$ in the collector circuit may be said to be proportional to the voltage across the *intrinsic* part of the equivalent circuit.

In keeping with the above observations if we had included C_{te} in the base branch of the h-equivalent circuit, the dependent current source βI_b should be defined in terms of the base current which goes only through the intrinsic h_{11e} (i.e. through the capacitance $g_{11}/(\beta_0\omega_\beta)$ and the resistance (β_0/g_{11}) and *not* through the junction capacitance C_{te} in the base circuit. (See Problem 6–9.)

6–13 EXTRINSIC PARAMETERS AND THE SHORT-CIRCUIT CURRENT GAIN

We have already calculated the short-circuit current gain for the intrinsic transistor. In this section we wish to reconsider the common base short-circuit current gain cut-off frequency including some extrinsic parameters. For the purpose of the present calculation we shall employ the so-called T-equivalent circuit of Fig. 6–13. Figure 6–13a is all set up to calculate the short-circuit current gain. Three specific extrinsic circuit elements taken into account now

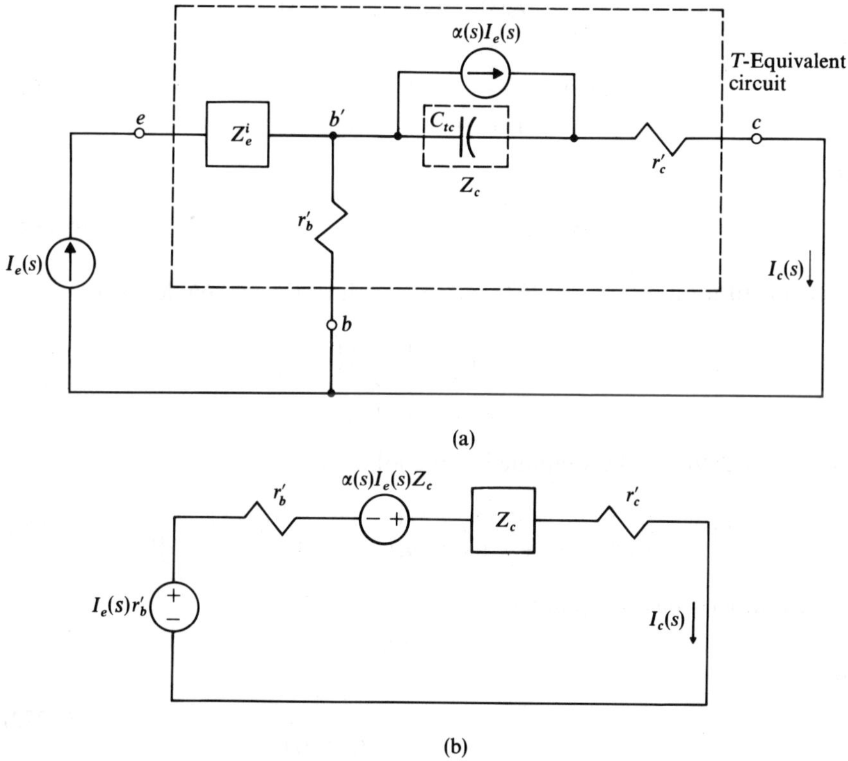

Fig. 6–13 Circuits for short-circuit current gain calculations including extrinsic parameters. (a) Circuit for calculating the short circuit current gain taking account of extrinsic parameters. (b) Figure 6–13a, redrawn after applying Thevenin's theorem to $I_e(s)$, r'_b, $Z_e(s)$ to the left of points b'-b, and to the dependent source and Z_c.

are r_b', r_c', and C_{tc}. The first two are the bulk base and collector resistances. It is here assumed that we are considering the case of those transistors whose extrinsic base impedance may be considered to be *real.*

If Thevenin's theorem is applied to the independent current source, $I_e(s)$, Z_e, and r_b' results in the two left-most circuit elements of Fig. 6–13b. Thevenin's theorem applied to the dependent current source and Z_c in Fig. 6–13a results in the dependent voltage source $\alpha(s)I_eZ_c$ and Z_c in Fig. 6–13b. Thus, the simple single loop circuit of Fig. 6–13b results. We may write the single loop equation for the circuit of Fig. 6–13b as follows:

$$I_c(s)\,[r_b' + r_c' + Z_c(s)] = I_e(s)r_b' + \alpha(s)I_e(s)Z_c(s). \tag{6–265}$$

Equation (6–265) allows us to solve for the current gain $A_i(s)$

$$A_i(s) = \frac{I_c(s)}{I_e(s)} = \frac{r_b' + \alpha(s)Z_c(s)}{r_b' + r_c' + Z_c(s)}. \tag{6–266}$$

Now

$$\alpha(s) = \frac{\alpha_0}{1 + \dfrac{s}{\omega_\alpha}} \tag{6–267}$$

and

$$Z_c(s) = \frac{1}{C_{te}s}. \tag{6–268}$$

Direct substitution of Eqs. (6–267) and (6–268) into Eq. (6–266) leads to

$$A_i(s) = \frac{r_b' + \left(\dfrac{\alpha_0}{1 + s/\omega_\alpha}\right)\left(\dfrac{1}{C_{tc}s}\right)}{r_b' + r_c' + 1/(C_{tc}s)}. \tag{6–269}$$

Equation (6–269) may be simplified to the following

$$A_i(s) = \frac{I_c(s)}{I_e(s)} = \frac{r_b'[s^2 + \omega_\alpha s + \alpha_0\omega_\alpha/(r_b'C_{tc})]}{(r_b' + r_c')(s + \omega_\alpha)\,\{s + 1/[C_{tc}(r_b' + r_c')]\}}. \tag{6–270}$$

The current gain has two poles at

$$s_1 = -\,\omega_\alpha \tag{6–271}$$

and

$$s_2 = -\frac{1}{C_{tc}(r_b' + r_c')}\,. \tag{6–272}$$

We may thus say that if

$$\omega_\alpha \gg \frac{1}{C_{tc}(r_b' + r_c')} \tag{6–273}$$

then

$$\omega_{\alpha co} = \frac{1}{C_{tc}(r'_b + r'_c)}, \tag{6–274}$$

where $\omega_{\alpha co}$ = cut-off frequency for the short-circuit current gain including extrinsic parameters.

If

$$\omega_\alpha \ll \frac{1}{C_{tc}(r'_b + r'_c)} \tag{6–275}$$

then

$$\omega_{\alpha co} = \omega_\alpha. \tag{6–276}$$

The above analysis invites the following conclusions. The short-circuit current gain cut-off frequency is ω_α only when the magnitude relationship (6–275) is valid. If, however, the magnitude relationship of (6–273) is true, the extrinsic parameters determine the dominant cut-off frequency. Because usually

$$r'_b \gg r'_c \tag{6–277}$$

the magnitude relationship (6–273) may be approximated by

$$\omega_{\alpha co} \doteq \frac{1}{C_{tc} r'_b}. \tag{6–278}$$

6–14 THE MAXIMUM FREQUENCY OF OSCILLATION—A FIGURE OF MERIT

If an amplifying device is to be used for high frequency application, it is useful to know a good figure of merit. A good figure of merit should have the following desirable attributes.

1. It should be exclusively a function of the device's own parameters, and relatively independent of the *external* circuit.
2. It should be a function of all the important high frequency parameters.

One possible candidate that could be considered is the gain bandwidth product. This, however, is not a good figure of merit for transistors as we shall presently see. The gain bandwidth product for voltage gain $(GB)_v$ may be written with the help of Eqs. (6–263) and (6–264) as follows:

$$(GB)_v = \frac{\beta_0 \omega_\beta R_L}{r'_b}. \tag{6–279}$$

Note that Eq. (6–279) fails both the criteria given above because it is proportional to an external circuit element, namely the load resistance R_L; and it does not contain such other important high frequency parameters as r'_b and C_{tc}.

The current gain bandwidth product $(GB)_{ic}$ for cascaded (two stage) identical amplifying stages and having a simple interstage coupling circuit consisting of a single shunt resistance R_1 gives the result (see Problem 6–11)

$$(GB)_{ic} = \frac{\beta_0 \omega_\beta R_1}{R_1 + r_b'}. \tag{6–280}$$

Again Eq. (6–280) fails the criterion for a good figure of merit for the same reasons as Eq. (6–279) failed.

A good figure of merit is the *maximum frequency of oscillation for transistors.* This is the maximum frequency of oscillation at which the transistor output current may be fed back to the input ideal conditions and may be made to oscillate. We shall utilize the h-equivalent circuit of Fig. 6–8 to calculate the maximum frequency of oscillation f_m. The h-parameters to be used at f_m, which is expected to be quite a high frequency, is obtained by taking limiting forms of Eqs. (6–245) through (6–248) as given below.

$$h_{11e} \doteq r_b', \tag{6–281}$$

$$h_{12e} \doteq 0, \tag{6–282}$$

$$h_{21e} \doteq \frac{\omega_\alpha}{j\omega} = -j\omega_\alpha/\omega, \tag{6–283}$$

$$h_{22e} \doteq \omega_\alpha C_{tc}. \tag{6–284}$$

Equation (6–283) is obtained by neglecting one in the denominator of h_{21e} in Eq. (6–247). The right hand side of Eq. (6–284) comes from considering the series R-C circuit of h_{22e} and neglecting the impedance of the capacitance $\beta_0 C_{tc}$. We have further neglected the following:

$$g_{22} \approx 0, \tag{6–285}$$

$$g_{\ell c}' \approx 0. \tag{6–286}$$

Figure 6–14 shows the equivalent circuit for the h-parameters, taking account of the above observations. The circuit is set up to enable us to calculate the maximum frequency of oscillations f_m. The ideal filter may consist of an ideal shunt inductor which is chosen so that its parallel combination with C_{tc} of the original h-equivalent circuit offers *zero* admittance at the maximum frequency of oscillation. For this reason C_{tc} is not shown in Fig. 6–14 to avoid confusion in the calculations which follow. The 90° phase shifter simply gets rid of the j in Eq. (6–283). The minus sign on the right hand side of Eq. (6–283) means that the *actual* current direction of the dependent source is opposite to that shown in Fig. 6–14. We shall adjust the turns ratio of the ideal transformer so as to maximize the frequency of oscillation. The necessary and sufficient condition of oscillation is that the current fed back to the input circuit, r_b', should equal the base current I_b. Thus,

$$I_f = I_b. \tag{6–287}$$

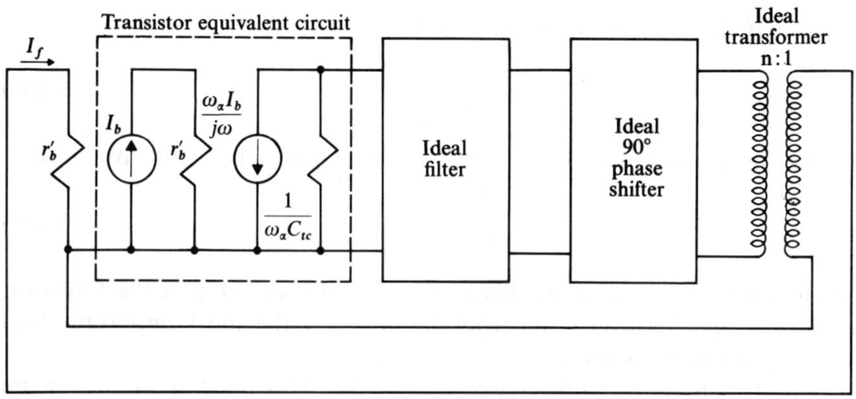

Fig. 6–14 Circuit for calculating the maximum frequency of oscillation for a junction transistor.

From Fig. 6–14 the feedback current I_f may be related through the circuit as follows:

$$I_f = \frac{n\omega_\alpha I_b}{\omega}\left[\frac{1/(\omega_\alpha C_{tc})}{1/(\omega_\alpha C_{tc}) + r'_b n^2}\right] = \frac{n\omega_\alpha I_b}{\omega(1 + n^2\omega_\alpha r'_b C_{tc})}. \tag{6–288}$$

The use of Eq. (6–287) and Eq. (6–288) yields the following

$$\omega = \frac{\omega_\alpha n}{1 + n^2\omega_\alpha r'_b C_{tc}}. \tag{6–289}$$

We require that the turns ratio must be adjusted to maximize the frequency of oscillation ω_m. To this end the following is done.

$$\frac{d\omega}{dn} = 0 = \omega_\alpha(1 + n^2\omega_\alpha r'_b C_{tc}) - 2\omega_\alpha^2 n^2 r'_b C_{tc} \tag{6–290}$$

or

$$\omega_\alpha + n^2\omega_\alpha^2 r'_b C_{tc} - 2\omega_\alpha^2 r'_b C_{tc} = 0 \tag{6–291}$$

$$\omega_\alpha - n^2\omega_\alpha^2 r'_b C_{tc} = 0. \tag{6–292}$$

We may now solve for n from Eq. (6–292) as follows:

$$n = \frac{1}{(\omega_\alpha r'_b C_{tc})^{\frac{1}{2}}}. \tag{6–293}$$

With n evaluated in Eq. (6–293) we may substitute it into Eq. (6–289) to obtain the maximum frequency of oscillation ω_m (rads/sec)

$$\omega_m = \frac{\omega_\alpha}{2}[1/(\omega r'_b C_{tc})^{\frac{1}{2}}]. \tag{6–294}$$

Equation (6–294) may be written alternatively as

$$\omega_m = \left[\frac{\omega_\alpha}{4r_b' C_{tc}}\right]^{\frac{1}{2}}. \qquad (6\text{–}295)$$

The maximum frequency of oscillation (in hertz) f_m may be readily obtained as

$$f_m = \left[\frac{f_\alpha}{8\pi r_b' C_{tc}}\right]^{\frac{1}{2}}. \qquad (6\text{–}296)$$

This is a good figure of merit because it is a function *only* of the transistor parameters and because it incorporates three of the most important high frequency transistor parameters ω_α, r_b', and C_{tc}.

The maximum available power gain (MAG) is defined as the maximum gain that is obtainable from an active device without external feedback and under the condition of complex conjugate matching both at the input terminals and output terminals. For step-junction transistors this gain is given by

$$\text{MAG} = \frac{f_\alpha}{8\pi f_m^2}\left(\frac{1}{r_b' C_{tc}}\right). \qquad (6\text{–}297)$$

It is to be noted that at the maximum frequency of oscillations MAG is unity

$$\text{MAG}(f_m) = 1. \qquad (6\text{–}298)$$

For linearly graded junctions

$$\text{MAG} = 0.2\left[\frac{f_\alpha}{f_m^3 r_b' C_{tc} r_e}\right]^{\frac{1}{2}}, \qquad (6\text{–}299)$$

where

$$r_e = \frac{kT}{qI_E}. \qquad (6\text{–}300)$$

We see that MAG diminishes at a rate of 20 db per decade for step junctions and 15 db per decade for linearly graded junctions.

Equations (6–296), (6–297), and (6–299) reveal the fact that those transistors are best in which the value of f_α is high and the values of r_b', C_{tc} are low.

PART C | TRANSIENT RESPONSE

6–15 TRANSIENT TIMES

So far we have considered only the *steady state time* dependent and time independent characteristics of the transistor. We now consider the transient response of junction transistors.

One of the most important characteristics of an electronic amplifying device is the speed with which it turns "on" or "off" in response to appropriate drives.

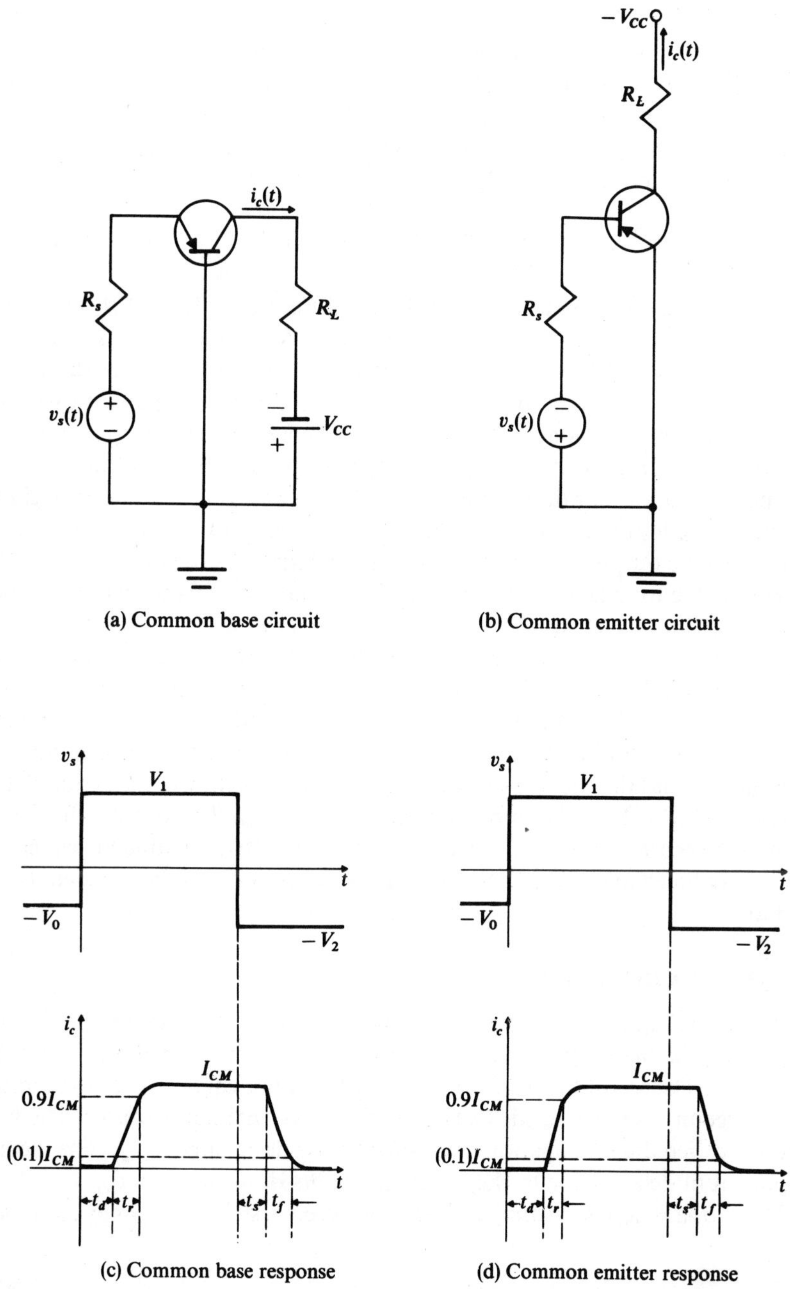

Fig. 6–15 Four transient times defined for common base and common emitter circuits.

The speed with which active devices switch "on" and "off" determines the rate at which, for example, computers can accept information or process it. The switching speeds determine at what frequency a monostable or bistable circuit can respond to an external trigger signal.

Figures 6–15a and 6–15b show a common base circuit and common emitter circuit in which the resistance R_s is chosen so that

$$r_{\text{in}}(\text{off}) \gg R_s \gg r_{\text{in}}(\text{on}). \tag{6–301}$$

With the choice given by Eq. (6–301) the combination of V_{in} and R_s looks like a voltage source of value V_{in} when the transistor is "off", and looks like a current source of value V_{in}/R_s when the transistor is "on". A comparison of the shapes of Figs. 6–15c and 6–15d show that the output current shape is a distorted version of the input voltage and/or current. There are four output transient times defined in Fig. 6–15d: delay time, t_d; rise time, t_r; storage time, t_s; and fall time, t_f. The output delay time is the time taken by the output to begin to respond to the input "on" drive in the absence of capacitive feed through from the input to the output. The output rise time is defined as the time taken by the output to reach from zero to 90 per cent of the steady-state value of the collector current I_{CM}. The $t = 0$, for the rise time, is tagged at the end of the delay time. *Output storage time* is defined as the time between the start of the "off" drive at the input and the *start* of the decrease in current at the output. This definition is consistant with Fig. 6–15d where the start of decrease in the collector current is sharp. If this point is not sharp, the *end* of storage time may be obtained by extrapolating the fall time curve towards I_{CM}; the intersection of the extrapolated curve with I_{CM} is the *end* of storage time. The *output fall time* is the difference in time between the output storage time and the instant of time at which the current is reduced to $(0.1\ I_{CM})$. We shall now examine each of the transient times in detail.

6–16 OUTPUT DELAY TIME

There are two quite unrelated causes which can contribute to the output delay time. The first is that minority carriers take a finite time to traverse the base region. The second is the time taken by the input capacitances to readjust their charge in response to the demands of the external input source. The first of these is called the propagation delay time, and is designated by t_{dp}. The second component of delay is due to the input capacitors designated by t_{dc}.

The propagation delay time will be considered first. If we begin by assuming that

$$\gamma = \alpha^* = 1 \tag{6–302}$$

then

$$\alpha = \beta^*. \tag{6–303}$$

We may now choose the fairly accurate expression for the base transport factor given by Eq. (6–43). In terms of Laplace transforms we may write

$$\frac{I_c(s)}{I_e(s)} = \operatorname{sech}\left[(w_0/L_p)(1 + s\tau_p)^{\frac{1}{2}}\right], \tag{6–304}$$

where $I_e(s)$ = Laplace transform of $i_e(t)$

$I_c(s)$ = Laplace transform of $i_c(t)$.

If the emitter current is a step function

$$I_e(s) = \frac{I_{e1}}{s} \tag{6–305}$$

and hence

$$i_c(t) = \mathcal{L}^{-1}\left[\frac{I_{e1}}{s} \operatorname{sech}\left[w_0/L_p)(1 + s\tau_p)^{\frac{1}{2}}\right]\right]. \tag{6–306}$$

If we had chosen the Taylor series expansion of the hyperbolic term retaining only two terms, the expression of the intrinsic short-circuit current gain would be that of Eq. (6–176) which may be written in terms of Laplace transforms as follows:

$$\frac{I_c(s)}{I_e(s)} = \frac{\alpha_N}{1 + \dfrac{s}{\omega_N}}, \tag{6–307}$$

where

$$\alpha_N = \alpha_0, \tag{6–308}†$$

$$\omega_N = \omega_\alpha. \tag{6–309}†$$

Note that in writing the right hand side of Eqs. (6–304) and (6–307) we have let

$$j\omega \to s \tag{6–310}$$

in the original Eqs. (6–43) and (6–176). Again if the emitter current is a step function of amplitude I_{e1}, the collector current may be expressed as follows:

$$i_c(t) = \mathcal{L}^{-1} \frac{I_{e1}}{s[1 + s/\omega_N]} = I_{e1}(1 - e^{-\omega_N t}). \tag{6–311}$$

Figure 6–16 shows the right hand sides of Eqs. (6–306) and (6–311) plotted

---

† We have introduced new notation here for the *normal* low frequency alpha α_N and the *normal* alpha cut-off frequency ω_N. The word normal here refers to the normally biased transistor with the emitter junction forward biased and the collector junction reverse biased. We shall have reason to refer to the *inverse* alpha α_I and *inverse* alpha cut-off frequency ω_I in connection with the transistor output *storage time* to be discussed in Section 6–17. The word *inverse* refers to the inverse biased transistor where the collector junction is forward biased and the emitter junction is reverse biased.

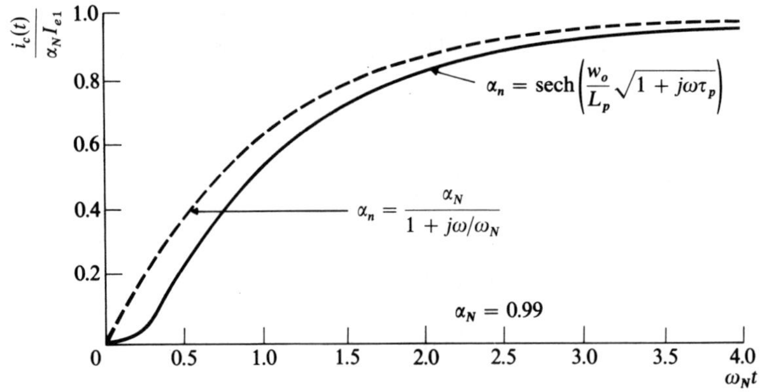

Fig. 6–16 Step current response corresponding to different expressions of α_n.

on a set of dimensionless axes. It can be seen from this figure that the plot of Eq. (6–306) shows that the collector current remains essentially zero up to approximately a time duration of $(1/5\omega_N)$. We interpret this as the time taken by the minority carriers to cross the base region from the emitter edge to the collector edge of the base region. Thus, the propagation delay time t_{dp} is given by

$$t_{dp} = 1/(5\omega_N). \tag{6–312}$$

It is worthwhile noting that the propagation delay time predicted by Eq. (6–311) is zero. This is, of course, an incorrect prediction, though not entirely as can be seen by the following reasoning. The expression of Eq. (6–307) was obtained by retaining only two terms in the Taylor series expansion of sech $(C_p w_0)$. As the higher order terms were assumed to be zero the information contained in them is not present in Eq. (6–307). *Very high frequency* information loss in the frequency domain corresponds to a loss of information for short times. That is, Eq. (6–307) is not expected to give valid information pertaining to very high frequencies, and Eq. (6–311) cannot be expected to give valid information for very short times. As the propagation delay time is a short time, the prediction of Eq. (6–307) is incorrect as expected.

To get a feel for numbers, a transistor having

$$f_\alpha = f_N = 80 \text{ MHz}, \; t_{pd} \doteq 0.4 \text{ nanoseconds},$$

$$f_\alpha = f_N = 800 \text{ MHz}, \; t_{dp} \doteq 40 \text{ picoseconds},$$

$$f_\alpha = f_N = 8 \text{ GHz}, \; t_{dp} = 4 \text{ picoseconds}.$$

These are extremely small times indeed. In the early 1950s, the propagation delay time was thought to be the reason for the existence of the output delay time. Nanavati and Johnson (1958) proposed that the major cause for the existence of the output delay time is the time needed to charge or discharge the input capacitances. That is, if there is an "off" voltage which holds the transistor in the cut-off condition, the input capacitances are charged and have voltages across them appropriate to the "off" voltage at the input. When the input current changes from "off" to "on" the voltages across the input capacitances *cannot* change instantaneously as only finite currents are flowing. It is not until the voltage across the emitter junction has changed from reverse bias to zero voltage (it is assumed that the cut-in voltage is zero), that the injection of minority carriers into the base region can begin. The delay due to this effect is designated t_{dc}. The *total* delay time is

$$t_d = t_{dp} + t_{dc}. \qquad (6\text{–}313)$$

We note before the evaluation of t_{dc} begins that the capacitances straddling each junction will consist of the linear stray capacitance inevitably present and the nonlinear junction capacitance. We shall first consider the *common base* circuit. The relevant equivalent circuit appropriate for the calculation of the common base circuit is as shown in Fig. 6–17. There is only one independent node pair in this circuit. Using ECL only one equation need be written as follows by equating the net current entering the node e to that leaving the same node. Thus, *after* the "on" drive has been applied at the input, the circuit equation from the conditions given above is:

$$\frac{V_1 - V_{eb}}{R_s} = C_{\ell e}\frac{dV_{eb}}{dt} + \frac{dQ_j}{dt} + I_{se}(e^{qV_{eb}/kT} - 1), \qquad (6\text{–}314)$$

where $C_{\ell e}$ = linear stray capacitance associated with the emitter junction,
Q_j = charge on one plate of the junction capacitance of the emitter junction,
I_{se} = back saturation current of the emitter junction,
V_1 = amplitude of the "on" input drive source,
R_s = resistance in series with the input voltage source.

The second term on the right hand side may be evaluated as follows:

$$\frac{dQ_j}{dt} = \frac{dQ_j}{dV_{eb}}\frac{dV_{eb}}{dt}. \qquad (6\text{–}315)$$

The first derivative on the right hand side of Eq. (6–315) depends on the kind of junction we are dealing with.

For *step junctions*

$$\frac{dQ_j}{dt} = \frac{k_{se}}{V_{eb}^{\frac{1}{2}}}\frac{dV_{eb}}{dt}. \qquad (6\text{–}316)$$

For *linearly graded junctions*

$$\frac{dQ_j}{dt} = \frac{k_{\ell e}}{V_{eb}^{\frac{1}{3}}}\frac{dV_{eb}}{dt}. \qquad (6\text{–}317)$$

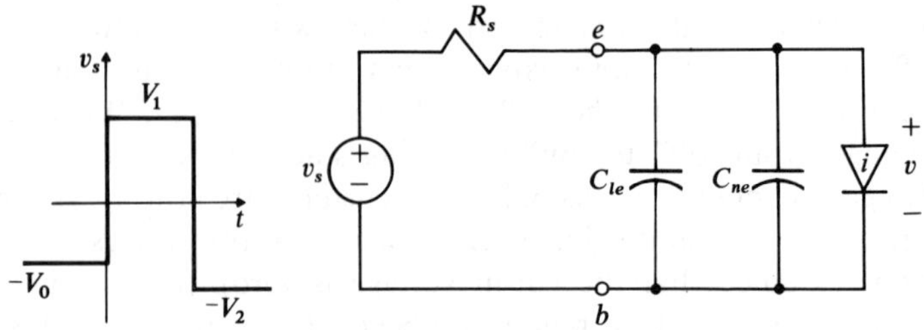

Fig. 6–17 Equivalent circuit appropriate to calculate the output delay time in common base circuits.

In writing Eqs. (6–316) and (6–317) we have assumed that the emitter built-in voltage satisfies

$$V_{BI}\,(\text{emitter}) \ll V_{eb}. \tag{6–318}$$

Thus, $V_{BI} - V_{eb}$, and if V_{eb} is reverse biased, as during delay time, the above quantity is V_{eb} itself. We may rewrite Eq. (6–314) with the help of Eq. (6–316) as follows:

$$\left(C_{\ell e} + \frac{k_{se}}{V_{eb}^{\frac{1}{2}}}\right)\frac{dV_{eb}}{dt} = \frac{V_1 - V_{eb}}{R} + I_{SE}. \tag{6–319}$$

We have neglected the exponential term in comparison with one in the last term on the right hand side of Eq. (6–314) because for most of the delay time V_{eb} will be quite large, compared to kT/q, and also negative. Equation (6–319) is a first order *nonlinear* differential equation. Happily it can be integrated as follows:

$$\int_0^{V_0} \frac{R_s C_{\ell e} dV_{eb}}{-R_s I_{SE} - V_1 + V_{eb}} + R_s k_{se} \int_0^{V_0} \frac{dV_{eb}}{V_{eb}^{\frac{1}{2}}(-R_s I_{SE} - V_1 + V_{eb})} = t_{dcb}, \tag{6–320}$$

where t_{dcb} = output delay time due to input capacitors in the common base circuit.

The above integrals may be evaluated to obtain

$$t_{dcb} = R_s C_{\ell e}\,\ell n(1 + \rho_d) + 2R_s C_{ne}(V_0)\,\rho_d^{\frac{1}{2}} \tan^{-1} \rho_d^{\frac{1}{2}}, \tag{6–321}$$

where

$$\rho_d \triangleq \frac{V_0}{V_1 + I_{SE} R_s}, \tag{6–322}$$

where V_0 = "off" voltage *before* the "on" pulse arrives as can be seen in Fig. 6–15c.

Equation (6–321) may be put in a dimensionless form as follows:

$$\frac{t_{dcb}}{R_s C_T(V_0)} = \left(\frac{1}{1+\eta}\right)\ell\,\mathrm{n}(1+\rho_d) + \left(\frac{2\eta\rho_d^{\frac{1}{2}}}{1+\eta}\right)\tan^{-1}\rho_d^{\frac{1}{2}} \text{ (step junctions)}, \quad (6\text{–}323)$$

where

$$C_T(V_0) = C_{\ell e} + C_{ne}(V_0). \quad (6\text{–}324)$$

Equation (6–324) represents the total input capacitance evaulated at the "off" voltage V_0 and

$$\eta \triangleq \frac{C_{ne}}{C_{\ell e}}, \quad (6\text{–}325)$$

where C_{ne} = nonlinear emitter junction capacitance,
$C_{\ell e}$ = linear emitter stray capacitance associated with the emitter junction.

The equation analogous to Eq. (6–320) for linearly graded junctions is

$$\int_0^{V_0} \frac{R_s C_{\ell e} dV_{eb}}{-R_s I_{SE} - V_1 + V_{eb}} + R_s k_{\ell e} \int_0^{V_0} \frac{dV_{eb}}{V_{eb}^{\frac{1}{3}}(-R_s I_{SE} - V_1 + V_{eb})} = t_{dcb}. \quad (6\text{–}326)$$

Upon carrying out the integration and after rearranging the terms, the following dimensionless expression results

$$\frac{t_{dcb}}{R_s C_T(V_0)} = \left(\frac{1}{1+\eta}\right)\ell\mathrm{n}\,(1+\rho_d) + \frac{\eta\rho_d^{\frac{1}{3}}}{1+\eta}\left\{\sqrt{3}\left[\left(\tan^{-1}\frac{2\rho_d^{\frac{1}{3}}-1}{3}\right)+\frac{\pi}{6}\right]+\right.$$
$$\left. + \tfrac{1}{2}\,\ell\mathrm{n}\left(\frac{1-\rho_d^{\frac{1}{3}}+\rho_d^{\frac{2}{3}}}{(1+\rho_d^{\frac{1}{3}})^2}\right)\right\}. \quad (6\text{–}327)$$

In both Eqs. (6–323) and (6–327) the first term on the right hand side represents the contribution due to the *linear capacitance $C_{\ell e}$ alone.*

In practice in Eq. (6–322)

$$I_{SE}R_s \ll V_1 \quad (6\text{–}328)$$

so that

$$\rho_d \doteq \frac{V_0}{V_1}. \quad (6\text{–}329)$$

Thus, ρ_d may be termed the input drive ratio. The linear stray capacitance $C_{\ell e}$ may be obtained by sufficiently reverse biasing the emitter junction, because

$$\lim_{V_0 \to \textit{sufficiently large}} [C_T(V_0)] = C_{\ell e}. \quad (6\text{–}330)$$

It may be the case, however, that a sufficiently high reverse voltage is not possible to be applied to the emitter junction because the required voltage may exceed the breakdown voltage of the junction. If both $C_{\ell e}$ and C_{ne} are known, then Eqs. (6–323) or (6–327) may be used to predict the delay time due to the input capacitances. But because of the above reasons, where it is not possible to determine the values of $C_{\ell e}$ and C_{ne}, the following procedure may be used. First we note that

$$\ln(1 + \rho_d) \leqslant \frac{t_{dcb}}{R_s C_T(V_0)} \leqslant 2\rho_d^{\frac{1}{2}} \tan^{-1} \rho_d^{\frac{1}{2}} \qquad \text{(step junctions)} \quad (6\text{–}331)$$

and

$$\ln(1 + \rho_d) \leqslant \frac{t_{dcb}}{R_s C_T(V_0)} \leqslant \rho_d^{\frac{1}{3}} \left\{ \sqrt{3} \left[\left(\tan^{-1} \frac{2\rho_d^{\frac{1}{3}} - 1}{\sqrt{3}} \right) + \frac{\pi}{6} \right] + \right.$$
$$\left. + \tfrac{1}{2} \ln \left[\frac{1 - \rho_d^{\frac{1}{3}} + \rho_d^{\frac{2}{3}}}{(1 + \rho_d^{\frac{1}{3}})^2} \right] \right\} \qquad \text{(linearly graded junctions).} \quad (6\text{–}332)$$

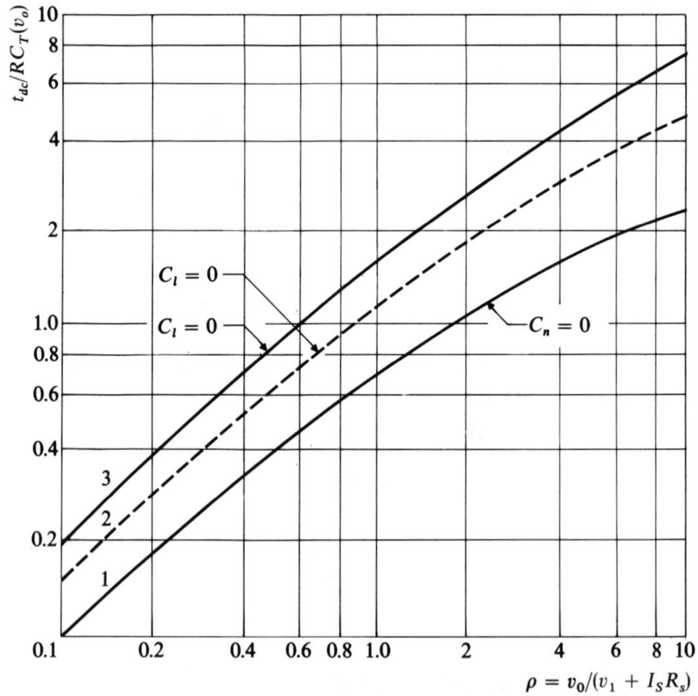

Fig. 6–18 Upper and lower bounds for output delay time. 1, lower bound, 2, upper bound for graded-junction transistors; 3, upper bound for step-junction transistors.

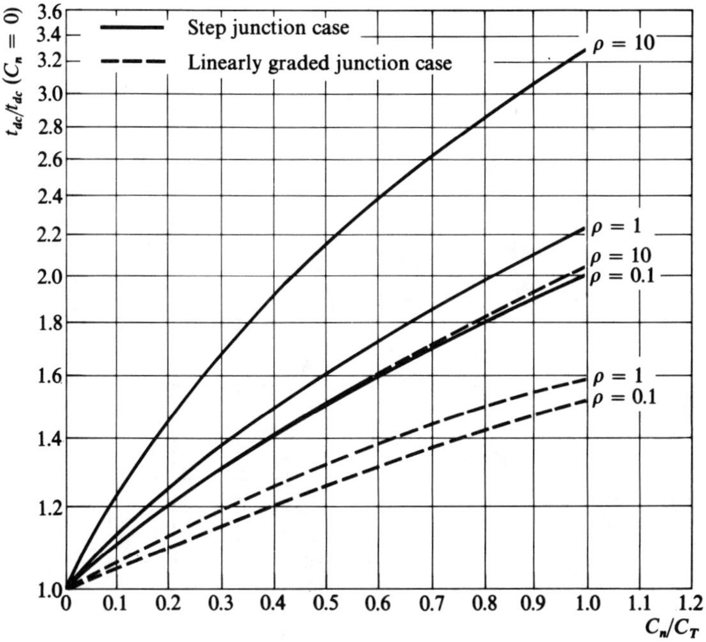

Fig. 6–19 Maximum theoretical variation in t_{dc} as a function of C_n/C_T.

Figure 6–18 shows plots of the predicted output delay time in two limiting circumstances when there is no $C_{\ell e}$ and when there is no C_{ne} for both step junction and linearly graded junction transistors. Figure 6–19 shows the maximum theoretical variation in t_{dc} over the *entire range* of values for parameter (C_n/C_T). The maximum variation in t_{dc} are not too large over a wide range (two orders of magnitude) of the drive parameter.

$$\frac{t_{dcb}}{RC_T} < 3.3 \qquad \text{(step junctions)} \tag{6–333}$$

$$\frac{t_{dcb}}{RC_T} < 2.05 \qquad \text{(linearly graded junctions).} \tag{6–334}$$

Now because of the magnitude relationships of Eqs. (6–333) and (6–334), and because of a lack of knowledge of the relative values of $C_{\ell e}$ and C_{ne} in some cases we may arbitrarily assume that

$$C_{ne} = C_{\ell e} = \tfrac{1}{2} C_T(V_0). \tag{6–335}$$

With the help of Eq. (6–335) the delay time may be predicted. As it is expected that neither limit $C_{\ell e} = 0$ or $C_{ne} = 0$ is practically feasible for real transistors,

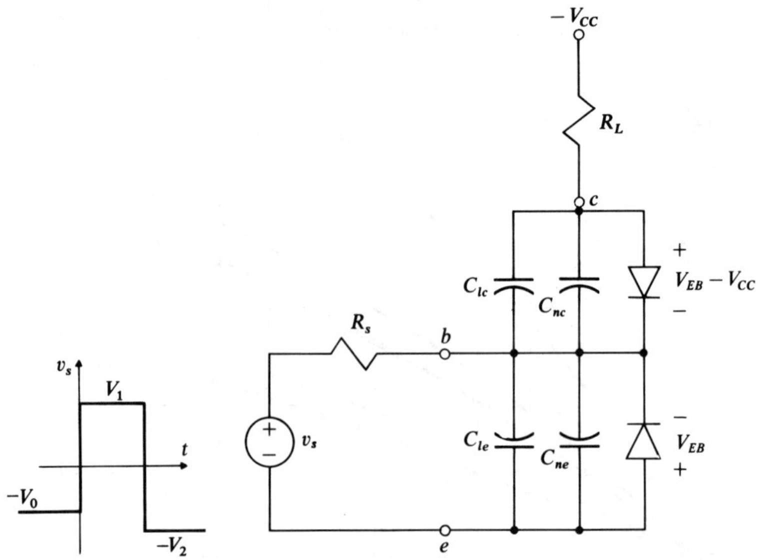

Fig. 6–20 Equivalent circuit appropriate to calculate the output delay time in common emitter circuits.

the predicted delay times will not be very close to the upper or lower bounds for delay time shown in Fig. 6–18. Hence Eq. (6–335) will lead to relatively small errors.

To calculate the delay time in the common emitter circuit configuration, the equivalent circuit of Fig. 6–20 needs to be utilized. Again we neglect the built-in voltage of *both* the junctions compared to the voltage due to external sources. The load resistance R_L is shown dotted because for the entire duration of the output delay time no collector current flows and the voltage drop across it is zero. The differential equation for this circuit is

$$(C_{\ell c} + C_{\ell e} + C_{nc} + C_{ne})\frac{dV_{eb}}{dt} + I_{se}(e^{qV_{eb}/kT} - 1) + $$

$$+ I_{sc}(e^{q(-V_{eb}-V_{CC})/kT} - 1) = \frac{V_1 - V_{cb}}{R_s}, \qquad (6\text{–}336)$$

where $C_{\ell c}$ = linear stray capacitance of the collector junction,

C_{nc} = junction capacitance of the collector junction,

I_{sc} = back saturation current of *only* the collector junction.

As both junctions are reverse biased both the exponential terms may be neglected in comparison with one. Thus Eq. (6–336) may be written as follows where the nonlinear capacitance expressions have been substituted. For *step-junctions,*

$$C_{\ell c} + C_{\ell e} + \left(\frac{k_{sc} + k_{se}}{V_{eb}^{\frac{1}{2}}}\right)\frac{dV_{eb}}{dt} = \frac{V_1 - V_{eb}}{R_s} + I_{SE} + I_{SC}, \tag{6–337}$$

for *linearly graded junctions,*

$$C_{\ell c} + C_{\ell e} + \left(\frac{k_{\ell c} + k_{\ell e}}{V_{eb}^{\frac{1}{3}}}\right)\frac{dV_{eb}}{dt} = \frac{V_1 - V_{eb}}{R_s} + I_{SE} + I_{SC}. \tag{6–338}$$

The nonlinear Eqs. (6–337) and (6–338) may be integrated in a straightforward way. Only the details of the step junction case will be carried out below. The result for the linearly graded junction will be stated. Equation (6–337) separates easily for integration as follows

$$\begin{aligned}
&R_s \int_{-V_0}^{0} \frac{C_{\ell e}\, dV_{eb}}{-R_s(I_{SE} + I_{SC}) - V_1 + V_{eb}} + R_s \int_{-V_0 - V_{CC}}^{-V_{CC}} \frac{C_{\ell c}\, dV_{eb}}{-R_s(I_{SE} + I_{SC}) - V_1 + V_{eb}} \\
&+ \int_{-V_0}^{0} \frac{k_{se} R_s\, dV_{eb}}{V_{eb}^{\frac{1}{2}}[-R_s(I_{SE} + I_{SC}) - V_1 + V_{eb}]} \\
&+ \int_{-V_{CC} - V_0}^{-V_{CC}} \frac{k_{sc} R_s\, dV_{eb}}{V_{eb}^{\frac{1}{2}} - R_s(I_{SE} + I_{SC}) - V_1 + V_{eb}} = \int_0^{t_{dce}} dt = t_{dce},
\end{aligned} \tag{6–339}$$

where t_{dce} = output delay time due to input capacitors in the common emitter circuit configuration. The integrals are readily evaluated to obtain

$$\begin{aligned}
t_{dce} = R_s(C_{\ell e} + C_{\ell c}) \ln(1 + \rho_{de}) &+ 2R_s C_{ne}(V_0)\rho_{de}^{\frac{1}{2}} \tan^{-1}(\rho_{de}^{\frac{1}{2}}) \\
&- 2R_s C_{ne}(V_0 + V_{CC})\rho_{dc}^{\frac{1}{2}} \tan^{-1}(\rho_{de}^{\frac{1}{2}}),
\end{aligned} \tag{6–340}$$

where

$$\rho_{de} \triangleq \frac{V_0 + V_{CC}}{V_1 + I_S' R_s}, \tag{6–341}$$

and

$$I_S' \triangleq I_{SE} + I_{SC}. \tag{6–342}$$

Equation (6–340) may be put in a dimensionless form as follows

$$\frac{t_{dce}}{R_s C_T(V_0)} = \frac{C_{\ell e} + C_{\ell c}}{C_{T1}(V_0)} \ln(1 + \rho_d) - \frac{C_{\ell c}}{C_{T1}(V_0)} \ln(1 + \rho_{de})$$

$$+ \frac{2[C_{ne}(V_0) + C_{nc}(V_0)]}{C_{T1}(V_0)} (\rho_d^{\frac{1}{2}} \tan^{-1} \rho_d^{\frac{1}{2}})$$

$$- \frac{2C_{nc}(V_0 + V_{CC})}{C_T(V_0)} (\rho_{de}^{\frac{1}{2}} \tan^{-1} \rho_{de}^{\frac{1}{2}}) \qquad \text{(step junctions).} \qquad (6\text{–}343)$$

$$\frac{t_{dce}}{R_s C_T(V_0)} = \frac{C_{\ell e} + C_{\ell c}}{C_{T1}(V_0)} \ln(1 + \rho_d) - \frac{C_{\ell c}}{C_T(V_0)} \ln(1 + \rho_{de}) +$$

$$+ \rho_d^{\frac{1}{3}} \left(\frac{C_{ne}(V_0) + C_{nc}(V_0)}{C_{T1}(V_0)} \right) \times \left\{ 3^{\frac{1}{2}} \left[\tan^{-1} \left(\frac{2\rho_d^{\frac{1}{3}} - 1}{3^{\frac{1}{2}}} \right) + \frac{\pi}{6} \right] \right\} +$$

$$+ \tfrac{1}{2} \ln \left(\frac{1 - \rho_d^{\frac{1}{3}} + \rho_d^{\frac{2}{3}}}{1 + \rho_d^{\frac{1}{3}}} \right) - \frac{C_{nc}(V_0 + V_{CC})}{C_{T1}(V_0)} \rho_{de}^{\frac{1}{3}}$$

$$\times \left\{ \sqrt{3} \left[\left(\tan^{-1} \frac{2\rho_{de}^{\frac{1}{3}} - 1}{3^{\frac{1}{2}}} \right) + \frac{\pi}{6} \right] + \tfrac{1}{2} \ln \left(\frac{1 - \rho_{de}^{\frac{1}{3}} + \rho_{de}^{\frac{2}{3}}}{1 + \rho_{de}^{\frac{1}{3}}} \right) \right\}$$

$$\text{(linearly graded junctions),} \qquad (6\text{–}344)$$

where

$$C_{T1}(V_0) \triangleq C_{\ell e} + C_{\ell c} + C_{ne}(V_0) + C_{nc}(V_0 + V_{CC}) \qquad (6\text{–}345)$$

and

$$\rho_{de} \triangleq \frac{V_0 + V_{CC}}{V_1 + I_S' R_s}. \qquad (6\text{–}346)$$

If all the parameters of both junctions are known, the delay time may be predicted from Eqs. (6–343) or (6–344). It is possible, however, that because of relatively low breakdown voltage of at least one of the junctions not all parameters are known. In such an event one way to proceed is to assume that the total input capacitance $C_{T1}(V_0)$ (with the collector shorted to the emitter) measured is made up of half nonlinear junction capacitance and half linear stray capacitance. The upper and lower bounds corresponding to the much simpler common base delay time with $n = \frac{1}{2}$ are applicable also to the common emitter circuit. Experimentally such a procedure does in fact give quite usable results. Figures 6–18 and 6–19 may now be used.

Lastly, we may note that junctions whose nonlinear junction capacitance follows a relationship given below:

$$C_n = \frac{k_1}{V_{eb}^n} \tag{6-347}$$

and having

$$n < \tfrac{1}{2} \tag{6-348}$$

the upper and lower bounds corresponding to the step junctions Eq. (6–331) will always give a pessimistic estimate of delay time. Similarly, if

$$n < \tfrac{1}{3} \tag{6-349}$$

the upper and lower bounds corresponding to the linearly graded junctions Eq. (6–332) will always give a pessimistic estimate of delay time.

Finally, we must remember that the total delay time is given by Eq. (6–313). It is possible that if the "on" drive is quite large that t_{dc} may be comparable to t_{dp}, in which case the latter cannot be neglected. In all other cases the neglect of t_{dp} does not lead to any great errors because t_{dc} usually is much bigger than t_{dp}.

6–17 RISE AND FALL TIMES

There are three major methods of analysis of transistor transients in current use.

I. The equivalent circuit approach.
II. The charge control approach.
III. The diffusion equation approach.

We shall look at all of the above. The equivalent circuit approach will be used for most of the analysis as it is the easiest to use. The charge control method will also be used. The diffusion equation approach will be used as an ideal approach to test the limits on the other two methods of analysis. We shall point out both the advantages as well as the disadvantages of each method.

I. The Equivalent Circuit Approach

This method consists simply of using the wideband equivalent circuits already obtained for the transistor in the circuit whose response needs to be calculated. The Laplace transform method of analysis will be used.

The major assumptions of this approach are the following:

1. The *small signal* wideband equivalent circuits derived in Part *B* of this chapter may be utilized in the essentially large signal transient analysis.

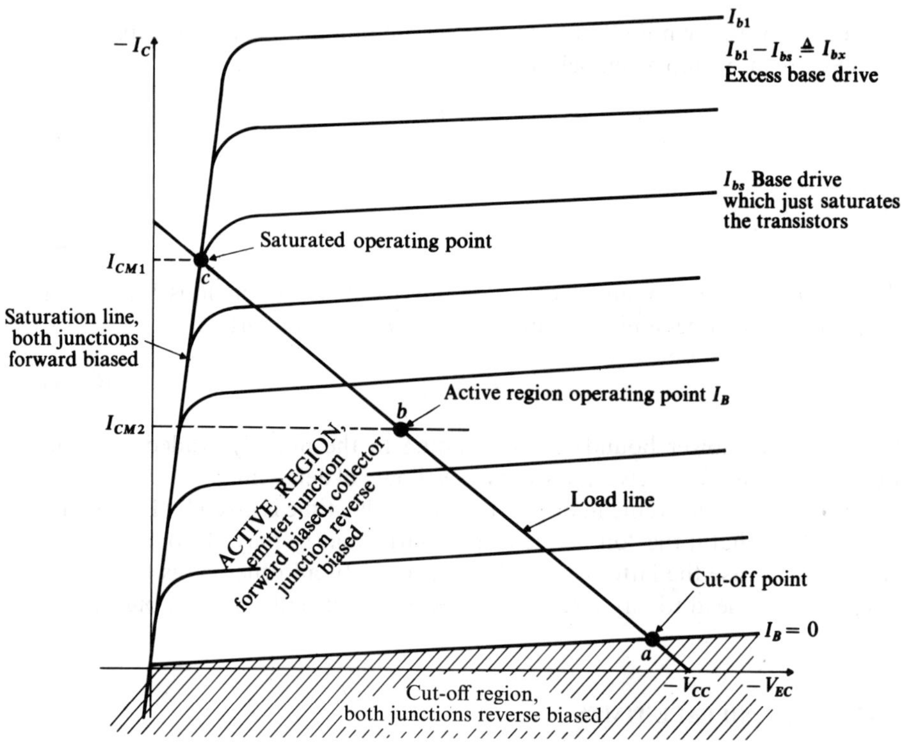

Fig. 6–21 Collector characteristics of common emitter transistor configuration with the three regions of operation identified.

Important conclusions of the above are that

a) low injection conditions are valid,

b) recombination-generation is negligible in the transition regions of both junctions.

c) All other assumptions and approximations necessary to obtain the wideband equivalent circuits of Part *A* are also assumed to be valid. On the face of it, it would appear as though this approach could not possibly work as seen by the following reasoning. Consider a common emitter circuit with a single load resistance R_L in the collector circuit. Figure 6–21 shows the collector characteristics of the common emitter configuration along with the load line. If the transistor is driven from the cut-off region point *a* through the *active* region point *b* to the *saturated region* point *c*, the *only* region for which the equivalent circuit is valid is the *active region.* There is no apriori reason to believe that the equivalent circuit which assumes *normal bias* should yield the correct prediction for any of the transient times.

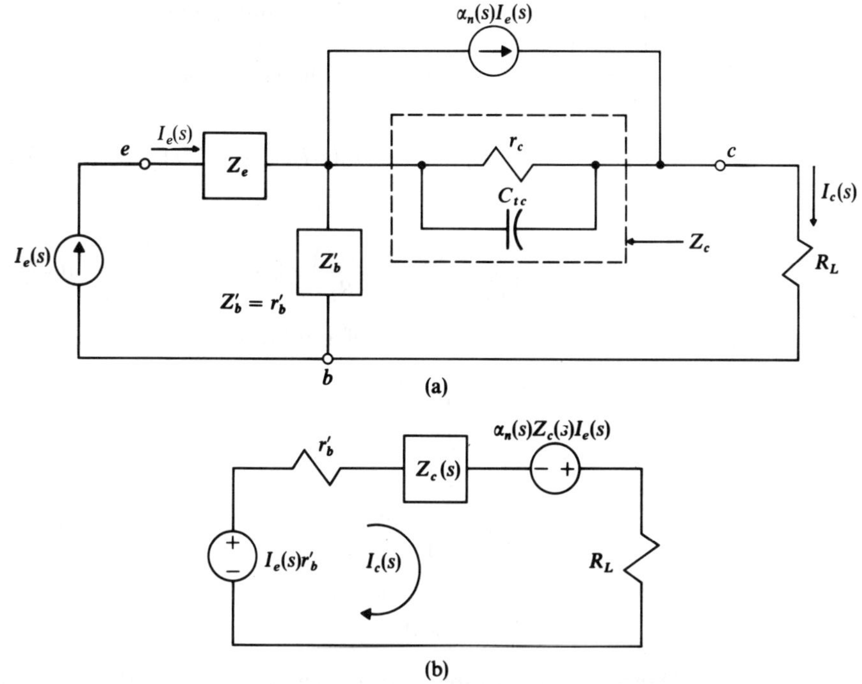

Fig. 6–22 The equivalent circuits used in the analysis of rise and fall times of junction transistors. (a) *T*-equivalent circuit for the transistor. (b) Circuit of Fig. 6–22a simplified.

Ebers and Moll (1954) boldly assumed that the above assumption 1 was valid, and their results did agree fairly well with experimental data they took for devices of their day. Here we follow their method. The T-equivalent circuit of Fig. 6–22a will be used in analyzing rise and fall times.

2. The input "on" and "off" drives are *step currents.*
3. The parameters α_N, ω_N, and C_{tc} are independent of bias.
4. We shall assume that Z'_b is real, that is r'_b.

The assumption of the collector junction capacitance being independent of bias seems like a drastic assumption. If this assumption is not made, we have a nasty nonlinear problem on our hands. It will be shown later that it is possible to find a constant *effective* C_{tc} which takes the nonlinearity into account in a simple way.

We shall begin by analyzing the common base circuit of Fig. 6–22a. We note that because the input is driven by an ideal current source, the series branch Z_e is trivial; it cannot possibly influence the collector current at all and may be removed. Next, Thevenin's theorem may be applied twice: first to the current source $I_e(s)$ and the parallel resistance r_b', and again to the current source $\alpha_n(s)I_e(s)$ and the parallel impedance $Z_c(s)$. The resulting single loop circuit is shown in Fig. 6–22b. Using KVL for the single loop we obtain the following in terms of Laplace transforms,

$$I_c(s)(r_b' + Z_c(s) + R_L) = \alpha_n(s)Z_c(s)I_e(s) + r_b'I_e(s) \tag{6–350}$$

or

$$I_c(s) = \left(\frac{\alpha_n(s)Z_c(s) + r_b'}{r_b' + Z_c(s) + R_L}\right) I_e(s). \tag{6–351}$$

As the emitter current is a step function of amplitude I_{e1},

$$I_e(s) = \frac{I_{e1}}{s}. \tag{6–352}\dagger$$

Furthermore,

$$\alpha_n(s) = \frac{\alpha_N}{1 + s/\omega_N} \tag{6–353}$$

$$Z_c(s) = \frac{r_c}{r_cC_{tc}s + 1}. \tag{6–354}$$

When Eqs. (6–352) through (6–354) are substituted into Eq. (6–351) the following is obtained:

$$I_c(s) = \frac{I_{e1}[\alpha_N r_c + r_b'(r_cC_{tc}s + 1)(1 + s/\omega_N)]}{s(1 + s/\omega_N)[(R_L + r_b')(r_cC_{tc}s + 1) + r_c]}. \tag{6–355}$$

The following magnitude relationships are valid in practical transistors,

$$(R_L + r_b') \ll \alpha_N r_c, \tag{6–356}$$

$$r_cC_{tc}\omega_N \gg 1. \tag{6–357}$$

By using the magnitude relationships of Eqs. (6–356) and (6–357) and by taking the inverse Laplace transform of the collector current, the time response of the collector current is obtained as

$$i_c(t) = \alpha_N I_{e1}\left\{1 - \left[\left(\frac{\omega_N C_{tc}(R_L + r_b')}{\omega_N C_{tc}(R_L + r_b') - 1} - \frac{r_b'}{\alpha_N(R_L + r_b')}\right) e^{-t/(R_L + r_b')C_{tc}} - \frac{e^{-\omega_N t}}{\omega_N C_{tc}(R_L + r_b') - 1}\right]\right\}. \tag{6–358}$$

† We shall follow the convention that the "on" drives will have the subscript 1 and the "off" drives will have the subscript 2 for rise, fall, delay, and storage times.

If a precisely similar procedure is used for the common emitter and common collector circuit configurations (with the load resistance R_L from the output to the common terminal and a current source from input to the common terminal), the output currents are obtained. For common emitter circuit

$$i_c(t) = -\left(\frac{\alpha_N I_{b1}}{1 - \alpha_N + R_L/r_c}\right)[1 - e^{-(1-\alpha_N)\omega_N t/(1+\omega_N C_{tc} R_L)}]. \quad (6\text{–}359)$$

For common collector circuit the output current, which is emitter current, is given by

$$i_e(t) = \left(\frac{I_{b1}}{1 - \alpha_N + R_L/r_c}\right)\left[1 - \left(\frac{\alpha_N + \omega_N C_{tc} R_L}{1 + \omega_N C_{tc} R_L}\right) e^{-(1-\alpha_N)\omega_N t/(1+\omega_N R_L C_{tc})}\right], \quad (6\text{–}360)$$

where I_{b1} = amplitude of the "on" base step drive.

In obtaining Eqs. (6–359) and (6–360) use was made of the magnitude relationships of Eqs. (6–356) and (6–357).

Having obtained the time response, we can easily obtain expressions for the rise times in all three circuits. The 0 to 90 per cent rise time is obtained by solving for the time t_r at which 0.9 times the *maximum* output current is reached.

The common base collector current response of Eq. (6–358) is unfortunately a transcendental equation. No closed form analytical expression for the rise time is obtainable. If one of the following two magnitude relationships is valid analytical solutions are possible.

$$\omega_N R_L C_{tc} \ll 1 \quad (6\text{–}361)$$

or

$$\omega_N R_L C_{tc} \gg 1. \quad (6\text{–}362)$$

If the magnitude relationship (6–361) is true and if

$$R_L \gg r_b'. \quad (6\text{–}363)$$

Equation (6–358) takes on the following simple form

$$i_c(t) = \alpha_N I_{e1}(1 - e^{-\omega_N t}). \quad (6\text{–}364)$$

Using the definition of rise time on Eq. (6–364) we have

$$0.9 I_{CM} = \alpha_N I_{e1}(1 - e^{-\omega_N t_{rb}}). \quad (6\text{–}365)\dagger$$

Equation (6–365) may be solved for t_{rb}

$$t_{rb} = \frac{1}{\omega_N} \ln\left(\frac{I_{e1}}{I_{e1} - 0.9 I_{CM}/\alpha_N}\right), \quad \omega_N R_L C_{tc} \ll 1. \quad (6\text{–}366)$$

† The second subscript on t indicates that this is a calculation for the common base circuit configuration.

If the magnitude relationship Eq. (6–362) is valid then Eq. (6–358) becomes

$$i_c(t) = \alpha_N I_{e1}(1 - e^{-t/R_L C_{tc}}). \tag{6–367}$$

Again using the definition of rise time we have

$$0.9I_{CM} = \alpha_N I_{e1}(1 - e^{-t_{rb}/R_L C_{tc}}). \tag{6–368}$$

Equation (6–368) may be solved for the rise time to obtain

$$t_{rb} = R_L C_{tc} \ln\left(\frac{I_{e1}}{I_{e1} - 0.9I_{CM}/\alpha_N}\right), \omega_N R_L C_{tc} \gg 1. \tag{6–369}$$

Fortunately, for the common emitter and common collector circuits the output currents given by Eqs. (6–359) and (6–360) involve a single exponential term, and a closed form analytic expression for the rise time is possible. Before stating the results we shall make use of another magnitude relationship which is valid in practical transistors,

$$R_L \ll r_c. \tag{6–370}$$

Usually R_L is at most a few thousand Ohms whereas r_c is at least 10–100 Meg-Ohms hence Eq. (6–370) is certainly valid.

Following a procedure just like the one used for common base circuit, the rise time in common emitter and common collector circuits is found to be

$$t_{re} = \frac{1 + \omega_N C_{tc} R_L}{\omega_N(1 - \alpha_N)} \ln\left(\frac{I_{b1}}{I_{b1} - 0.9I_{CM}(1 - \alpha_N)/\alpha_N}\right), \tag{6–371}$$

$$t_{rc} = \frac{1 + \omega_N C_{tc} R_L}{\omega_N(1 - \alpha_N)} \ln\left(\frac{I_{b1}(\alpha_N + \omega_N C_{tc} R_L)/(1 + \omega_N C_{tc} R_L)}{I_{b1} - 0.9I_{EM}(1 - \alpha_N)}\right). \tag{6–372}$$

As

$$\frac{\alpha_N + \omega_N C_{tc} R_L}{1 + \omega_N(1 - \alpha_N)} \approx 1 \tag{6–373}$$

Equation (6–372) may be written more simply as follows:

$$t_{rc} = \frac{1 + \omega_N C_{tc} R_L}{\omega_N(1 - \alpha_N)} \ln\left(\frac{I_{b1}}{I_{b1} - 0.9I_{EM}}\right). \tag{6–374}$$

Note that I_{CM} and I_{EM} are the maximum collector and emitter currents in the common emitter and common collector circuits, respectively. If the transistor saturates,

$$I_{CM}, I_{EM} \doteq V_{CC}/R_L, V_{EE}/R_L. \tag{6–375}$$

If the transistor does not saturate I_{CM} may be obtained from the collector characteristics by graphical analysis. Thus, for example as seen in Fig. 6–19, if the base drive is only I_b, the transistor *does not saturate* and I_{CM} is I_{CM2}.

A precisely similar procedure to the one used in obtaining the rise times may also be utilized to obtain the analytic expressions for the *fall times.* The output current at the beginning of fall time will be assumed to be the same current which flows just before the "off" drive is applied. This will be I_{CM} or I_{EM} in common emitter or common collector circuits, if the transistor saturates.

Recall that fall time ends when the output current has been reduced to 10 per cent of its initial value. The "off" drive is a current step of amplitude I_{e2} in the common base circuit, and I_{b2} in the common emitter and the common collector circuits.

For common base circuit,

$$t_{fb} = \frac{1}{\omega_N} \ln \left(\frac{I_{CM} + \alpha_N I_{e2}}{0.1 I_{CM} + \alpha_N I_{e2}} \right), \; \omega_N R_L C_{tc} \ll 1, \tag{6-376}$$

$$t_{fb} = R_L C_{tc} \ln \left(\frac{I_{CM} + \alpha_N I_{e2}}{0.1 I_{CM} + \alpha_N I_{e2}} \right), \; \omega_N R_L C_{tc} \gg 1. \tag{6-377}$$

For the common emitter circuit,

$$t_{fe} = \left(\frac{1 + \omega_N R_L C_{tc}}{\omega_N (1 - \alpha_N)} \right) \ln \left(\frac{I_{b2} + I_{CM}(1 - \alpha_N)}{I_{b2} + 0.1 I_{CM}(1 - \alpha_N)/\alpha_N} \right). \tag{6-378}$$

For the common collector circuit,

$$t_{fc} = \left(\frac{1 + \omega_N R_L C_{tc}}{\omega_N (1 - \alpha_N)} \right) \ln \left(\frac{I_{b2} + I_{\epsilon M}(1 - \alpha_N)}{I_{b2} + 0.1 I_{\epsilon M}(1 - \alpha_N)} \right). \tag{6-379}$$

In writing Eq. (6–379) we have utilized Eq. (6–373).

The equations derived for rise and fall times involve the collector junction capacitance which we know is a nonlinear function of voltage. Recall that we had assumed in the above derivation that this capacitance was independent of the bias voltage.

How can these two facts be reconciled? We shall define an average linear capacitance which displaces the same charge as does the actual nonlinear junction capacitance between the initial and final voltage swing for rise and fall times as follows:

$$C_{av} = \frac{\int_{V_I}^{V_F} C_{tc}(V) dV}{V_F - V_I}, \tag{6-380}$$

where V_F = final voltage across C_{tc},

V_I = initial voltage across C_{tc}.

From the definition of junction capacitance we have

$$C_{tc} dV = dQ, \tag{6–381}$$

where Q = instantaneous charge on the junction C_{tc}.

Thus, Eq. (6–380) takes the following form with the help of Eq. (6–381)

$$C_{av} = \frac{\int_{V_I}^{V_F} dQ}{V_F - V_I} = \frac{Q(V_F) - Q(V_I)}{V_F - V_I}. \tag{6–382}$$

For step or abrupt junctions,

$$\frac{dQ}{dV} = k_a V^{-\frac{1}{2}} \Rightarrow Q = 2k_a V^{\frac{1}{2}}. \tag{6–383}$$

For linearly graded junctions

$$\frac{dQ}{dV} = k_\ell V^{-\frac{1}{3}} \Rightarrow Q = \tfrac{3}{2} k_\ell V^{\frac{2}{3}}. \tag{6–384}$$

Note that Eq. (6–382) is general and is valid for all kinds of semiconductor junctions.

For 0 to 90 per cent rise time calculations,

$$V_I = V_{CC} \tag{6–385}$$

and

$$V_F = 0.1 V_{CC}. \tag{6–386}$$

With the use of Eqs. (6–385), (6–386), and (6–383), Eq. (6–382) gives

$$C_{av} = 2k_a \cdot \frac{V_{CC}^{\frac{1}{2}} - (0.1 V_{CC})^{\frac{1}{2}}}{0.9 V_{CC}} = 1.52 C_{tc}(V_{CC}) \quad \text{(abrupt junctions)}. \tag{6–387}$$

A similar calculation for linearly graded junctions gives

$$C_{av} = 1.31 C_{tc}(V_{CC}) \qquad \text{(linearly graded junctions)}. \tag{6–388}$$

A formulation for the calculation of the average capacitance for rise times is possible for a *general junction* (Nanavati, 1959) of the following kind.

$$C_{tc} = k_n v^{-n/m}, \qquad n < m \tag{6–389}$$

$$Q = \int C_{tc} dV = k_n V^{-n/m} dV = \frac{m k_n V^{(m-n)/m}}{m - n}. \tag{6–390}$$

To calculate 0 to 90 per cent rise time we may still use Eqs. (6–385), (6–386), and (6–382).

$$C_{av} = \frac{mk_n}{(m-n)} \cdot \frac{V_{CC}{}^{(m-n)/m} - (0.1V_{CC})^{(m-n)/m}}{0.9V_{CC}}, \tag{6–391}$$

$$C_{av} = \left(\frac{m}{m-n}\right)\left(\frac{1-(0.1)^{(m-n)/m}}{0.9}\right) C_{tc}(V_{CC}) \qquad \text{(rise time)}. \tag{6–392}$$

To calculate 0 to 90 per cent *fall time*

$$V_I = 0 \tag{6–393}$$

$$V_F = 0.9V_{CC}. \tag{6–394}$$

Equations (6–382) and (6–390) now lead to the following result for 0 to 90 per cent fall time.

$$C_{av} = \left(\frac{m}{m-n}\right)(0.9)^{-n/m} C_{tc}(V_{CC}) \qquad \text{(fall time)}. \tag{6–395}$$

Equations (6–392) and (6–395) are quite general. They can be easily specialized to the abrupt junction and the linearly graded junction cases by letting $n = 1/2$ and $n = 1/3$, respectively.

Thus, for abrupt junctions 0 to 90 per cent fall time

$$C_{av} = 2.11C_{tc}(V_{CC}). \tag{6–396}$$

For linearly graded junctions 0 to 90 per cent fall time

$$C_{av} = 1.55C_{tc}(V_{CC}). \tag{6–397}$$

The values given in Table 6–3 should be used in the expressions for rise and fall times already obtained. The values will enable us to properly calculate these transient times. We must keep in mind that even though the correct rise and fall

Table 6–3 The Value of Collector Capacitance to Be Used in Expressions for Rise and Fall Times

	Rise time, 0–90 per cent	Fall time, 0–90 per cent
linearly graded junctions $C_{tc} = k_\ell V^{-\frac{1}{3}}$	$1.31C_{tc}(V_{CC})$	$1.55C_{tc}(V_{CC})$
Step junctions $C_{tc} = k_s V^{-\frac{1}{2}}$	$1.52C_{tc}(V_{CC})$	$2.11C_{tc}(V_{CC})$
General junctions $C_{tc} = k_n V^{-n/m}$	$\left(\frac{m}{m-n}\right)\frac{1-(0.1)^{(m-n)/m}}{0.9}C_{tc}(V_{CC})$	$\frac{m}{m-n}(0.9)^{-n/m}C_{tc}(V_{CC})$

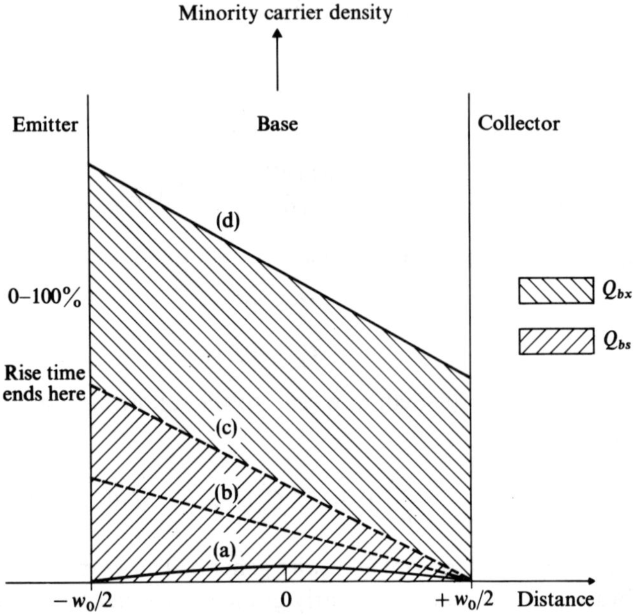

Fig. 6–23 Minority carrier density build-up as the transistor is driven from cut-off to saturation. (a) Cut-off. (b) Input drive not enough for saturation, for example, operating point *b* in Fig. 6–21. (c) Output drive just enough to saturate the transistor, for example $I_b = I_{bs}$ in Fig. 6–21. (d) Input drive more than enough to saturate transistor, for example $I_b = I_{b1} > I_{bs}$ in Fig. 6–21.

times are calculated by the above procedure, it will not predict the proper shape between the 0 to 90 per cent points in rise and fall times. In most applications the rise and fall times are adequate information. When it is necessary to predict the detailed shape difficult nonlinear equations would need to be solved.†

6–18 STORAGE TIME

To understand the storage time it is very useful to look at the changes in the minority carrier density in the base region. Figure 6–23 shows the build-up of minority carrier charge density as the input current is increased from cut-off to saturation.

It is worthwhile noting that the minority carrier charge density is *not* through building up when the rise time is over. The steady state distribution corresponding to the base drive current I_{b1} is obtained some time later.

We shall assume that when storage time begins that the minority carrier

† See Nanavati (1963b).

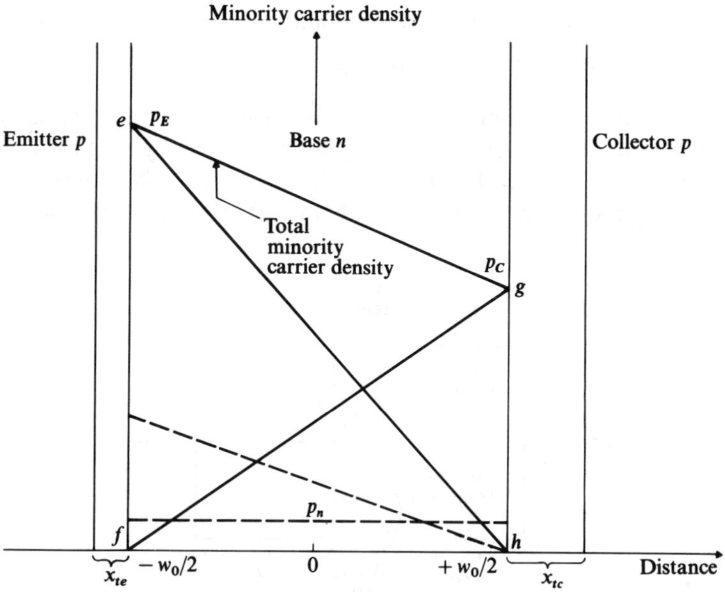

Fig. 6–24 The figure shows how a saturated transistor is equivalent to two unsaturated transistors, one operating in the normal and the other operating in the inverse mode. Triangle *efh* represents the normal transistor. Triangle *gfh* represents the inverse transistor.

density in the base region is at its steady state value corresponding to the build-up during the turn-on drives.

Ebers and Moll made a major contribution to the understanding of storage time when they suggested that the initial distribution of minority carrier density in the base region may be thought to be made up of a sum of individual distributions as shown in Fig. 6–24. First, the total minority carrier density is approximated by a *straight line.* The total minority carrier density in the base region may be thought to consist of two seperate triangular distributions *efh* and *ghf.* From elementary geometrical considerations it is evident that the sum of areas of the two triangles indeed equals the area *efgh.* The triangular distribution *efh* looks just like the distribution for a transistor in Fig. 4–5 curve *b*, which is normally biased (emitter junction forward biased, collector junction reverse biased). The triangular distribution *ghf* is the distribution we would expect from a sort of backward transistor--an *inverse transistor. A transistor in which the collector junction is forward biased and the emitter junction is reverse biased is said to be operated in the inverse mode.* Thus, the saturated transistor is equivalent to two transistors back to back, one normal and one inverse, each operating in its active region. Figure 6–25 shows this model. Note that because the minority carrier density at both junctions is significantly greater than the thermal equilibrium minority carrier density, both junctions are forward biased because

$$p(-w_0/2) = p_n e^{qV_E/kT}, \; p(+w_0/2) = p_n e^{qV_C/kT}.$$

This implies that both the emitter and collector are injecting minority carriers into the base region. The storage time may be calculated by requiring that the excess base charge Q_{bx} (see Fig. 6–23) be zero. Alternatively, storage time may be obtained by requiring that the current emitted by the collector be zero.

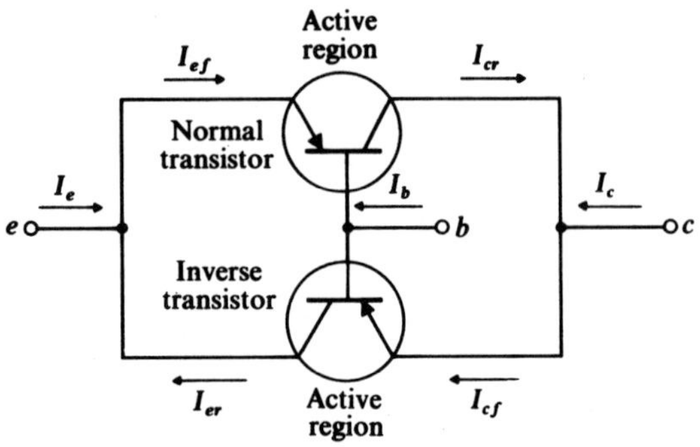

Fig. 6–25 A model of a saturated *p-n-p* transistor.

In terms of the model of Fig. 6–25, we may write the following expressions for the total emitter and collector currents:

$$I_e = I_{ef} - I_{er}, \tag{6–398}$$

$$I_c = I_{cf} - I_{cr}. \tag{6–399}$$

For the normal transistor the normal gain may be written as:

$$I_{cr}/I_{ef} \triangleq \alpha_n(s) = \frac{\alpha_N}{1 + s/\omega_N}. \tag{6–400}$$

For the inverse transistor the inverse gain may be written analogously

$$\frac{I_{er}}{I_{cf}} \triangleq \alpha_i(s) = \frac{\alpha_I}{1 + s/\omega_I}, \tag{6–401}$$

where α_I = low frequency common base current gain for the inverse transistor,
ω_I = the intrinsic common base current gain cut-off frequency for the inverse transistor.

We shall assume that a step excitation is used in turning off the transistor such that the emitter base and collector currents change their values from I_{e1}, I_{b1}, I_{c1} to I_{e2}, I_{b2}, I_{c2}.

With the help of Eqs. (6–398) through (6–401) and some algebra we obtain

$$I_{cf}(s) = -\frac{(I_{c2} - I_{c1})(1 + s/\omega_N)(1 + s/\omega_I) + \alpha_N(I_{e2} - I_{e1})(1 + s/\omega_I)}{(s/\omega_N\omega_I)[s^2 + s(\omega_N + \omega_I) + \omega_N\omega_I(1 - \alpha_N\alpha_I)]}, \quad (6\text{–}402)$$

$$I_{er}(s) = -\frac{\alpha_I(I_{c2} - I_{c1})(1 + s/\omega_N) + \alpha_I\alpha_N(I_{e2} - I_{e1})}{(s/\omega_N\omega_I)[s^2 + s(\omega_N + \omega_I) + \omega_N\omega_I(1 - \alpha_N\alpha_I)]}. \quad (6\text{–}403)$$

Taking the inverse transform of Eqs. (6–402) and (6–403) we obtain

$$\begin{aligned} i_{cf}(t) = {} & -\frac{I_{c2} + \alpha_N I_{e2}}{1 - \alpha_N\alpha_I} \\ & + \frac{\left[(I_{c2} - I_{c1})\left(\frac{\omega_B}{\omega_I} - \frac{\omega_B}{\omega_A}\right)\left(1 - \frac{\omega_A}{\omega_N}\right) + \alpha_N(I_{e2} - I_{e1})\left(\frac{\omega_B}{\omega_I} - \frac{\omega_B}{\omega_A}\right)\right]e^{-\omega_A t}}{(1 - \alpha_N\alpha_I)\left(1 - \frac{\omega_B}{\omega_A}\right)} \\ & + \frac{\left[(I_{c2} - I_{c1})\left(1 - \frac{\omega_B}{\omega_N}\right)\left(1 - \frac{\omega_B}{\omega_I}\right) + \alpha_N(I_{e2} - I_{e1})\left(1 - \frac{\omega_B}{\omega_I}\right)\right]e^{-\omega_B t}}{(1 - \alpha_N\alpha_I)\left(1 - \frac{\omega_B}{\omega_A}\right)} \end{aligned} \quad (6\text{–}404)$$

and

$$\begin{aligned} i_{er}(t) = {} & -\frac{\alpha_I(I_{c2} + \alpha_N I_{e2})}{1 - \alpha_N\alpha_I} \\ & + \frac{\left[\alpha_I(I_{c2} - I_{c1})\left(\frac{\omega_B}{\omega_N} - \frac{\omega_B}{\omega_A}\right) - \frac{\omega_B}{\omega_A}\alpha_N\alpha_I(I_{e2} - I_{e1})\right]e^{-\omega_A t}}{(1 - \alpha_N\alpha_I)\left(1 - \frac{\omega_B}{\omega_A}\right)} \\ & + \frac{\left[\alpha_I(I_{e2} - I_{e1})\left(1 - \frac{\omega_B}{\omega_N}\right) - \alpha_N\alpha_I(I_{e2} - I_{e1})\right]e^{-\omega_B t}}{(1 - \alpha_N\alpha_I)\left(1 - \frac{\omega_B}{\omega_A}\right)}, \end{aligned} \quad (6\text{–}405)$$

where ω_A and ω_B are the larger and smaller roots respectively of the following equation:

$$\omega^2 - \omega(\omega_N + \omega_I) + \omega_N\omega_I(1 - \alpha_N\alpha_I) = 0. \quad (6\text{–}406)$$

By solving Eq. (6–406) ω_A and ω_B may be obtained as follows:

$$\omega_A = \omega_N + \omega_I - \frac{\omega_N\omega_I(1 - \alpha_N\alpha_I)}{\omega_N + \omega_I} \quad (6\text{–}407)$$

$$\omega_B = \frac{\omega_N \omega_I (I - \alpha_N \alpha_I)}{\omega_N \omega_I}. \tag{6–408}$$

Generally

$$\omega_B \ll \omega_I. \tag{6–409}$$

Comparing Eqs. (6–407) and (6–408) it may be concluded that

$$\omega_A \gg \omega_B. \tag{6–410}$$

During storage time both junctions are forward biased and hence the terminal currents will be determined primarily by external circuit elements. This implies in common emitter and common base circuits that

$$I_{c2} = I_{c1} \doteq V_{CC}/R_L. \tag{6–411}$$

Finally the requirement that

$$I_{cf}(t_s) = 0 \tag{6–412}$$

and algebraic manipulations lead to the following expressions for the storage times.

$$t_{sb} = \frac{\omega_N + \omega_I}{\omega_N \omega_I (1 - \alpha_N \alpha_I)} \ln \left(\frac{I_{e2} - I_{e1}}{I_{e2} - I_{es}} \right) \tag{6–413}$$

$$t_{se} = \frac{\omega_N + \omega_I}{\omega_N \omega_I (1 - \alpha_N \alpha_I)} \ln \left(\frac{I_{b2} - I_{b1}}{I_{b2} - I_{bs}} \right). \tag{6–414}$$

By setting $I_{e2} = I_{e1}$ the storage time for the common collector circuit is obtained

$$t_{sc} = \frac{\omega_N + \omega_I}{\omega_N \omega_I (1 - \alpha_N \alpha_I)} \ln \left(\frac{I_{b2} - I_{b1}}{I_{b2} - I_{bs}} \right), \tag{6–415}$$

where I_{es} and I_{bs} = the emitter and base currents necessary to just saturate the transistor.

In Eqs. (6–413) through (6–415) the input drives are algebraic quantities. If the "off" drive is opposite in direction to that of the "on" drive, as is often the case, then it is itself negative and the magnitudes of the "on" and "off" drives *add*.

It is worthwhile noticing that the factor outside the logarithm in all three above equations is the same. This factor, a time constant, is

$$T_s^* = \frac{\omega_N + \omega_I}{\omega_N \omega_I (1 - \alpha_N \alpha_I)}. \tag{6–416}$$

A more fundamental analysis of storage time indicates that the storage time constant T_s should be used in place of T_s^* as follows:

$$T_s = 1.22\, T_s^*. \tag{6–417}$$

Thus, the storage time for the common emitter circuits, for example, should be more accurately expressed as

$$t_{se} = T_s \ln\left(\frac{I_{b2} - I_{b1}}{I_{b2} - I_{bs}}\right). \tag{6–418}$$

The storage time constant T_s is one of the fundamental high frequency parameters of the transistor.

The factor of 1.22 appearing in Eq. (6–417) has its counterpart in the expressions for rise and fall times also. Table 6–4 shows the expressions of rise, fall, and storage transient times where the factor of 1.22 is shown where it belongs. Table 6–3 shows the expressions for rise time. This numerical factor has its origin in relating the minority carrier lifetime in the base region to ω_N, α_N when the diffusion equation is used as the starting point.

6–19 THE CHARGE CONTROL APPROACH

Another method of analyzing transistor transients is the *charge control analysis.* Beaufoy and Sparkes (1957) first suggested that the transistor ought to be considered a *charge controlled device.* A major assumption of the analysis is the following.

1. The *total* minority carrier charge in the base region determines the behavior of the transistor.

The starting point of this analysis is the continuity equation written in terms of excess minority carrier density p^* which stands for either electrons or holes. We may write the continuity equation in terms of p^* as follows:

$$\nabla \cdot J_n = -\nabla \cdot J_p = \frac{q\partial p^*}{\partial t} + \frac{qp^*}{\tau}, \tag{6–419}$$

where $p^* = p - p_n$ or $n - n_p$,

J_n, J_p = electron and hole current densities,

τ = minority carrier lifetime in the base.

We may integrate Eq. (6–419) over the *field-free base* region as follows:

$$\int_{\text{base}} \nabla \cdot J_n dV = -\int_{\text{base}} \nabla \cdot J_p dV = \frac{\partial Q_b(t)}{\partial t} + \frac{Q_b(t)}{\tau}, \tag{6–420}$$

where Q_b = total excess base charge of either kind of carriers in the field free base region.

dV = volume element in the field free base region.

Table 6–4

Quantity	Common base	Common emitter	Common collector
t_r (rise time)	For $\omega_N R_L C_{tc} \ll 1$ $\frac{1.22}{\omega_N} \ell n \frac{I_{e1}}{I_{e1} - 0.9 I_{CM}/\alpha_N}$ For $\omega_N R_L C_{tc} \gg 1$ $R_L C_{tc} \ell n \frac{I_{e1}}{I_{e1} - 0.9 I_{CM}/\alpha_N}$	$\frac{1.22 + \omega_N R_L C_{tc}}{\omega_N (1 - \alpha_N)} \ell n \frac{I_{b1}}{I_{b1} - (0.9 I_{CM}/\alpha_N)(1 - \alpha_N)}$	$\frac{1.22 + \omega_N R_L C_{tc}}{\omega_N (1 - \alpha_N)} \ell n \frac{I_{b1}}{I_{b1} - 0.9 I_{CM}(1 - \alpha_N)}$
t_s (storage time)	$\frac{(\omega_N + \omega_I)\, 1.22}{\omega_N \omega_I (1 - \alpha_N \alpha_I)} \ell n \frac{I_{e2} - I_{e1}}{I_{e2} - I_{es}}$	$\frac{(\omega_N + \omega_I)\, 1.22}{\omega_N \omega_I (1 - \alpha_N \alpha_I)} \ell n \frac{I_{b1} - I_{b2}}{I_{bs} - I_{b2}}$	$\frac{(\omega_N + \omega_I)\, 1.22}{\omega_N \omega_I (1 - \alpha_N \alpha_I)} \ell n \frac{I_{b1} - I_{b2}}{I_{bs} - I_{b2}}$
t_f (fall time)	For $\omega_N R_L C_{tc} \ll 1$ $\frac{1.22}{\omega_N} \ell n \frac{I_{CM} + \alpha_N I_{e2}}{0.1 I_{CM} + \alpha_N I_{e2}}$ For $\omega_N R_L C_{tc} \gg 1$ $R_L C_{tc} \ell n \frac{I_{CM} + \alpha_N I_{e2}}{0.1 I_{CM} + \alpha_N I_{e2}}$	$\frac{1.22 + \omega_N R_L C_{tc}}{\omega_N (1 - \alpha_N)} \ell n \frac{I_{b2} + \frac{I_{CM}(1 - \alpha_N)}{\alpha_N}}{I_{b2} + (0.1 I_{CM}/\alpha_N)(1 - \alpha_N)}$	$\frac{1.22 + \omega_N R_L C_{tc}}{\omega_N (1 - \alpha_N)} \ell n \frac{I_{b2} + I_{CM}(1 - \alpha_N)}{I_{b2} + 0.1 I_{CM}(1 - \alpha_N)}$

The two volume integrals on the left hand side may be converted into surface integrals by the use of the divergence theorem (also known as Gauss's theorem†). The evaluation of these surface integrals over the surface which encloses the volume V gives the following general basic equation of the charge control analysis.

$$i_{in}(t) = \frac{dQ_b(t)}{dt} + \frac{Q_b(t)}{\tau}, \tag{6–421}$$

where $i_{in}(t)$ = *net* current of either kind of carriers into the base region.

Note that going from the right hand side of Eq. (6–420) to the right hand side of Eq. (6–421) we have converted the partial derivatives to total derivatives since Q_b is a function only of time t.

It is instructive to note that in Table 6–4 there are only three time constants (not including $R_L C_{tc}$). One interesting interpretation of these time constants is in terms of Eq. (6–421). If there is no *net* current going into the base region, any excess charge which may be present will reduce itself to zero. Furthermore, in *steady state*

$$\frac{dQ_b}{dt} = 0 \tag{6–422}$$

and

$$i_{in} = \frac{Q_b}{\tau}. \tag{6–423}$$

If we assume that the collector current and emitter current are essentially the same,‡ the net current into the base region is the *base* current I_B. Then from Eq. (6–423) we have

$$\frac{Q_b}{I_B} \triangleq T_b = \tau. \tag{6–424}$$

The charge Q_b is called the *base control charge*. The time constant T_b can be expressed in terms of ω_N and α_N as follows:

$$T_b = \tau = \frac{1}{\omega_\beta} = \frac{1.22}{\omega_N(1 - \alpha_N)}. \tag{6–425}$$

The time constant just obtained may be interpreted as the ratio of the base

† This theorem is not to be confused with Gauss's theorem in the electro-magnetic field theory discussed in Chapter 9.

‡ This also implies that the minority carrier density distribution in the base region is a straight line.

control charge to the current necessary to maintain it.† The following new time constants may be defined.

$$T_e \triangleq \frac{Q_b}{I_E}, \tag{6–426}$$

$$T_c \triangleq \frac{Q_b}{I_C}. \tag{6–427}$$

The direct integration of the excess minority carrier base charge density in the base region under normal bias condition gives:

$$Q_b = \frac{w_0^2 I_E}{2D_p} = \frac{1.22 I_E}{\omega_N} \tag{6–428}$$

or

$$T_e = \frac{Q_b}{I_E} = \frac{1.22}{\omega_N}. \tag{6–429}$$

To calculate T_c we note first that

$$I_C = \alpha_N I_E. \tag{6–430}$$

With the help of Eq. (6–430), Eq. (6–429) may be written

$$T_c = \frac{Q_b}{I_C} = \frac{Q_b}{\alpha_N I_E} = \frac{1.22}{\alpha_N \omega_N}. \tag{6–431}$$

The storage time constant is defined and obtained as follows:

$$T_s \triangleq \frac{Q_{bx}}{I_{BX}} = \frac{1.22(\omega_N + \omega_I)}{\omega_N \omega_I (1 - \alpha_N \alpha_I)}, \tag{6–432}$$

where Q_{bx} = base charge in excess of what is present when the transistor is just saturated,

I_{BX} = excess base current, the current in excess of the current which just saturates the base current I_{bs} (see Fig. 6–21).

If $i_{\text{in}}(t)$ is a step function, the first order differential equation, Eq. (6–421) has a solution of the following form:

$$Q_b(t) = Q_{bss} + (Q_{bi} - Q_{bss}) e^{-t/\tau_{\text{eff}}}, \tag{6–433}$$

where $Q_{bss} = Q_b(\infty)$ = steady state base control charge,

$Q_{bi} = Q_b(0)$ = initial base control charge,

τ_{eff} = effective time constant.

† The time constants may also be defined in terms of incremental quantities, e.g.

$$T_b = \frac{dQ_b}{dI_B}$$

Let us calculate the rise time to illustrate the method in the simple case where the magnitude relationship Eq. (6–361) is true. In this case:

$$\tau_{\text{eff}} = \tau = \frac{1.22}{\omega_N(1 - \alpha_N)}. \tag{6–434}$$

In this case

$$Q_{bi} = 0 \tag{6–435}$$

$$Q_{bss} = I_b(\infty)T_b = I_{b1}T_b. \tag{6–436}$$

Rise time ends when the collector current reaches $0.9I_{CM}$ at which time

$$Q_b(t_r) = 0.9\, I_{CM}T_c. \tag{6–437}$$

Substituting Eqs. (6–435) through (6–437) into Eq. (6–433), we have

$$Q_b(t_r) = 0.9\, I_{CM}T_c = I_{b1}T_b + (0 - I_{b1}T_b)\,e^{-t_{re}/\tau} \tag{6–438}$$

or

$$\frac{0.9\, I_{CM}T_c}{I_{b1}T_b} = (1 - e^{-t_{re}/\tau}) \tag{6–439}$$

or

$$e^{-t_{re}/\tau} = \frac{I_{b1}T_b - 0.9\, I_{CM}T_c}{I_{b1}T_b}. \tag{6–440}$$

Now solving for t_{re} we obtain

$$t_{re} = \tau \ln\left(\frac{I_{b1}}{I_{b1} - 0.9\, I_{CM}T_c/T_b}\right). \tag{6–441}$$

Note that

$$\frac{T_c}{T_b} = \frac{1.22/(\alpha_N\omega_N)}{1.22/[\omega_N(1 - \alpha_N)]} = \frac{1 - \alpha_N}{\alpha_N}. \tag{6–442}$$

With the use of Eqs. (6–434) and (6–442), Eq. (6–441) becomes

$$t_{re} = \frac{1.22}{\omega_N(1 - \alpha_N)} \ln\left[\frac{I_{b1}}{I_{b1} - 0.9\, I_{CM}[(1 - \alpha_N)/\alpha_N]}\right]. \tag{6–443}$$

It is to be noted that Eq. (6–443) is the same as Eq. (6–371) under the condition given by the magnitude relationship (6–361). If instead of using $\tau_{\text{eff}} = \tau$, the following is used

$$\tau_{\text{eff}} = T_{bv} = \frac{1.22 + R_L C_{tc}\omega_N}{\omega_N(1 - \alpha_N)} \tag{6–444}$$

we obtain

$$t_{re} = \left(\frac{1.22 + R_L C_{tc}\omega_N}{\omega_N(1 - \alpha_N)}\right) \ln\left[\frac{I_{b1}}{I_{b1} - 0.9\, I_{CM}[(1 - \alpha_N)/\alpha_N]}\right] \tag{6–445}$$

which is precisely the same as Eq. (6–371).

Thus, we see that the same results are obtained by the charge control method as by the equivalent circuit method. This is a general observation true for all transient times. Their fundamental limitations are essentially the same. Both approaches make comparably drastic approximations. One assumes that the large signal prediction can be made from small signal equivalent circuits. The other assumes that the total minority carrier charge is adequate to determine the transient response. Both of these are drastic assumptions.

As a final example we shall examine the influence of "on" pulse width on storage time. We have already seen that the steady state distribution of minority carriers is not reached when the rise time ends. If the input pulse width does not last long enough for the steady state minority carrier distribution to be reached, the storage time will be reduced because Q_{bx} is less.

We shall now obtain the expression for storage time in terms of the initial charge $Q_b(0)$ when storage time begins.

$$Q_b(t) = Q_{bss} + [Q_b(0) - Q_{bss}]\, e^{-t/T_s}, \tag{6–446}$$

where

$$Q_{bss} = I_{b2} T_b, \tag{6–447}$$

where I_{b2} = "off" base drive.

The storage time ends when the total base charge takes on the value appropriate to the edge of saturation. Thus,

$$Q_b(t_{se}) = I_{CM} T_c. \tag{6–448}$$

The substitution of Eqs. (6–448) and (6–447) into Eq. (6–446) leads to

$$t_{se} = T_s \ln\left[\frac{Q_b(0) - I_{b2} T_b}{I_{CM} T_c - I_{b2} T_b}\right]. \tag{6–449}$$

Now we shall calculate $Q_b(0)$. To do this we shall need to write the appropriate equation valid during the build-up of charge while the transistor is being saturated.

$$Q_b(t') = Q_{bss} + [Q_b(0') - Q_{bss}]\, e^{-(t' - t_d - t_r)/T_s}, \tag{6–450}$$

where t' = time begins when the transistor saturates,

$t_d = t_{dc} + t_{dp}$ delay time,

t_r = 0 to 100 per cent rise time.

Note that 0 to 100 per cent rise time is easily written as

$$t_{re} = T_{bv} \ln\left[\frac{I_{b1}}{I_{b1} - I_{CM}(1 - \alpha_N)/\alpha_N}\right], \tag{6–451}$$

where I_{CM} replaces (0.9) I_{CM}.

The base charge at the beginning of saturation is

$$Q(0') = I_{CM}T_c \tag{6–452}$$

$$Q_{bss} = I_{b1}T_b. \tag{6–453}$$

If the input pulse duration is of T seconds, Eqs. (6–452) and (6–453) substituted into Eq. (6–450) give us

$$Q_b(T - t_d - t_{re}) = I_{b1}T_b + (I_{CM}T_c - I_{b1}T_b)\,\mathrm{e}^{-(T-t_d-t_r)/T_s}. \tag{6–454}$$

The right hand side can serve as the initial charge $Q_b(0')$ in Eq. (6–449). When this substitution is made

$$t_{se} = T_s \ln\left[\frac{I_{b1} - I_{b2} + (I_{bs} - I_{b1})\,\mathrm{e}^{-(T-t_d-t_r)/T_s}}{(I_{bs} - I_{b2})}\right]. \tag{6–455}$$

We note that when

$$T - t_d - t_r \geqslant 5T_s \tag{6–456}$$

the exponential term inside the logarithm is negligibly small and the storage time expression reduces to the previously obtained result in Eq. (6–418). At the other extreme if

$$T = t_d + t_r \tag{6–457}$$

then

$$t_{se} = T_s \ln\left[\frac{-I_{b2} + I_{bs}}{-I_{b2} + I_{bs}}\right] = 0. \tag{6–458}$$

This result agrees with our understanding that when the excess charge Q_{bx} is zero, the time necessary to remove it, t_s, is also zero.

The equations derived for storage time in the last section are correct provided that the magnitude relationship Eq. (6–456) is valid.

6–20 THE DIFFUSION EQUATION AS THE STANDARD TO DETERMINE THE LIMITS OF THE EQUIVALENT CIRCUIT AND THE CHARGE CONTROL ANALYSES

The most fundamental approach to the transient analysis is the solution of the diffusion equation itself. Very few authors have tried this approach only because the mathematics is more difficult and because of widespread belief, first expressed in the papers of Ebers and Moll, that this approach gives essentially the same result as the much simpler equivalent circuit approach. Nanavati (1960b, 1961) has shown that this approach is worth considering not only because it puts limits on the validity of the other methods of analyses but also because it predicts *new* results not obtainable from other methods of analysis.

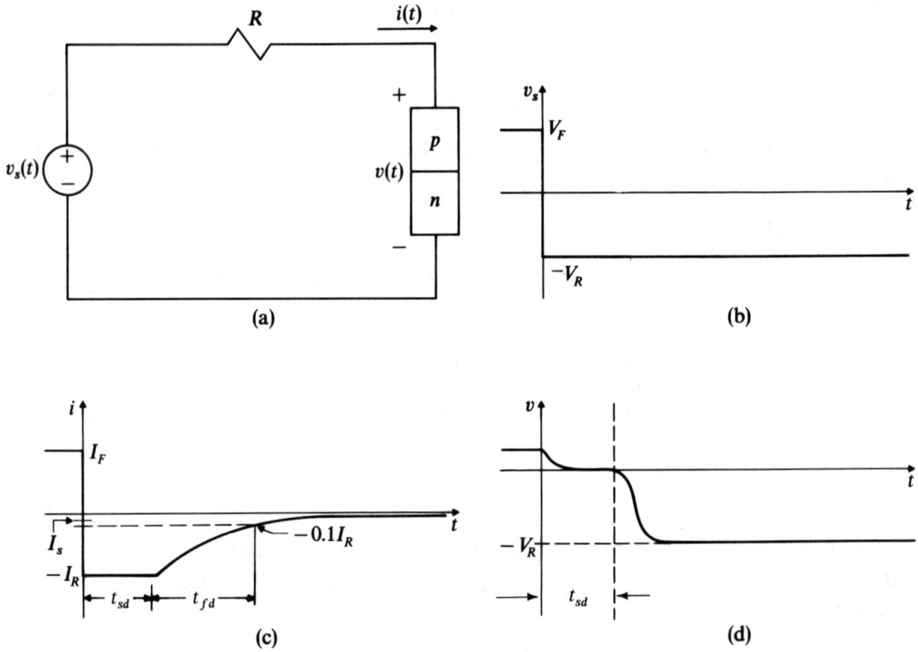

Fig. 6–26 A *p-n* junction diode being switched from saturation (fully "on") to cut-off.

If the diffusion equation approach is the most fundamental, then we conclude that the transistor in reality is neither a current controlled device, nor is it a charge controlled device. It is fundamentally a *minority carrier density distribution controlled device.* That is, its response depends essentially on how the minority carrier charge is distributed in the base region at any given time.

Before considering the transistor let us take a brief look at the junction diode and calculate its storage time from the diffusion equation point of view. Figure 6–26 shows the diode circuit, the voltage source waveform and the resulting current through and voltage across the diode as a function of time. The diode storage time is t_{sd} and the diode fall time is t_{fd}.

Assuming that

$$\sigma_p \gg \sigma_n \tag{6–459}$$

the major current crossing the junction is hole current and the events of interest are occurring on the *n*-side. Figure 6–27 shows the decay of minority carriers from saturation to cut-off. The initial distribution is the exponentially decaying function discussed in Chapter 3. The current is

$$i(0^-) = I_f = \frac{V_f}{R}. \tag{6–460}$$

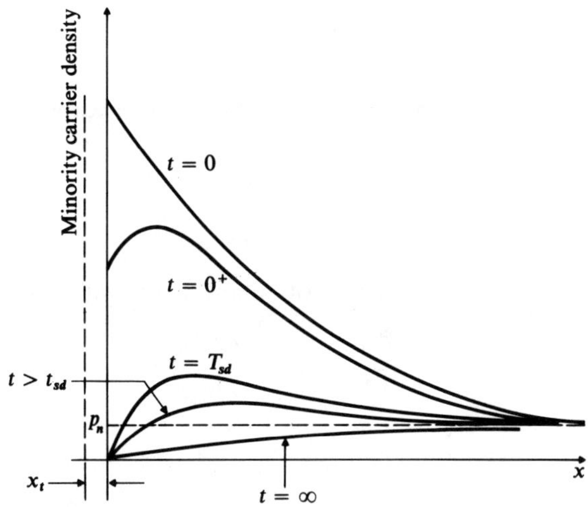

Fig. 6–27 The decay of minority carrier density during storage and fall times in *p-n* junction diode.

When the source voltage reverses, so does the current

$$i(0^+) = -I_r = -\frac{V_r}{R}. \tag{6–461}$$

As diffusion is the only transporting mechanism for minority carriers in the bulk of the *n*-region, the transport equation is simply

$$j_p = -qD_p\frac{\partial p}{\partial x}. \tag{6–462}$$

Because of the magnitude relationship (6–459)

$$i_p = i, \tag{6–463}$$

where i = total current crossing the junction.

From Eqs. (6–463) and (6–462) we have

$$\frac{\partial p}{\partial x} = -\frac{i}{AqD_p}. \tag{6–464}$$

Equation (6–464) makes the fact clear that when i changes sign, $\partial p/\partial x$ must also change sign, hence, the change in slope of the curve at $t = 0^+$. During diode storage time the minority carrier density is greater than p_n and the boundary conditions for the decay phase are

$$\left.\frac{\partial p}{\partial x}\right|_{x=0} = +\frac{I_r}{AqD_p} \quad \text{during storage time} \tag{6–465}$$

and
$$p(x \text{ large}) = p_n. \tag{6–466}$$

When the carrier density at $x = 0$ passes through p_n and quickly goes to zero, diode storage time ends, and diode fall time commences with the boundary conditions.

$$p(x = 0) = 0 \tag{6–467}$$

$$p(x \text{ large}) = p_n. \tag{6–468}$$

The diffusion equation may be written in the following form

$$\frac{\partial^2 p^*}{\partial x^2} - p^* = \frac{\partial p^*}{\partial T_p}, \tag{6–469}$$

where

$$p^* \triangleq p(x, t) - p_n \tag{6–470}$$

$$X \triangleq x/L_p \tag{6–471}$$

$$T_p = t/\tau_p. \tag{6–472}$$

The solution of Eq. (6–469) for diode storage time is obtained as

$$\operatorname{erf}\left(\frac{t_{sd}}{\tau_p}\right)^{\frac{1}{2}} = \frac{1}{1 + I_r/I_f}. \tag{6–473}$$

The left hand side is the error function which is defined by

$$\operatorname{erf}(x) = \frac{2}{\pi^{\frac{1}{2}}} \int_0^x e^{-y^2}\, dy. \tag{6–474}$$

The diode fall time is obtained as

$$0.1\left(\frac{I_r}{I_f}\right) = \operatorname{erf}\left(\frac{T_{fd} - t_{sd}}{\tau_p}\right)^{\frac{1}{2}} + \frac{e^{-(t_{fd} - t_{sd})/\tau_p}}{\left[\frac{\pi}{\tau_p}(t_{fd} - t_{sd})\right]^{\frac{1}{2}}} - 1 \tag{6–475}$$

for

$$t > t_{sd} \quad \text{and} \quad t \gg \tau_p. \tag{6–476}$$

Note that the important drive parameter is the reverse to forward current, and the device parameter is the minority carrier lifetime† τ_p. Indeed, Eq. (6–473) is often used to measure the lifetime of the lightly doped side of the *p-n* junction.

The transistor case is quite similar to the diode case except that now there are two junctions. Assuming that the transistor is initially staturated and that the equilibrium minority carrier density distribution is prevailing in the

† If $\sigma_n \gg \sigma_p$, the lifetime would be τ_n.

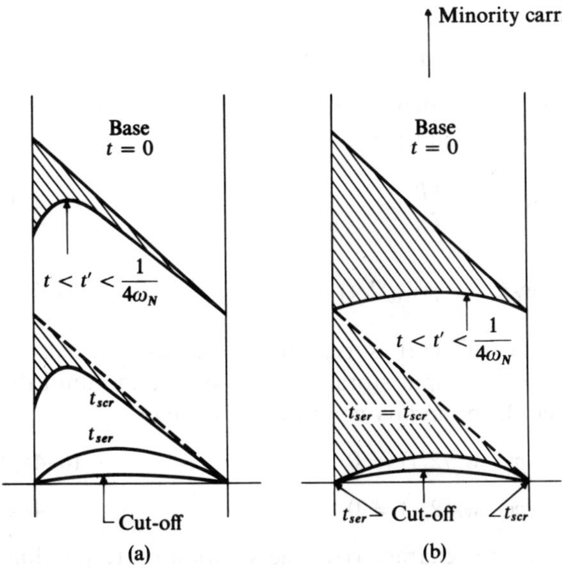

Fig. 6–28 Minority carrier density in the base region of a symmetric transistor as it decays during turn-off times. The shaded area represents the amount by which the equivalent circuit and charge-control methods of analysis are in error for calculating storage and fall times.

(a) Collector recovers first	(b) Emitter and collector recover at the same time in a symmetric device	(c) Emitter recovers first
$\lvert I_{e2}\rvert \ll \lvert I_{CM}\rvert$ or $\lvert I_{b2}\rvert \ll 2\lvert I_{CM}\rvert$	$\lvert I_{e2}\rvert = \lvert I_{CM}\rvert$ or $\lvert I_{b2}\rvert = 2\lvert I_{CM}\rvert$	$\lvert I_{e2}\rvert > \lvert I_{CM}\rvert$ or $\lvert I_{b2}\rvert > 2\lvert I_{CM}\rvert$

base region when the "off" drive is applied, the minority carrier decay can follow one of three different sequences shown in Fig. 6–28.

Case I

Figure 6–28a shows the case where the "off" drive is small or moderate in magnitude, i.e.

$$|I_{e2}| \ll |I_{CM}| \tag{6–477}$$

$$|I_{b2}| \ll |I_{CM}| \tag{6–478}$$

In such a case the collector junction recovers first. During the storage time the boundary conditions for common base circuit are

$$\text{for} \qquad x = -w_0/2, \quad \frac{\partial p}{\partial x} = -\frac{I_{e2}}{qD_pA} \tag{6–479}$$

for $$x = +w_0/2, \quad \frac{\partial p}{dx} = -\frac{I_{CM}}{qD_pA}. \tag{6–480}$$

When the collector minority carrier density reaches zero $t = t_{scr}$ (collector recovers) the boundary conditions change to

for $$x = -w_0/2, \quad \frac{\partial p}{\partial x} = -\frac{I_{e2}}{qD_pA} \tag{6–481}$$

for $$x = +w_0/2, \quad p(w_0/2, t) = 0. \tag{6–482}$$

The boundary condition in Eq. (6–482) is reasonable because the *carrier density can never go below zero.* The input storage time does not end until the emitter junction recovers when the boundary conditions change to

$$p(-w_0/2, t) = 0 \tag{6–483}$$

$$p(+w_0/2, t) = 0. \tag{6–484}$$

A different dominant time constant characterizes the solution corresponding to different boundary conditions. This is why *we need to pay particular attention to changes in the boundary conditions.*

Case II

Figure 6–28c shows the decay of minority carrier density under the condition that the emitter junction recovers before the collector junction. This happens when the "off" drives follow the following magnitude relationships

$$|I_{e2}| > |I_{CM}| \tag{6–485}$$

$$|I_{b2}| > 2|I_{CM}|. \tag{6–486}$$

Right after the "off" drive is applied the boundary conditions are the same as in Case I, i.e. Eqs. (6–479) and (6–480). When the emitter junction recovers, the emitter storage time, t_{ser}, is over and the boundary conditions now become

$$p(-w_0/2, t) = 0 \tag{6–487}$$

for $$x = +w_0/2, \quad \frac{\partial p}{\partial x} = -\frac{I_{CM}}{qD_pA}. \tag{6–488}$$

The boundary conditions change again when the collector junction recovers at the end of collector storage time t_{scr}, the boundary conditions during the collector fall time are the same as Eqs. (6–483) and (6–484).

It should be pointed out that the storage time and fall time calculations done by the use of equivalent-circuit approach or the charge-control approach assumes that the decay is *only* that of Case I. *Thus, the sequence of Case II is completely ignored.* In sequence II, when both junctions have recovered, the two junctions are reverse biased. Hence, the external drives can no longer influence the turn-off. Thus, the device must turn-off *by itself.* We, therefore, *expect that there will exist a minimum fall time which cannot be improved.* That is, *there exists a fundamental lower bound on the fall time.*

An analysis by Navavati (1960b, 1961) indicates that a minimum fall time indeed does exist. Because the analysis is fairly involved, only the result is stated below.

$$t_{f\,\min} = \left(\frac{1}{4\omega_N}\right) \ln \left[\frac{(\pi A_1/w_0)}{qD_p \left.\dfrac{\partial p_{ss}}{\partial x}\right|_{x=w_0} - 0.1\, I_{CM}}\right] \quad (p\text{-}n\text{-}p). \qquad (6\text{–}489)$$

where

$$A_1 = \frac{2}{w_0} \int_0^{w_0} [p_i(x) - p_{ss}(x)] \sin\left(\frac{\pi x}{w_0}\right) dx, \qquad (6\text{–}490)$$

where $p_i(x)$ = minority carrier density when the collector hole density reaches zero under the condition that the emitter hole density already has reached zero,

$p_{ss}(x)$ = steady state minority carrier density given below.

$$p_{ss}(x) \triangleq p_n + p_n\,[\coth(w_0/L_p) - \operatorname{csch}(w_0/L_p)] \sinh(x/L_p) - p_n \cosh(x/L_p). \qquad (p\text{-}n\text{-}p) \quad (6\text{–}491)$$

The $p_i(x)$ is hard to calculate because it depends on the past history of the decay process. We can obtain reasonable estimates of $t_{f\,\min}$ by assuming that $p_{ss}(x) = 0$, and $p_i(x)$ is a perfect half sine wave spanning the base width w_0 and having the slope at the collector corresponding to I_{CM}. The result for the intrinsic device is

$$t_{f\,\min} = \frac{0.575}{\omega_N}. \qquad (6\text{–}492)$$

It is interesting to observe that the data presented by Ebers and Moll (1954) in their paper does in fact show the existence of a minimum fall time even though the authors make no comment on this fact.

In Fig. 6–28b is shown the unusual case where the emitter and collector junctions recover precisely at the same time. In a perfectly symmetric device (that is, $\omega_N = \omega_I$ and $\alpha_N = \alpha_I$) this happens when $I_{e2} = I_{CM}$ or $I_{b2} = 2\,I_{CM}$.

In realistic asymmetric devices when the "off" drives are significantly less than this case the collector junction recovers first, and when the "off" drives exceed this case, the emitter junction recovers first. In the symmetric case the input and output storage times are equal to each other. The fall times at the input and the output are the minimum fall time already calculated in Eq. (6–489).

It may be noted that the dominant time constant for $t_{f\,\min}$ of $1/(4\omega_N)$ is comparable to $1/(5\omega_N)$ which was found to be the propagation delay time. The second curve from the top in all three figures of Fig. 6–26 is a distribution which occurs in a time of approximately $1/(4\omega_N)$ duration as determined by the solution to the diffusion equation. This time may be considered to be the *redistribution time* for minority carriers, i.e. the time taken by the minority carriers to redistribute themselves according to the demands of the external constraints.

Finally, we may summarize the shortcomings of the equivalent circuit and the charge control approaches. First, the transient times predicted by these approaches of the order of $1/(4\omega_N)$ or less are not believable. Second, the transient times as a result of Case II are not predictable. Hence, for example the existence of minimum fall time is impossible to predict. Third, in Case I, the fall time should be at least a *two* time constant fall time, one in each time segment as the boundary conditions change (except when the second time constant occurs when i_c is less than 0.1 I_{CM}), and in Case II the collector storage time is fundamentally at least a two time constant (one in each time segment) process. The assumption that the minority-carrier distribution is the same as that obtained when the transistor is biased statically at the edge of saturation results in an error. This error is shown by the shaded areas in Figs. 6–28a, b, and c. The storage time is under-estimated by the shaded area. The fall time is over-estimated by this *same* area. This also explains the experimentally found fact that the time constant associated with the rise time exceeds that for fall times. More charge needs to be established in the base region during rise time than needs to be removed during fall time.

6–21 SUMMARY

We have studied the time dependent diffusion equation in this chapter. Both steady state and transient solutions were considered. First order perturbation theory needed to be used to obtain steady state time dependent solutions taking account of the variation of the location of the collector boundary position with time (the Early effect). Wide band equivalent circuits were obtained for the bipolar junction transistor in terms of the fundamental device parameters and bias conditions.

The common base equivalent circuits are characterized in terms of the dominant cut-off frequency ω_N, whereas the common emitter circuit is characterized by the dominant cut-off frequency ω_β. The equivalent circuits are valid within a finite bandwidth of frequencies. A good figure for merit, for the transistor, is the *maximum frequency of oscillations.* The gain-band-width product was seen to be *not* a good figure of merit.

The transient times for the transistor were analytically obtained both as functions of the fundamental device parameters and in terms of the input drives. The equivalent-circuit, charge-control, and diffusion-equation approaches were used in the transient analysis. The last of these was used to determine the limits of the first two, as well as to predict a new result—the minimum fall time. The diffusion equation approach though is a cumbersome analytical method, more involved than the other two approaches.

PROBLEMS

6–1 In a saturated transistor the perturbation terms are expected to be negligible. Keeping this in mind, obtain the expressions for the common base intrinsic admittance parameters for a saturated transistor.

6–2 By the use of Taylor series expansions of Y^i_{ee} and Y^i_{cc} for a saturated transistor abtained in Problem 6–1, synthesize these admittances.

6–3 Obtain an analytic expression for Z_{11e} for common emitter connected transistor in the common base Y^i's and extrinsic parameters excluding C_{tc}. Assume $Z'_b = r'_b$.

6–4 Obtain the common base admittance parameters for the inverse transistor (collector forward biased, and emitter reverse biased) *Hint:* use symmetry and the results already obtained in the present chapter to arrive at the answer.

6–5 In calculating the fall times we shall assume that the collector junction recovers substantially *before* the emitter junction recovery. We define the recovery of the junction to have occurred when the minority carrier density at the junction is equal to or less than its thermal equilibrium value in the base region. Calculate the 10 to 90 per cent fall time in the common emitter circuit.

6–6 You are to compare two *p-n-p* transistors having numerically the same parameters C_{te}, C_{tc}, w_0, L_p, L_{ne}, $k_s = k_\ell$, D_p, at the same bias (normal bias) point. If one transistor A has step junctions and the other transistor B has linearly graded junctions, how do ω_α, ω_β, α_0, β_0 compare in these devices if the collector junction reverse bias is doubled?

6–7 Two semiconductor materials A and B are being considered for making transistors out of them. They have the same, L_p, L_{ne}, L_{nc}, w_0, N_a (emitter), N_d (base), N_a (collector). The mobilities are

$$\mu_n(A) = 25{,}000 \text{ cm}^2/(\text{Volt-sec.}),\ \mu_n(B) = 25{,}000 \text{ cm}^2/\text{Volt-sec.}$$
$$\mu_p(\text{A}) = 12{,}500 \text{ cm}^2/(\text{Volt-sec.}),\ \mu_p(B) = 1{,}250 \text{ cm}^2/\text{Volt-sec.}$$

Which is a better high frequency device? Why?

6–8 If C_{te} is to be taken into account properly in calculating h_{11e} use the appropriate method to do the calculation.

6–9 Modify the common emitter h-equivalent circuit by adding C_{te} in parallel with h_{11e} to take account of those transistors in which C_{te} is non-negligible. Keeping the remarks of Section 6–12 in mind calculate:

a) a-c current gain $= \dfrac{\text{Output current in } R_L}{\text{Input current}}$;

b) a-c voltage gain $= \dfrac{\text{Voltage across } R_L}{\text{Input voltage}}$,

when there is a load resistance R_L in series with a battery V_{CC} in the collector to emitter circuit.

6–10 Using the equivalent circuit of Fig. 6–11 driven at the input by an independent voltage source V_s, and the voltage V_0 across the load resistance R_L in the collector to emitter circuit, show that the low frequency voltage gain and the cut-off frequency are given by Eqs. (6–263) and (6–264) respectively.

6–11 Consider a two stage common emitter amplifiers having an interstage coupling network as a simple single shunt resistance R_1. Using the h-equivalent circuit, designating the input current by I_1, and the current into the base of the second transistor by I_2, solve for the gain bandwidth product of the current gain I_2/I_1.

6–12 Show that in 0 to 100 per cent rise time, calculation of C_{av} which should be used for step junctions is

$$C_{av} = 2\,C_{tc}(V_{CC}).$$

6–13 Show that in 0 to 100 per cent fall time calculation of C_{av} which should be used for linearly graded junctions is

$$C_{av} = 1.5\,C_{tc}(V_{CC}).$$

6–14 Show that for the general junction given by Eq. (6–389) the C_{av} which should be used for 0 to 100 per cent rise time is

$$C_{av} = \left(\frac{m}{m-n}\right) C_{tc}(V_{CC}).$$

6–15 Calculate the common collector amplifier intrinsic current gain cut-off frequency.

6–16 A resistance R_e is placed in the common emitter amplifier circuit having a value of R_L in the collector circuit such that h_{12e} and h_{22e} may be neglected. Calculate the upper cut-off frequency for voltage gain. What is the low frequency voltage gain?

6–17 If h_{11e} is experimentally plotted as a function of frequency, what information does this give on the value of r_b' (assume that the extrinsic base impedance is real)?

6–18 Five silicon common emitter amplifiers are connected in cascade. The first amplifier is driven by a current source. The last amplifier has $R_L = 500\ \Omega$. In the intermediate stages the collector of each stage is connected directly into the base of the next stage. Assume that the d-c bias arrangement is proper so that all transistors are operated in their respective normal active region. The following transistor parameters are given, calculate current gain cut-off frequency.

$$\frac{w_0}{L_p} = 0.05 \qquad C_{tc} = 5\ \mu\mu f$$

$$\tau_p = 500\ \mu\text{-sec.} \qquad C_{te} = 10\ \mu\mu f$$

$$I_E = 2\ \text{mA} \qquad r_b' = 50\ \text{Ohm}$$

6–19 Calculate the common emitter current gain when the extrinsic elements r_b', r_c', and C_{tc} must be taken account of. When will only the extrinsic parameters determine the cut-off frequency?

$$w_0/L_p = 0.05 \qquad Z_b' = r_b' = 50\ \Omega$$

$$\tau_p = 5\ \mu\text{-sec.} \qquad I_E = 26\ \text{mA}$$

$$C_{tc} = 1\ \text{pF}$$

Calculate (a) the upper cut-off frequency for the overall cut-off frequency, and (b) low frequency overall current gain.

6–20 Obtain an expression for the 10 to 90 per cent rise time in common emitter circuit.

6–21 What value of C_{tc} should be used in the expression for Problem 6–20?

6–22 In a common emitter switching circuit all "on" and "off" currents and the supply voltages are kept fixed. Only the load resistance is increased. How are the four transient times expected to change?

6–23 Give an example of a minority carrier distribution in the base region for which *the major assumption* of the charge control analysis is not valid.

6–24 Consider a germanium *p-n-p* transistor which has the following parameters:

$\tau_p = 0.1\ \mu$-sec. $\quad$ $C_{tc} = 2$ pF at $V_{CB} = -1$ Volt

$w_0/L_p = 0.1$ $\quad$ built-in voltages are negligible

$V_{CC} = 12$ Volt

The transistor is to be used in a common emitter switching circuit with a load resistance of 10 kΩ. The input circuit is a voltage source in series with $R_s = 10$ kΩ. The source voltage is a step of 1 V. Calculate for step junction device.

a) the rise time of the collector current, and

b) the rise time of the V_{EC}.

6–25 Consider a step junction diode whose *p*-side is very much more heavily doped than the *n*-side. Starting from first principles prove that the transition region is primarily on the *n*-side.

6–26 Derive Eqs. (6–90) and (6–91).

REFERENCES

Baker, A. N., and May, W. G., "Charge Analysis of Transistor Operation Including Delay Effects", *IRE Trans. on Electron Devices,* **ED-8,** 152–155, March (1961).

Bashkow, T. R., "Effect on Nonlinear Collector Capacitance on Collector Current Rise Time," *IRE Trans. on Electron Devices,* **ED-3,** 167–172, Oct. (1956).

Beaufoy, R., and Sparkes, J., "The Junction Transistor as a Charge-Controlled Device," *ATE Journal,* **13,** No. 4, 310–327.

Early, J. M., "Design Theory of Junction Transistors," *Bell System Technical Journal,* **32,** 1271–1312, Nov. (1953).

Early, J. M., "Stored Charge Analysis of Transistors," *IRE WESCON Conv. Rec.,* **pt. 2,** 127 (1959).

Easley, J. W., "The Effect of Collector Capacity on the Transient Response of Junction Transistors," *IRE Trans. on Electron Devices,* pp. 6–14, Jan. (1957).

Ebers, J. J., and Moll, J. L., "Large-signal Behavior of Junction Transistors," *Proc. IRE,* **42,** 1761–1772, Dec. (1954).

Giacoletto, J. L., "Study of *p-n-p* Alloy Junction Transistor from d-c through Medium Frequencies," *RCA Review,* **15,** 506–562 (1959).

Middlebrook, R. D., *An Introduction to Junction Transistor Theory,* Chapters 10,11, John Wiley & Sons, New York.

Moll, J. L., "Large-signal Transient Response of Junction Transistors," *Proc. IRE,* **42,** 1173–1183, Dec. (1954).

Nanavati, R. P., *An Introduction to Semiconductor Electronics,* Chapters 7–10, McGraw-Hill Book Co., New York (1963).

Nanavati, R. P., and Johnson, R. A., "Turn-on Delay Time and Its Predication," *Proc. National Electronics Conference,* **14,** 25–37 (1958).

Nanavati, R. P., "The Influence of Nonlinear Junction Capacitance on Transistor Rise and Fall Times," *Proc. National Electronics Conference,* **15,** 1–9 (1959).

Nanavati, R. P., "Correlation between Input and Output Transient Times for Junction Transistors," *Proc. NEC,* **16,** 729–734 (1960).

Nanavati, R. P., "A Fundamental Lower Bound for Junction Transistor Fall Time," *Proc. NEC,* **17,** 178–183 (1961).

Nanavati, R. P., "Charge Control Analysis of Transistor Storage Time Dependence on Input 'On' Pulse-width," *IEEE Trans. Proc. Tech. Gr. Electron Devices,* **ED-10,** No. 4, July (1963).

Nanavati, R. P., and Wilfinger, R., "Transistor Storage Time in Circuits With Speedup Capacitors," *Proc. IRE,* Dec. (1961).

Nanavati, R. P., and Wilfinger, R., "Predicting Transistor Storage Time for Non-Stop, Quasi-voltage Inputs," *IRE Trans.,* PGED, Nov. (1962).

Nanavati, R. P., and Wilfinger, R., "The Relationship Between Emitter Junction Recovers and 'Wiggle Effect'," *Proc. IRE,* Dec. (1961).

Pritchard, R. L., "Frequency Variation of Current Amplification Factor of Junction Transistors," *Proc. IRE,* **40,** 1477 (1952).

Pritchard, R. L., and Coffey, W. N., "Small Signal Parameters of Grown-Junction Transistors at High Frequencies," *IRE WESCON Conn. Rec.,* **Part 3,** 90–98 (1954).

Sparkes, J. J., "A Study of the Charge Control Parameters of Transistors," *Proc. IRE,* **48,** 1696–1705, Oct. (1960).

CHAPTER 7

TRANSISTORS WITH FIELDS IN THE BASE REGIONS DUE TO NONUNIFORM DOPING

7-1 INTRODUCTION

So far we have considered only those transistors which have a *uniformly doped base region.* These are called *diffusion transistors* because the major transport process which carries the minority carriers across the base region is simply diffusion under low injection conditions. In the previous chapter we saw that the transit time, or the propagation delay time for minority carriers to traverse the base region, is inversely proportional to the cut-off frequency of the transistor. If this time can be reduced, the intrinsic cut-off frequencies could be higher and the transient times will be shorter. Both of these are desirable high frequency characteristics. Thus anything which helps to transport the minority carriers across the base region in a shorter time will tend to improve the high frequency performance of the transistors. If, for example, an electric field can be introduced into the bulk base region for example of a *p-n-p* transistor directed from the emitter to the collector, it will aid the transport of holes across the base region.

It turns out that if the doping in the base region is nonuniform in the base region, an electric field will exist because of this fact as shown in Sec.2–16. Furthermore if the doping density is high at the emitter end and low at the collector end of the base region, the electric field due to nonuniform doping will aid the diffusion flow of minority carriers across the base region. Such fields are called *aiding fields.* Fields which oppose the flow of minority carriers across the base region are called *opposing fields.* If the base is fabricated by the solid state *diffusion process* a nonuniform doping results. Thus it is not hard to fabricate devices which have the required impurity grading (nonuniform impurity density as a function of distance). Almost all the devices made today are of this kind. Such transistors are variously called *diffused-base-transistors* or *drift-transistors.* The latter designation arises from the fact that the carriers are aided in their transport by the drift electric field that is built into the bulk region as a result of nonuniform doping.

Many of the methods of analysis of these transistors are the same as for the diffusion transistors considered in detail in Chapter 6, though the mathematical equations are somewhat different. We shall make maximum use of the experience gained in Chapter 6 to understand the somewhat more general results of the present chapter.

7-2 THE BUILT-IN VOLTAGE ACROSS THE BULK BASE REGION AND THE FIELD FACTOR

We have already seen in Sec. 2–16 that the nonuniform doping gives rise to a built-in field and a built-in voltage across the bulk of the base region (excluding the emitter and collector transition regions).

We consider the case of *p-n-p* transistor, though the conclusions will be generally stated. We wish to address ourselves to the question of the *optimum choice* of impurity grading in the base region. The first choice is the selection of impurity distribution as a function of distance. We note that this requirement must be compatible with the necessity of the biggest field as this improves the high frequency response. In Chapter 2 we saw that the electric field is given by

$$\mathscr{E} = -\left(\frac{kT}{q}\right)\frac{\nabla n}{n} \doteq -\left(\frac{kT}{q}\right)\frac{\nabla N_d(x)}{N_d(x)}. \qquad (7\text{–}1)$$

The extreme right hand side of Eq. (7–1) follows from the assumptions of all donors being ionized. Thus we wish to maximize the right hand side of Eq. (7–1) under the constraint that a finite upper bound exists on this field (as will be shown later). In a simple minded way we require then that

$$\frac{d(\nabla N_d/N_d)}{dx} = 0. \qquad (7\text{–}2)$$

The function $N_d(x)$ which satisfied the above equation must be such that the gradient of the function divided by the function itself gives a constant. For we know that the derivative of a constant is zero. The only function which gives back the function times a constant when differentiated is the exponential function. Thus we choose the following function.

$$N_d(x) = N_d(0)\, e^{-x/L_d}, \qquad (7\text{–}3)$$

where $N_d(0)$ = the donor density at the emitter end of the base region,

L_d = distance constant associated with the impurity grading in the base region.

Let us calculate the electric field in the base region due to chemical impurity grading in the base region given by Eq. (7–3).

$$\mathscr{E} = -\left(\frac{kT}{q}\right)\frac{\nabla N_d(x)}{N_d(x)} = -\frac{kT}{q}\left(-\frac{1}{L_d}\right)\frac{N_d(0)\,e^{-x/L_d}}{N_d(0)\,e^{-x/L_d}} = \frac{kT}{qL_d}. \tag{7-4}$$

We note from Eq. (7-4) that the electric field is constant and is in the positive x-direction, i.e. from the emitter to the collector direction. A field in this direction will indeed aid the transport of minority carriers across the base region. Thus the impurity grading according to Eq. (7-3) does indeed produce an aiding field.

We may now use another result from Chapter 2, namely that the potential difference between two planes of constant x is a function *only* of the values of carrier densities at these planes and is given by

$$V(0) - V(w_0) \triangleq V_d = \frac{kT}{q}\,\ell n\left(\frac{n(0)}{n(w_0)}\right). \tag{7-5}$$

Even though a field is present in the base region the space charge neutrality is still essentially valid in the base region. The presence of the electric field in the base region implies that there must be sources (i.e. net charges) of the field in the base region. It can be shown, however, from quantum mechanical calculations that very small percentage departures from space-charge neutrality result in extremely high electric fields—enough to break down the material. As no one has ever experimentally detected such large fields in the bulk semiconductors, we are forced to conclude that *to a good approximation the space charge neutrality must obtain in the bulk of the base region.* Thus in the n-type base region

$$n(x) = N_d(x) + p(x). \tag{7-6}$$

Thus Eq. (7-5) becomes

$$V_d = \frac{kT}{q}\,\ell n\left[\frac{N_d(0) + p(0)}{N_d(w_0) + p(w_0)}\right] \tag{7-7}$$

under *low injection conditions*

$$N_d(0) \gg p(0) \tag{7-8}$$

and

$$N_d(w_0) \gg p(w_0) \tag{7-9}$$

and Eq. (7-7) becomes

$$V_d = \frac{kT}{q}\,\ell n\left[\frac{N_d(0)}{N_d(w_0)}\right] \tag{7-10}$$

for a fixed base width the electric field due to impurity grading in the base region will be maximized by maximizing V_d. Thus the largest $N_d(0)$ and the smallest $N_d(w_0)$ will maximize V_d. Clearly the minimum $N_d(w_0)$ is

$$N_d(w_0) = n_i. \tag{7-11}$$

If $N_d(w_0)$ gets any smaller than n_i the collector edge of the n-type base region would have to be p-type—a clear violation of the integrity of the *n-type* base region. The limit on $n(0)$ is set by the requirement that the injection efficiency must not be too low. For present purposes we shall use the expression for injection efficiency γ from Chapter 6 given below.

$$\gamma = \frac{1}{1 + n_n D_n C_{en}[\tanh(C_p w_0)]/(p_{pe} D_p C_p)} \tag{7-12}$$

The reason why it is permissible to consider Eq. (6–45) here, even though it was derived for diffusion transistor, is that the dependence of γ of drift transistors on n_n and p_{pe} is identical to that of Eq. (7–12). To make sure that γ is not too low,

$$p_{pe} = N_a(\text{emitter}) \quad \gg n_n. \tag{7-13}$$

There are limits to how high p_{pe} could be made. First is the *solid* solubility of acceptors in the emitter which is about $10^{21}/\text{cm}^3$. At a doping density of about $10^{19}/\text{cm}^3$ the emitter becomes *degenerate.*† In this case tunneling currents would be important in the emitter base junction. For transistor action tunneling currents are unwelcome. Thus, the upper limit of p_{pe} is about $10^{19}/\text{cm}^3$ in Ge, Si, and GaAs. This implies that the maximum minority carrier density at the emitter edge of the base region is

$$n(0) \doteq N_d(0) \doteq 10^{18}/\text{cm}^3. \tag{7-14}$$

Thus optimum V_d is

$$V_d = \frac{kT}{q} \ell\text{n}\left(\frac{10^{18}}{n_i}\right) \tag{7-15}$$

and the corresponding field for the constant field case is

$$\mathscr{E} = \frac{V_d}{w_0} = \frac{kT}{q w_0} \ell\text{n}\left(\frac{10^{18}}{n_i}\right). \tag{7-16}$$

For Ge, Si, and GaAs at room temperature

$$V_d(\text{max}) = 0.276 \text{ Volt} \qquad \text{(Ge)} \tag{1}$$
$$V_d(\text{max}) = 0.476 \text{ Volt} \qquad \text{(Si)} \tag{2}$$
$$V_d(\text{max}) = 0.685 \text{ Volt} \qquad \text{(GaAs)} \tag{3}$$

† A degenerate semiconductor is one in which the Fermi level is within the conduction or valence band.

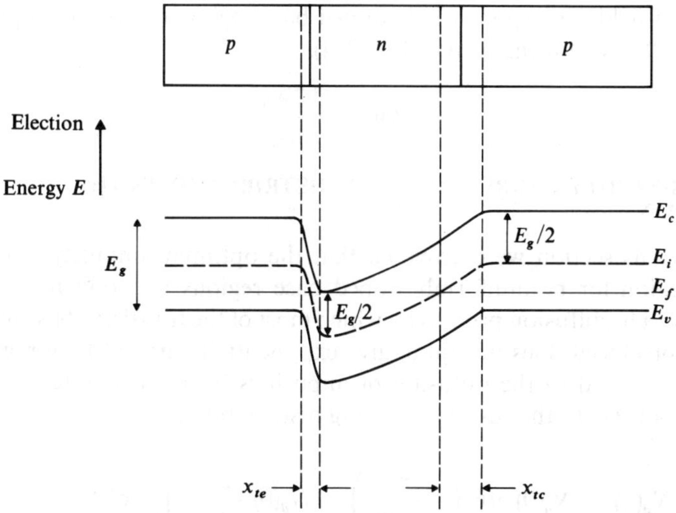

Fig. 7–1 Energy band picture for a *p-n-p* transistor with optimum impurity grading in the base region.

An absolute upper bound on V_d is

$$V_d(\max) = V_g/2, \tag{7–17}$$

where V_g = voltage corresponding to band gap energy E_g.

To see the legitimacy of Eq. (7–17) let us consider the following. At the emitter edge of the base region the Fermi level cannot be higher than the bottom edge of the conduction band, otherwise the base is degenerate and the emitter base junction is a tunneling junction. The Fermi level is at essentially the middle of the band gap at the collector end of the base region. Under these circumstances the intrinsic energy E_i (corresponding to $n = n_i$) changes by $E_g/2$ over the width of the base region. This proves Eq. (7–17). Figure 7–1 shows the band picture for this case.

A parameter which turns out to be important in further considerations of the drift transistors is the *field factor* f given below

$$f \triangleq \frac{V_d}{kT/q}. \tag{7–18}$$

By the use of Eq. (7–10), Eq. (7–18) may be rewritten

$$f = \ell n\left[\frac{N_d(0)}{N_d(w_0)}\right]. \tag{7–19}$$

For constant field, i.e. exponentially doped base region, the voltage V_d may be written as follows with the help of Eq. (7–4).

$$V_d = \mathscr{E} w_0 = \frac{kTw_0}{qL_d} \tag{7–20}$$

7–3 D-C MINORITY-CARRIER-DENSITY DISTRIBUTION IN THE BASE REGION

In the previous section we discovered that the optimum impurity grading in the base region for nonuniformly doped base regions is exponential doping distribution. The diffusion process by which most of the transistor base impurity grading is produced does not quite give exponentially graded base regions. If the base is exposed to the diffusion of impurities from an infinite source, i.e. donors, for a time T_d, the resulting doping distribution is given by

$$N_d(x) = N_d(0) \operatorname{erfc}\left(\frac{x}{2\sqrt{DT_d}}\right) = N_d(0) \frac{2}{\sqrt{\pi}} \int_{x/(2\sqrt{DT_d})}^{\infty} e^{-x^2} dx. \tag{7–21}$$

The complementary error function in Eq. (7–21) is related to the error function by the following

$$\operatorname{erfc}(x) = 1 - \operatorname{erf}(x). \tag{7–22}$$

The electric field due to a doping density given by Eq. (7–21) is

$$\mathscr{E}(x) = \frac{kT}{q\sqrt{\pi DT_d}} \cdot \frac{e^{-x^2/(4DT_d)}}{\operatorname{erfc}[x/(2\sqrt{DT_d})]}, \tag{7–23}$$

where D = diffusion constant which characterizes the diffusion of impurities into the base region for the purpose of impurity grading.

Our present aim is to obtain the minority carrier density as a function of distance x in the base region. The starting point is the transport equation which now *must* include both the field and the diffusion terms. Thus

$$j_p(x) = q\mu_p p(x)\mathscr{E} - qD_p \nabla p(x) \tag{7–24}$$

and

$$j_n(x) = q\mu_n n(x)\mathscr{E} + qD_n \nabla n(x). \tag{7–25}$$

We are considering *p-n-p* transistors. We noted in Chapter 6 that the electron component of current in the base region of diffusion transistors was very small. In drift transistors having an aiding field, the *loss of minority carriers* as they cross the base region is expected to be significantly reduced from that of the diffusion transistors. Hence the electron current in the base region, which was small in diffusion transistors, *must be even smaller in the drift transistors.* Hence we may set the electron current component equal to zero, as an approximation, so that the electric field may be calculated. Thus Eq. (7–25) becomes

$$0 = q\mu_n n(x)\,\mathscr{E} + qD_n \nabla n(x). \tag{7–26}$$

Therefore

$$\mathscr{E} = -\left(\frac{D_n}{\mu_n}\right)\frac{\nabla n(x)}{n(x)}. \tag{7–27}$$

But from the Einstein relationship we may conclude that the following is true

$$\frac{D_n}{\mu_n} = \frac{D_p}{\mu_p}. \tag{7–28}$$

Equation (7–27) may be rewritten with the help of Eq. (7–28) as follows

$$\mathscr{E} = -\left(\frac{D_p}{\mu_p}\right)\frac{\nabla n(x)}{n(x)}. \tag{7–29}$$

This expression for the electric field may be substituted into Eq. (7–24) to obtain

$$j_p(x) = -q\mu_p p(x)\left(\frac{D_p}{\mu_p}\right)\frac{\nabla n(x)}{n(x)} - qD_p \nabla p(x) \tag{7–30}$$

$$i_p(x) = j_p(x)A = -qAD_p\left[\frac{p(x)}{n(x)}\nabla n(x) + \nabla p(x)\right]. \tag{7–31}$$

Space charge neutrality in the bulk of the base region implies that

$$n(x) = p(x) + N_d(x). \tag{7–32}$$

When Eq. (7–32) is substituted into Eq. (7–31), the following is obtained

$$i_p(x) = -qD_pA\left[\frac{p(x)[\nabla p(x) + \nabla N_d(x)]}{p(x) + N_d(x)} + \nabla p(x)\right]. \tag{7–33}$$

Under low injection conditions

$$p(x) \ll N_d(x) \tag{7–34}$$

$$\nabla p(x) \ll \nabla N_d(x). \tag{7–35}$$

Using the magnitude relationships Eqs. (7–34) and (7–35) we obtain

$$\frac{i_p(x)}{qD_pA} \doteq -\left[\frac{p(x)}{N_d(x)}\cdot \nabla N_d(x) + \nabla p(x)\right]. \tag{7–36}$$

The one-dimensional version of Eq. (7–36) is

$$\frac{i_p(x)}{qD_pA} = -\frac{p(x)}{N_d(x)}\frac{dN_d(x)}{dx} - \frac{dp(x)}{dx}. \tag{7–37}$$

We have just argued the case for the electron current component in the base region being negligibly small because of the low recombination loss. We may hence conclude that to a good approximation $i_p(x)$ is essentially constant in the base region.

$$i_p(x) = \text{constant} = I_E. \tag{7-38}$$

Using this as one condition, and the other condition for reverse biased collector junction

$$p(w_0) = 0 \tag{7-39}$$

the solution to Eq. (7–37) is found to be

$$p(x) = \frac{I_E}{qD_p A_e N_d(x)} \int_x^{w_0} N_d(x)dx. \tag{7-40}$$

Equation (7–40) is general in that it is valid for any spatial variation of the doping density (donor density). We may hence look at the special cases of (a) exponential grading and (b) complimentary-error-function grading in the base region.

For exponential grading, i.e. constant field case, we have in the base region

$$p(x) = \frac{P_0}{f}(1 - e^{-f(1-x/w_0)}), \tag{7-41}$$

where

$$P_0 \triangleq \frac{I_E w_0}{qA_e D_p}, \tag{7-42}$$

where A_e = emitter junction area,

f = field factor given by Eq. (7–19).

For the special case of $f = 0$, Eq. (7–41) takes the indefinite form which may be evaluated by l'Hospital's rule to the following in the base region

$$p(x) = P_0(1 - x/w_0); \quad f = 0. \tag{7-43}$$

For complementary-error-function grading the following minority carries density in the base region is obtained

$$p(x) = \frac{P_0}{w_0/[2(DT_d)^{\frac{1}{2}}]} \times \left\{ \frac{\{w_0/[2(DT_d)^{\frac{1}{2}}]\} \operatorname{erfc}\{w_0/[2(DT_d)^{\frac{1}{2}}]\} - \{x/[2(DT_d)^{\frac{1}{2}}]\} \operatorname{erfc}\{x/2[DT_d]^{\frac{1}{2}}\}}{\operatorname{erfc}\{x/[2(DT_d)^{\frac{1}{2}}]\}} + \frac{1}{\pi^{\frac{1}{2}}}\left[\frac{e^{-x^2/(4DT_d)} - e^{-w_0^2/(4DT_d)}}{\operatorname{erfc}\{x/[2(DT_d)^{\frac{1}{2}}]\}}\right] \right\}. \tag{7-44}$$

Figure 7–2 shows the minority carrier density in the base region as a function of position in the base region for various values of the field factor f. Most of the curves are for the exponential grading. Two curves for the complementary error function grading are also shown in Fig. 7–2 as are several things worthy of note. First, for a sufficiently high field factor ($f = 6$), the minority carrier density has no appreciable gradient for roughly seventy per cent of the base

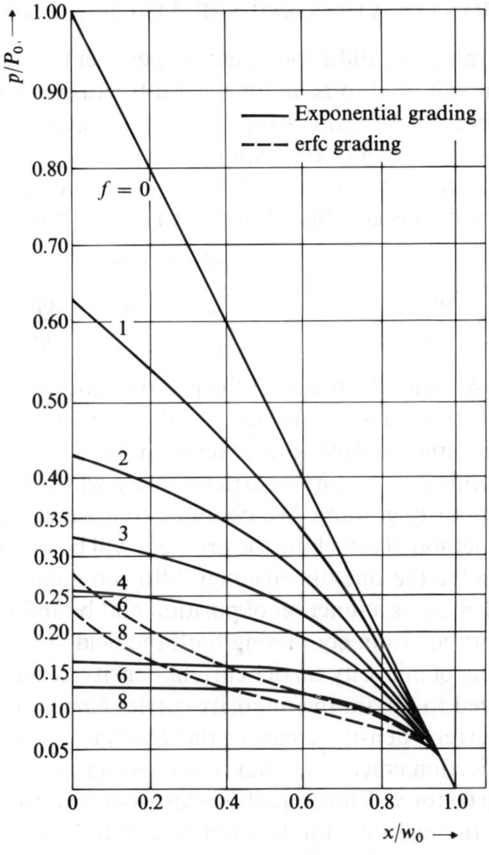

Fig. 7–2 Minority carrier density in the base region given by Eq. (7–41) for low injection condition.

region. This fact implies that for roughly seventy per cent of the base width, starting at the emitter edge, the transport mechanism is essentially field aided transport. Carrier gradient becomes appreciable only in the twenty per cent of the base width at the collector end. At this end the diffusion transport process would be contributing significantly. Second, it is to be noted that in the case of the complimentary-error-function grading, the gradient of minority carriers is almost everywhere greater than that in the exponentially graded base†. Thus, at least in this case, the drift field is more efficiently used in the latter case confirming the fact that it is an optimum grading scheme. Third, when the field factor is zero the minority carrier distribution is perfectly linear. The actual distribution will not be quite a straight line because of the presence of recombination which was ignored in obtaining both Eqs. (7–41) and (7–44).

7–4 A-C DIFFUSION EQUATION AND THE WIDE BAND MODEL

We shall now solve the a-c diffusion equation pertinent to drift transistor. The diffusion equation will need to be more general because the *built-in field due to impurity grading in the base region must be taken into account.* We shall, for the purpose of analysis, treat only the exponentially graded base transistor. For this not only gives an optimum device but it also is relatively easier to analyze because the electric field is constant. For this case the diffusion equation is

$$D_p \frac{\partial^2 p(x,t)}{\partial x^2} - \mu_p \mathscr{E} \frac{\partial p(x,t)}{\partial x} - \frac{p(x,t) - \dfrac{n_i^2 \, e^{q(x - w_0/2)/(kT)}}{N_d(-w_0/2)}}{\tau_p} = \frac{\partial p}{\partial t} \quad (p\text{-}n\text{-}p) \qquad (7\text{–}45)$$

Equation (7–45) presumes the origin of the position coordinate system is at the middle of the average base width as was also done in Chapter 6. Note that the second term in the numerator of the third term on the left hand side of Eq. (7–45) is the thermal equilibrium minority carrier density which is now a function of position due to impurity grading. We continue to assume that G_p is zero.

Our general method of attacking the problem will be the same as in Chapter 6. We shall first solve the one-dimensional diffusion equation and obtain the minority carrier density as a function of position and boundary conditions. The use of the transport equation containing both the field and the diffusion terms allows the obtaining of minority carrier current density as a function of position. It is then specialized to obtain the minority carrier current crossing each junction. The *total* current density crossing the emitter junction is obtained by dividing by the injection efficiency γ and that crossing the collector junction by multiplying by the factor α^*. Thus, equations are obtained where the a-c currents crossing each junction are found to be a function of the a-c voltages across each of the junctions. The intrinsic common base admittance parameters are readily obtained from these equations. These are expanded retaining only a few terms. Networks are then synthesized corresponding to the intrinsic admittances. The extrinsic parameters are then added to obtain the "complex" wide band models. We shall take account of the perturbation effect of the collector boundary of the

† Thus, for a given emitter current level, the diffusion component of current is greater in complementary error function than in the exponentially graded case. Note that in both types of grading the field induced current flow is expected to be greater than diffusion induced flow.

base region varying with time, but shall assume that the emitter edge of the base region remains fixed as was also assumed in Chapter 6.

We shall leave out many routine steps as similar steps have been gone into in considerable detail in the previous chapter in connection with the diffusion transistor.

The total solution to the more general diffusion equation under consideration here (Eq. 7–45) is the following:

$$p_t(x,t) = \left\{ p_{ne}\, e^{2(x+w_0/2)/g} - p_{ne}\, e^{(x+w_0/2)/g} \left[\operatorname{csch}(w_0/\delta_0)\right] \times \left[(e^{qV_E/(kT)} - 1)\sinh\left(\frac{(x - w_0/2)}{\delta_0}\right) + e^{w_0/g}\sinh\left(\frac{x + w_0/2}{\delta_0}\right)\right]\right\}$$
$$+ \left\{ \frac{qV_e\, e^{j\omega t}}{kT}(-p_{ne})\, e^{qV_E/(kT)}\, e^{(x+w_0/2)/g} \cdot \frac{\sinh\left(\dfrac{x + w_0/2}{\delta}\right)}{\sinh(w_0/\delta)} \right\}$$
$$+ \left\{ \frac{\partial w_0}{\partial V_C} V_c p_{nc}\, e^{j\omega t} \left[\frac{1}{\delta_0}\left(\frac{(e^{qV_E/(kT)} - 1)\, e^{w_0/g}}{\sinh(w_0/\delta)} + \coth(w_0/\delta_0)\right) - \frac{1}{g}\right] \times \left(\frac{\sinh\left(\dfrac{x + w_0/2}{\delta}\right)}{\sinh(w_0/\delta)}\right) e^{(x+w_0/2)/g} \right\}, \tag{7–46}$$

where the following parameters are defined as

$$g \triangleq \frac{2kT}{q\mathscr{E}} \tag{7–47}$$

$$\frac{1}{\delta_0^2} \triangleq \frac{1}{g^2} + \frac{1}{L_p^2} \tag{7–48}$$

$$\frac{1}{\delta^2} \triangleq \frac{1}{g^2} + C_p^2 \tag{7–49}$$

where p_{ne} = thermal equilibrium value of minority carrier density at the emitter edge of the base region,

p_{nc} = thermal equilibrium value of the minority carrier density at the collector edge of the base region.

The first { } is the d-c, the second { } is the a-c unperturbed, and the third { } is the a-c perturbed solutions to Eq. (7–45). The first { } is easily recognized as the d-c solution as it is independent of frequency. The last is also easily recognized as the a-c perturbed solution because of the presence of $\partial w_0/\partial V_C$. The middle { }

is frequency dependent but does not have the perturbation factor in it, hence it must be the a-c unperturbed solution. The total current density then is obtained by substituting Eq. (7–46) into the general one-dimensional transport equation given below

$$j_{pt}(x,t) = q\mu_p p_t(x,t)\,\mathscr{E} - qD_p\,\frac{\partial p_t(x,t)}{\partial x}. \tag{7–50}$$

The method of obtaining the intrinsic admittances will be outlined below.

1. After Eq. (7–46) is substituted into the transport equation x is specialized to $x = -w_0/2$ and to $x = +w_0/2$ to obtain the total minority carrier current density crossing the emitter and collector junctions. When these quantities are multiplied respectively by the emitter junction area A_e and the collector junction area A_c, the emitter and collector current densities are obtained as functions of the voltages across the emitter and collector junctions.
2. Only the total a-c components of currents are retained as we only want the a-c wideband models.
3. All a-c currents and voltages will be assumed to be sinusoidal quantities characterized by the radian frequency ω.
4. The collector current direction is now reversed so that the reference directions of currents and voltages obtained from the device theory are compatible with the reference directions appropriate to the admittance parameters defined in circuit theory.
5. The emitter current is divided by the injection efficiency and the collector current is multiplied by the collector multiplication factor to obtain the total currents (i.e. electron and hole currents together) crossing the junctions.
6. The resulting equations allow us to recognize the common base intrinsic admittance parameters, Y's.

By following the above first five steps the following common base intrinsic parameters (Kromer, 1954; Almond and McIntype, 1957) are obtained

$$Y^i_{eed} = \left(\frac{q}{kT}\right)\left(\frac{[I_{Ep} - I_{SE}]}{\gamma}\right)\left[\frac{1 + (g/\delta)\coth(w_0/\delta)}{1 + (g/\delta_0)\coth(w_0/\delta_0)}\right] \tag{7–51}$$

$$Y^i_{ecd} = \frac{I_{Cp}}{\delta\gamma}\left(\frac{\partial w_0}{\partial V_C}\right)\left[\frac{e^{-w_0/g}}{\delta\sinh(w_0/\delta)}\right] \tag{7–52}$$

$$Y^i_{ced} = -\frac{q\alpha^*(I_{Ep} - I_{SE})}{kT\delta}\left[\frac{e^{w_0/g}}{[\sinh(w_0/\delta)][(1/g) + (1/\delta_0)\coth(w_0/\delta)]}\right] \tag{7–53}$$

$$Y^i_{ccd} = -\left[\frac{I_{Cp}}{\delta}\tanh(w_0/\delta) - I_D/L_p\right]\alpha^*\,\frac{\partial w_0}{\partial V_C}. \tag{7–54}$$

The d-c currents I_{SE}, I_{Cp}, I_{Ep}, and I_D are given by the following:

$$I_{Ep} = \frac{qA_eD_pp_{ne}}{\delta_0}[(e^{qV_E/(kT)} - 1)(\coth(w_0/\delta_0) + (\delta_0/g)) + e^{w_0/g}\operatorname{csch}(w_0/\delta_0)] \tag{7–55}$$

$$I_{Cp} = -\frac{qA_cD_pp_{nc}}{\delta_0}[(e^{qV_E/(kT)} - 1)e^{-(w_0/g)}\operatorname{csch}(w_0/\delta_0) + \coth(w_0/\delta_0) - (\delta_0/g)] \tag{7–56}$$

$$I_D = \frac{qA_eD_pp_{ne}}{L_p} \tag{7–57}$$

$$I_{SE} \triangleq \frac{qA_eD_pp_{ne}}{\delta_0}[\coth(w_0/\delta_0) + (\delta_0/g) - e^{(w_0/g)}\operatorname{csch}(w_0/\delta_0)]. \tag{7–58}$$

Some observations are pertinent here. The parameter δ_0 is a generalized *effective diffusion length* taking the presence of electric field into account. The parameter δ is the generalized effective reciprocal C_p including the influence of the electric field.

Note that the presence of the electron current components has been taken into account by the presence of γ and α^* both of which are extremely close to unity indicating that the electron current components are negligibly small. They will need to be taken into account only in those situations where two almost equal admittances are to be subtracted, one from the other. Generally α^* is a lot closer to one than either γ or β^*. For this reason from now on we let

$$\alpha^* \doteq 1. \tag{7–59}$$

We seek a one-dependent source admittance equivalent circuit. This requires differences between some of the admittances. Because Y_{ee}^i and Y_{ec}^i are of comparable magnitude, γ will be retained in Y_{ee}^i. In all other admittances γ will be ignored with relatively negligible error.

Figures 7–3 and 7–4 show the form of the one-dependent source admittance equivalent circuit for the common base and the common emitter circuits. Each of the admittances of Figs. 7–3 and 7–4 can be expressed in terms of the intrinsic admittances of Eqs. (7–51) through (7–54) as given below:

$$Y_e^i = Y_{eed}^i + Y_{ecd}^i \tag{7–60}$$

$$Y_c^i = Y_{ccd}^i + Y_{ecd}^i \tag{7–61}$$

$$Y_f^i = -Y_{ecd}^i \tag{7–62}$$

$$Y_m^i = Y_{ced}^i + Y_{ecd}^i \tag{7–63}$$

$$Y_s^i = Y_{eed}^i + Y_{ced}^i - Y_{ecd}^i. \tag{7–64}$$

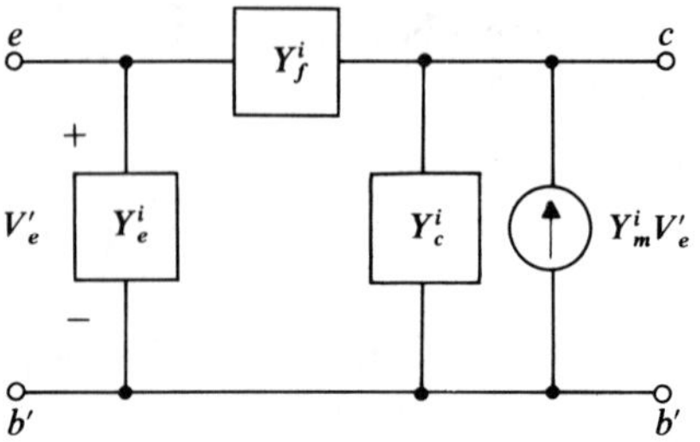

Fig. 7–3 The form of the common base equivalent circuit.

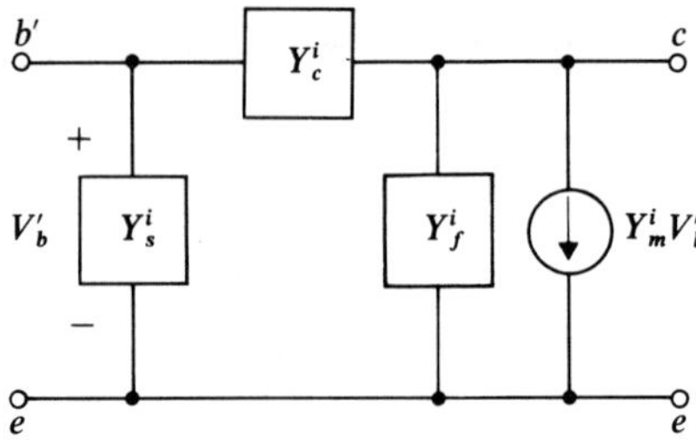

Fig. 7–4 The form of the common emitter equivalent circuit.

Each of the admittances Y_e^i, Y_c^i, Y_f^i, Y_s^i is a driving point admittance and may be synthesized only if each of them is a *positive real function* which generally they are not. Taylor series expansions will be made and terms up to and including the third power will be retained. Then a three-element admittance network will be synthesized for each of the intrinsic admittances in Figs. 7–3 and 7–4.

Before proceeding with the above scheme we shall find the following useful.

$$\varepsilon_g \triangleq w_0/g \tag{7–65}$$

$$\varepsilon_\delta \triangleq w_0/\delta_0. \tag{7–66}$$

The ratio of $\varepsilon_\delta/\varepsilon_g$ is given by

$$\frac{\varepsilon_\delta}{\varepsilon_g} = \frac{g}{\delta_0} = \sqrt{1 + \frac{w_0^2}{L_p^2\,\varepsilon_g^2}}. \tag{7–67}$$

Equation (7–65) is expressible in another way

$$\varepsilon_g = qV_d/2kT. \tag{7–68}$$

For most practical devices

$$\frac{w_0^2}{L_p^2\,\varepsilon_g^2} \ll 1 \tag{7–69}$$

The magnitude relationship Eq. (7–69) allows us to conclude that

$$\varepsilon_g = \varepsilon_\delta. \tag{7–70}$$

Another parameter of considerable interest is

$$\Omega \triangleq \sqrt{1 + \frac{j\omega\delta_0^2}{D_p}}. \tag{7–71}$$

The admittance Y_{ee}^i may be expressed in terms of the above defined quantities as

$$Y_{ee}^i = \left(\frac{qI_{Ep}}{2kT}\right)\left[\frac{1 + (\varepsilon_\delta/\varepsilon_g)\,\Omega(1 + e^{-2\varepsilon_g\Omega})/(1 - e^{-2\varepsilon_g\Omega})}{1 + (\varepsilon_\delta/\varepsilon_g)(1 + e^{-2\varepsilon_\delta})/(1 - e^{-2\varepsilon_\delta})}\right]. \tag{7–72}$$

The exponential terms in Eq. (7–72) may be neglected if

$$\varepsilon_\delta > 2$$

The error in neglecting the exponential terms in Eq. (7–72) is about 5 per cent. The error is $< 1\%$ if $\varepsilon_\delta = 3$. If in addition to Eq. (7–69) and $\varepsilon_\delta > 2$ we neglect I_{SE} and I_D, which are quite small compared to I_{Ep} and I_{Cp}, the intrinsic admittances simplify to

$$Y_{ee}^i = \frac{qI_{Ep}}{\gamma kT}\left(\frac{1+\Omega}{2}\right) \tag{7–73}$$

$$Y_{ec}^i = \left(\frac{2I_{Cp}\varepsilon_\delta}{w_0}\right)\left(\frac{\partial w_0}{\partial V_C}\right)\Omega\, e^{-(\varepsilon_g - \varepsilon_\delta \Omega)} \tag{7–74}$$

$$Y_{ce}^i = \frac{-qI_{Ep}\Omega\, e^{-\varepsilon_g - \varepsilon_\delta \Omega}}{kT} \tag{7–75}$$

$$Y_{cc}^i = \left(\frac{I_{Cp}}{w_0}\right)\left(\frac{\partial w_0}{\partial V_C}\right)(2\varepsilon_\delta \Omega\, e^{-2\varepsilon_\delta \Omega} + \varepsilon_\delta \Omega - \varepsilon_g). \tag{7–76}$$

Because other terms on the right hand side of Eq. (7–76) almost cancel, an additional term had to be used.

The admittances on the left hand side of Eqs. (7–60) through (7–64) may be written in the following form with the help of the simplified expressions for the intrinsic admittances of Eqs. (7–73) through (7–76).

$$Y_e^i = g_e + \frac{j\omega C_e}{1 + j\omega C_e/g_e} \tag{7–77}$$

$$Y_e^i \doteq g_e\left(1 + j\frac{\omega C_e}{g_e} + \frac{\omega^2 C_e^2}{g_e^2}\right) \tag{7–78}$$

if

$$\frac{\omega C_e}{g_e} < 1. \tag{7–79}$$

The conductance g_e and capacitance C_e are given by

$$C_e \doteq \delta_0^2 g_e/(4D_p) \tag{7–80}$$

$$g_e \doteq qI_{Ep}/(\gamma_0 kT). \tag{7–81}$$

The admittance Y_e^i is not hard to synthesize. The admittance of a conductance g in series with a capacitance C shown in Fig. 7–5a, is seen to be

$$Y(s) = \frac{g(Cs)}{g + Cs} = \frac{Cs}{1 + (Cs/g)}. \tag{7–82}$$

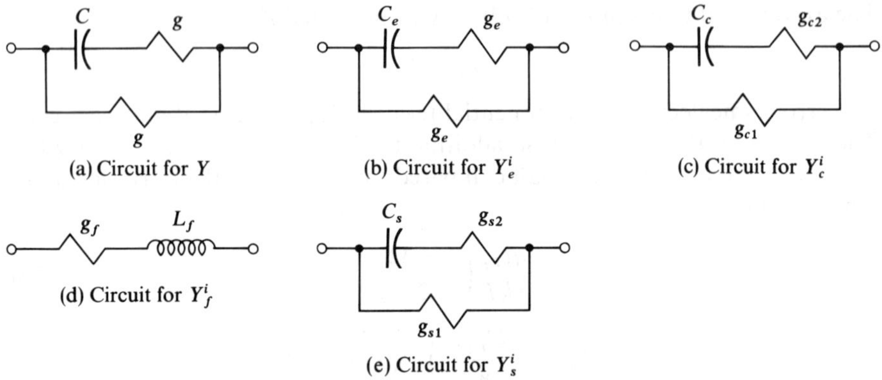

Fig. 7–5 Individual circuits synthesized for admittances Y, Y_e^i, Y_c^i, Y_s^i, and Y_f^i.

When s is replaced by $j\omega$ in Eq. (7–82) it is seen to have the same form as the second term on the right hand side of Eq. (7–77). Thus, the second term on the right hand side of Eq. (7–77) may be synthesized as a series combination of a conductance g_e and a capacitance C_e. Figure 7–5b shows the synthesized admittance Y_e^i.

Similarly by the use of Eqs. (7–61), (7–74), and (7–76), Y_c^i is obtained

$$Y_c^i = -\frac{I_C}{w_0}\left(\frac{\partial w_0}{\partial V_C}\right)(\varepsilon_\delta - \varepsilon_g)\left[1 + \frac{j\omega\delta_0^2}{2D_p}\left(\frac{\varepsilon_\delta}{\varepsilon_g - \varepsilon_\delta}\right) + \frac{\omega^2\delta_0^4}{8D_p^2}\left(\frac{\varepsilon_\delta}{\varepsilon_\delta - \varepsilon_g}\right)\right]. \quad (7\text{–}83)$$

The admittance Y_c^i may also be synthesized as a series R-C network in parallel with a conductance as shown in Fig. 7–5c. The individual elements of this subnetwork are given by

$$g_{c1} = -\left(\frac{I_C}{w_0}\right)\left(\frac{\partial w_0}{\partial V_C}\right)(\varepsilon_\delta - \varepsilon_g) \quad (7\text{–}84)$$

$$g_{c2} = 2\left(\frac{\varepsilon_\delta}{\varepsilon_\delta - \varepsilon_g}\right)g_{c1} \quad (7\text{–}85)$$

$$C_c = \frac{g_{c1}\delta_0^2}{2D_p}\left(\frac{\varepsilon_\delta}{\varepsilon_\delta - \varepsilon_g}\right). \quad (7\text{–}86)$$

The admittance Y_s^i is obtained by the use of Eqs. (7–64), (7–73), (7–74), and (7–75) as follows:

$$Y_s^i = \frac{qI_E(1-\gamma_0)}{kT\gamma_0}\left[1 + \frac{j\omega\delta_0^2}{4D_p}\left(\frac{1+2\gamma_0(\varepsilon_g - 1)}{1-\gamma_0}\right) + \frac{(j\omega)^2\delta_0^4}{16D_p^2} \times \left(\frac{1+2\gamma_0(\varepsilon_g^2 - \varepsilon_g - 1)}{1-\gamma_0}\right)\right]. \quad (7\text{–}87)$$

The Y_s^i can also be synthesized as a three-element network shown in Fig. 7–5d with

$$g_{s1} = \frac{qI_{Ep}}{kT}\left(\frac{1-\gamma_0}{\gamma_0}\right) \tag{7–88}$$

$$g_{s2} = \frac{[1+2\gamma_0(\varepsilon_g-1)]^2 g_{s1}}{[1+2\gamma_0(\varepsilon_g^2-\varepsilon_g-1)](1-\gamma_0)} \tag{7–89}$$

and

$$C_s = \frac{\delta_0^2}{4D_p}\left[\frac{1+2\gamma_0(\varepsilon_g-1)}{1-\gamma_0}\right] g_{s1}. \tag{7–90}$$

With the use of Eqs. (7–62) and (7–74) an admittance of a series conductance g_f and inductance L_f are obtained with

$$g_f = -\left(\frac{2I_C\varepsilon_g\, e^{-2\varepsilon_g}}{w_0}\right)\frac{\partial w_0}{\partial V_C} \tag{7–91}$$

$$L_f = \frac{\delta_0^2(\varepsilon_g-1)}{2D_p g_f}. \tag{7–92}$$

The subnetwork of Y_f is shown in Fig. 7–5e.

The admittance Y_m^i maybe expressed by the use of Eqs. (7–63), and (7–74), and (7–75) in terms of the fundamental device parameters as follows:

$$Y_m^i = \frac{qI_{Ep}}{kT}\,\Omega\, e^{\varepsilon_g-\varepsilon_\delta\Omega} = \Omega\, g_{md}\, e^{\varepsilon_\delta(1-\Omega)}, \tag{7–93}$$

where

$$g_{md} = \left(\frac{qI_{Ep}}{kT}\right) e^{\varepsilon_g-\varepsilon_\delta}. \tag{7–94}$$

Now Ω is expanded in power series and only terms upto the ones involving ω^2 are kept. The expression for Y_m^i becomes

$$Y_m^i \doteq \frac{g_{md}\, e^{-[j\omega\delta_0^2/(2Dp)][\varepsilon_\delta-1-(\varepsilon_\delta-2)^{\frac{1}{2}}]}}{1+j\omega[\delta_0^2/(2D_p)](\varepsilon_\delta-2)^{\frac{1}{2}}}. \tag{7–95}$$

With all the individual admittances synthesized we obtain the one generator equivalent circuit, or the hybrid-π drift transistor equivalent circuit for both the common base as well as common emitter circuits shown in Figs. 7–6 and 7–7. It may be seen that these are complete equivalent circuits in that they include the extrinsic elements as well. Note that the dependent source depends upon the internal voltage and not the terminal voltage. Note also that

$$\gamma \doteq \gamma_0. \tag{7–96}$$

Equation (7–96) is justified by virtue of the fact that γ is very close to one and its

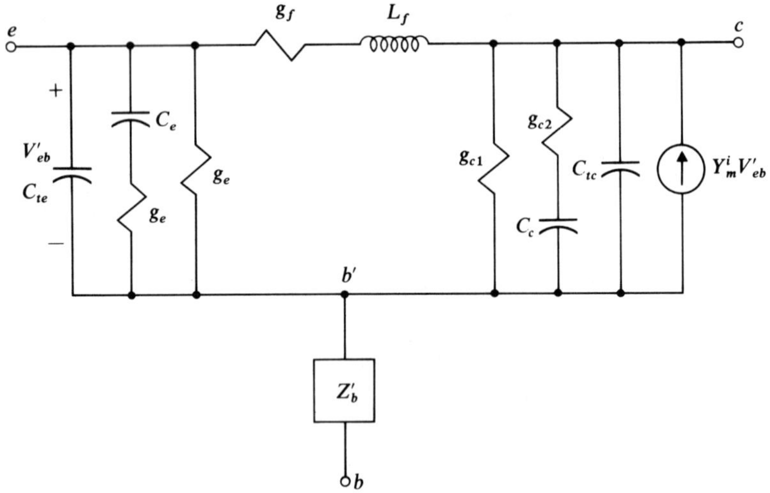

Fig. 7–6 Common base equivalent circuit for drift transistor.

frequency dependence is expected to have a negligible influence on the predicted performance of the transistor.

If the magnitude of the load impedance is low, as is often the case, the admittance between the collector and common terminals in the equivalent circuit may be neglected. Further simplifications may result for specific devices.

It is of interest to note that if R_L is of normal value, i.e. a few hundred or a thousand ohms, in the common emitter equivalent circuit both g_f and L_f may

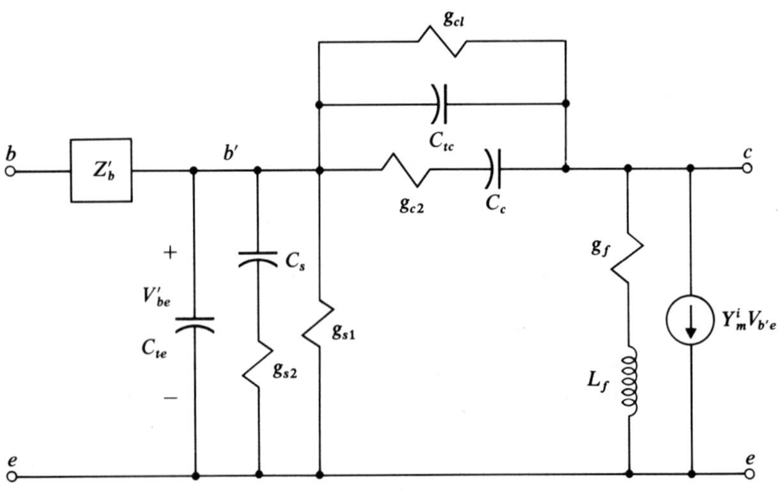

Fig. 7–7 Common emitter equivalent circuit for drift transistor.

be neglected. Furthermore, g_{c2}, C_c and g_{c1} may also be neglected. If in addition C_{te} is negligible and g_{s2} is neglected the resulting hybrid-π equivalent circuit is precisely of the same form as that of Fig. 6–10. Indeed this form of equivalent circuit is in widespread use both for diffusion transistors as well as drift transistors.

Fundamentally the extrinsic parameters, such as the extrinsic base resistance, collector junction capacitance, collector leakage admittance, are distributed parameters. A more accurate model of the device should show the distributed character of these parameters which we have not incorporated in our models. With the aid of a digital computer a better model including the distributed extrinsic parameters can be analyzed. Several attempts have been made at such modeling (Nanavati, 1963; Pritchard, 1959; Poon, 1969).

A more moderate approach to this problem is to represent the collector junction capacitance and extrinsic base resistance as a ladder like network consisting of just three to four elements, e.g. C_{tc1}, C_{tc2}, r_b' or C_{tc}, r_{b1}', r_{b2}' may be a compromise between analytic simplicity and accuracy of the model.

7–5 INTRINSIC SHORT-CIRCUIT CURRENT GAIN

As the short-circuit current gain, and its upper cut-off frequency are among the most important transistor parameters, we shall calculate these for the drift transistor having a *constant* electric *field.*

We shall assume that for the purpose of calculating the intrinsic short-circuit current gain α_d,

$$\gamma_d \approx 0 \tag{7–97}$$

$$\alpha_d^* \approx 0. \tag{7–98}$$

The current gain α_d may be expressed in terms of the intrinsic admittances the same as for diffusion transistors,

$$\alpha_d = -\frac{Y_{ce}^i}{Y_{ee}^i} \doteq \frac{2\Omega\, \mathrm{e}^{\varepsilon_g - \varepsilon_\delta \Omega}}{1 + \Omega}. \tag{7–99}$$

Use was made of Eqs. (7–73) and (7–75) in arriving at Eq. (7–99).

The frequency dependence of the quantity $2\Omega/(1 + \Omega)$ is quite weak in practical transistors. Indeed at almost all frequencies of interest it is very close to unity. Hence the following approximation is well justified.

$$\alpha_d \doteq \mathrm{e}^{\varepsilon_g - \varepsilon_\delta \Omega} \tag{7–100}$$

or

$$\alpha_d = \mathrm{e}^{-w_0 g/(2L_p^2)}\, \mathrm{e}^{-j\omega\, w_0 \delta_0/(2D_p)}\, \mathrm{e}^{-\omega^2 w_0 \delta_0^3/[2(2Dp)^2]}. \tag{7–101}$$

The first exponential factor on the right hand side of Eq. (7–101) is the low frequency value α_{d0} of α_d.

$$\alpha_{d0} = e^{-w_0 g/(2L_p^2)} \doteq 1 - (\tfrac{1}{2})\frac{w_0 g}{L_p^2}. \tag{7–102}$$

From Eq. (7–101) the following expressions for upper intrinsic cut-off frequency $\omega_{\alpha d}$ and phase angle ϕ_{dd} as a function of frequency are obtained

$$\omega_{\alpha d} = \frac{2D_p}{(w_0\delta_0^3)^{\frac{1}{2}}} = \left(\frac{2D_p}{w_0^2}\right)\left(\frac{w_0}{\delta_0}\right)^{\frac{3}{2}} \doteq \omega_\alpha\left(\frac{w_0}{\delta_0}\right)^{\frac{3}{2}} \tag{7–103}$$

and

$$\phi_{\alpha d} = \frac{\omega w_0 \delta_0}{2D_p} = \left(\frac{\omega}{\omega_{\alpha d}}\right)\left(\frac{w_0}{\delta_0}\right)^{\frac{1}{2}}. \tag{7–104}$$

From Eq. (7–103) it may be concluded that the intrinsic short-circuit current gain cut-off frequency for the drift transistor with a constant field is higher than that of a corresponding diffusion transistor by a factor of $(w_0/\delta_0)^{\frac{3}{2}}$. It is important to note that the phase angle of the short-circuit current gain at $\omega_{\alpha d}$ is greater than expected from a simple single pole expression for gain by the factor $(w_0/\delta_0)^{\frac{1}{2}}$. This excess phase is important in the case of cascaded amplifiers. The excess phase can lead to instability in cascaded amplifiers.

Example 7–1. Calculate at room temperature the ratio of $(\omega_{\alpha d}/\omega_\alpha)$ in Ge and Si transistors given that

$$L_p \gg g. \tag{1}$$

Solution:

From Eq. (7–103) we may solve for the ratio

$$\left(\frac{\omega_{\alpha d}}{\omega_\alpha}\right) = \left(\frac{w_0}{\delta_0}\right)^{\frac{3}{2}}. \tag{7–105}$$

Equation (7–48) may be rewritten

$$\delta_0^2 = \frac{g^2 L_p^2}{g^2 + L_p^2} \doteq g^2. \tag{7–106}$$

The approximation in Eq. (7–106) results from the use of the magnitude relationship of (1). Thus

$$\delta_0 = g. \tag{7–107}$$

Hence Eq. (7–105) may be rewritten for the case at hand as follows:

$$\left(\frac{\omega_{\alpha d}}{\omega_\alpha}\right) = \left(\frac{w_0}{g}\right)^{\frac{3}{2}}. \tag{7–108}$$

The parameter g may be rewritten by the use of Eq. (7–47) as follows:

$$g = \frac{2kT}{q\mathscr{E}} = \frac{2kTw_0}{qV_d}, \tag{7–109}$$

where V_d = base built-in voltage due to impurity grading.

The extreme right hand side of Eq. (7–109) assumes constant electric field in the base region. When Eq. (7–109) is substituted into Eq. (7–108) we obtain:

$$\left(\frac{\omega_{\alpha d}}{\omega_\alpha}\right) = \left(\frac{w_0 q V_d}{2kTw_0}\right)^{\frac{3}{2}} = \left(\frac{qV_d}{2kT}\right)^{\frac{3}{2}}. \tag{7–110}$$

If the optimum voltages V_d for Ge and Si obtained in Section 7–2 are used the following numerical results are obtained. Recall that $kT/q = 0.026$ V at room temperature. Thus,

$$\left(\frac{\omega_{\alpha d}}{\omega_\alpha}\right) = \left(\frac{276}{52}\right)^{\frac{3}{2}} = 12.23 \qquad \text{Ge} \tag{2}$$

$$\left(\frac{\omega_{\alpha d}}{\omega_\alpha}\right) = \left(\frac{476}{52}\right)^{\frac{3}{2}} = 27.7 \qquad \text{Si} \tag{3}$$

Thus, it is seen that a fairly substantial improvement in the upper cut-off frequency is possible in drift transistors.

7–6 PHYSICAL STRUCTURES AND SOME DEVICE CONSIDERATIONS IN HIGH FREQUENCY (MICROWAVE) TRANSISTORS

Two geometries often used in high frequency transistors are the *stripe base* and the *ring base* geometries. The stripe base device geometry was shown in Fig. 4–9. A more realistic configuration of the stripe base device is shown in Fig. 7–8 and that of a ring-base is shown in Fig. 7–9. Both these figures show merely the top view of the plannar-passivated expitaxial transistors.

We have already discussed the fact that the optimum doping profile (i.e. the variation of the doping density starting from the emitter and going through the base region) is the exponential grading. The practical doping profile is not truly exponential and is probably more nearly the complimentary-error-function type. Figure 7–10 shows a typical doping profile in a silicon double diffused microwave transistor cross-section. Note that the emitter is extremely heavily doped (approaching the solid solubility of the chemical impurity). The epitaxial layer is quite broad. Ideally this layer should be intrinsic. In practice it is lightly doped. The substrate is another heavily doped region. The heavily doped emitter increases the injection efficiency. The heavily doped collector gives a low value of collector bulk resistance and hence a low value of

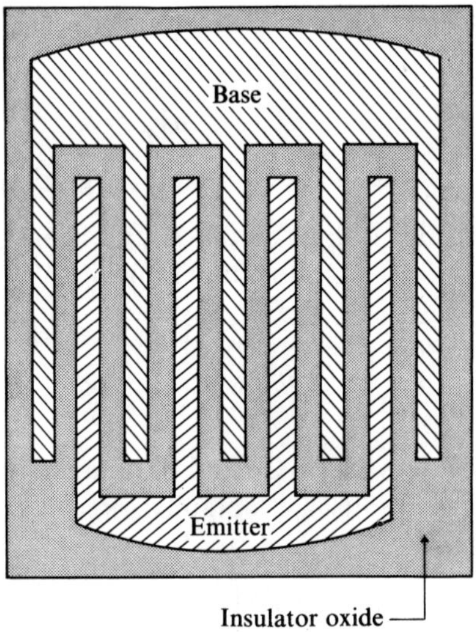

Fig. 7–8 Top view of a stripe base geometry plannar passivated epitaxial transistor. The collector contact is at the bottom of devices (out of view).

voltage across a saturated transistor. The presence of the relatively high resistivity epitaxial region enables us to obtain

(a) high collector breakdown voltage,
(b) low value of C_{tc},
(c) high doping density in the substrate.

The state-of-the-art dimensions in microwave transistors (very high frequency) are given in Table 7–1.

The actual doping profile looked at in detail shows a *retarding field* over a relatively short distance of the base region. It is possible to take this into account. For most purposes it is not worth the trouble to do this; so we omit its treatment here. We should be aware of its existence.

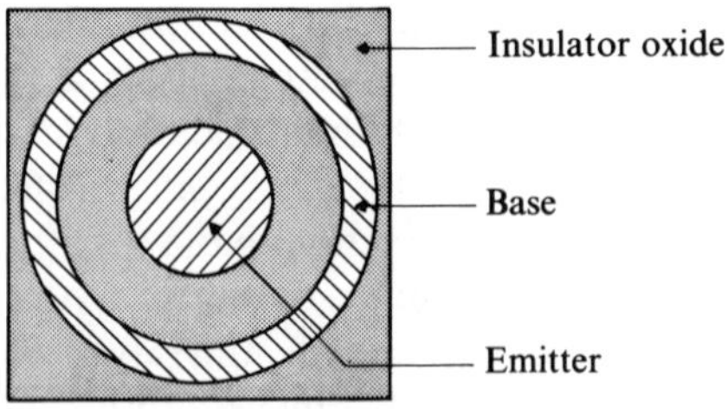

Fig. 7–9 Top view of a ring base geometry plannar passivated epitaxial transistor. The collector contact is at the bottom of devices (out of view).

Table 7–1 State of the Art Transistor Parameters

	Base width w_B (μ)	0.15
	Epitaxial region width w_C (μ)	1.0
	Stripe width (μ)	1.0
Theory (GHz) f_T	Ge (*p-n-p*)	6.7
	Ge (*n-p-n*)	10.4
	Si (*p-n-p*)	5.2
	Si (*n-p-n*)	8.6
	GaAs (*p-n-p*)	5.0
	GaAs (*n-p-n*)	18.5
Expt'l (GHz) f_T	Ge (*p-n-p*)	> 6
	Ge (*n-p-n*)	> 7
	Si (*p-n-p*)	> 5
	Si (*n-p-n*)	> 8
	GaAs (*p-n-p*)	> 0.5
	GaAs (*n-p-n*)	> 1.4
$r_b C_c$ (p-sec)	Ge (*p-n-p*)	5
	Ge (*n-p-n*)	15
	Si (*p-n-p*)	5
	Si (*n-p-n*)	10

A figure of merit that has been widely used is the cut-off frequency f_T. This frequency is *defined to be the frequency at which the common emitter short-circuit current gain is unity.*

In evaluating it the following arbitrary relationship is assumed to be valid.

$$f_T = \frac{1}{2\pi \tau_{ec}}, \qquad (7\text{–}111)$$

where τ_{ec} = total delay time experienced by minority carriers as they travel from the emitter across the base into the collector.

The total delay time consists of *four* delay times. In each case the delay is either equated to a time constant or to an actual transit time through a region.

a) *Emitter Charging Times* (τ_E)

This is a misnamed time as it is meant to represent the delay in the input of a common emitter circuit.

$$\tau_E = r_{ed}[C_{te} + C_{tc} + C_\ell] = \frac{kT}{qI_E}[C_{te} + C_{tc} + C_\ell], \qquad (7\text{–}112)$$

where r_{ed} = emitter diffusion resistance,

C_ℓ = linear stray capacitance between the base and emitter terminals.

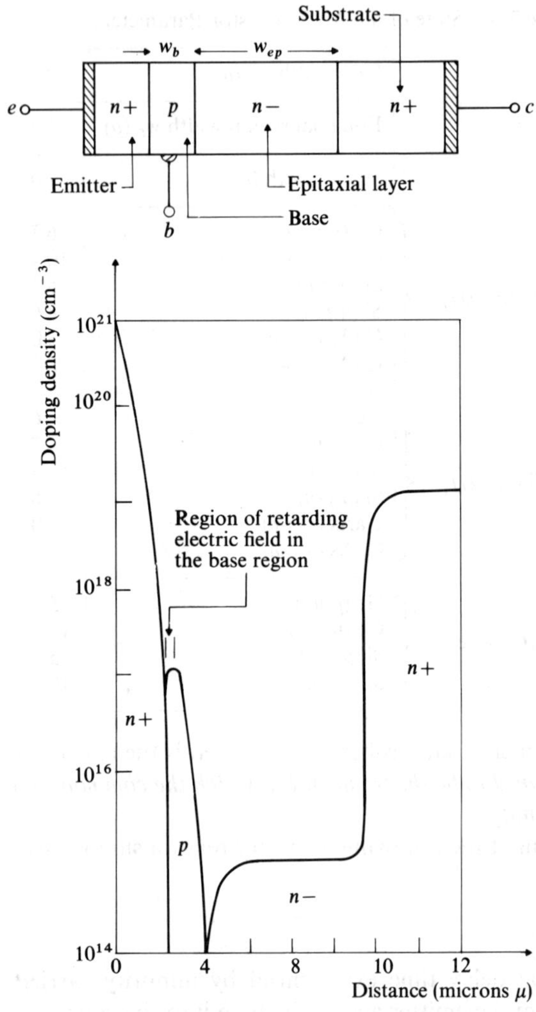

Fig. 7–10 Typical doping profile in high frequency transistors. (Poon, 1969.)

b) *Base Transport Time* (τ_B)

This too is misleadingly named and is simply the reciprocal of ω_N

$$\tau_B = \frac{w_0^2}{\eta D_B}, \tag{7–113}$$

where D_B = diffusion constant for minority carriers in the base region,

η = numerical factor which equals 2.43 for diffusion transistors and can be significantly higher for drift transistors (see Example 7–1).

c) *Collector Depletion Region Transit Time* (τ_C)

This is the transit time for minority carriers to traverse the collector transition region under scattering limited condition usually obtaining in the collector of the very high frequency transistors

$$\tau_C = \frac{x_{tc}}{2v_\ell}. \tag{7–114}$$

Occasionally, a fourth delay time is considered.

d) *Collector Delay Time* (τ_{CB})

$$\tau_{CB} = r'_c C_{tc}, \tag{7–115}$$

where r'_c = collector bulk resistance.

Here τ_{CB} is clearly a time constant, involving the collector bulk.

From the above times f_T is calculated as follows:

$$f_T = \frac{1}{2\pi\,\tau_{ec}} = \frac{1}{2\pi[\tau_E + \tau_B + \tau_C]} \tag{7–116}$$

$$f_T = \frac{1}{2\pi\left[\dfrac{kT(C_{te} + C_{tc} + C_\ell)}{qI_E} + \dfrac{w_0^2}{\eta D_B} + \dfrac{x_{tc}}{2v_\ell}\right]}. \tag{7–117}$$

A better value of f_T would involve the fourth time τ_{CB} in the denominator of Eq. (7–116), so that

$$f_T^* = \frac{1}{2\pi\left[\dfrac{kT(C_{te} + C_{tc} + C_\ell)}{qI_E} + \dfrac{w_0^2}{\eta D_B} + \dfrac{x_{tc}}{2v_\ell} + r'_c C_{tc}\right]} \tag{7–118}$$

The rather important high frequency parameter left out of f_T is the extrinsic base resistance r'_b. This is a major short-coming of this figure of merit. It is to be noted that f_T includes both *intrinsic and extrinsic parameters.*

7–7 MAXIMUM FREQUENCY OF OSCILLATIONS AND STRIPE WIDTH

We had already derived the expression for the maximum frequency of oscillation in the previous chapter which is repeated below:

$$f_{\max} = \sqrt{\frac{f_\alpha}{8\pi r'_b C_{tc}}} \quad \text{(diffusion transistor).} \tag{7–119}$$

As the above was obtained for the diffusion transistor, the following may be written for drift transistors

$$f_{\max,\,d} = \sqrt{\frac{f_{\alpha d}}{8\pi r'_b C_{tc}}} \quad \text{(drift transistors),} \tag{7–120}$$

where $f_{\alpha d}$ has replaced f_α.

It can be shown (see Problem 7–4) that if α_d can be expressed as a single pole expression then

$$f_{\alpha d} \approx f_T. \tag{7–121}$$

Furthermore, if the *stripe width* is S and the length is L then we may write for a single stripe geometry device

$$r_b' = \rho_0 S/L \tag{7–122}$$

$$C_{tc} = C_{c0} SL. \tag{7–123}$$

Then $f_{\max}$ becomes

$$f_{\max,\,d} = \frac{1}{S}\left(\frac{f_T}{8\pi\rho_0 C_{c0}}\right)^{\frac{1}{2}} \tag{7–124}$$

Eq. (7–124) reveals the fact that the stripe width is a most important high frequency parameter. The smaller the S is, the better the high frequency performance is expected to be. Thus, we want as large a number of stripes that the fabrication procedure will permit.

7–8 TRANSIENT TIME

The simple minded equivalent circuit approach used in Chapter 6 may also be applied here. The expressions for all four transient times may also be used for drift transistors where now the cut-off frequency ω_N and gain α_N should be replaced by ω_{Nd} and α_{Nd}. That is

$$\omega_N \rightarrow \omega_{Nd} \tag{7–125}$$

$$\alpha_N \rightarrow \alpha_{Nd}, \tag{7–126}$$

where ω_{Nd} is given by Eq. (7–103) and
α_{Nd} is given by Eq. (7–102).

The storage time expression in terms of T_s is experimentally found to work quite well for drift transistors as well.

7–9 SUMMARY

We considered in this chapter the transistors which have a built-in field in the bulk of the base region due to impurity grading. We have seen how this base field helps the overall high frequency performance of the transistor. Both higher gain and higher cut-off frequency are obtained from such devices. Current–voltage equations were derived and wide band models were obtained for diffused-base or drift transistors. Only those features which were different from the diffusion transistor covered in Chapter 6 were covered in this chapter. The method of analysis used in this chapter was the same as that in Chapter 6.

PROBLEMS

7–1 Discuss how the use of a wider band gap semiconductor such as GaAs as compared with Ge for fabricating a diffused base transistor is expected to influence the maximum base electric field, low-frequency current gain, and current-gain cut-off frequency.

7–2 Obtain a low frequency incremental equivalent circuit for the common base configuration of drift transistors.

7–3 Obtain a low frequency incremental equivalent circuit for the common emitter configuration of drift transistors.

7–4 You are to assume that

$$\alpha_d(j\omega) = \frac{\alpha_{d0}}{1 + j\omega_{\alpha d}} \tag{1}$$

and that the common emitter short circuit current gain is given by

$$\beta_d(j\omega) = \frac{\beta_{d0}}{1 + j\omega_{\beta d}}. \tag{2}$$

Furthermore

$$\beta_{d0}\omega_{\beta d} = \omega_{\alpha d} \tag{3}$$

and

$$\beta_d = \frac{\alpha_d}{1 - \alpha_d}. \tag{4}$$

Show that

$$f_T \doteq f_{\alpha d}.$$

7–5 Compare the low frequency incremental equivalent circuits of exponentially graded transistors with those of the homogenous base transistors.

7–6 Obtain analytic expresssions for β_{d0} and $\omega_{\beta d}$ defined in Problem 7–4.

7–7 How is the transport of minority carriers different in an exponentially graded transistor compared with complimentary error function graded transistor?

7–8 A detailed look at the actual doping profile, such as that of Fig. 7–10, indicates that there is a retarding field at the emitter edge of the base region. Discuss what major influence this retarding field has on the transistor characteristics.

7–9 If there is a retarding field at the collector end of the base region due to some cause, discuss how you would expect the minority carrier distribution to be altered by this. Assume that the aiding field is present for most of the base region.

7–10 How do Y_{ee}^i and Y_{cc}^i of diffusion transistors compare with similar admittances for the exponentially graded base transistor.

7–11 You are given an exponentially graded transistor which is used as a common emitter amplifier. The load resistance in the collector circuit is low enough so that h_{12e} and h_{22e} may be neglected. In this case the C_{te} may not be neglected. As an approximation assume that C_{te} may be included in the input circuit by being merely put in parallel with h_{11e} and that no other h-parameters are influenced by C_{te}. Calculate (a) the voltage gain A_v, (b) what is the upper cut-off frequency for A_v?

7–12 Estimate the collector current at which the injected minority carrier density at the emitter edge of the base region is 10 per cent of the majority carrier density there. Discuss the relationship between the above calculation and the current level up to which the full advantage of the built-in base field is effective.

7–13 Compare the d-c h_{21e} for a diffusion transistor with the same parameter in an exponentially graded transistor.

7–14 Discuss the relative advantages of the reciprocal transit time and the maximum frequency of oscillation as a figure of merit for the transistor.

7–15 Under the most simplifying approximations calculate the 0 to 90 per cent rise time for the common emitter configuration and a step of base current drive.

REFERENCES

Almond, J., and McIntyre, R. J., "The Equivalent Circuit of the Drift Transistor," *RCA Rev.,* **18,** No. 3, 361–384, September (1957).

Becke, H., Flatley, D., and Stolnitz, D., "Double Diffused Gallium Arsenide Transistors," *Solid State Electron,* **8,** 255 (1965).

Carslaw, H. S., and Jaeger, J. C., *Conduction of Heat in Solids,* Oxford University Press, 2nd ed. (1959).

Das, M. B., and Boothroyd, A. R., "Determination of Physical Parameters of Diffusion and Drift Transistors," *IRE Trans. on Electron Devices,* **ED-8,** No. 1, 15–30, January (1961).

Daw, A. N., Mitra, R. N., and Choudhury, N. K. D., "Cutoff Frequency of a Drift Transistor," *Solid State Electron.,* **10,** 359 (1967).

Edwards, R., "Fabrication Control is Key to Microwave Performance," *Electronics,* 109–113, Feb. (1968).

Edwards, R., and Pritchard, R. L., "Planner Ge Microwave Transistors," *NEREM Record,* 246 (1965).

Emeis, R., Herlet, A., and Spenke, E., "The Effective Emitter Area of Power Transistors," *Proc. IRE,* **46,** 1220–1229, June (1958).

Fletcher, N. H., "Self Bias Cut-off Effect in Power Transistors," *Proc. IRE,* **43,** 1669, November (1955).

Gansauge, P., "Double Diffused High-Speed Germanium Transistors," *IEEE, PGED,* **ED-15,** No. 10, 728–731.

Giacoletto, L. J., "Junction Capacitance and Related Characteristics Using Graded Impurity Semiconductors," *IRE Trans. on Electron Devices,* **ED-4,** 22–24, July (1957).

Gudmundesen, R. A., "The Reduction of Effective Emitter Voltage Due to Base Current," presented at the AIEE-IRE Semiconductor Device Research Conference, Pennsylvania State University, University Park, Pa., July 7 (1953).

Gummel, H. K., and Scharfetter, D. L., "Depletion-Layer Capacitance of *p-n* Step Junction," *J. Appl. Phys.,* **38,** 2148 (1967).

Hall, R. N., and Dunlap, W. C., "*p-n* Junctions Prepared by Impurity Diffusion," *Phys. Rev.,* **80,** 467 (1950).

Hoerni, J. A., "Planar Silicon Transistor and Diodes," IRE Electron Devices Meeting, Washington, D.C. (1960).

Johnson, E. O., "Physical Limitations of Frequency and Power Parameters of Transistors," *IEEE. Interc. Conv. Record,* **Part 5,** 27 (1965).

Kesterbaum, A. L., and Ditrick, N. H., "Design Construction and High Frequency Performance of Drift Transistors," *RCA Rev.,* **28,** 12–23, March (1957).

Kromer, H., "Zur Theorie des Diffusions und des Driftransistors, Parts I-III," *Arch. electrischem Ubertragung,* **8,** 223–228, May (1954); 363–369, August (1954); 499–504, November (1954).

Kromer, H., "Theory of a Wide-Gap Emmiter for Transistors," *Proc. IRE,* **45,** 1535 (1957).

Lee, C. A., "A High Frequency Diffused Base Germanium Transistor," *Bell Syst. Tech. J.,* **35,** 23 (1956).

Moll, J. L., and Ross, I. M., "The Dependence of Transistor Parameters on the Distribution of Base Layer Resistivity," *Proc. IRE,* **44,** 72 (1956).

Nanavati, R. P., *An Introduction to Semiconductor Electronics,* Chapter 11, McGraw-Hill Book Co., New York (1963).

Poon, H. C. et al, "High Injection in Epitaxial Transistors," *IEEE Prof. Group on Electron Devices,* 455–458, May (1969).

Prince, M. B., "Diffused *p-n* Junction Silicon Rectifiers," *Bell Syst. Tech. J.,* **35,** 661 (1956).

Pritchard, R. L., "Transistor Equivalent Circuits, presented at the International Convention on Transistors and Associated Semiconductor Devices," London, May 22 (1959).

Pritchard, R. L., Angell, J. B., Adler, R. B., Early, J. M., and Webster, W. M., "Transistor Internal Parameters of Small-Signal Representation," *Proc. IRE,* **49,** 725 (1961).

Statz, H., "Double Diffused *p-n-p* GaAs Transistor," *Solid State Electron.,* **8,** 827 (1965).

Sze, S. M., and Gummel, H. K., "Appraisal of Semiconductor-Metal-Semiconductor Transistors," *Solid State Electron,* **9,** 751 (1966).

Tanenbaum, M., and Thomas, D. E., "Diffused Emitter and Base Silicon Transistors," *Bell Syst. Tech. J.,* **35,** 1 (1956).

Theuerer, H. C., Kleimack, J. J., Loar, H. H., and Christenson, H., "Epitaxial Diffused Transistors," *Proc. IRE,* **48,** 1642 (1960).

Thomas, D. E., and Moll, J. L., "Junction Transistor Short-circuit Current Gain and Phase Determination," *Proc. IRE,* **46,** 1177–1184 (1958).

Von Munch, W., "Gallium Arsenide Planar Technology," *IBM Journal,* 348, Nov. (1966).

Von Munch, W., Statz, H., and Blakeslee, A. E., "Isolated GaAs Transistors on High-Resistivity GaAs Substrate," *Solid State Electron,* **9,** 826 (1966).

CHAPTER 8

THE JUNCTION FIELD EFFECT TRANSISTOR

8–1 INTRODUCTION

The junction field effect transistor was first proposed by Shockley (1952). Dacey and Ross (1953, 1955) were the first to physically realize this device. The junction field effect transistor, often called simply the field effect transistor, is abbreviated as JFET or simply FET. This is a majority carrier device in that only the majority carrier participates in the conduction mechanism. Thus, because only *one* kind of carrier, i.e. electrons or holes but not both, is an active participant it is also called the *unipolar* transistor. In contrast the regular junction transistor has carriers of both types participating in the conduction mechanism, and is hence called the *bipolar* transistor.

The operation of the FET consists of the variation of the resistance of the channel, along which the majority carriers travel, by a variation of the cross-sectional area of the channel by a change in the transition region width of a reverse biased *p-n* junction. In this chapter we shall study the physics of this device. Terminal *I-V* characteristics will be obtained in terms of the physical parameters of the FET.

Figure 8–1a shows the physical structure of an FET. It can be seen that the device has a source region, a drain region, and a gate region. (The two gate regions which are tied together may be considered to be one extended gate region. There are geometries other than that of Fig. 8–1 also in use where the gate is more clearly a single region.)

The source serves as the source of majority carriers which travel from the source to the drain along the channel. There are *n*-channel FETs and also *p*-channel FETs. Figures 8–1b and 8–1c show the symbols of the *n*- and *p*-channel FETs respectively. The arrow is associated only with the source

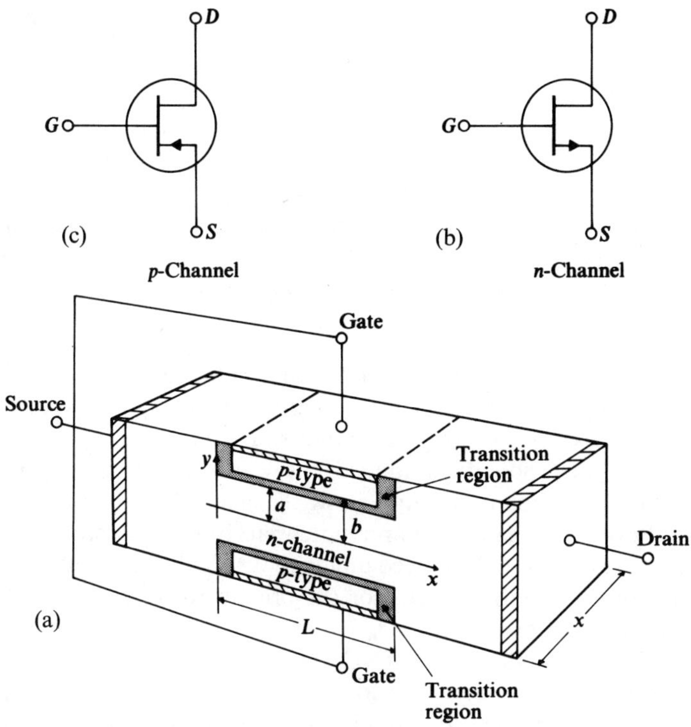

Fig. 8–1 Structure of the idealized symmetric junction FET.

terminal.† The arrow is directed along the direction of conventional current flow due to majority carriers going from the source to the drain. Recall that the direction of conventional current flow is along the direction of hole flow and opposite the direction of electron flow. That is why in a *p*-channel device the arrow is pointed into the device, whereas it is pointed out of the device in an *n*-channel device.

8–2 IDEALIZED SYMMETRIC FET

In Fig. 8–1a is shown the structure of an idealized symmetric FET. The following is assumed in the analysis given below.

1. The *IR* drop in the source and drain end regions are negligibly small. (These voltage drops will be easily accounted for later on in the complete model of the device.)

†Another convention which is also used shows the arrow associated with the gate terminal, and points into the device for *n*-channel and out of the device for the *p*-channel FETS. We prefer the convention used here as it identifies the three terminals automatically.

2. The current between the drain and source has only an x component.
3. The gate region is very much more heavily doped compared with the channel region.
4. The space charge region in the transition region is characterized by a step junction or abrupt junction distribution of fixed charge density.
5. The rate of change with distance of the field $\mathscr{E}_x$ is much less than the rate of change of $\mathscr{E}_y$ with distance. This is also called the *gradual channel approximation*. Thus,

$$\frac{d\mathscr{E}_x}{dx} \ll \frac{d\mathscr{E}_y}{dy}. \tag{8–1}$$

Figure 8–1a shows the transition region straddling the p-n junction to be of uniform thickness. Usually there is potential applied not only between gate and source but also between the drain and source.

With both these potentials applied, the width of the transition region is expected to be a function of position along the channel. For simplicity we shall first consider only the gate to source potential. In such a case the transition region width is expected to be indeed uniform. We shall consider the application of drain to source potential later. For the present we assume that that drain terminal has been left open-circuited.

We shall assume that the gate channel p-n junction is a step junction (assumption 4). Thus, the equation for the width of the transition region corresponding to the step junction given in Eq. (3–98) is valid here. That equation will be repeated here for convenience.

$$x_t = \left(\frac{2\varepsilon}{q}\right)^{\frac{1}{2}} \left\{\left(\frac{N_a/N_d}{N_a + N_d}\right)^{\frac{1}{2}} + \left(\frac{N_d/N_a}{N_a + N_d}\right)^{\frac{1}{2}}\right\} V_t^{\frac{1}{2}}. \tag{8–2}$$

Because of assumption 3

$$N_a \gg N_d. \tag{8–3}$$

With the use of the magnitude relationship (8–3) we simplify Eq. (8–2) as follows:

$$x_t = \left(\frac{2\varepsilon}{q}\right)^{\frac{1}{2}} \left\{\left(\frac{1}{N_d}\right)^{\frac{1}{2}} + \left(\frac{N_d}{N_a^2}\right)^{\frac{1}{2}}\right\} V_t^{\frac{1}{2}} \tag{8–4}$$

or

$$x_t = \left(\frac{2\varepsilon}{qN_d}\right)^{\frac{1}{2}} V_t^{\frac{1}{2}}. \tag{8–5}$$

According to the discussion in Section 3–7 almost all of the transition region is in the channel region because the gate is extremely heavily doped. Thus, in terms of the parameters of Fig. 8–1 and Eq. (8–5) the following may be written:

$$b - a = \left(\frac{2\varepsilon}{qN_d}\right)^{\frac{1}{2}} (V_{BI} - V_{GS})^{\frac{1}{2}}. \tag{8–6}$$

In the usual case where the gate channel *p-n* junction is reverse biased and where V_{BI} is the built-in voltage of the gate-channel junction

$$|V_{GS}| \gg V_{bi}. \tag{8–7}$$

Eq. (8–6) takes on the simpler form

$$b - a = -\left(\frac{V_{GS} 2\varepsilon}{qN_d}\right)^{\frac{1}{2}}. \tag{8–8}$$

The width of the conducting channel is (2*a*). The specific gate to source voltage V_{GS} at which the conducting channel is choked off, called *pinch-off* voltage V_p, may be easily calculated from Eq. (8–8) by setting *a* equal to zero.

$$V_{GS}\Big|_{a=0} = V_P = -\frac{b^2 qN_d}{2\varepsilon}. \tag{8–9}$$

Equation (8–8) may be rewritten in terms of V_P more compactly as follows:

$$a = b\left[1 - \left(\frac{V_{GS}}{V_P}\right)^{\frac{1}{2}}\right]. \tag{8–10}$$

In the schematic of Fig. 8–1, the channel represents a rectangular resistance of length *L*, width 2*a*, and height *h*. Neglect the end segments of the source and drain which is consistent with assumption 1. Then the conductance of the channel is given by

$$G = \frac{\sigma A}{L} = \frac{(q\mu_n n_n)(2ah)}{L}, \tag{8–11}$$

where n_n = thermal equilibrium density of electrons in the *n* channel and

μ_n = mobility of electrons.

In evaluating the channel conductivity, we have used Eq. (2–39) and neglected the contribution due to minority carrier–holes. By the use of Eq. (8–10) for *a* Eq. (8–11) may be written

$$G = \frac{2qhbq\mu_n n_n}{L}\left[1 - \left(\frac{V_{GS}}{V_P}\right)^{\frac{1}{2}}\right] \tag{8–12}$$

$$G = G_0\left[1 - \left(\frac{V_{GS}}{V_P}\right)^{\frac{1}{2}}\right], \tag{8–13}$$

where

$$G_0 = \frac{2qhbq\mu_n n_n}{L}. \tag{8–14}$$

The conductance G_0 is the conductance when gate to source voltage is zero.

It is worthwhile noting that Eqs. (8–13), (8–6), and (8–8) were derived under the assumption that the transition region width *was constant* along the channel. This is precisely true when the drain to source voltage V_{DS} is zero. It is a *good approximation* when

$$V_{DS} \ll V_{GS}. \tag{8–15}$$

When the magnitude relationship given by Eq. (8–15) is valid, the drain current I_D and V_{DS} are linearly related to each other as follows:

$$I_D = GV_{DS} = G_0\left[1 - \left(\frac{V_{GS}}{V_P}\right)^{\frac{1}{2}}\right]V_{DS}. \tag{8–16}$$

When Eq. (8–15) is not satisfied it is to be expected that the width of the transition region will be a function of x because of the externally applied voltage due to the presence of sources in the gate-source and the drain-source circuits. For now, we wish to consider drain to source voltage less than pinch-off.

$$V_{DS} \leqslant -V_P. \tag{8–17}$$

The magnitude relationship Eq. (8–7) assumes that the channel is not pinched off.

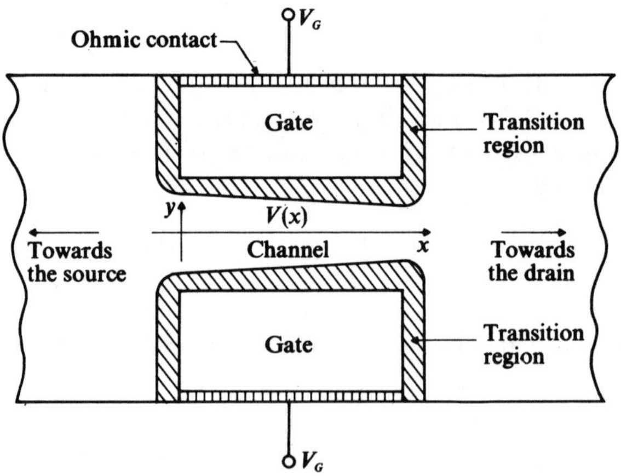

Fig. 8–2 The gate-channel transition region below pinch-off.

We shall assume that the channel width varies gradually with x so that the preceding one-dimensional analysis is valid. *This is the gradual channel approximation.* The only difference is that the half channel width a is now a function of x. Figure 8–2 shows the transition region in this case.

The analysis starts with the writing of the transport equation in the absence of concentration gradients. Thus, for the n-channel

$$J_n(x) = q\mu_n n_n \mathscr{E}, \tag{8–18}$$

where μ_n = mobility of electrons in the channel,

n_n = thermal equilibrium density of electrons in the channel.

Replacing the electrostatic field by the negative gradient of the potential, we obtain the following equation for current

$$I_n(x) = -q\mu_n n_n A(x)\frac{dV(x)}{dx}, \tag{8-19}$$

where the cross sectional area $A(x)$ is a function of the x coordinate as follows:

$$A(x) = h2a(x). \tag{8-20}$$

Thus, Equation (8–19) may be rewritten:

$$I_n(x) = -2hq\mu_n n_n a(x)\frac{dV(x)}{dx}. \tag{8-21}$$

The half channel width $a(x)$ may be obtained from Eq. (8–10) in which the following substitution is used.

$$V_{GS} \rightarrow [V_G - V(x)], \tag{8-22}$$

where $V(x)$ is the voltage of the channel as a function of x.
The right hand side of Eq. (8–22) stands for the reverse voltage across the transition region at x.

Then

$$a(x) = b\left[1 - \left(\frac{V_G - V(x)}{V_P}\right)^{\frac{1}{2}}\right]. \tag{8-23}$$

Equation (8–23) may now be substituted into Eq. (8–21) to obtain

$$I_n(x) = -2hq\mu_n n_n b\left[1 - \left(\frac{V_G - V(x)}{V_P}\right)^{\frac{1}{2}}\right]\frac{dV(x)}{dx}. \tag{8-24}$$

Multiplying both sides of Eq. (8–24) by dx, and recognizing that by Kirchhoff's current law the channel current $I_n(x)$ must be continuous and equal to the drain current $-I_D$, we obtain the following (we are neglecting any gate current by virtue of assumption 2 which is experimentally found to be valid, the drain current is several orders of magnitude higher than the gate current).

$$\int_{x=0}^{x=L} (-I_D)dx = -2ahq\mu_n n_n \int_{V_S}^{V_D}\left[1 - \left(\frac{V_G - V(x)}{V_P}\right)^{\frac{1}{2}}\right]dV. \tag{8-25}$$

The limits of integration are from the source end of the channel to the drain end of the channel. The voltage drop in the bulk of semiconductor in the source and drain have not been considered by virtue of assumption 1.

$$I_D L = 2ahq\mu_n n_n \left\{\int_{V_S}^{V_D} dV - \left(\frac{V_G}{V_P}\right)^{\frac{1}{2}} \int_{V_S}^{V_D}\left[1 - \frac{V(x)}{V_G}\right]^{\frac{1}{2}} dV\right\}. \tag{8-26}$$

The last integral on the right hand side of Eq. (8–26) may be evaluated by using the following:

$$\int (fx + g)^{\frac{1}{2}}\,dx = \frac{2}{3f}(fx + g)^{\frac{3}{2}}. \tag{8–27}$$

Comparing the last integral on the right hand side of Eq. (8–26) with Eq. (8–27) we see that

$$g = 1 \tag{8–28}$$

and

$$f = -1/V_G. \tag{8–29}$$

Using the integral of Eq. (8–27) to evaluate Eq. (8–26) with the help of Eqs. (8–28) and (8–29) we obtain:

$$I_D L = 2ahq\mu_n n_n \left\{(V_D - V_S) - \left(\frac{V_G}{V_P}\right)^{\frac{1}{2}}\left[-\frac{2V_G}{3}\left(1 - \frac{V(x)}{V_G}\right)\right]_{V_S}^{V_D}\right\} \tag{8–30}$$

$$I_D L = 2ahq\mu_n n_n \left\{(V_D - V_S) - \left(\frac{V_G}{V_P}\right)^{\frac{1}{2}}\left(-\frac{2V_G}{3}\right)\left[\left(1 - \frac{V_D}{V_G}\right)^{\frac{3}{2}} - \left(1 - \frac{V_S}{V_G}\right)^{\frac{3}{2}}\right]\right\}. \tag{8–31}$$

Dividing Eq. (8–31) by L and taking V_G out common from the last two square brackets on the right hand side of this equation we obtain the following:

$$I_D = G_0\left\{(V_D - V_G + V_G - V_S) + \tfrac{2}{3}\left[\frac{(V_G - V_P)^{\frac{3}{2}}}{V_P^{\frac{1}{2}}} - \frac{(V_G - V_S)^{\frac{3}{2}}}{V_P^{\frac{1}{2}}}\right]\right\}. \tag{8–32}$$

But

$$V_D - V_G \triangleq V_{DG} \tag{8–33}$$

$$V_G - V_S \triangleq V_{GS}. \tag{8–34}$$

With the use of Eqs. (8–33) and (8–34), Eq. (8–32) may be rewritten.

$$I_D = G_0\left[V_{GS}\left[1 - \frac{2}{3}\left(\frac{V_{GS}}{V_P}\right)^{\frac{1}{2}}\right] - V_{GD}\left[1 - \frac{2}{3}\left(\frac{V_{GD}}{V_P}\right)^{\frac{1}{2}}\right]\right]. \tag{8–35}$$

Equation (8–35) is valid from $I_D = 0$ up to the pinch-off point. It will be useful to write the equation for drain current in terms of drain to source voltage V_{DS} and gate to source voltage V_{GS} because the common source configuration is the most widely used amplifier configuration for which the input voltage is V_{GS} and the output voltage is V_{DS}. Thus, to obtain the desired result we shall use the following:

$$V_{GD} = V_{GS} - V_{DS}. \tag{8–36}$$

With the help of Eq. (8–36), Eq. (8–35) may be rewritten

$$I_D = G_0\left[V_{GS}\left[1 - \frac{2}{3}\left(\frac{V_{GS}}{V_P}\right)^{\frac{1}{2}}\right] - (V_{GS} - V_{DS})\left[1 - \frac{2}{3}\left(\frac{V_{GS} - V_{DS}}{V_P}\right)^{\frac{1}{2}}\right]\right]. \tag{8–37}$$

Regrouping leads to the following:

$$I_D = G_0\left[\frac{2}{3}V_{GS}\left[\left(\frac{V_{GS} - V_{DS}}{V_P}\right)^{\frac{1}{2}} - \left(\frac{V_{GS}}{V_P}\right)^{\frac{1}{2}}\right] + V_{DS}\left[1 - \frac{2}{3}\left(\frac{V_{GS} - V_{DS}}{V_P}\right)^{\frac{1}{2}}\right]\right]. \tag{8–38}$$

Equation (8–38) represents the drain characteristics of the ideal FET up to the point of pinch-off. These characteristics are plotted on Fig. 8–3 which also shows the linear region where the slopes are designated on the figure.

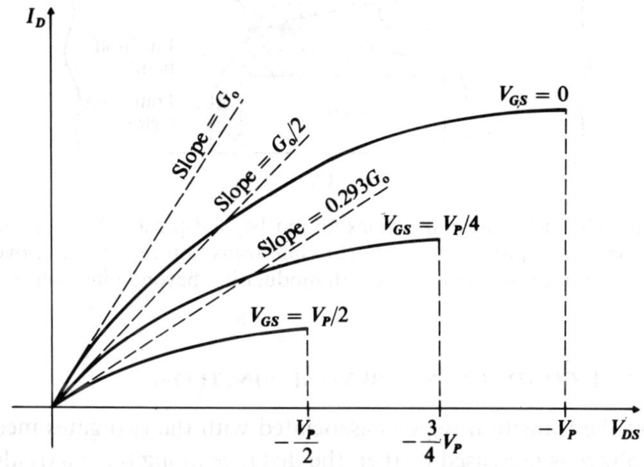

Fig. 8–3 Common source output characteristics together with the initial slope of the $I_D - V_{DS}$ characteristics.

It should be recognized that as the device approaches pinch-off at the drain end, the current density increases dramatically as the channel cross-sectional area, through which the current passes, diminishes almost to the vanishing point. At pinch-off the two transition regions meet as shown in Fig. 8–4 and further increase in current is expected to cease. Experimentally, however, the current *does* continue to increase although at a low rate. A mechanism which can account for the continuing increase in drain current after pinch-off is the shortening of the channel length which is discussed in some detail in the next section.

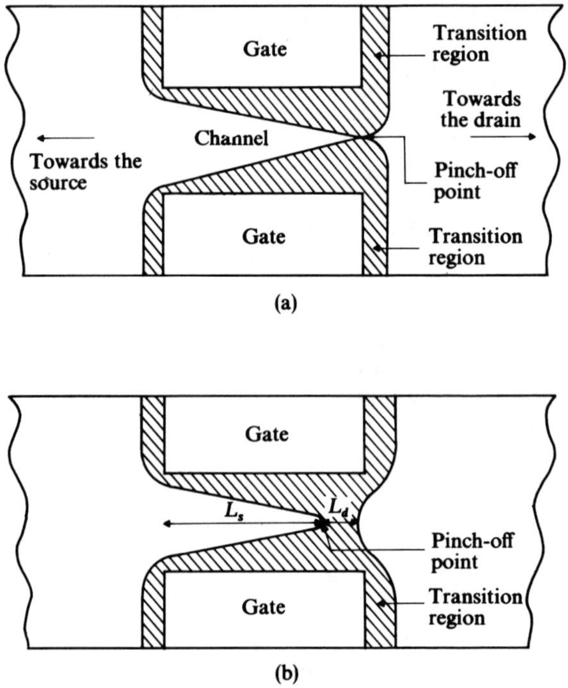

Fig. 8–4 Gate-channel transition regions at and beyond pinch-off regions. (a) The gate transition regions meet just at pinch-off. (b) Gate transition regions meet over a greater length showing channel length modulation beyond pinch-off.

8–3 CHANNEL CONDUCTANCE BEYOND PINCH-OFF

At pinch-off the transition regions associated with the two gates meet. If drain to source voltage is increased further, the distance along the y-axis along which the transition regions meet increases leading to a net *decrease* in the *effective channel length.* This phenomenon is referred to as *channel length modulation,* and was predicted by Shockley (1952). Equation (8–38) may be rewritten now to explicitly show its dependence on the channel length L as follows:

$$I_D = \frac{2ahq\mu_n n_n}{L_s}\left\{\frac{2}{3}V_{GS}\left[\left(\frac{V_{GS}-V_{DS}}{V_P}\right)^{\frac{1}{2}}-\left(\frac{V_{GS}}{V_P}\right)^{\frac{1}{2}}\right] + V_{DS}\left[1-\frac{2}{3}\left(\frac{V_{GS}-V_{DS}}{V_P}\right)^{\frac{1}{2}}\right]\right\}, \tag{8–39}$$

where L_s = effective channel length.

It is obvious from Eq. (8–39) that as the effective channel length L_s decreases, the drain current increases. This is one of the important reasons why the *drain characteristics in actual devices do not truly saturate.* The channel conductance *beyond pinch-off* may be written as

$$g_D = \frac{\partial I_D}{\partial V_{DS}} = \frac{\partial I_D}{\partial L_s}\frac{\partial L_s}{\partial V_{DS}}, \tag{8-40}$$

where $L_s =$ is the effective channel length.
Furthermore,

$$L = L_s + L_d, \tag{8-41}$$

where $L_d =$ distance over which the transition regions meet at the drain end (see Fig. 8-4b).

From Eq. (8-41) we may write

$$\frac{\partial L_s}{\partial V_{DS}} = -\frac{\partial L_d}{\partial V_{DS}}. \tag{8-42}$$

With the help of Eq. (8-42) we may rewrite Eq. (8-40) as follows:

$$g_D = -\frac{\partial I_D}{\partial L_s}\cdot\frac{\partial L_d}{\partial V_{DS}}. \tag{8-43}$$

Note that L_d depends on the transition region widening effects. Equations (3-98) and (3-103) convince us that $\partial L_d/\partial V_{DS}$ is *independent of current.* Let us designate

$$\frac{\partial L_d}{\partial V_{DS}} \triangleq \varepsilon_d. \tag{8-44}$$

Next we shall calculate the first derivative on the right hand side of Eq. (8-43). In preparation for this calculation, we rewrite Eq. (8-39) as follows:

$$I_D = \frac{K_1}{L_s}, \tag{8-45}$$

where

$$K_1 \triangleq 2ahq\mu_n n_n \left\{\frac{2}{3} V_{GS}\left[\left(\frac{V_{GS} - V_{DS}}{V_P}\right)^{\frac{1}{2}} - \left(\frac{V_{GS}}{V_P}\right)^{\frac{1}{2}}\right]\right.$$
$$\left. + V_{DS}\left[1 - \frac{2}{3}\left(\frac{V_{GS} - V_{DS}}{V_P}\right)^{\frac{1}{2}}\right]\right\}. \tag{8-46}$$

The derivative may now be calculated

$$\frac{\partial I_D}{\partial L_s} = -\frac{K_1}{L_s^2}. \tag{8-47}$$

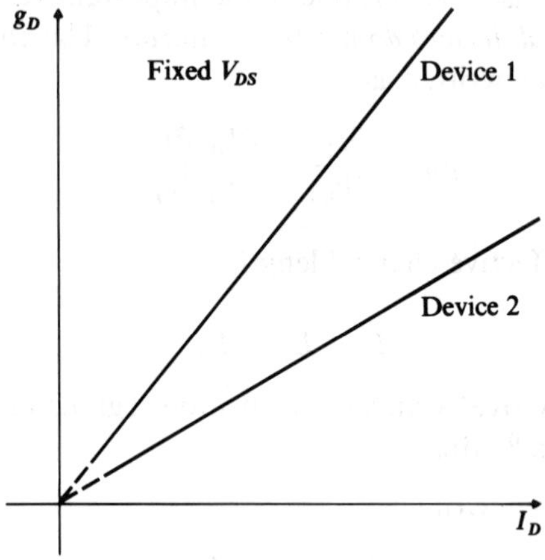

Fig. 8–5 Experimentally obtained incremental drain conductance as a function of drain current when V_{DS} is held fixed.

The right hand side of Eq. (8–47) may be written

$$-\frac{K_1}{L_s^2} = -\left(\frac{K_1}{L_s}\right)\left(\frac{1}{L_s}\right). \tag{8–48}$$

The first bracket on the right hand side of Eq. (8–48) may be recognized as L_D. This fact, together with Eqs. (8–47) and (8–48), yields

$$\frac{\partial I_D}{\partial L_s} = -\frac{I_D}{L_s}. \tag{8–49}$$

Equation (8–49) may be combined with Eqs (8–43) and (8–44). Thus

$$g_D = \frac{\varepsilon_d I_D}{L_s}. \tag{8–50}$$

Recall that ε_d is independent of current. Thus, we conclude that the *drain conductance on the basis of channel length modulation model is expected to be proportional to drain current and inversely proportional to the effective channel length.* When V_{DS} is held fixed g_D is indeed found to be proportional to drain current I_D as shown in Fig. 8–5. Shockley's theory, which used the gradual channel approximation, applies to those junction FET structures where the gate electrode width W_g is significantly greater than the channel width b.

$$W_g \gg b. \tag{8–51}$$

The theory which we have been using also assumes that the mobility for mobile carriers is a constant, i.e. independent of the electric field. Thus, *the above theory is not expected to apply to those devices in which* (a) *the gate width is narrow compared to channel width and/or in which* (b) *field dependent mobility needs to be taken into account.*

8-4 MECHANISMS LEADING TO SATURATION

Several different theories have been proposed to explain the mechanism which leads to saturation in the drain characteristics. Shockley's model suggested that saturation occurred when the channel got pinched-off. He did not discuss in any detail what accounted for the flow of drain current through the pinched-off region. He simply observed that the channel must have a sufficient concentration of mobile carriers, even while pinched-off, to allow the drain current to flow. The material of Section 8-3 states that the point along the channel which gets pinched-off first (the *pinch-off point*) moves toward the source end as the voltage V_{DS} is increased. Grebene and Ghandhi (1968) modified the Shockley model and proposed that in the pinched-off section of the channel, the current flow is confined to a rather narrow path surrounding the center of the channel. Because this model allows the gradual movement of the pinch-off point along the channel, it predicts the so called *soft saturation.*

The devices which fit the theory of the last section do in fact show soft saturation. There are other junction FET, which show the *hard saturation*—meaning the onset of saturation occurs more rapidly than in the soft saturation devices. Thus, it is clear that the hard saturation devices could not be adequately explained by the Shockley model.

Dacey and Ross (1955) were the first to propose that the hard pinch-off could be explained by a model which asserts that the pinch-off point occurs at the *drain end of gate junctions and remains there essentially immobile, independent of the applied voltage.* They claimed that saturation in the *I-V* characteristics was due to the phenomenon of *saturation velocity* reached by mobile carriers in the constricted channel region where the longitudinal electric field is large enough to produce this phenomenon. When the carrier velocity ceases to be independent of the electric field, we may say that the *mobility begins to be a function of the electric field.*

Figure 2-26 and the concept of saturation carrier velocity is seen to be consistent with the observation of saturation of the drain current. Indeed Grosvalet et. al (1963) showed that the onset of current saturation *does not* occur when the channel gets pinched-off but occurs *only* when the carriers within the channel reach *saturation velocity.* The experimental evidence for the above conclusion was the observation of the temperature dependence of the drain current. In *n*-channel Ge devices before saturation, the current varied as $T^{-\frac{3}{2}}$; after onset of saturation it varied as $T^{-\frac{1}{2}}$. The latter dependence is consistent with electron mobility in *n*-type germanium.

Computer studies (Kennedy and O'Brien, 1970) have shown that saturation velocity limited transport along the channel leads to regions of accumulation and depletion of majority carriers located symmetrically along the longitudinal axis down the center of the channel. The regions of accumulation and depletion constitute an electrostatic bilayer which exists along that position of the channel normally assumed to be pinched-off by the gate transition regions. *The existence of this additional electrostatic potential due to the bilayer results in keeping the pinch-off point essentially static with variation in drain to source voltage as is required by Dacey and Ross hypothesis.* This bilayer alters the potential variation along the channel which in turn influences the transition region width of the gate junctions. Dacey and Ross did not discuss the mechanics which would keep the pinch-off point immobile with variation in drain to source voltage.

It turns out that the bilayer is more pronounced in *narrow gate* devices. One major effect of the bilayer is a substantial contraction of the gate channel transition region. As a result, these devices *never completely* pinch-off the channel; only a narrowing of the channel cross-sectional area occurs. In such devices saturation seems to occur at externally applied potential substantially less than that predicted for pinch-off according to the Shockley model. Of course the Shockley model is not expected to work in these devices as they violate one of the important assumptions (wide gate) of the Shockley model.

8–5 EQUIVALENT CIRCUIT FOR THE JUNCTION FET

The wide band model of the junction FET may be rather readily obtained from the structure of the device and the knowledge of *p-n* junctions. Thus, from phenomenological considerations the equivalent circuit of Fig. 8–6 may be obtained from the physical structure of Fig. 8–7. Capacitances C_{gs} and C_{gd} are junction capacitances; C_{ds} is linear stray capacitance. The leakage conductances $g'_{s\ell}$ and $g'_{d\ell}$ are usually quite low in value, and for most calculations may be left out. Thus, the equivalent circuit of Fig. 8–8 is obtained. Resistances r'_s and r'_d represent the bulk resistances of the source and drain *end segments* and are usually designed to be quite small. The resistance r_D is the incremental drain resistance.

The dependent current source arises from the following considerations. The voltage across the capacitance C_{gs} controls both the value of this capacitance as well as the width of the transition region of the gate channel *p-n* junction at the source end. The transition region width in turn controls the cross-sectional area through which the channel current, which is essentially also the drain current, passes. Thus, the voltage across C_{gs} is related to drain current, hence the dependent source $Y_m V'_{gs}$. It is to be noted that the dependent source depends on V'_{gs} and *not* on V_{gs}. With this choice the parameter Y_m turns out not to be a function of frequency. If the dependent source were written in terms of V_{gs}, it would be a function of frequency. Typical values of the parameters of the equivalent circuit will be given in a table in the next chapter where they are compared with the corresponding elements of MOSFETS. The form of the wide-band

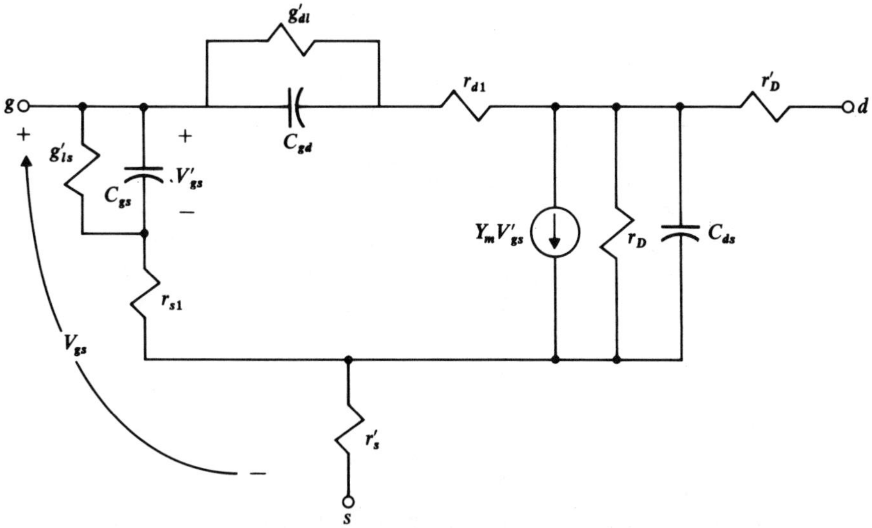

Fig. 8–6 A wide-band equivalent circuit for junction FET in pinch-off.

equivalent circuit proposed in Fig. 8–6 has been experimentally verified (Rego, 1964).

Voltage step function response of the junction FET with a resistive load in the drain circuit using the model of Fig. 8–8 (r'_s and r'_D were neglected in this

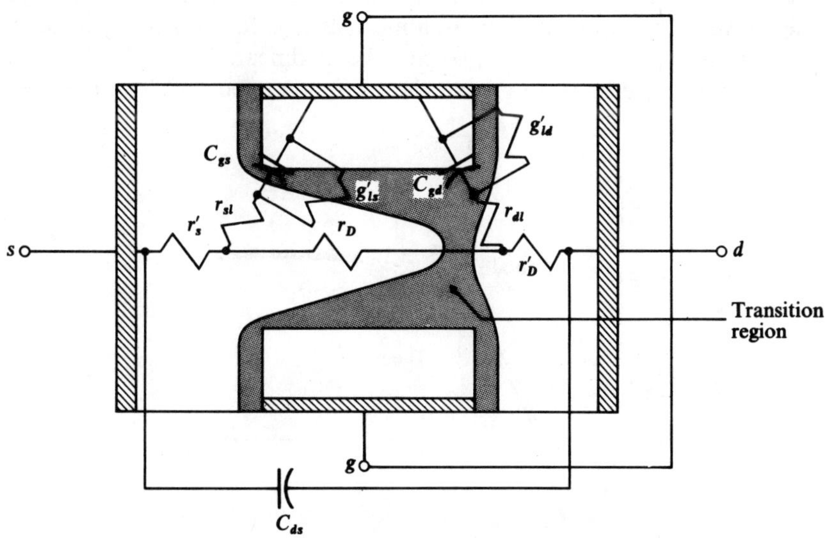

Fig. 8–7 The elements of Fig. 8–6 related to physical structure of the junction FET in pinch-off.

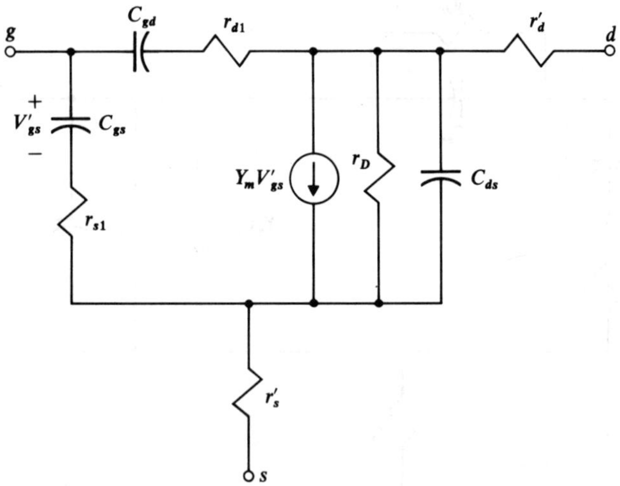

Fig. 8–8 Equivalent circuit of FET modified from that of Fig. 8–6.

particular case) gives quite a good correlation (error $\leqslant$ 10 per cent) in the general shape as well as the rise time between the theoretical and experimental results (Rego, 1964). Figure 8–9 indicates this. Experimentally the time constants associated with the rise and fall times were found to be approximately equal which, in fact, is indicative of a linear circuit confirming the validity of the linear analysis which was used in predicting the step response. The input gate capacitance, however, is expected to be nonlinear as it is fundamentally a junction capacitance which we have already seen to be nonlinear. For the "on" and "off" drives used to experimentally drive the FET "on" and "off" were evidently

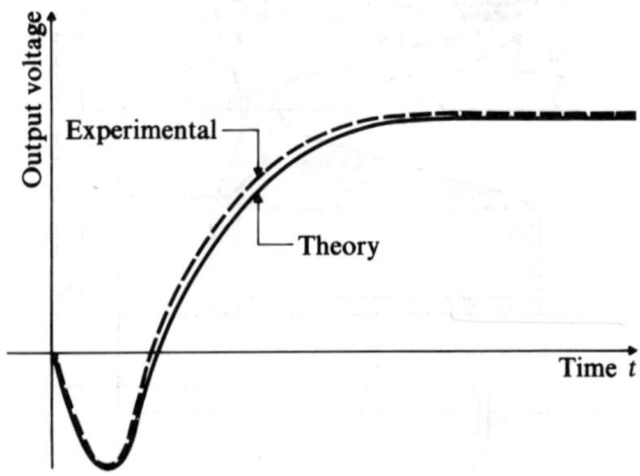

Fig. 8–9 Step response of the junction FET based on the model of Fig. 8–8 (without including r'_s and r'_d) common source configuration.

sufficiently small as to make the linear analysis valid. If the step drive is big enough, the nonlinearity of the input gate capacitance will need to be taken into account.

8–6 CUT-OFF FREQUENCY

The cut-off frequency of an amplifier is one of its most important high frequency parameters. As the junction FET may be used as an amplifier, it is useful to know its cut-off frequency. There are *two* different fundamental causes for the existence of the cut-off frequency in the junction FET. In the last section we saw that a fairly good wide-band model for this device consists of capacitance, conductance, and a dependent current source. It is possible to analyze such a circuit in an amplifier configuration (i.e. driven by a source with its own impedance and the FET driving a load impedance) by the methods of circuit analysis and obtain the cut-off frequency. This will not be done here. We simply note that it *can* be done by fairly straightforward methods. The other source of frequency limitation is the *transit time effect*. That is, it takes a finite time for a carrier to traverse the length of the channel region. This effect has been left out of our models thus far.

We shall now try to calculate the frequency limit due to this latter effect. We shall start out by assuming that the transit time τ_t corresponds to a half period of frequency f_{tc}. Thus,

$$f_{tc} = \frac{1}{2\tau_t}. \tag{8–52}$$

The time τ_t taken by a carrier to traverse the length L of the channel in a straight line is given by:

$$\tau_t = \frac{L}{u}, \tag{8–53}$$

where u = drift velocity of the carrier.

If the majority carriers are electrons (as is the case for an n-channel device) Eq. (8–53) may be written

$$\tau_t = \frac{L}{u} = -\frac{L}{\mu_n \mathscr{E}} = \frac{L}{\mu_n V_{DS}/L} = \frac{L^2}{\mu_n V_{DS}}, \tag{8–54}$$

where $\mathscr{E}$ = electric field in the channel.

We shall evaluate the transit time and the corresponding cut-off frequency under the just pinched-off condition for zero gate bias, i.e. when

$$V_{DS} = V_P = \frac{b^2 q N_d}{2\varepsilon}. \tag{8–55}$$

Now for constant mobility

$$\sigma_n = qN_d\mu_n \tag{8-56}$$

or

$$\mu_n = \frac{\sigma_n}{qN_d} \tag{8-57}$$

Equations (8–55) and (8–57) may be substituted into Eq. (8–54) to obtain the following:

$$\tau_t = \frac{L^2}{\dfrac{\sigma_n}{qN_d}\left(\dfrac{b^2qN_d}{2\varepsilon}\right)} = \frac{L^2 2\varepsilon}{\sigma_n b^2}. \tag{8-58}$$

Finally Eq. (8–58) may be substituted into Eq. (8–52) to obtain the following:

$$f_{ct} = \frac{\sigma_n b^2}{4\varepsilon L^2}. \tag{8-59}$$

If the carriers are traveling the entire channel under scattering limited conditions, the saturation velocity v_{sat} must be taken into account as follows:

$$\tau_t = \frac{L}{v_{\text{sat}}}. \tag{8-60}$$

Then

$$f_{tc} = \frac{v_{\text{sat}}}{2L}. \tag{8-61}$$

Example 8–1. Calculate the cut-off frequency in Ge and Si junction FET of *n*-channel length 10 μ and whose frequency is limited primarily by transit time under scattering limited carrier flow (saturation velocity).

Solution

$$f_{tc} = \frac{v_{\text{sat}}\ \text{cm/sec}}{2 \times 10^{-3}\ \text{cm}}. \tag{1}$$

Using the saturation velocities given in Table 2–9 we obtain for germanium

$$f_{tc} = \frac{6.2 \times 10^6}{2 \times 10^{-3}} = 3.1 \times 10^9 \quad \text{Hz} \tag{2}$$

and for silicon

$$f_{tc} = \frac{9.0 \times 10^6}{2 \times 10^{-3}} = 4.5 \times 10^9 \quad \text{Hz} \tag{3}$$

8–7 NONLINEAR INPUT GATE CAPACITANCE

The analysis of the previous section dealt with linear analysis. If the gate terminal is identified as the input terminal, the gate-channel being a *p-n* junction, the input capacitance will necessarily be nonlinear. We shall now attempt to obtain its voltage dependence. The junction capacitance is given by:

$$C = \frac{dQ}{dV_{GS}}. \tag{8-62}$$

Let

B_C = volume of the channel, i.e. charge neutral region.
B = volume of the charge depleted transition region.

Then

$$Q = h\rho_c \int_{-L}^{0} 2a(x)\,dx, \tag{8-63}$$

where ρ_c = majority carrier charge density in the channel.

Here the origin of the x coordinate system is at the drain end. Shockley in his original paper gave the following relationship.

$$x = -\frac{6LI_m}{I(p)}(k_v^2/2 - k_v^3/3), \tag{8-64}$$

where

$$k_v \triangleq \frac{a(x)}{b} \tag{8-65}$$

$$I(p) = I_m(1 - 3p + 2p^{\frac{3}{2}}) \tag{8-66}$$

$$I_m = \frac{2b\sigma_c V_P}{3L} \tag{8-67}$$

where σ_c = channel conductivity, and

$$p \triangleq \frac{V_{GS}}{V_P}. \tag{8-68}$$

From Eq. (8–64) the differential is obtained as

$$dx = -\frac{6LI_m}{I(p)}(k_v - k_v^2)\,dk_v. \tag{8-69}$$

Now when the FET is pinched-off, Eq. (8–9) allows us to write the following at the source end of the channel $x = -L$.

$$k_v(-L) = \frac{a(-L)}{b} = [1 - p^{\frac{1}{2}}]. \tag{8-70}$$

At the drain end $x = 0$ because of the pinched-off condition.

$$k_v(0) = (0) \tag{8-71}$$

Substituting Eqs. (8–66) through (8–71) into Eq. (8–63) we obtain:

$$Q = \frac{12\rho_c LbZI_m}{I(p)} \int_0^{1-p^{1/2}} k_v^2(1 - k_v)\,dk_v. \tag{8–72}$$

When the above indicated integration is carried out, Q may be expressed as:

$$Q = \rho_c LbZI_m \left[1 - \frac{3p}{1 + 2p^{\frac{1}{2}}}\right]. \tag{8–73}$$

Finally, the use of Eqs. (8–62) and (8–68) applied to Eq. (8–73) gives us

$$C(V) = \rho_c LbZ \frac{d}{dV_{GS}}\left[1 - \frac{3p}{1 + 2p^{\frac{1}{2}}}\right] \tag{8–74}$$

$$C(V) = -\rho_c \frac{LbZ}{V_P}\frac{d}{dp}\left(\frac{3p}{1 + 2p^{\frac{1}{2}}}\right)$$

$$C(V) = 6\varepsilon Z\left(\frac{L}{b}\right)\frac{1 + p^{\frac{1}{2}}}{[1 + 2(p)^{\frac{1}{2}}]^2} \tag{8–75}$$

Equation (8–75) agrees well with experimental data on FET devices whose gate channel junctions are step junctions (Riecher, 1963; Lindholm and Latham, 1963).

If the gate channel is a linearly graded junction the following expression for the gate input capacitance is obtained.

$$C(v) = [(45/28)\,\varepsilon Z(L/b)]\,(4/3)\cdot\frac{(3 - 7p^{2/3} + 7p^{5/3} - 3p^{7/3})(1 - p^{2/3})}{(2 - 5p + 3p^{5/3})^2}. \tag{8–76}$$

Thus, the gate input capacitance is indeed a nonlinear capacitance as amply evidenced by Eqs. (8–75) and (8–76).

8–8 SUMMARY

We considered the physics of junction FET in this chapter. This enabled us to obtain the theoretical current-voltage drain characteristics. We examined the competing theories which attempt to explain the current flow in current saturation condition. Channel length modulation effects were thought to be important in the soft saturation devices. In the hard saturation devices the pinch-off point remains essentially static with current and voltage.

Wide band incremental models for the FET were obtained from phenomological considerations. Frequency limits of the device were examined in which the scattering limited saturation carrier velocity was of importance. Finally, the gate input nonlinear capacitance expressions were obtained.

The major advantage of the JFET is that among semiconductor junction amplifying devices it has a very high input resistance. The metal-oxide-semiconductor MOS device has an input impedance even several orders of magnitude higher than that of JFET. The insights gained in the present chapter will be valuable in obtaining a better understanding of the MOSFETS which we shall consider in the next chapter.

PROBLEMS

8–1 You are given an abrupt junction *n*-channel FET having the structure of Fig. 8–1 and having the following parameters:
$N_a = 10^{17}/\text{cm}^3$, $N_d = 10^{14}/\text{cm}^3$, $b = 1\,\mu$, $L = 20\,\mu$, $h = 100\,\mu$, $n_i = 10^{10}/\text{cm}^3$, $\mu_n = 10^3\text{cm}^2/\text{Volt-sec.}$, $\mu_p = 500\ \text{cm}^2/\text{Volt-sec.}$, $\varepsilon = 10^{-12}$ Farad/cm.
Calculate (a) pinch-off voltage, (b) built-in voltage of the gate-channel junction, (c) conductance G_0.

8–2 All parameters of the JFET are the same as given in Problem 8–1 except that this is *p*-channel device. Calculate the same quantities as in Problem 8–1.

8–3 Given that the common source (source terminal is common to input and output) amplifier circuit is driven with an ideal voltage source between gate and source at the input terminals and the load is a resistance R_L between the drain and source, (a) calculate the incremental voltage gain, (b) how important are r_d' and r_s' to the answer?

8–4 If the current flow in the device of Fig. 8–1 along the *x*-axis may be assumed to be constant and equal to $-I_D$, obtain an analytic expression for the variation of voltage along the channel.

8–5 Obtain an expression for the maximum drain current in an *n*-channel FET at pinch-off.

8–6 Obtain the value of the incremental resistance when $V_{DS} = V_P/2$ and when $V_{GS} = 0$, $V_P/2$, $V_P/4$.

8–7 Obtain the voltage step response of the common source amplifier with a resistor load R_L if in the equivalent circuit of Fig. 8–8 r_s', r_d', r_{s1} and C_{ds} are all considered zero and r_{d1} is infinite.

8–8 Figure P8–8 shows an application of the FET as a variable attenuator. Calculate the attenuator voltage ratio V_2/V_1 for $V_{GS} = 0$, $V_{P/2}$ and $V_{P/4}$.

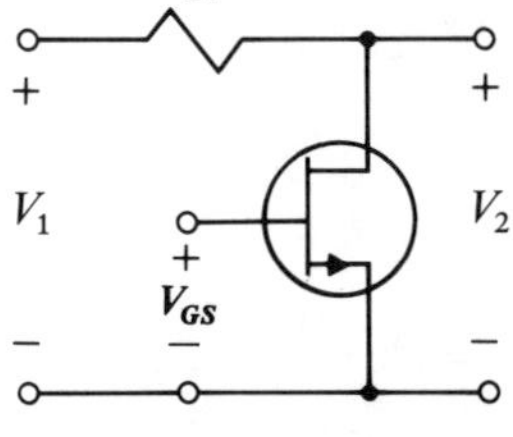

Fig. P8–8

REFERENCES

Biehl, B. L., "Nonlinear JFET Model for Computer-Aided Circuit Analysis," *IEEE Journal of Solid State Circuits,* **SC-6,** No. 1, Feb. (1971).

Dacey, G. C., and Ross, I. M., "Unipolar Field Effect Transistor," *Proc. IRE,* **41,** 970 (1953).

Dacey, G. C., and Ross, I. M., "The Field-Effect Transistor," *Bell System Tech. J.,* **34,** 1149 (1955).

Gray, P. E., and Searle, C. L., *Electronic Principles,* Chapter 10, John Wiley & Sons, New York (1969).

Grebene, A. B., and Ghandhi, S. K., "The Behavior of Junction-Gate Field-Effect Transistors Beyond Pinch-Off," Int. Solid-State Circuits Conf. Digest, IEEE, Philadelphia (1968).

Grosvalet, J. et al, "Physical Phenomenon Responsible for Saturation Current in Field Effect Devices," *Solid State Electronics,* **6,** 65 (1963).

Kennedy, D. P., and O'Brien, R. R., "Computer Aided Two-Dimensional Analysis of the Junction Field-Effect Transistor," *IBM, J. Res. Develop.,* **14,** No. 2, 95, March (1970).

Lindholm, F. A., and Latham, D. C., "Junction Capacitance of Field Effect Transistors," *Proc. IEEE,* 404–405, Feb. (1963).

Rego, J. L. G., "Equivalent Circuit of the Field Effect Transistor," M. S. Thesis, Electrical Engineering Department, Syracuse University (1964).

Riecher, I., "Input Capacitance of Field-Effect Transistors," *Proc. IEEE,* 1249–1250, Sept. (1963).

Shockley, W., "A Unipolar Field-Effect Transistor," *Proc. IRE,* **40,** 1365 (1952).

Sze, S. M., *Physics of Semiconductor Devices,* Chapter 7, John Wiley & Sons, New York (1969).

CHAPTER 9

METAL-OXIDE-SEMICONDUCTOR FIELD EFFECT TRANSISTOR

9-1 INTRODUCTION

In the previous chapter we studied a three terminal device consisting of gate, source, and drain. As observed in Section 8-8, it has a very high input impedance among amplifying junction devices. In this chapter we shall consider another device which also has a gate, source, and drain and which has input impedance several orders of magnitude higher than that of the JFET. The general name for such a device is the *insulated gate field effect transistor*, abbreviated *IGFET*. In such a device the channel region is insulated from the gate by an insulator. An important special subclass of these devices is the *metal-oxide-semiconductor field effect transistor*, abbreviated *MOSFET*.† In this latter device, the gate electrode is a metal, the insulator is an oxide and the channel region is within a semiconductor. It is this latter device we shall examine in some detail in this chapter.

In the JFET the high input resistance is obtained because of a reverse biased *p-n* junction. In the MOSFET the extremely high input resistance ($\sim 10^{14}\Omega$) is made possible by the insulator.

If one were to look at the drain characteristics of JFET and MOSFET, they look quite similar. The underlying mechanisms which are responsible for them in the two devices are, however, quite different. We shall study these mechanisms in the section following the next in which we examine the structure of the MOSFET.

9-2 THE STRUCTURE AND FABRICATION OF MOSFETS

There are two basic types of MOSFETS, (a) the *depletion type* and (b) the *induced type*. Figure 9-1a shows the depletion type. The major difference

† MOSFET is also written briefly as MOS.

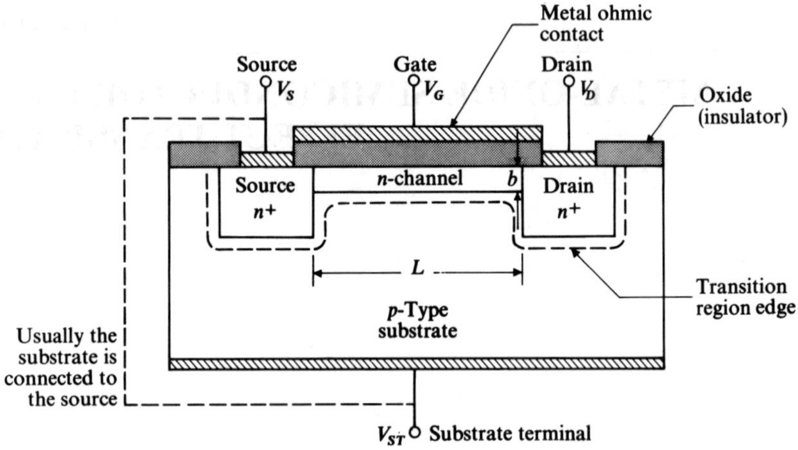

(a)

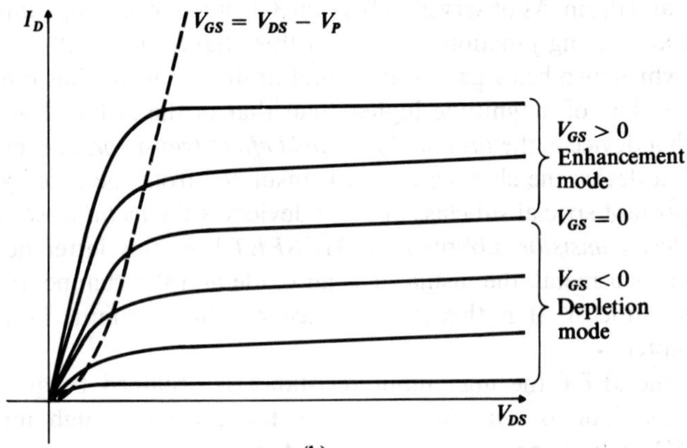

(b)

Fig. 9–1 Depletion-type *n*-channel MOSFET. (a) Cross-section of the structure of depletion-type MOSFET (*n*-channel). (b) Drain characteristics for the above device.

between the two is that with the terminals of the device open-circuited, the depletion type has a conducting channel which links the drain to the source, whereas the induced type has a region of opposite type to that of the drain and the source linking the two regions. In Figs. 9–1 and 9–2, *n*-channel devices are shown. Devices having a *p*-channel have *p*-type source and drain regions. In Figs. 9–1b and 9–2b are shown the drain characteristics of the depletion type and induced type, respectively.

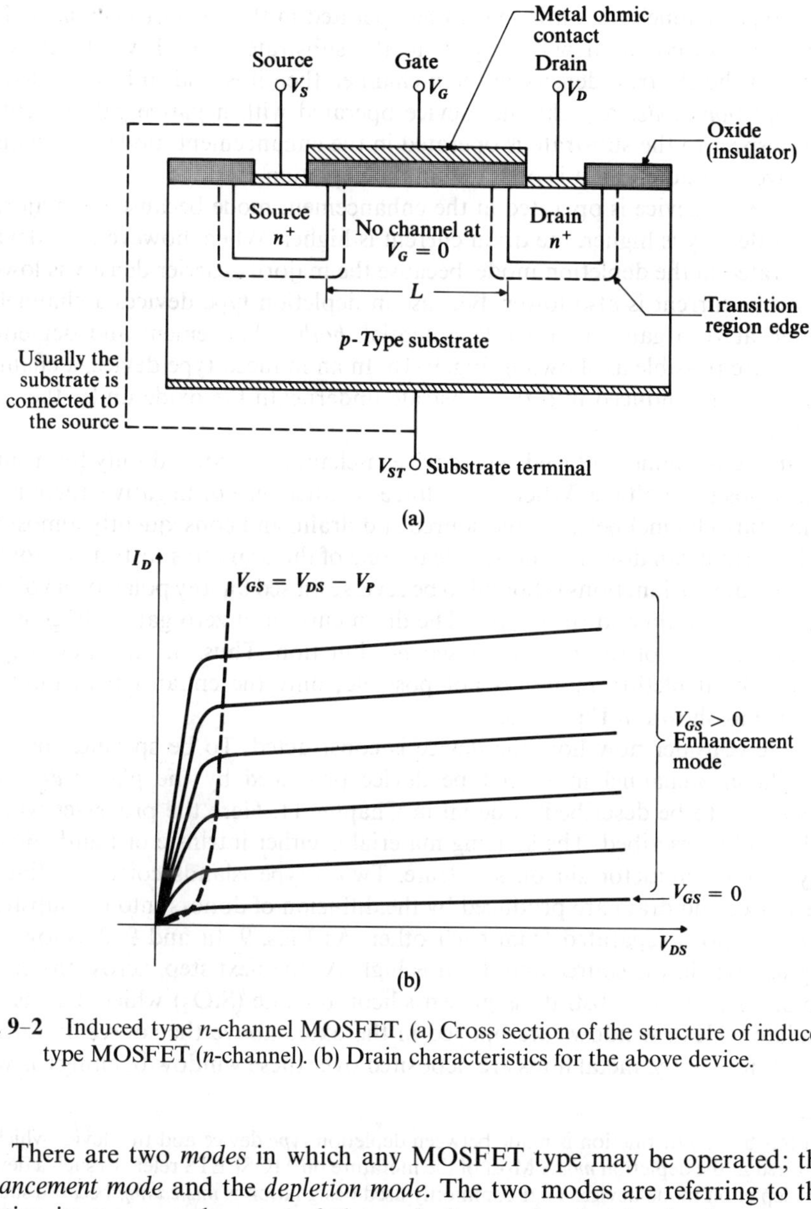

Fig. 9–2 Induced type *n*-channel MOSFET. (a) Cross section of the structure of induced type MOSFET (*n*-channel). (b) Drain characteristics for the above device.

There are two *modes* in which any MOSFET type may be operated; the *enhancement mode* and the *depletion mode*. The two modes are referring to the relative increase or decrease of the majority carrier density in the channel connecting the source to the drain. If a given gate bias tends to increase the majority carrier density in the channel, the device is said to be operated in *enhancement mode*. Thus if in an *n*-channel device, a positive gate potential (with respect to the substrate) is used which tends to draw more electrons into the *n*-type channel, then it is said to be operated in the *enhancement mode.* If a negative gate potential with respect to the substrate is used which tends to diminish the electron density in the *n*-channel, then it is said to be operated in the *depletion mode.* A *p*-channel device operated with negative gate potential with respect to the substrate is operated in the enhancement mode; and with a negative gate potential it is operated in the depletion mode.

When a device is operated in the enhancement mode because the majority carrier density is higher, the drain current is higher. When, however, the device is operated in the depletion mode, because the majority carrier density is lower, the drain current is also lower. Because in depletion type devices a channel *is present* at zero gate to substrate potential, *both* enhancement and depletion modes are possible as shown in Fig. 9–1b. In an induced type device, a channel has to be first induced into the substrate underneath the oxide under the gate electrode.

In an *n*-channel induced type device, a channel is induced only for a sufficiently positive voltage. When the voltage is either zero or negative, there is no connecting channel between the source and drain, and consequently almost no drain current can flow. For now, at least, one of the drain to substrate or source to substrate *p*-*n* junctions is bound to be reverse biased for any polarity of voltage between the drain and the source. The drain current at zero gate voltage is the leakage current of the reverse biased *p*-*n* junction. Thus, in an induced type device, the depletion mode† is not possible; only the enhancement mode is possible as shown in Fig. 9–2b.

We consider now how the device is constructed. To be specific, consider the silicon *n*-channel induced type device produced by the *planar diffusion technology* to be described in detail in Chapter 11. Here the processes will be only briefly described. The starting material is either intrinsic or lightly doped *p*-type semiconductor silicon substrate. Two *n*-type islands corresponding to the source and drain are produced by the diffusion of donors into the substrate in two regions separated from each other. As Figs. 9–1a and 9–2a show, the doping level in the source and drain is high. As the next step, across the entire top surface of the substrate is grown silicon dioxide (SiO_2) which acts as the insulator. Next, windows are opened in the oxide above the source and drain diffusions. Lastly, metal films are deposited over these window openings as well

† In this book a distinction is made between depletion *type* device and the device which is operated in the depletion *mode.* Most of the literature on MOSFETS refers to such a device as a depletion mode device. That commonly used designation is *misleading* because such a device can be operated in the enhancement mode also. To avoid this ambiguity we have adopted the terminology whereby the depletion type and induced type devices are determined by whether a conducting channel physically links the drain to the source at zero gate voltage (with respect to substrate) or not. Depletion and enhancement modes are *ways* of operating the device. Thus, the *type* of device indicates the physical nature of the device, and the *mode* indicates the manner in which it is operated.

as over the oxide region between the source and drain. These metallic films serve as the ohmic contacts to the outside world. An ohmic contact is also made to the bottom of the substrate. Note that while the source and drain regions are contacted directly by the metallic film, the gate film can communicate with the underlying semiconductor only through the insulating oxide. For this reason the device is also called an *insulated gate field effect transistor*. The device just described is shown in Fig. 9–3. Its cross section is shown in Fig. 9–2a. Hence it is an *induced type* device. If the region between the drain and the source is also made *n*-type by, for example, a shallow diffusion of lower density of donors than in the drain and the source, we get a depletion type device whose cross section is shown in Fig. 9–1a.

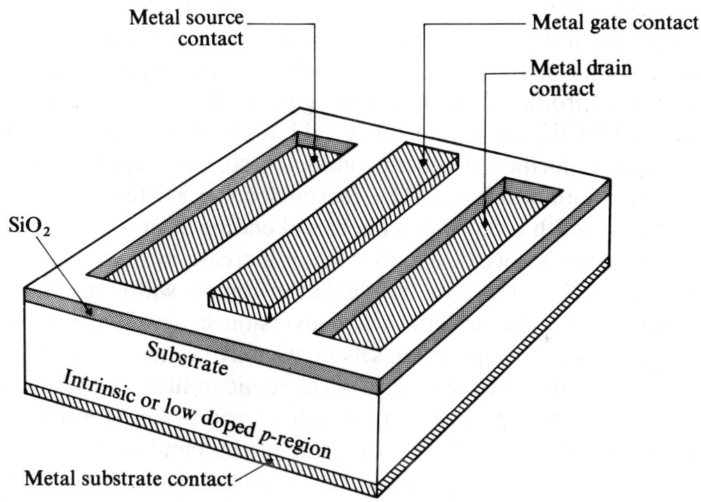

Fig. 9–3 Physical structure of the MOSFET.

Typical dimensions of the device are as follows, substrate area $\sim 200\ \mu^2$, oxide thickness ~ 1000 Å, source and drain diffusions are $\sim 7\ \mu$ deep, channel length $\sim 10\ \mu$.

9–3 ENERGY BANDS AND ACCUMULATION, DEPLETION, AND INVERSION LAYERS UNDER THE GATE

Because in the MOSFET the gate potential controls the *mode* in which the device is operated, and because in the induced type device the gate potential is responsible in inducing a channel, it is important to examine the band picture of the MOSFET so that we can more readily understand how an accumulation or depletion of majority carriers is obtained or how the inversion layer comes about. If a vertical slice of the device through the metal gate electrode, the oxide, and the semiconductor is considered as in Fig. 9–4a, its band picture looks as shown in Fig. 9–4b when the gate voltage is zero. If a small positive gate bias voltage V_{G1} is applied, a depletion region close to the oxide-semiconductor interface is formed where essentially no free mobile carriers exist. The ionized acceptors in the depletion region provide the negative charge Q_i. This charge is given by

$$Q_i = -qN_A x_t, \tag{9–1}$$

where x_t is the width of the depletion region.

If a greater positive gate bias voltage is applied, the further bending of the bands in the semiconductor results in a situation shown in Fig. 9–4e, where the intrinsic energy *crosses* the Fermi level. At this point there is a sudden increase in the free electron density close to the oxide–semiconductor interface. Thus, the original *p*-type material will now look like an *n*-type material, and an *inversion layer* is produced just below the gate oxide.

In the MOSFET an *n*-channel is said to have been *induced.* When an inversion layer is formed, the semiconductor surface depletion region soon reaches a maximum width x_{tm}. The reason for the existence of a maximum depletion region width is that when the band bending has progressed far enough for strong inversion to occur, a rather small increase in band bending corresponding to a small increase in the depletion region width results in a quite large increase in the charge within the inversion layer. As a measure of the point at which this strong inversion sets in, we may use the reasonable criterion that the electron density at the surface of the semiconductor closest to the oxide equals the substrate doping density which at room temperature equals the hole density (for *p*-type substrate). In terms of the band picture this criterion translates to the requirements that at the interface of the semiconductor and oxide the Fermi level be as far above E_i as the Fermi level is below E_i in the bulk of the substrate. Recall that the electron and hole densities will be equal to each other when the difference $(\psi_i - \varphi_f)$ in the bulk equals $(\varphi_f - \psi_i)$ at the interface, as can be clearly seen from Eqs. (3–13) and (3–14). Taking the above discussion into account a look at Fig. 9–4 allows us to write

$$(E_f - E_i)_{\text{surface}} = (E_i - E_f)_{\text{substrate}}. \tag{9–2}$$

But

$$E_{f_{\text{surface}}} = E_{f_{\text{substrate}}} \tag{9–3}$$

because the Fermi level must be flat, as discussed in Chapter 2. Thus with the use of Eq. (9–3), Eq. (9–2) may be rewritten as follows:

$$2E_f = E_{i_{\text{substrate}}} - E_{i_{\text{surface}}} \tag{9–4}$$

$$E_f = (1/2)(E_{i_{\text{substrate}}} - E_{i_{\text{surface}}}). \tag{9–5}$$

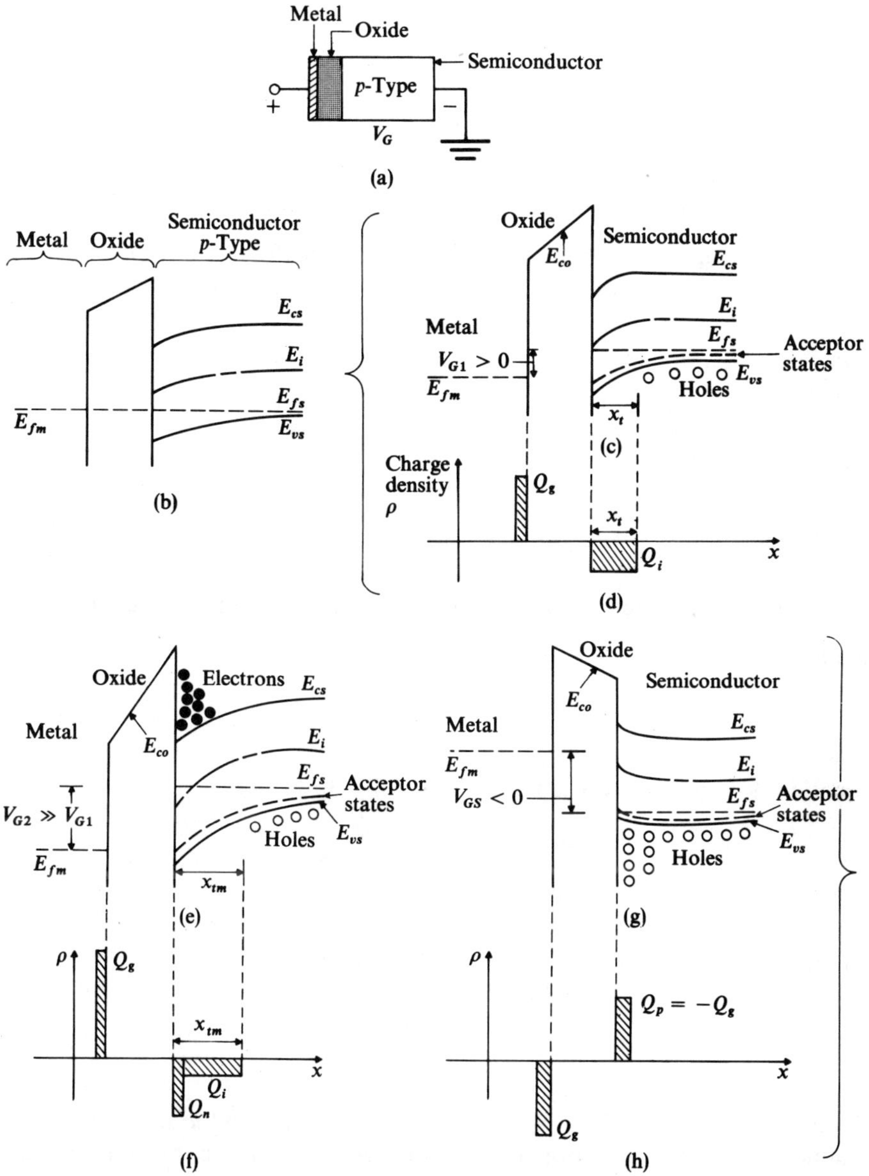

Fig. 9–4 Energy band bending due to external bias in the MOS device to produce depletion or accumulation of majority carriers or to produce an inversion layer in the semiconductor close to the oxide. (a) MOS. (b) No bias (thermal equilibrium). (c) Depletion of majority carriers from oxide—semiconductor interface. (d) Charge density in (c). (e) Inversion of the semiconductor region close to the oxide. (f) Charge density in (e). (g) Accumulation of majority carriers near the oxide-semiconductor interface. (h) Charge density in (g).

Dividing Eq. (9–5) by $-q$, we obtain

$$\frac{-E_f}{q} = -\frac{1}{2q}(E_{i_{\text{substrate}}} - E_{i_{\text{surface}}}) \qquad (9\text{–}6)$$

or

$$\varphi_f = (1/2)\psi_{\text{band bending}}, \qquad (9\text{–}7)$$

where φ_f = is the Fermi potential corresponding to Fermi level,

$\psi_{\text{band bending}}$ = potential difference between the intrinsic potential in the substrate (away from the oxide-semiconductor interface) and that at the interface.

If Eq. (9–7) is utilized as the criterion which applies at the onset of strong inversion, we can estimate the maximum depletion region width $x_{t\,\max}$. First if Poisson's equation is integrated twice under the depletion assumption, the potential ψ_s is given by

$$\psi_s = \psi_{s0}\left(1 - \frac{x}{x_t}\right)^2, \qquad (9\text{–}8)$$

where ψ_{s0} = the potential at the surface with respect to that in the bulk.

In a p-type substrate

$$\psi_{s0} = \frac{qN_a x_t^2}{2\varepsilon}. \qquad (9\text{–}9)$$

Equation (9–9) may be solved for x_t to obtain

$$x_t = \sqrt{\frac{2\varepsilon\psi_{s0}}{qN_a}} = \sqrt{\frac{2\varepsilon\psi_{\text{band bending}}}{qN_a}} \qquad (9\text{–}10)$$

To solve for $x_{t\,\max}$ we simply use Eq. (9–7) to finally obtain

$$x_{t\,\max} = \sqrt{\frac{4\varepsilon\varphi_f}{qN_a}}. \qquad (9\text{–}11)$$

Figure 9–5 shows how x_t varies with the charge induced in the semiconductor under the depletion assumption. We note that the transition region (depletion region) width varies initially linearly with the induced charge and then saturates to $x_{t\,\max}$. Thus the depletion region width first increases with increasing induced charge in the semiconductor. When the semiconductor surface, however, gets inverted and an induced junction is formed, a maximum depletion region width, given by Eq. (9–11), is attained. Note that $x_{t\,\max}$ is a function of the substrate doping level. The depletion approximation turns out

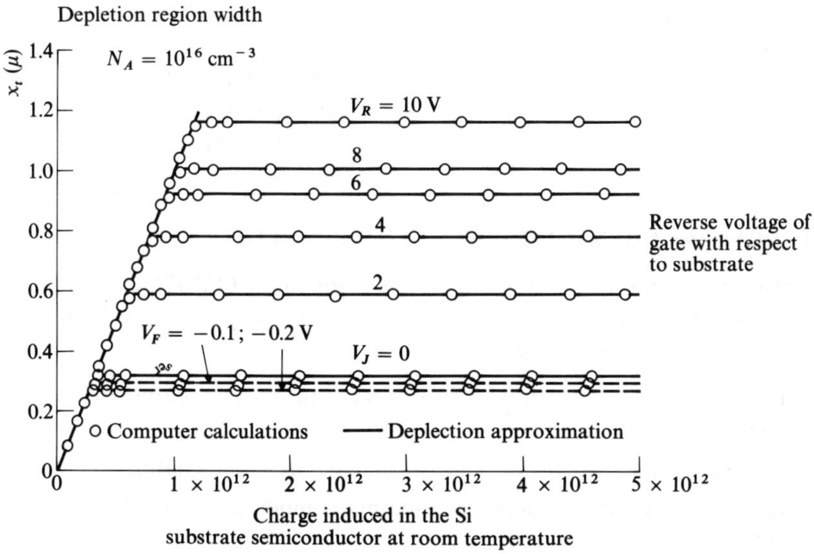

Fig. 9–5 Depletion region width as a function of induced charge in the silicon substrate at room temperature. (After Grove and Fitzgerald, 1966.)

to be an excellent one here. The charge in the depletion region can now be calculated under a strong inversion condition as

$$Q_i = -qN_A x_{t\,\max}. \tag{9–12}$$

The total charge Q_{tot} in the strongly inverted semiconductor is given by

$$Q_{tot} = Q_i + Q_n, \tag{9–13}$$

where Q_n = free electron charge in the inverted region.

If a negative gate bias voltage is applied the bands bend in the opposite direction to the previous case and the band picture looks as shown in Fig. 9–4g. Here the Fermi level is a lot closer to the valence band edge, and the free hole density close to the oxide-semiconductor interface increases or the *accumulation* of holes is said to have occurred. The charge distribution in this case is shown in Fig. 9–4h.

From the above discussion, we see that a conducting channel can indeed be induced between the source and drain in an induced type device by an appropriate choice of gate voltage. Thus for a *p*-substrate MOSFET with *n*-type source and drain regions, a sufficiently positive gate voltage with respect to the substrate will induce an *n-channel* in the semiconductor under the oxide. If the substrate were to be *n*-type and the source and gate were *p*-type, a sufficiently negative gate voltage will induce a *p*-type channel under the oxide. In the depletion type device where a channel already exists between the source and drain regions, a positive gate voltage with respect to the substrate will enhance the electron density in the n-channel so that it operates in the *enhancement mode.* A negative gate voltage will tend to reduce the electron density in the n-type channel, and the device operates in the *depletion mode.*

9–4 DRAIN *I-V* CHARACTERISTICS UP TO PINCH-OFF

We shall now obtain the I-V characteristics for the MOSFET up to pinch-off. Pinch-off here has the same meaning as in JFET. The following things will be assumed in the analysis to be used.

1. The IR voltage drops in the end source and drain regions are negligibly small.
2. There is no recombination-generation current in the channel.
3. The diffusion current along the channel is much less than the current due to the field along the channel.
4. There is no d-c current entering or leaving the gate terminal.
5. The gradual channel approximation is valid, i.e.

$$\frac{\partial \mathscr{E}_y}{\partial y} \gg \frac{\partial \mathscr{E}_x}{\partial x} \tag{9–14}$$

6. The substrate is nondegenerate.†
7. The charge in the electronic states at the semiconductor-oxide interface is to be considered as a fixed charge having a uniform surface charge γ_{SS} per unit area.‡ This charge exists independent of the gate voltage.
8. The channel depth b (see Fig. 9–6) is small compared to the oxide thickness x_0.
9. The bulk of the oxide contains no volume charge density.
10. Majority carrier mobility is constant.

We begin by writing Ohm's law.

$$J_x = \rho_{\text{total}}\, u_x, \tag{9–15}$$

---

†A *degenerate* semiconductor is one in which the doping level is such that the Fermi level is within the conduction band in n-type and valence band in p-type materials.

‡In much of the literature this surface *charge density* is designated by Q_{SS}. We choose a different symbol here so as not to confuse it with *total charge* which is also designated by the symbol Q.

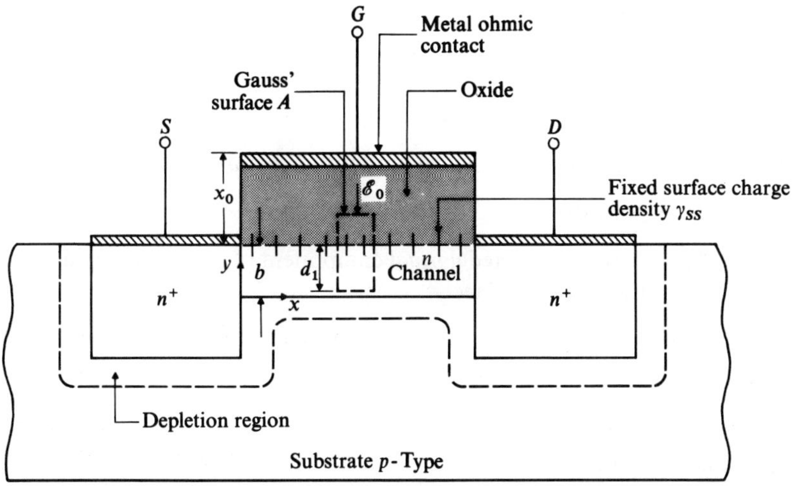

Fig. 9–6 *n*-Channel MOSFET showing a Gaussian surface.

where J_x = current density along the channel,

ρ_{total} = total mobile charge density in the channel,

u_x = velocity of mobile carriers along the channel.

Equation (9–15) permits us to write the current in the channel as follows:

$$I_x = \rho_{\text{total}}\, u_x A, \tag{9–16}$$

where A = cross-sectional area of the channel through which the current flows.

We may use the definition of mobility to rewrite Eq. (9–16) for an *n*-channel MOSFET.

$$I_x = -\rho_{\text{total}}\, \mu_n \mathscr{E} b\, h, \tag{9–17}$$

when b = depth of the channel

h = width of the channel

We need to evaluate the total mobile charge density ρ_{total}. In doing this calculation we wish to consider the general case where there maybe a fixed surface charge density γ_{SS} at the oxide-semiconductor interface. We shall presently see that if the surface charge density γ_{SS} is positive, where the substrate is *p*-type, it gives rise to an induced *n*-type channel, even if the gate is open-circuited!

We shall use Gauss's law to calculate ρ_{total}. Gauss's law is given below.

$$\oint_A \varepsilon \mathscr{E}_n dA = Q, \tag{9–18}$$

where Q = total charge enclosed within the *closed surface* of area A.

ε = permittivity

$\mathscr{E}_n$ = *outward* directed *normal* component of the electric field $\mathscr{E}$ crossing the closed surface A.

In order to apply Gauss's law we must choose the closed surface A. Figure 9–6 shows the cross section of the device along with surface A. Note that the *bottom* surface is a distance d_1 from the oxide-semiconductor interface such that the entire channel region is *within* the Gaussian surface A but the depletion region is not, and the electric field is *zero* along the bottom face of the Gaussian surface. Furthermore, the Gaussian surface is narrow, so that the variation of the electric field in the x-direction with distance is neglected consistent with assumption 5—the gradual channel assumption. Thus, the contribution of the electric field *entering* the face of the Gaussian surface closest to the drain is cancelled by an equal and opposite contribution of the electric field *leaving* the face closest to the source. There is no field normal to the closest and farthest faces of the Gaussian surface and hence there is no contribution to the integral of Eq. (9–18) from these faces. The only remaining contribution is from the top surface due to the field in the oxide due to the gate potential. Thus, Eq. (9–18) applied to the Gaussian surface gives

$$\oint_A \varepsilon \mathscr{E}_n dA = Q \tag{9–18}$$

$$-\varepsilon \mathscr{E}_0 A_t = \gamma_{SS} A_t + \int_0^b \rho(y) A_t dy, \tag{9–19}$$

when $\rho(y)$ = volume charge density in the channel,

A_t = top surface area of the Gaussian surface.

Furthermore,

$$\rho(y) = q(\Delta p - \Delta n), \tag{9–20}$$

where Δp = hole density in the channel in *excess* of the thermal equilibrium value in the bulk of the semiconductor,

Δn = electron density in the channel in *excess* of the thermal equilibrium value in the bulk of the semiconductor.

The reason for the use of excess free carrier densities in Eq. (9–20) is that it is only the *excess* carrier density which gives rise to the presence of *net* charge in the semiconductor. The free carrier densities corresponding to the thermal equilibrium values *do not give rise to net charge densities.* The reason for this is that the majority carrier charge density is precisely balanced by the ionized acceptor or donor charge densities.

If the channel is strongly inverted

$$\Delta n \gg \Delta p. \tag{9–21}$$

With the use of Eq. (9–21) we may rewrite Eq. (9–9) as follows:

$$-\varepsilon\mathscr{E}_0 A_t = \gamma_{SS} A_t + \int_0^b \rho(y) A_t dy = \gamma_{SS} A_t - A_t q \int_0^b \Delta n\, dy. \tag{9–22}$$

Cancelling the common factor A_t we obtain

$$q \int_0^b \Delta n\, dx = \varepsilon\mathscr{E}_0 + \gamma_{SS}. \tag{9–23}$$

It is worthwhile observing that even if $\mathscr{E}_0 = 0$, so long as γ_{SS} is positive, an inversion layer exists in the semiconductor under the oxide. It follows that if a positive charge density exists at the oxide–semiconductor interface then we have a *depletion type* device because it satisfies the requirements of the definition that a conducting channel exist at zero gate bias. In such a device a *positive* bias produces a stronger inversion layer and hence a higher electron density, and higher drain current. If the gate is made *negative*, the inversion layer is made weaker, the channel electron density is reduced and the drain current is reduced. This is consistent with the experimental observations in the drain characteristics of depletion type devices as seen in Fig. 9–1b.

Thus, there is a second way other than the one described in Section 9–2 to obtain a depletion type *n*-channel device. It is to somehow introduce the *positive* surface charge density at the oxide-semiconductor interface. However, the details of how such a surface charge density is produced is proprietary information zealously guarded by different corporations. It is known that when SiO_2 is grown in an atmosphere of dry hydrogen, a positive charge density is formed at the oxide-semiconductor interface.

Now we shall obtain the drain *I-V* characteristics of the ideal MOSFET starting with Ohm's law.

$$J_x = \sigma\mathscr{E}_x = q\mu_n n(y)\mathscr{E}_x, \tag{9–24}$$

where J_x = channel current density along the x-direction,

$n(y)$ = free electron density,

$\mathscr{E}_x$ = electric field along the length of the channel to obtain the channel current we need to integrate the channel current density over the channel cross-sectional area.

$$I_{CH}(x) = \int_0^b q\mu_n(y)\mathscr{E}_x h\,dy, \tag{9–25}$$

where hdy = differential area through which the channel current flows.

Now in view of assumption 8, it may be stated that the electric field $\mathscr{E}_x$ does not vary significantly over the range of integration. Equation (9–25) may hence be rewritten

$$I_{CH}(x) = q\bar{\mu}_n\mathscr{E}_x h \int_0^b n(y)\,dy, \tag{9–26}$$

where the *average* mobility $\bar{\mu}_n$ is defined as

$$\bar{\mu}_n = \frac{\int_0^b \mu_n n(y)dy}{\int_0^b n(y)dy}, \tag{9–27}$$

b = channel depth.

Thus, the problem of finding the channel current along the x-direction reduces to the evaluation of the integral of Eq. (9–26).

In attempting to evaluate the integral of Eq. (9–26) we shall find the concept of overall *charge neutrality* in the MOSFET useful. Mathematically we may state

$$\gamma_G + \gamma_{CH} + \gamma_D + \gamma_{SS} = 0, \tag{9–27a}$$

where γ_G = charge per unit area in the metal gate,

γ_{CH} = charge per unit area in the channel,

γ_D = charge per unit area in the depletion region,

γ_{SS} = charge per unit area at the oxide–semiconductor interface.

The surface charge density γ_{SS} is assumed to be constant. This is the effective surface charge density which takes into account the oxide traps, surface states, ionic charge within the oxide, if any, and work function differences between the different substances of the MOSFET.

We shall use Gauss's law to a narrow box of thickness dx, located at distance x from the source end, and enclosing the oxide and the channel as shown in Fig. 9–7. The lower end of the box is in the substrate just outside of

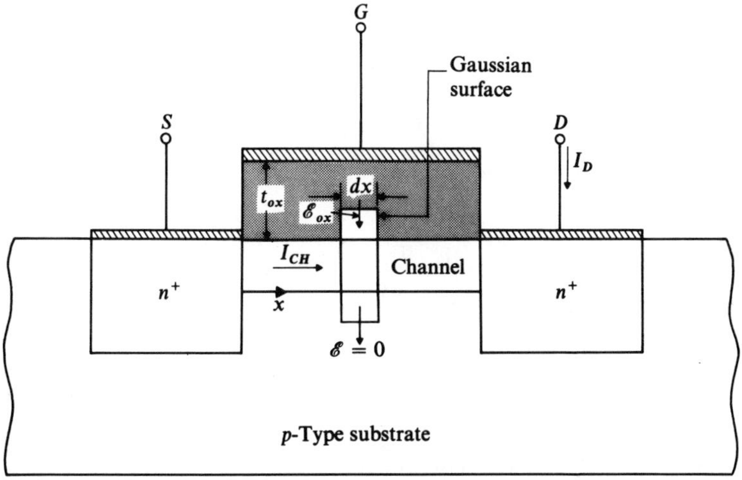

Fig. 9–7 MOSFET showing the Gaussian surface appropriate to Eq. (9–28).

the channel. Thus

$$\oint \varepsilon \mathscr{E}_{ox} dA = (\gamma_{SS} + \gamma_D + \gamma_{CH}) h dx. \tag{9–28}$$

The gradual channel approximation allows us to say that the left hand and right hand areas of the Gaussian surface have no field coming out of (or going into) them, so their contribution to the integral is zero. The bottom surface is in the substrate where the field is essentially zero and hence it does not contribute anything to the integral. The only contribution is to the top surface where the field $\mathscr{E}_{ox}$ is *into* the surface. Thus, Eq. (9–28) becomes

$$-\varepsilon_{ox} \mathscr{E}_{ox}\, h dx = (\gamma_{SS} + \gamma_D + \gamma_{CH})\, h dx. \tag{9–29}$$

The negative sign on the left hand side is due to the fact that the field is directed *into* the Gaussian surface. The oxide field at a distance x from the source may be written as

$$\mathscr{E}_{ox} = \frac{V_{GS} - V(x)}{t_{ox}}, \tag{9–30}$$

where $V(x)$ = potential with respect to the source in the channel at a distance x from the source end of the channel,

t_{ox} = oxide thickness.

Substituting Eq. (9–30) into Eq. (9–29) and cancelling the common factor hdx, we obtain

$$-\frac{\varepsilon_{ox}}{t_{ox}}[V_{GS} - V(x)] = \gamma_{SS} + \gamma_D + \gamma_{CH}. \tag{9–31}$$

The capacitance of the oxide is obtainable from the parallel plate capacitance relationship

$$C_{ox} = \frac{\varepsilon_{ox} A}{t_{ox}}. \tag{9–32}$$

Thus, the *oxide capacitance per unit area* c_{ox} is seen to be

$$\frac{C_{ox}}{A} \triangleq c_{ox} = \frac{\varepsilon_{ox}}{t_{ox}}. \tag{9–33}$$

Substituting Eq. (9–32) into (9–31) and solving for the channel mobile charge density γ_{CH} we obtain

$$\gamma_{CH} = -q \int_0^b n(y)dy = -\left[-c_{ox}[V_{GS} - V(x)] - (\gamma_{SS} + \gamma_D)\right]. \tag{9–34}$$

The integral of Eq. (9–34) may now be substituted into Eq. (9–26) to obtain the channel current I_{CH}.

$$I_{CH}(x) = -h\bar{\mu}_n\left(-\frac{dV}{dx}\right)\left[c_{ox}[V_{GS} - V(x)] + (\gamma_{SS} + \gamma_D)\right]. \tag{9–35}$$

Now Eq. (9–35) may be integrated from the source to the drain end of the channel to obtain the following.

$$\int_0^L I_{CH}(x)dx = -h\bar{\mu}_n \int_0^{V_{DS}} \left[c_{ox}[V_{GS} - V(x)] + (\gamma_{SS} + \gamma_D)\right] dV. \tag{9–36}$$

By assumption 4 and the use of Kirchhoff's continuity of current law we recognize the channel current I_{CH} to be the drain current $-I_D$. Note that I_{CH} is positive when it is directed towards the drain, whereas the drain current I_D is positive when it enters the drain terminal. This is the reason for (see Fig. 9–7)

$$I_{CH} = -I_D. \tag{9–37}$$

Substituting Eq. (9–37) into Eq. (9–36) we have

$$\int_0^L I_D dx = h\bar{\mu}_n c_{ox} \int_0^{V_{DS}} \left[V_{GS} - V(x) + \frac{1}{c_{ox}}(\gamma_{SS} + \gamma_D)\right] dV \tag{9–38}$$

$$I_D L = h\bar{\mu}_n c_{ox} \int_0^{V_{DS}} \left[V_{GS} + \frac{1}{c_{ox}}(\gamma_{SS} + \gamma_D)\right] dV - \int_0^{V_{DS}} V(x)dV. \tag{9–39}$$

Carrying out the integration and dividing through by L we obtain the following.

$$I_D = \frac{h\bar{\mu}_n c_{ox}}{L}\left[V_{GS} + \frac{1}{c_{ox}}(\gamma_{SS} + \gamma_D)\right] V_{DS} - \frac{V_{DS}^2}{2}. \tag{9–40}$$

We define the following quantity as the threshold voltage V_T.

$$V_T \triangleq -\frac{1}{c_{ox}}(\gamma_{SS} + \gamma_D). \tag{9–41}$$

Equation (9–40) may be rewritten in terms of V_T as follows

$$I_D = \frac{h\bar{\mu}_n c_{ox}}{L}[(V_{GS} - V_T)V_{DS} - V_{DS}^2/2]. \tag{9–41a}$$

If the *total* oxide capacitance C_{ox} is to be used in Eq. (9–40) instead of the oxide capacitance per unit area, the following may be obtained.

$$C_{ox} = c_{ox}hL. \tag{9–42}$$

Equation (9–40) may be rewritten in terms of C_{ox} as follows.

$$I_D = \frac{\bar{\mu}_n C_{ox}}{L^2}\left[(V_{GS} - V_T)V_{DS} - \frac{V_{DS}^2}{2}\right] \tag{9–43}$$

Equation (9–43) is valid only *up to the point* where the channel is *pinched-off*. That is, Eq. (9–43) is valid in the range

$$V_{DS} \leqslant (V_{GS} - V_T); (V_{GS} - V_T) \geqslant 0. \tag{9–44}$$

For *low* V_{DS} the drain characteristics of Eq. (9–43) look linear. When

$$V_{DS} \ll (V_{GS} - V_T) \tag{9–45}$$

the squared term V_{DS}^2 may be dropped in Eq. (9–43) which then becomes

$$I_D = \bar{\mu}_n \frac{C_{ox}}{L^2}(V_{GS} - V_T)V_{DS}. \tag{9–46}$$

Thus, the drain current is linearly proportional to both V_{DS} and $(V_{GS} - V_T)$. For fixed V_{GS} the drain conductance is linear and constant.

$$g_D = \frac{I_D}{V_{DS}} = \frac{\bar{\mu}_n C_{ox}}{L^2}(V_{GS} - V_T). \tag{9–47}$$

It should be noted that Eqs. (9–46) and (9–47) are valid only provided that the magnitude relationship of Eq. (9–45) is satisfied.

The fact that the drain conductance under the above conditions is proportional to the effective gate to source voltage $(V_{GS} - V_T)$ may be utilized to obtain a variable attenuator. Figure 9–8 shows such a circuit.

Example 9–1. Calculate the attenuation ratio V_2/V_1 for (a) $V_{GS} = -2V_T$ (b) $V_{GS} = -15\ V_T$ (c) $V_{GS} = +0.5\ V_T$. Average mobility $\bar{\mu}_n = 1000\ \text{cm}^2/\text{V-sec.}$, $C_{ox} = 10$ pF, channel length $= 10\ \mu$, $V_T = -2$ V. The MOSFET is a depletion type device. (d) At what voltage V_{GS} will the channel current be turned off?

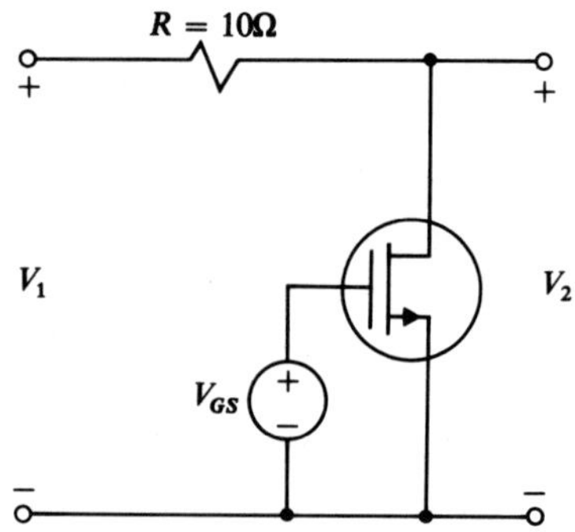

Fig. 9–8 Circuit for Example 9–1.

Solution. (a) With the use of the voltage divider idea, we may immediately write the attenuation ratio in terms of g_D as follows.

$$\frac{V_2}{V_1} = \frac{1/g_D}{\dfrac{1}{g_D} + R} = \frac{r_D}{r_D + R} = \frac{1}{1 + \dfrac{R}{r_D}} \tag{1}$$

Using Eq. (9–41) for g_D or $1/r_D$ we obtain

$$\frac{V_2}{V_1} = \frac{1}{1 + \dfrac{R\bar{\mu}_n C_{ox}}{L^2}(V_{GS} - V_T)} \tag{2}$$

$$= \frac{1}{1 + \dfrac{10(1000)\,10^{-11}}{(10 \times 10^{-4})^2}(-3\,V_T)} \tag{3}$$

$$= \frac{1}{1 + 10^{-1}(+6)} \tag{4}$$

$$\frac{V_2}{V_1} \doteq 0.625 \tag{5}$$

b) $$V_{GS} = -15\,V_T = -15(-2) = 30$$

$$\frac{V_2}{V_1} = \frac{1}{1 + 10^{-1}(+16 \times 2)} \doteq 0.238. \tag{6}$$

c) $$V_{GS} = +0.5\,V_T$$

$$\frac{V_2}{V_1} = \frac{1}{1 + 10^{-1}[0.5\,V_T - V_T]} \tag{7}$$

$$\frac{V_2}{V_1} = \frac{1}{1 + 10^{-1}(0.5)(2)} \doteq 0.91. \tag{8}$$

d) The channel current should theoretically cease when

$$V_{GS} = V_T = -2V. \tag{9}$$

That is the n-channel will cease to exist at the above voltage.

We have already noted that the drain I-V characteristics are valid from the origin up to the pinch-off point. Figure 9–9 shows the drain characteristics as predicted by Eq. (9–43) in the region of their validity. Also shown on these curves is the tangent to each curve at the origin as predicted by Eq. (9–46). The dotted extension of the drain characteristics beyond pinch-off point in each gate curve shows what the predicted curves would be if the value of drain current is assumed to remain constant beyond pinch-off. This implies that the incremental drain conductance beyond pinch-off is zero. This is not experimentally confirmed. It is not difficult to calculate the value of drain current at pinch-off where presumably it saturates. We simply require that the derivative $\partial I_D/\partial V_{DS}$ be zero just at the onset of *saturation*. These saturation values of I_D and V_{DS} will be designated by I_{DO}, V_{DSO} respectively. Thus, the partial derivative of Eq. (9–43) is found and set equal to zero for n-channel devices.

$$\frac{\partial I_D}{\partial V_{DS}} = \frac{\overline{\mu}_n C_{ox}}{L^2}\left[(V_{GS} - V_T) - V_{DSO}\right] = 0. \tag{9–48}$$

Thus, at the onset of saturation

$$V_{DSO} = (V_{GS} - V_T), \tag{9–49}$$

where V_{DSO} = drain to source voltage at the onset of saturation.

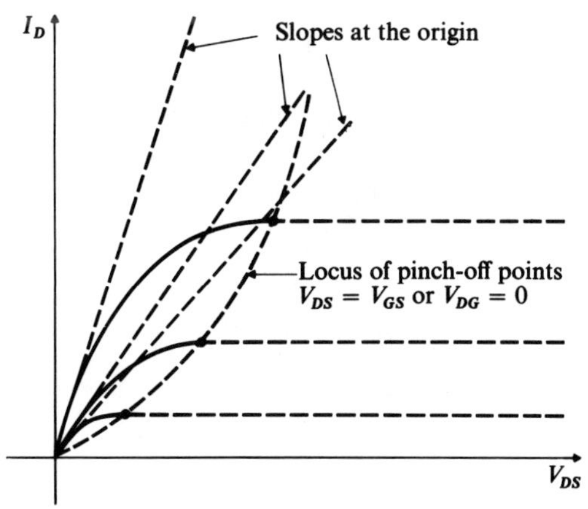

Fig. 9–9 Drain characteristics showing the slopes at the origin and locus of pinch-off points.

This value of drain to source voltage may be substituted into Eq. (9–43) to obtain

$$I_{DO} = \frac{\bar{\mu}_n C_{ox}}{L^2}\left(\frac{V_{DSO}^2}{2}\right). \tag{9–50}$$

With the help of Eq. (9–49), Eq. (9–50) may also be rewritten as

$$I_{DO} = \frac{\bar{\mu}_n C_{ox}}{2L^2}(V_{GS} - V_T)^2. \tag{9–51}$$

Another form of the drain current at the onset of saturation in terms of the oxide capacitance per unit area c_{ox}, is

$$I_{DO} = \frac{\bar{\mu}_n c_{ox} h}{L}(V_{GS} - V_T)^2. \tag{9–52}$$

9–5 INTERPRETATION OF THE THRESHOLD VOLTAGE

It is worthwhile to pause and really try to understand and interpret the *threshold* voltage. It is defined by Eq. (9–41), which tells us that the threshold voltage depends on the oxide-semiconductor interface charge density and on the depletion region charge density. Through Eqs. (9–34) and (9–41) another more revealing observation may be made. That is, whenever the gate to channel voltage $[V_{GS} - V(x)]$ equals V_T, the *mobile* charge density in the channel *disappears*, and the channel current is reduced to zero as indicated by Eq. (9–35). We may use this criterion to conclude that $V_T = -(\gamma_{SS} + \gamma_D)$ for the depletion type device.

Equation (9–43) is equally valid for both *depletion type* as well as *induced type* devices. As seen from the last Example 9–1, if V_T is *negative*, i.e. if

$$\gamma_{SS} + \gamma_D > 0. \tag{9–53}†$$

then V_T represents the minimum gate voltage required to *turn-off* the conduction path from the source to the drain in the *depletion type* device. Recall that in such devices a conducting channel exists even at $V_{GS} = 0$. This then is the *turn-off* gate to source voltage for depletion type devices. In the induced type devices, where

$$\gamma_{SS} + \gamma_D < 0, \tag{9–54}†$$

then V_T is positive, and represents the minimum gate voltage necessary to

---

† The depletion type device is said to be *normally* "*on*" because the drain to source currents can flow even at zero gate voltage.

The induced type device is said to be *normally* "*off*" because no drain to source currents except leakage currents can flow at zero gate voltage.

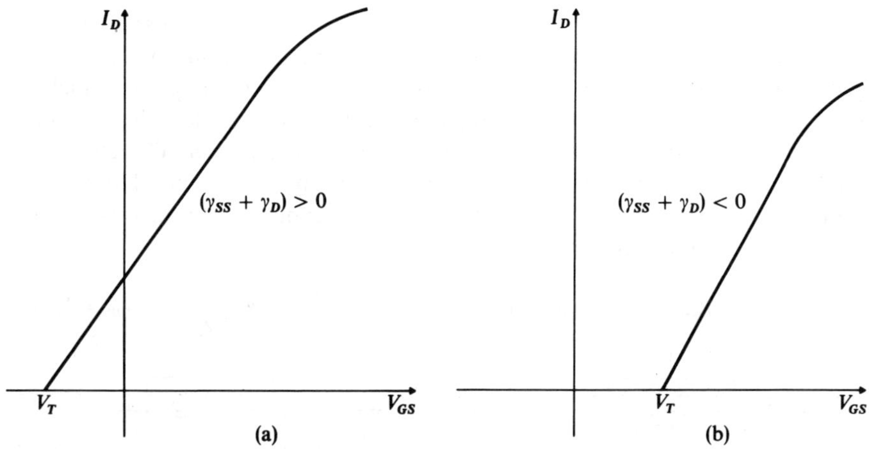

Fig. 9–10 Threshold voltage V_T and the two types of MOSFETS. (a) Depletion type n-channel devices ($V_T < 0$). (b) Induced type n-channel devices ($V_T > 0$).

induce an inversion layer so that a conducting channel can exist. This is then the *turn-on* voltage for induced type devices. Figure 9–10 shows threshold voltage for both types of devices.

The above remarks are valid for n-channel devices. Analogous remarks can be made for p-channel devices. See Problem 9–9.

9–6 DRAIN CHARACTERISTICS FOR $V_{DS} > V_{DSO}$ AND THE PHENOMENON OF "SATURATION"

When the drain to source voltage exceeds V_{DSO} two approximations used in obtaining Eq. (9–43) are no longer valid. First, the gradual channel approximation given by the magnitude relationship Eq. (9–14) is no longer valid. Second, the diffusion current is no longer negligible compared to the current due to the longitudinal drift field along the length of the channel.

To understand this latter point, consider the situation in the MOSFET for the three cases shown in Fig. 9–11. In Fig. 9–11a the drain to source voltage is rather small, i.e.

$$V_{DS} \ll V_{DSO}$$

and the channel depth looks essentially uniform. The main carrier transport mechanism is drift due to an electric field. Figure 9–11b shows the situation just at the onset of saturation. Note that the channel depth has been reduced to essentially zero at the drain end of the channel. The channel length is still essentially L. The major transport mechanism for the majority carriers is still drift due to the electric field for almost the entire channel length except right at the pinch-off point where the carrier density is zero. Note that a nonzero carrier density is necessary to explain the current contribution due to drift field as is seen from Eq. (9–17). As V_{DS} is increased byond V_{DSO}, the pinch-off point moves toward the source and the effective channel length is reduced analogous to a similar phenomenon discussed in Chapter 8. The channel is pinched-off over a channel length L_d at the drain end of the channel. The channel is *open*, or not pinched-off over the channel length L_s towards the source end as can be seen in Fig. 9–11c.

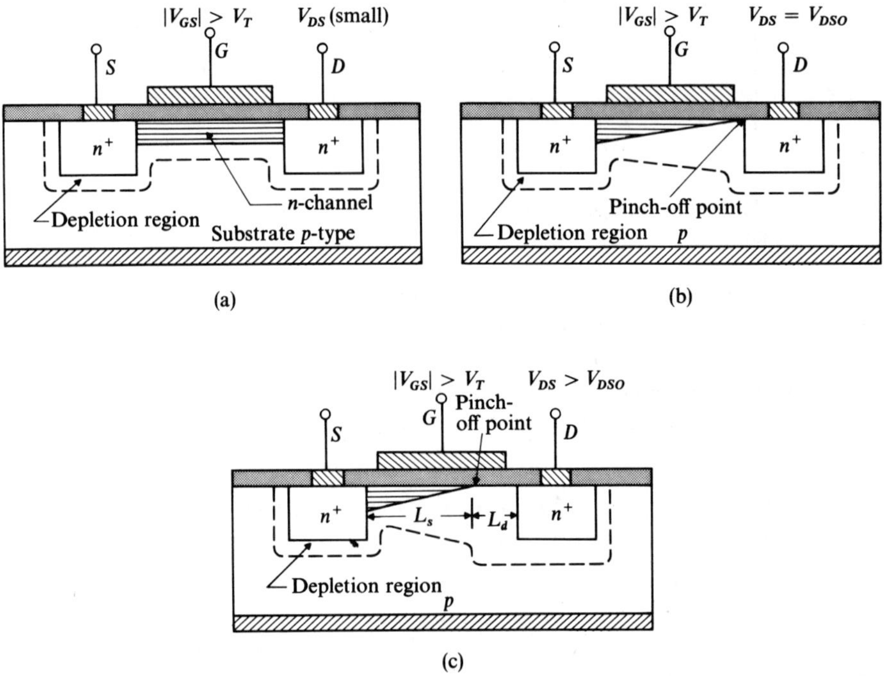

Fig. 9–11 MOSFET (*n*-channel) showing variation in the channel and depletion regions. (a) Very low V_{DS} and hence considerably below pinch-off. (b) Just at the onset of saturation. (c) In the saturation region where the channel length shrinkage is shown.

Let us consider the question of current continuity over the entire channel length L. As the channel is open at the source end for a distance L_s the major transport mechanism is drift of electrons from the source towards the drain due to the electric field. At a distance L_s the free carrier density is rather small as this is the beginning of the *depletion* region of the drain-substrate reverse biased *p-n* junction. The free carrier density is rather low in the channel depletion region. Hence, the drift current cannot account for the entire saturation current in this region unless the field is extremely large (i.e. $> 10^6$ volts/cm).

Here the rest of the current, other than the drift current, is due to *diffusion*. Recall that the diffusion mechanism is fundamentally dependent on the carrier *density gradient* and *not* on the carrier density itself. Thus, even though the carrier density is low in the pinched-off region of the channel L_d, the *gradient* of the electron density is quite significant and hence diffusion (Pao and Sah, 1966) can and does account for the rest of the current, and the current continuity is satisfied. Thus, we see that the assumption of diffusion current being negligible is *not valid* in the pinched-off region. The carriers going through the depletion region and being collected by the drain is quite similar to the carriers going across the collector depletion (transition) region in the regular bipolar transistor.

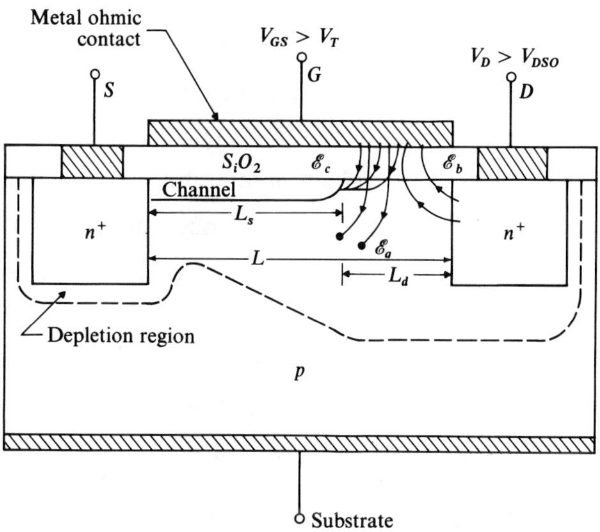

Fig. 9–12 MOSFET showing three fields $\mathscr{E}_a$, $\mathscr{E}_b$, $\mathscr{E}_c$ which contribute to the finite drain conductance in saturation.

The *finite* conductivity which is experimentally observed in the saturation region of MOSFETS is due primarily to the channel length shrinkage already mentioned. As the drain current here as in the JFET is inversely proportional to the effective channel length, it may seem reasonable to assume that the channel conductance may be given by Eq. (8–50). Some MOSFETS do indeed follow this behaviour (Crawford, 1967). Equation (8–50), however, does not take account of the fringing fields associated with the gate electrode, nor of the substrate bias, nor the oxide thickness, all of which are found experimentally to influence the saturation conductance in a significant way (Frohman-Bentchkowsky and Grove, 1969). Thus, in general, we must tackle this problem again in the MOSFET case taking account of the above mentioned parameters. In Fig. 9–12 are shown the fringing fields associated with the gate electrode at the *drain* end. Note that three different field components emanating from the gate electrode are shown in this figure. We shall take account of all of them after judicious approximations have been made to keep the problem tractable.

First note that the drain-to-source voltage *in excess of* V_{DSO} appears across L_d. The voltage V_{DSO} itself appears across L_s. Thus, the *total average* electric field $\mathscr{E}_{TL}$ component along the length of the channel may be easily calculated by simply dividing the voltage across depletion region width at the drain end by the region width. Thus,

$$\mathscr{E}_{TL} = \frac{V_{DS} - V_{DSO}}{L_d} \tag{9–55}$$

or

$$L_d = \frac{V_{DS} - V_{DSO}}{\mathscr{E}_{TL}}, \tag{9–56}$$

where V_{DS} = drain-to-source voltage *in* saturation,

V_{DSO} = drain-to-source voltage at the onset of saturation,

$\mathscr{E}_{TL}$ = total electric field along the direction of the length of the channel taking the fields $\mathscr{E}_a$, $\mathscr{E}_b$, and $\mathscr{E}_c$ of Fig. 9–12 in this direction and adding them.

The total field along the channel may be expressed as follows

$$\mathscr{E}_{TL} = \mathscr{E}_{aL} + \mathscr{E}_{bL} + \mathscr{E}_{cL}, \tag{9–57}$$

where $\mathscr{E}_{aL}$, $\mathscr{E}_{bL}$, and $\mathscr{E}_{cL}$ = the components of fields $\mathscr{E}_a$, $\mathscr{E}_b$, and $\mathscr{E}_c$ along the direction of the length of the channel.

We may obtain average $\mathscr{E}_{aL}$ simply from our previous knowledge of the depletion region width, its dependence on the net voltage across the depletion (transition) region as follows for a reversed biased *p-n* junction.

$$\mathscr{E}_{aL} = \frac{V_{DS} - V_{DSO}}{(2\varepsilon_s/qN_s)^{\frac{1}{2}}\,(V_{DS} - V_{DSO})^{\frac{1}{2}}}. \tag{9–58}$$

In writing Eq. (9–58) we have neglected the drain substrate built-in voltage, we have assumed that the doping of the drain region is much more than that of the substrate and Eq. (9–58) may be simplified to the following:

$$\mathscr{E}_{aL} = \frac{1}{K_a}(V_{DS} - V_{DSO})^{\frac{1}{2}}, \tag{9–59}$$

where

$$K_a = \left(\frac{2\varepsilon_s}{qN_s}\right)^{\frac{1}{2}}, \tag{9-60}$$

where ε_s = dielectric constant of the bulk semiconductor,

N_s = density of the doping atoms in the substrate (in p-type substrate $N_s = N_a$).

To evaluate $\mathscr{E}_{bL}$ and $\mathscr{E}_{cL}$ we shall make the simplifying assumption that these fields *are proportional to the respective normal electric fields across the oxide.* The constants of proportionality shall be the field-fringing factors f_b and f_c. Thus, the field $\mathscr{E}_{bL}$ which is due to the voltage from drain to gate may be written

$$\mathscr{E}_{bL} = f_b\left(\frac{\varepsilon_o}{\varepsilon_s}\right)\left(\frac{V_{DS} - V_{GS\text{eff}}}{x_o}\right), \tag{9-61}$$

where ε_o = dielectric constant of the oxide,

ε_s = dielectric constant of the semiconductor,

t_o = oxide thickness,

f_b = fringing factor,

$V_{GS\text{eff}}$ = effective gate to source voltage given below

$$V_{GS\text{eff}} \triangleq V_{GS} - V_T. \tag{9-62}$$

The average field $\mathscr{E}_{cL}$ is due to the voltage difference between the gate and the end of the channel on the drain side as seen in Fig. 9–12. We may hence write for the field $\mathscr{E}_{cL}$.

$$\mathscr{E}_{cL} = f_c\left(\frac{\varepsilon_o}{\varepsilon_s}\right)\left(\frac{V_{GS\text{eff}} - V_{DSO}}{x_o}\right) \tag{9-63}$$

Equations (9–59), (9–61), and (9–63) may be substituted into Eq. (9–57) to obtain

$$\mathscr{E}_{TL} = \frac{1}{K_a}(V_{DS} - V_{DSO})^{\frac{1}{2}} + f_b\left(\frac{\varepsilon_o}{\varepsilon_s}\right)\left(\frac{V_{DS} - V_{GS\text{eff}}}{x_o}\right) + f_c\left(\frac{\varepsilon_o}{\varepsilon_s}\right)\left(\frac{V_{GS\text{eff}} - V_{DSO}}{x_o}\right) \tag{9-64}$$

Equation (9–64) may be introduced into Eq. (9–56) and the reciprocal of L_d may be solved for as follows

$$\frac{1}{L_d} = \frac{\mathscr{E}_{TL}}{V_{DS} - V_{DSO}} = \frac{1}{K_a(V_{DS} - V_{DSO})^{\frac{1}{2}}} + \left(\frac{\varepsilon_o}{\varepsilon_s x_o}\right)\left[\frac{f_b(V_{DS} - V_{GS\text{eff}}) + f_c(V_{GS\text{eff}} - V_{DSO})}{(V_{DS} - V_{DSO})}\right] \tag{9-65}$$

It is instructive to write the above equation in the following form

$$\frac{1}{L_d} = \frac{1}{L_{ds}} + \frac{1}{L_{dg}}, \tag{9-66}$$

where

$$\frac{1}{L_{ds}} = \frac{1}{K_a(V_{DS} - V_{DSO})^{\frac{1}{2}}} \tag{9-67}$$

$$\frac{1}{L_{dg}} = \left(\frac{\varepsilon_o}{\varepsilon_s x_o}\right)\left[\frac{f_b(V_{DS} - V_{GSeff}) + f_c(V_{GSeff} - V_{DSO})}{(V_{DS} - V_{DSO})}\right]. \tag{9-68}$$

Note that L_d represents the amount by which the saturated channel shrinks when the device is operated in saturation region. Part of the channel shrinkage is due to the field associated with the substrate and given by the term of Eq. (9–67). The term given in Eq. (9–68) is the contribution to the channel shrinkage due to the field associated with the gate. *We see that the reciprocals of the channel shrinkages due to the substrate and gate are additive.*

The channel conductance g_D in saturation is defined as

$$g_D \triangleq \frac{\partial I_D}{\partial V_{DS}} = \frac{\partial I_D}{\partial L_d} \cdot \frac{\partial L_d}{\partial V_{DS}}. \tag{9-69}$$

In order to obtain $\partial I_D/\partial L_d$ consider the following. The drain current, as given in Eq. (9–41a), is inversely proportional to the channel length L. In the saturation region of the drain characteristics the unpinched-off channel length L should be replaced by the effective channel length L_s. The drain current *in saturation* may be related to the drain current just at the onset of saturation (pinch-off) by multiplying the latter with the ratio (L_s/L). Thus,

$$I_D = I_{DO}\left(\frac{L}{L_s}\right) = I_{DO}\left(\frac{L}{L - L_d}\right) = I_{DO}\left(\frac{1}{1 - \frac{L_d}{L}}\right) \tag{9-70}$$

Differentiation of Eq. (9–70) yields the following

$$\frac{\partial I_D}{\partial L_d} = \frac{I_{DO}}{L\left(1 - \frac{L_d}{L}\right)^2}. \tag{9-71}$$

To obtain $\partial L_d/\partial V_{DS}$ we simply differentiate Eq. (9–65) to obtain

$$\frac{\partial L_d}{\partial V_{DS}} = \frac{\left(\frac{1}{2K_a}\right)(V_{DS} - V_{DSO})^{\frac{1}{2}} + \left(\frac{\varepsilon_o}{\varepsilon_s x_o}\right)(f_c - f_b)(V_{GSeff} - V_{DSO})}{\left(\frac{1}{K_a^2}\right)(V_{DS} - V_{DSO}) + \left(\frac{\varepsilon_o}{\varepsilon_s x_o}\right)^2 G^2 + \left(\frac{2\varepsilon_o}{\varepsilon_s x_o K_a}\right)(V_{DS} - V_{DSO})^{\frac{1}{2}} G}, \tag{9-72}$$

$$G \triangleq f_b(V_{DS} - V_{GSeff}) + f_c(V_{GSeff} - V_{DSO}). \tag{9–73}$$

Finally by the use of Eqs. (9–71), (9–73), and (9–69) the channel conductance of the saturated device is given by the following

$$g_D = \frac{I_{DO}}{L\left(1 - \frac{L_d}{L}\right)^2} \cdot \left\{ \frac{\frac{1}{2K_a}(V_{DS} - V_{DSO})^{\frac{1}{2}} + \frac{\varepsilon_o}{\varepsilon_s t_o}(f_c - f_b)(V_{GSeff} - V_{DSO})}{\frac{1}{K_a^2}(V_{DS} - V_{DSO}) + \left(\frac{\varepsilon_o}{\varepsilon_s t_o}\right)^2 G^2 + \left(\frac{2\varepsilon_o}{\varepsilon_s t_o K_a}\right)(V_{DS} - V_{DSO})^{\frac{1}{2}} G} \right\}. \tag{9–74}$$

The dependence of the incremental drain conductance g_D in saturation on the oxide thickness t_o and on the substrate doping N_s through K_a (Eq. 9–60) can be seen in Eq. (9–74).

We may see what form Eq. (9–74) takes under two sets of limiting conditions.

Case A
$$x_o \gg K_a(V_{DS} - V_{DSO})^{\frac{1}{2}}. \tag{9–75}$$

The magnitude relationship Eq. (9–75) is the case which applies for thick oxides and/or heavily doped substrate. Then L_d takes the following simple form from Eq. (9–65).

$$L_d \doteq L_{ds} = K_a(V_{DS} - V_{DSO})^{\frac{1}{2}}. \tag{9–76}$$

The incremental drain conductance g_D becomes from Eq. (9–74)

$$g_D \doteq \frac{K_a I_{DO}}{2L\left(1 - \frac{L_d}{L}\right)^2 (V_{DS} - V_{DSO})^{\frac{1}{2}}}. \tag{9–77}$$

Substituting Eq. (9–76) into Eq. (9–77) we obtain

$$g_D \doteq \frac{K_a I_{DO}}{2[L - K_a(V_{DS} - V_{DSO})^{\frac{1}{2}}]^2 (V_{DS} - V_{DSO})^{\frac{1}{2}}}. \tag{9–78}$$

Case B
$$x_o \ll K_a(V_{DS} - V_{DSO})^{\frac{1}{2}}. \tag{9–79}$$

This case corresponds to *thin* oxides and/or low substrate doping density. From Eq. (9–68) L_d is obtained as

$$L_d \doteq L_{dg} = \left(\frac{\varepsilon_o x_o}{\varepsilon_o}\right) \frac{V_{DS} - V_{DSO}}{f_b(V_{DS} - V_{GSeff}) + f_c(V_{GSeff} - V_{DSO})}. \tag{9–80}$$

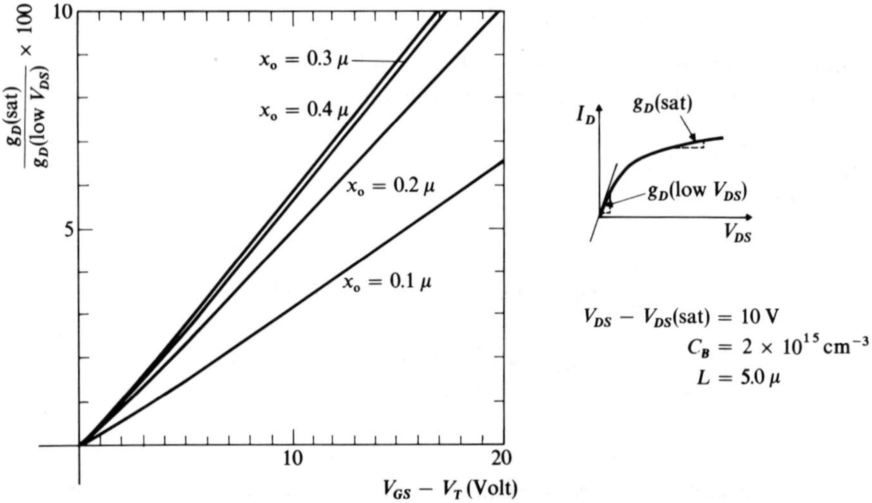

Fig. 9–13 Normalized saturation conductance as a function of effective gate voltage for different values of oxide thickness. (After Froham-Bentchkowsky and Grove, 1969.)

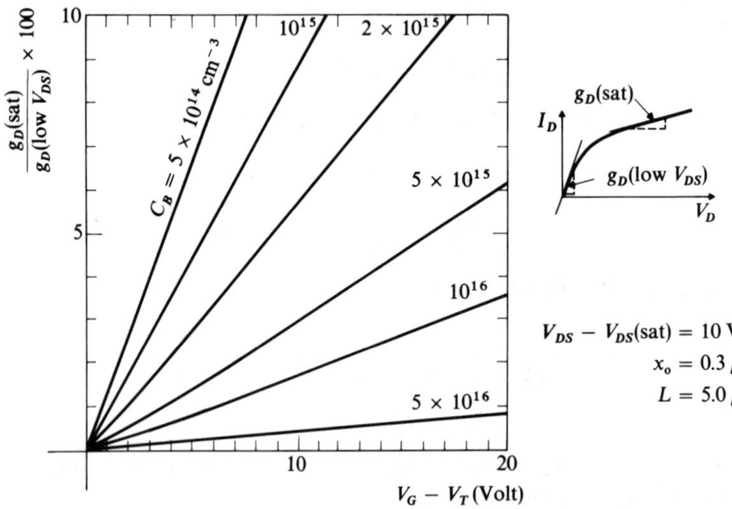

Fig. 9–14 Normalized saturation conductance as a function of effective gate voltage for different values of substrate impurity concentration. (After Frohman-Bentchkowsky and Grove, 1969.)

The incremental drain conductance g_D becomes from Eq. (9–74)

$$g_D = \left(\frac{\varepsilon_s x_o}{\varepsilon_o L}\right)\frac{(f_c - f_b)(V_{GSeff} - V_{DSO})}{[f_b(V_{DS} - V_{GSeff}) + f_c(V_{GSeff} - V_{DSO})]^2}\cdot\left[\frac{I_{DO}}{1-\left(\frac{L_d}{L}\right)^2}\right] \tag{9–81}$$

When neither the conditions of Case A nor case B are satisfied, the more general equation (Eq. 9–74) must be used. If we wish to get a clearer picture of the dependence of g_D on the oxide thickness and on the substrate doping in *saturation region* we may normalize this general g_D by the value of g_D for low value of V_{DS} (Eq. 9–47). This procedure will eliminate the influence of those parameters which affect the device performance below saturation. Figures 9–13 and 9–14 show the dependence of the normalized drain conductance on the oxide thickness and on substrate doping density. It is seen that the normalized saturation drain conductance increases with increased oxide thickness and decreases with increasing substrate doping density.

The saturated drain conductance as predicted by the general Eq. (9–74) has been compared with experimental data on both n-channel and p-channel devices having oxide-thickness varying between 0.1 μ and 1.0 μ and devices having substrate doping density (Frohman-Bentchkowsky and Grove, 1969) varying between 5×10^{14} and 2×10^{16}/c.c. The experimental results check quite well† with the predictions of Eq. (9–74).

In some devices the carriers traverse the drain depletion region under conditions of saturated carrier velocity discussed in Section 2–15. In such devices taking the saturation velocity into account gives better results (Baum and Bencking, 1970).

For a fixed gate to source voltage and for sufficiently high V_{DS} the electric field in the drain substrate reverse biased p-n junction may be sufficiently big to cause breakdown usually due to the avalanche effect. In such a case the drain characteristics show a sharply rising drain current, as shown in Fig. 9–15a.

9-7 WIDE BAND MODELS FOR THE MOSFETS

In Fig. 9–15 is shown a wide frequency band model for the MOSFET including the substrate. The two diodes represent the p-n junctions of the drain and substrate, and source and substrate. The diode direction is correct for an n-type channel device with a p-type substrate. If the channel is p-type and the substrate is n-type, the orientation of these two diodes should be reversed. The resistances r'_S and r'_D represent *the bulk resistances of the source and drain regions*

†An exception to this good agreement occurs in devices with quite low substrate doping density. Here the drain depletion region reaches the source, and the phenomenon of punch through, which was not considered in the above theory presented, takes over.

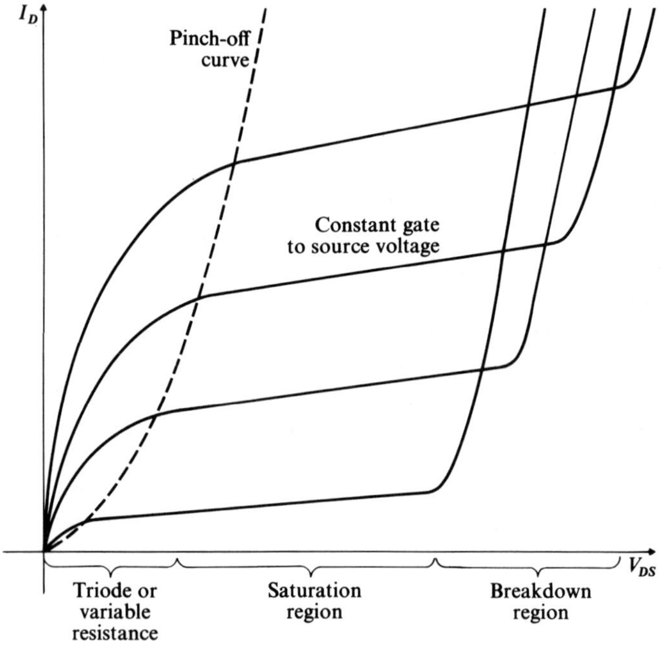

Fig. 9–15a MOSFET drain characteristics showing three different regions of possible operation.

which were neglected upto this point. The capacitances C_{gs}, C_{gd}, and C_{ds} are the interelectrode capacitances between the three terminals source, drain, and gate. For example, C_{gs} is the gate to source capacitance. The resistances r_{gs}, r_{gd} represent the resistances between the gate and source, and between the gate and drain. These are expected to be quite high because these are the resistances of the oxide insulator under the gate electrode. The series branch consisting of C_{gch} and r_{ch} represents, in an approximate way, the actual distributed gate to channel capacitance and channel resistance. The resistance r_D represents the incremental drain resistance obtainable from the drain characteristics. The dependent current source $g_m v_1$ represents the influence the voltage across the gate to channel capacitance has on the drain current. Note that g_m is the transconductance, and the voltage v_1 is only across C_{gch} and is *not* the gate to source terminal voltage.

The equivalent circuit of Fig. 9–15b can be simplified to that of Fig. 9–16 under certain circumstances. If the substrate is tied to the source, which is often the case, diode D_{SS} is shorted out and so is C_{js}. The diode D_{SD} is reverse biased and may be left out. The resistances r'_D and r'_S are usually quite small, usually a few Ohms or less, and for many calculations, though not all, may be neglected.

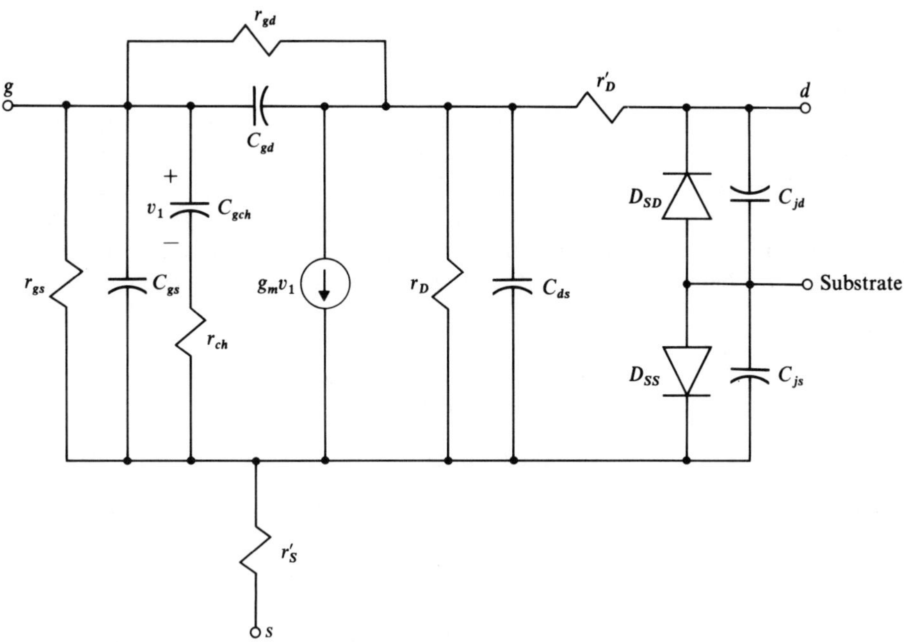

Fig. 9–15b A wideband equivalent circuit for the MOSFET convenient for the common source circuit configuration (n-channel).

The resistances r_{gd} and r_{gs} are extremely high, about 10^{14} to $10^{15}\,\Omega$, and may be left out. The channel resistance r_{ch} which is of the order of $50\,\Omega$ also may be left out.

The equivalent circuit of Fig. 9–16 results after the elements mentioned above are neglected, and is useful in many circuit applications. The capacitance C_i is the sum total of C_{gch} and C_{gs}.

$$C_i \triangleq C_{gs} + C_{gch}. \tag{9–82}$$

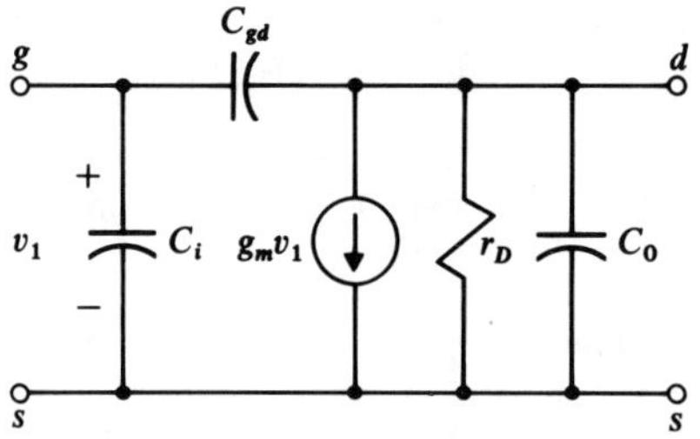

Fig. 9–16 Simplified equivalent circuit for the MOSFET.

Capacitance C_0 is obtained as

$$C_0 \triangleq C_{ds} + C_{js}. \tag{9-83}$$

The equivalent circuit of Fig. 9–15b is a fairly complete lumped equivalent circuit for the MOSFET. Figure 9–16 shows a practical, simple equivalent circuit for MOSFET useful and valid for many calculations. Table 9–1 gives typical values of MOSFET equivalent circuit elements. It should be remembered that the parameters such as r_D, g_m and the junction capacitances C_{js} and C_{jd}, are in fact nonlinear. Their values will vary with different choices of bias point.

Table 9–1 Typical Values of the Parameters of MOSFETS Operated in the Pinched-Off (Saturation) Region

Parameter	Typical value
r_{gs}, r_{gd}	10^{14}, $10^{15}\Omega$
C_{gs}	1 pF
C_{ds}	1 pF
r_{ch}	50Ω
C_{gch}	4 pF
r'_S, r'_D	few Ω
C_{js}	3 pF
C_{jd}	1 pF
r_D	40 kΩ
g_m	10 mA/Volt

9–8 FIGURE OF MERIT

Just as the maximum frequency of oscillation turned out to be a good figure of merit for the bipolar transistors, it is also a good figure of merit for the MOSFET.

Thus, we seek the maximum frequency under *ideal circumstances* at which the MOSFET can be made to oscillate. We shall use the equivalent circuit of Fig. 9–16 to deduce the maximum frequency of oscillation. We see in Fig. 9–17 that we have neglected the capacitance C_{gd} (i.e. assumed it is zero) and the drain

Table 9–2 Comparison of Key Parameter Values in JFET and MOSFETS.

Parameter	JFET	MOSFET
g_D	5 mA/V	10 mA/V
r_D	300 kΩ	30 kΩ

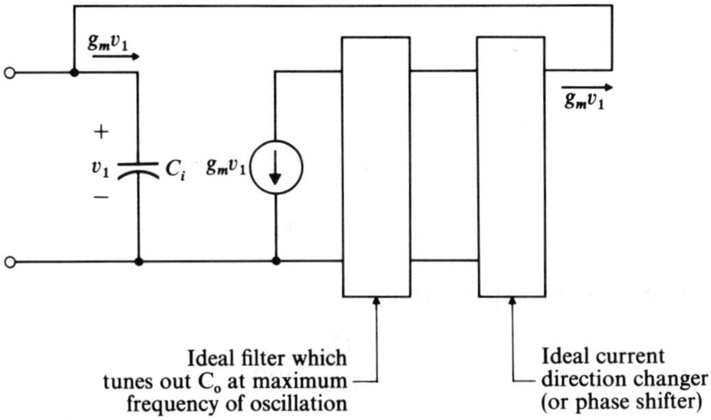

Fig. 9–17 Circuit for the calculation of the MOSFET maximum frequency of oscillation—a figure of merit.

resistance r_D (i.e. assumed to be infinite). This is part of the "ideal circumstances". The ideal filter of Fig. 9–17 may consist of an ideal inductor such that the admittance of the C_0 and the ideal inductance in parallel is zero at the frequency of maximum oscillation. The ideal current direction changer (or 180° phase shifter) makes the output current of the dependent source feed *down* through capacitance C_i such that the voltage V_1 produced by it is positive. When this is the case the voltage across C_i is

$$|V_1| = \frac{g_m|V_1|}{C_i\omega_m}, \tag{9–84}$$

where V_1 = amplitude of the sinusoidally varying voltage v_1 across C_i,
ω_m = maximum frequency of oscillation (radians/sec).

From Eq. (9–84) we may solve for ω_m

$$\omega_m = \frac{g_m}{C_i} \tag{9–85}$$

or

$$f_m = \frac{g_m}{2\pi C_i}, \tag{9–86}$$

where the units of f_m are in Hertz.

We may now go a step farther and express g_m in terms of the fundamental device parameters from the drain characteristics derived in Section 9–6.

Just at pinch-off

$$I_{DO} = \frac{\overline{\mu}_n C_{ox}}{2L^2}(V_{GSO} - V_T)^2. \tag{9–87}$$

We may now use the definition of g_m

$$g_m = \left.\frac{\partial I_D}{\partial V_{GS}}\right|_{V_{DS} = \text{constant}} = \frac{\bar{\mu}_n C_{ox}}{L^2}(V_{GSO} - V_T). \tag{9–88}$$

At pinch-off

$$V_{GSO} - V_T = V_{DSO}. \tag{9–89}$$

Thus Eq. (9–88) may be expressed as

$$g_m = \frac{\bar{\mu}_n C_{ox}}{L^2} V_{DSO}. \tag{9–90}$$

Equation (9–90) may be substituted into Eqs. (9–85) and (9–86) to obtain the following,

$$\omega_m = \frac{\bar{\mu}_n C_{ox}}{C_i L^2} V_{DSO} \tag{9–91}$$

$$f_m = \frac{\bar{\mu}_n C_{ox}}{2\pi C_i L^2} V_{DSO}. \tag{9–92}$$

If the channel length modulation may be neglected and if the incremental drain conductivity may be ideally considered to be infinite throughout the saturation region, Eqs. (9–91) and (9–92) may be written in terms of V_{DS} rather than V_{DSO}. Thus

$$\omega_m = \frac{\bar{\mu}_n C_{ox}}{C_i L^2} V_{DS} \tag{9–93}$$

$$f_m = \frac{\bar{\mu}_n C_{ox}}{2\pi C_i L^2} V_{DS}. \tag{9–94}$$

It is interesting and instructive to see that the maximum frequency of oscillations is intimately related to the *transit time* τ_t which is the time that a carrier takes to traverse the length L of the channel region. Thus,

$$\tau_t = \frac{L}{u} = \frac{L}{\bar{\mu}_n \mathscr{E}_x}, \tag{9–95}$$

where u = average velocity of carriers,

τ_t = transit time.

If the field along the x-direction is constant we may write the following *at* pinch-off.

$$\mathscr{E}_x = \frac{V_{DSO}}{L}. \tag{9–96}$$

Equation (9–96) may be substituted into Eq. (9–95) to obtain

$$\tau_t = \frac{L^2}{\bar{\mu}_n V_{DSO}}. \tag{9–97}$$

Comparing Eqs. (9–91) and (9–97) we conclude that

$$\tau_t = \frac{1}{\omega_m}\left(\frac{C_{ox}}{C_i}\right). \tag{9–98}$$

Provided that

$$C_{gch} \gg C_{gs} \tag{9–99}$$

and

$$C_{gch} \approx C_{ox} \tag{9–100}$$

then

$$C_{ox}/C_i = 1. \tag{9–101}$$

Equation (9–98) takes the following simple form

$$\tau_t = \frac{1}{\omega_m}. \tag{9–102}$$

Note that this relationship is true *only* if both Eqs. (9–99) and (9–100) are valid and the *just* pinched-off condition is obtained. In the saturation region *other than just pinched-off*, the channel length should be the *effective channel length* L_s and not L which should be used in Eqs. (9–95) and (9–96).

The above n-channel results may be readily converted to the p-channel results. For example, for p-channel devices the equations corresponding to Eqs. (9–91) and (9–92) are the following:

$$\omega_m = \frac{\bar{\mu}_p C_{ox}}{C_i L^2} V_{SDO} \tag{9–103}$$

$$f_m = \frac{\bar{\mu}_p C_{ox}}{2\pi C_i L^2} V_{SDO}, \tag{9–104}$$

where $\bar{\mu}_p$ = average hole mobility in the channel,

V_{SDO} = d-c source to drain voltage *just at* pinch-off.

9-9 ELECTRON HOLE MOBILITIES IN INVERSION LAYERS

We have used the effective average mobilities in many equations for the MOSFETS in this chapter. It is appropriate to wonder what values should be used for $\bar{\mu}_n$ and $\bar{\mu}_p$ in these equations. Leistiko et al (1965) have studied this question. They experimentally examined the $\bar{\mu}_n$ and $\bar{\mu}_p$ in silicon inversion layers and found that these effective mobilities were *constant* up to surface fields of about 1.5×10^5 Volts/cm and $|\gamma_s/q| = 10^{12}$ cm^{-2}, where γ_s is the charge per unit area induced in the semiconductor under the silicon oxide. (Beyond this point the $\bar{\mu}_n$ and $\bar{\mu}_p$ reduced a little with increasing γ_s.) They found that

$$\bar{\mu}_n \doteq \frac{\mu_n}{2} \text{ bulk} \tag{9–105}$$

$$\bar{\mu}_p \doteq \frac{\mu_p}{2} \text{ bulk.} \tag{9–106}$$

Furthermore it was found that the mobilities calculated by Eqs. (9–105) and (9–106) were independent of structural and gematrical parameters (e.g. channel length and shape, type of oxide and its thickness, γ_{SS}) as well as bulk impurity density.

Example 9–2. Calculate (a) the transit time and (b) the maximum frequency of oscillation for a silicon n-channel MOSFET whose channel length is 10 μ, $V_{DSS} = 5$ Volts and for which Eqs. (9–99) and (9–100) are valid.

Solution: Equation (9–102) is

$$\tau_t = \frac{1}{\omega_m}.$$

Because of Eq. (9–99) and (9–100), Eq. (9–101) is true, so that Eqs. (9–91) and (9–92) become

$$\omega_m = \frac{\bar{\mu}_n V_{DSO}}{L^2} \tag{9–107}$$

$$f_m = \frac{\bar{\mu}_n V_{DSO}}{2\pi L^2}. \tag{9–108}$$

Using Eq. (9–105) we get for Eq. (9–108) the following

$$f_m = \frac{\mu_{n\ \text{bûlk}} V_{DSO}}{4\pi L^2} \tag{9–109}$$

$$f_m = \frac{1350 \dfrac{\text{cm}^2}{\text{Volt-sec}} \cdot (5) \text{ Volts}}{4\pi \left[10 \times 10^{-6} \text{ meters} \times \dfrac{100 \text{ cm}}{1 \text{ meter}} \right]^2} \tag{1}$$

$$f_m = 5.36 \times 10^8 \text{ Hz.} \tag{2}$$

$$\tau_t = \frac{1}{\omega_m} = \frac{1}{2\pi f_m} = \frac{1}{2\pi(5.36 \times 10^8)} \tag{3}$$

$$\tau_t = 2.97 \times 10^{-10} \text{ sec.} \tag{4}$$

These calculations indicate the *best that the device can possibly do under the most ideal circumstances.* An actual device would yield a practical maximum frequency of oscillation which could be considerably less than theoretically predicted.

It is also worthwhile to note that on the basis of Eqs. (9–103) and (9–104) and the fact that the electron mobilities are generally greater than hole mobilities in all materials, *that* n-*channel devices seem to offer better ultimate high frequency performance compared to* p-*channel devices for devices with identical doping densities and geometry.* Furthermore, Eqs. (9–107) and (9–108) turn out to be good figures of merit for JFET as well.

9–10 MOS CAPACITANCE

It is useful to consider the capacitance of an MOS structure such as Fig. 9–18a which is the same structure as that of Fig. 9–4a but in a different orientation. It is not hard to see that the *total* capacitance of the metal–oxide–semiconductor (MOS) structure is the same as the oxide capacitance in series with the depletion (transition) capacitance due to the depletion region adjoining the oxide region. Thus, the total capacitance is given by the following,

$$C_t = \frac{dQ_G}{dV_G} = \frac{1}{\dfrac{1}{C_{ox}} + \dfrac{1}{C_j}} = \frac{C_{ox}C_j}{C_{ox} + C_j}, \tag{9–110}$$

where C_t = total MOS capacitance,
C_{ox} = oxide capacitance,
C_j = depletion capacitance.

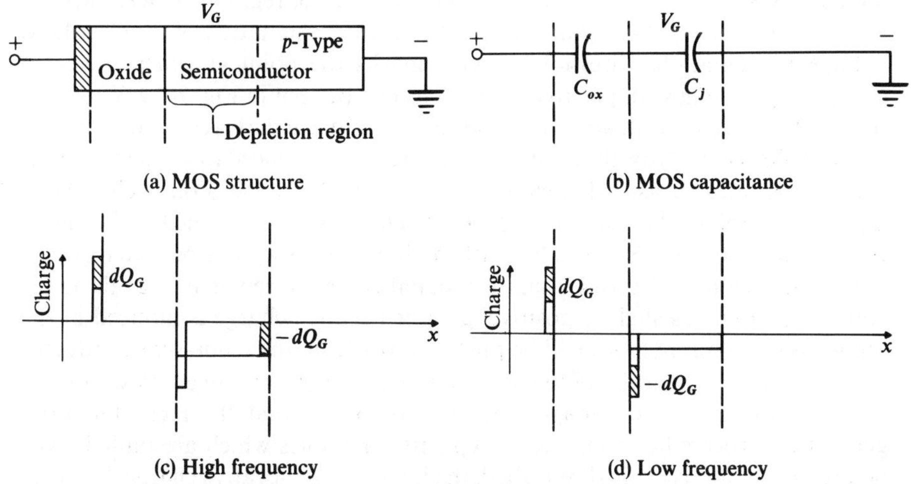

Fig. 9–18 MOS structure and charge distribution at high and low frequencies.

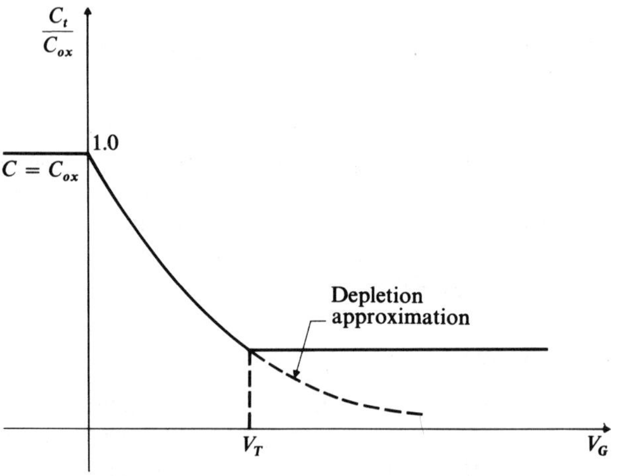

Fig. 9–19 MOS capacitance as a function of V_G.

It could be shown (Grove, 1967) that

$$\frac{C_t}{C_{ox}} = \frac{1}{\sqrt{1 + \dfrac{2\varepsilon_o^2 V_G}{qN_a\varepsilon_s t_{ox}^2}}}. \tag{9–111}$$

Equation (9–111) is valid up to the threshold voltage V_T. The equation shows that as the top of the *p*-type semiconductor is being depleted of free carriers the total capacitance drops with the voltage V_G. As the voltage is made more positive and a strong inversion layer forms, the transition region reaches a maximum width. Thus, the capacitance ratio C_t/C_{ox} falls to a minimum value and stays there as illustrated in Fig. 9–19. The start of the level region is determined by the condition of Eq. (9–7) which also led to the determination of the threshold voltage V_T. Thus, the threshold voltage marks the point at which the ratio C_t/C_{ox} stops falling with positive V_G at V_T as can be seen in Fig. 9–19. When the voltage V_G is negative an accumulation (of majority carriers) region to the right of the oxide occurs. Now the semiconductor region is reduced to just a resistance. The capacitance measured then will be that of the oxide alone. The above arguments explain the curve of Fig. 9–19. This curve is obtained if the signal voltage across the MOS is of sufficiently high frequency as may be seen from the following reasoning. Consider that the signal is a high frequency signal. As the voltage V_G becomes slightly positive, greater negative charge is induced in the semiconductor, more holes will be pushed away from the oxide-semiconductor interface. Hence, the *width* of the depletion region will increase as shown in Fig. 9–18c. If, however, the voltage frequency is low so that the recombination-generation process has a chance to replenish the holes which are pulled away before the voltage is again diminished, the incremental negative charge due to an incremental voltage will appear at the oxide-semiconductor interface as shown in Fig. 9–18d. The result is that the capacitance seen should be essentially the oxide capacitance C_{ox}. Thus, we expect the MOS capacitance to change with frequency as shown in Fig. 9–20.

As the final item of discussion we wish to consider the influence of *surface states* on MOS capacitance. In Chapter 2 we have already noted that any interruption in the perfectly periodic arrangement of the crystal surface can lead to the introduction of additional permissible electronic states not otherwise present. From this observation, it is not difficult to accept the fact that at surfaces of crystals additional energy states will exist because the surface is a major interruption of the periodic arrangement. This theoretical expectation has been experimentally verified. These additional states which are quite high in density, of the order of surface atom density, exist primarily in the band gap.

The surface states have been traditionally classified as *fast states* and *slow states.* The fast states are thought to be at the interface between the insulator (oxide) and the semiconductor. These states can quickly and easily exchange their charge with the conduction or the valence band, hence they are called *fast states.*†

The *slow states* are thought to exist at the interface of the insulator and air or vacuum. These therefore are expected to take a longer time to exchange their charge with the conduction or valence bands. We shall here consider only the *fast states.* The charge associated with the surface states is expected to change with the extent of band bending. It is not hard to see that the charge in the surface states will change because the surface charge will change as the Fermi level passes through the surface state. Fig. 9–21 shows how the charge occupancy of the fast surface states roughly in the middle of the band gap changes with

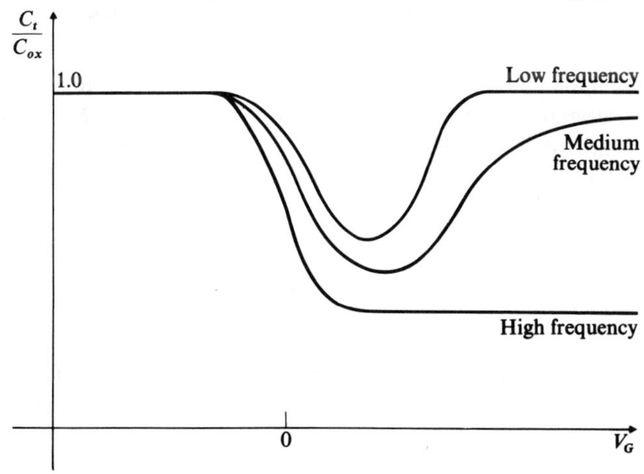

Fig. 9–20 MOS capacitance as a function V_G at various frequencies.

† This is a somewhat misleading terminology because at low temperatures the time constant associated with these states can be quite long.

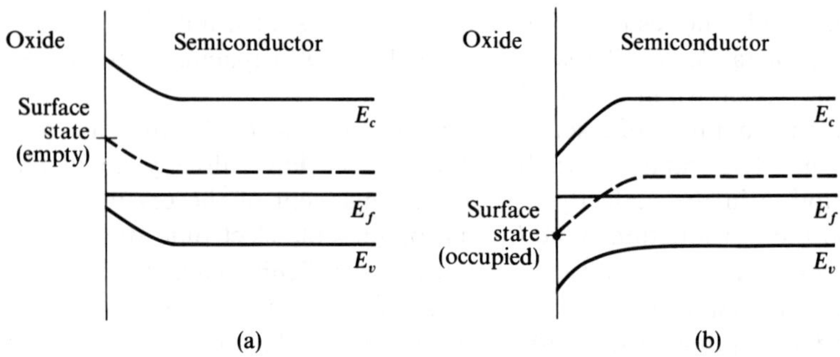

Fig. 9–21 Fast surface state occupancy changes with band bending. (a) $V_G < 0$, (b) $V_G > 0$.

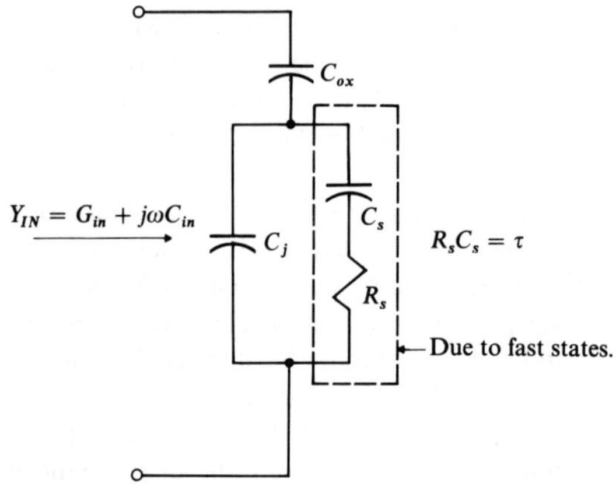

Fig. 9–22 Equivalent circuit of MOS including the influence of fast surface states.

band bending. This change of charge on the fast surface states will contribute to the MOS capacitance. Figure 9–22 shows the modified MOS capacitance equivalent circuit of Fig. 9–18 to include the contribution of the fast surface states. The input admittance of this circuit is given by

$$Y_{IN} = G_{IN} + j\omega C_{IN}. \tag{9–112}$$

when

$$G_{IN} = \frac{\omega^2 C_s \tau C_{ox}}{(C_j + C_{ox} + C_s)^2 + \omega^2 \tau^2 (C_{ox} + C_j)^2} \tag{9–113}$$

and

$$C_{IN} = \frac{C_{ox}}{C_{ox} + C_j + C_s}\left[C_j + C_s\left(\frac{(C_{ox} + C_j + C_s)^2 + \omega^2\tau^2 C_j(C_{ox} + C_j)}{(C_{ox} + C_j + C_s)^2 + \omega^2\tau^2(C_{ox} + C_j)^2}\right)\right]. \tag{9–114}$$

It is possible with the help of Eqs. (9–113) and (9–114) and experimental data, to extract valuable information about the fast surface states. (See Nicollian and Getzberger, 1965.)

9–11 SUMMARY

In this chapter we looked at the physics of the MOSFET. The MOSFET was conceived as early as 1929 but it was not physically realized until the early 1960s when the new silicon planar technology made this device possible. Because the number of processing steps in obtaining the MOSFET are less than those needed in the bipolar transistor, the former device is inherently less expensive and hence more attractive to produce. The currently available cut-off frequencies of MOSFETS, however, are less than those of the bipolar transistors.

The static drain characteristics were obtained by considering the influence of the gate voltage and the field due to it which can produce either an accumulation, a depletion, or an inversion layer in the semiconductor under the oxide. Two basic types of devices, the induced type and the depletion type were described. Two modes of operation are possible in the latter device—the enhancement mode and/or the depletion mode.

The gate voltage determines the carrier density in the channel. Hence, the drain current turns out to be a function of the gate to source voltage and the drain to source voltage. For low V_{DS} the drain characteristics appear linear. For higher values of V_{DS} the drain current saturates. In the pinch-off region the experimentally observed finite conductance is due to the channel length modulation. An analytical expression was derived for it.

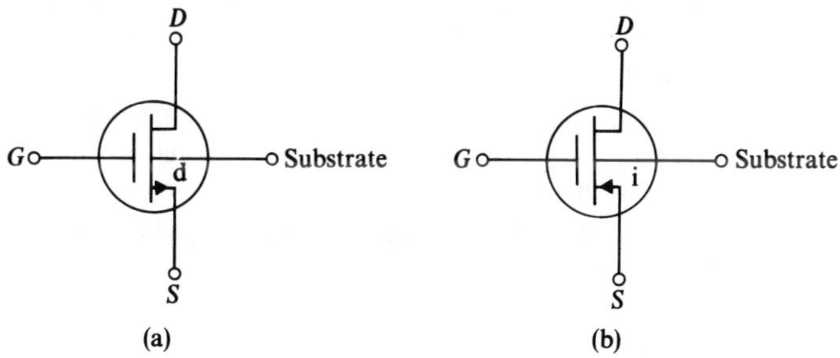

Fig. 9–23 Symbols of MOSFET or IGFET. (a) Depletion type n-channel MOSFET or IGFET. (b) Induced type p-channel MOSFET or IGFET.

A wide band equivalent circuit was obtained for the MOSFET. A simplified model was used to obtain a figure of merit—the maximum frequency of oscillation which was found to be a sensitive function of the channel length, and the oxide capacitance.

The effective mobility of carriers in the inversion layers to be used in analytic expressions for the silicon MOSFET is approximately *half* the bulk mobilities for electrons and holes. The MOS capacitance is found to be a function of both gate voltage and frequency. Most of the expressions derived are for *n*-channel devices. The corresponding expressions for *p*-channel devices are trivially different from those of the *n*-channel devices.

Finally, a symbol is proposed for the MOSFET or IGFET in Fig. 9–22. The letter *d* within the circle implies a *depletion* type device. The letter *i* within the circle implies an *induced* type device. This way you can tell from the symbol what kind of a device you have.

PROBLEMS

9–1 Calculate the surface field on a *p*-type semiconductor necessary to make the surface intrinsic.

9–2 Calculate the surface field on a *p*-type semiconductor to produce a strong inversion layer.

9–3 Obtain an expression for the minimum MOS capacitance.

9–4 Calculate g_m when (a) $V_{GS} = V_T/2$ and $V_{DS} = V_{DSO}/2$. (b) $V_{GS} = 3V_T/4$ and $V_{DS} = 3V_{DSO}/4$.

9–5 Write expressions for the drain *I-V* characteristics of a *p*-channel device including an expression for the threshold voltage V_T.

9–6 If $V_T = 0$ in a given MOSFET, and $V_{GS} = V_{DS} = 4$ Volts when $I_D = 3$ mA, is the transistor operating in the saturation part of its characteristics? Why?

9–7 Calculate the small signal 0 to 90 per cent rise time of the drain to source voltage to a step voltage of amplitude V_g applied between the gate and source. Between the drain and source terminals is connected a load resistance R_L in series with V_{DD} battery. You may use the simple model.

9–8 Calculate the small signal fall time of the drain to source voltage to a step voltage of 1 Volt applied between the gate and source. The drain terminal is connected to a load resistance R_L in series with a V_{DD} battery returned to the source.

9–9 Interpret the threshold voltage for *p*-channel induced type and depletion type devices as was done in Section 9–5.

REFERENCES

Baum, G., and Beneking, H., "Drift Velocity Saturation in MOS Transistors," *IEEE Transactions on Electron Devices,* Vol. ED 17, No. 6, June 1970, pp. 481–2.

Crawford, R. H., *MOSFET in Circuit Design,* McGraw-Hill Book Co., p. 36 (1967).

Frohman-Bentchkowsky, D., and Grove, A. S., "Conductance of MOS Transistors in Saturation," *IEEE Transactions on Electron Devices,* **ED. 16,** No. 1, 108–112, Jan. (1969).

Frohman-Bentchkowsky, D., and Vodasz, L., "Computer-Aided Design and Characterization of Digital MOS Integrated Circuits," *J. IEEE Solid-State Circuits,* **SC-4,** No. 2, April (1969).

Gray, P. E., and Searle, C. L., *Electronic Principles,* John Wiley & Sons, Chapter 9 (1969).

Grove, A. S., *Physics and Technology of Semiconductor Devices,* Chapter 9, John Wiley & Sons (1967).

Grove, A. S., and Fitzgerald, D. J., "Surface Effects on *p-n* Junctions: Characteristics of Surface Space-Charge Regions Under Nonequilibrium Conditions," *Solid State Electrons,* **9,** 783 (1966).

Hamilton, et al, *Principles and Applications of Semiconductor Device Modeling*, Chapter 5, Holt Rinehart & Winston, New York (1971).

Ihantola, H. K. J., "Design Theory of a Surface Field-Effect Transistor," *Stanford Electronics Labs., Stanford, Calif., Tech. Rept.,* 1661–1 (1961).

Leistiko, O., Jr., et al, "Electron and Hole Probabilities in Inversion Layers on Thermally Oxidized Silicon Surfaces," *IEEE Transactions on Electron Devices,* **ED. 12,** No. 5, 248–253, May (1965).

Nicollian, E. H., and Getzberger, A., "MOS Conductance technique for Measuring Surface State Parameters," *Appl. Phys. Letters,* Vol. 7, 216 (1965).

Pao, H. C., and Sah, C. T., "Effects of Diffusion Current on Characteristics of Metal-Oxide (Insulator)—Semiconductor Transistors," *Solid State Electronics,* **9,** 927–937 (1966).

Sah, C. T., "Characteristics of the MOS transistors," *IEEE Transactions on Electron Devices,* **ED. 11,** No. 7, 324-344, July (1964).

CHAPTER 10

SOME ADDITIONAL SEMICONDUCTOR DEVICES

10–1 INTRODUCTION

In Chapters 1 through 9, starting with the ideas of modern physics, device theory was developed for the bipolar transistor, the junction field effect transistor JFET, and the metal oxide semiconductor field effect transistor MOSFET. As this device theory was developed from first principles, it should be possible to use the insight developed from the analysis of the preceding chapters to understand other semiconductor devices including, hopefully, many which may come along in the years ahead.

We shall consider several devices which we shall analyze with the methods developed so far, as concrete illustrations of the use of these methods.

10–2 UNIJUNCTION TRANSISTORS

As the name suggests, the unijunction transistor UJT consists of a single *p-n* junction. Figure 10–1a shows the schematic of a unijunction transistor along with the normal biasing network. The device was first proposed by Shockley et al (1949) and was called the *filamentary transistor*. The device has also been called the *double base diode* (Less and Mathis, 1953). Now, however, the term *unijunction transistor* (Clark, 1965, 1967; Trofimenkoff and Huffs, 1966; Bluhn and Sylvan, 1964) has been widely accepted and used.

The major mechanism which explains the *I-V* characteristics of UJT is *conductivity modulation*. In Fig. 10–1a is shown the normal biasing of the UJT. The resistance of the *n*-type bar between the two bases is r_{BB}. The resistance between base one and B is r_{B1}; and that between B and base two is r_{B2}.

The input circuit consists of a voltage source V_S in series with R_s which is chosen such that

$$r_\ell \gg R_s \gg r_{on}, \tag{10–1}$$

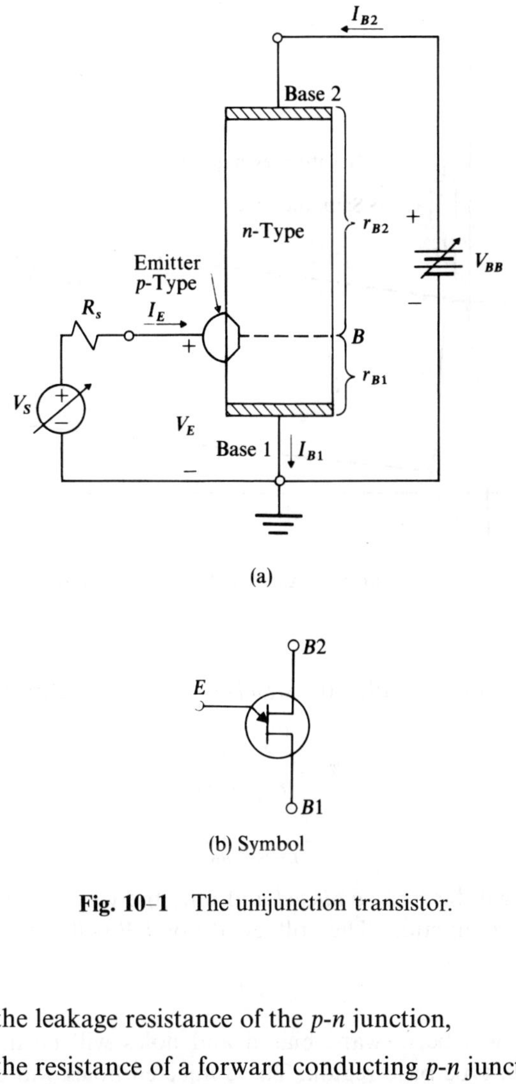

Fig. 10–1 The unijunction transistor.

where r_ℓ = the leakage resistance of the *p-n* junction,

r_{on} = the resistance of a forward conducting *p-n* junction.

Thus the combination of V_S and R_s behaves like a voltage source when the *p-n* junction is reverse biased, and it behaves like a current source when the *p-n* junction is forward biased. It is relatively easy to pick R_s which satisfies the magnitude relationship Eq. (10–1). The voltage at point *B* due to the battery V_{BB} alone is

$$V_{BO} = \frac{V_{BB} r_{B1}}{r_{B1} + r_{B2}}. \qquad (10\text{–}2)$$

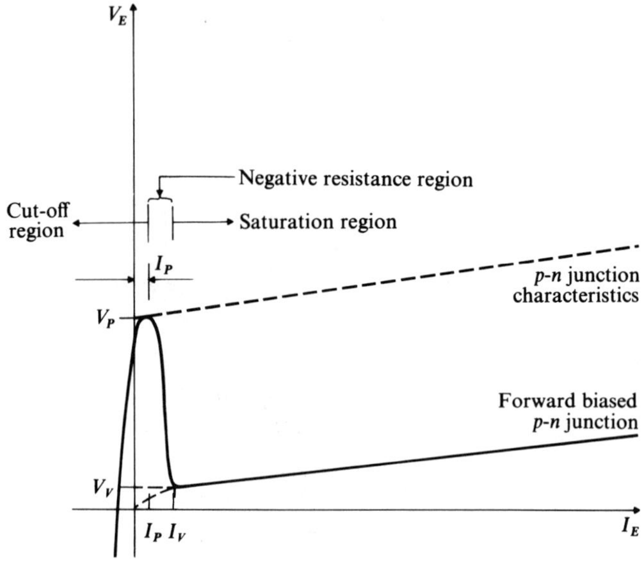

Fig. 10–2 *I-V* characteristics for the unijunction transistor.

A useful device parameter called the *stand-off-ratio* $\mathcal{R}$ is defined from the above as follows

$$\mathcal{R} = \frac{r_{B1}}{r_{B1} + r_{B2}}. \tag{10–3}$$

Thus, when

$$V_E < V_{BB} \tag{10–4}$$

the *p-n* junction will be *reverse biased* and only the very small *leakage* current will flow in the *p-n* junction. The voltage at point B will then be given by V_{BO}.

If

$$V_E > V_{BB} \tag{10–5}$$

the *p-n* junction will be forward biased and holes will be injected from the emitter into the *n*-type base region. These injected holes will drift down from the emitter towards the base one contact due to the electric field on account of battery V_{BB}. The presence of the injected holes in the region between the plane B, shown dotted in Fig. 10–1a, and the base one contact will *increase* the conductivity of that region. This increase is due not only to the injected holes but also due to the increase of majority carriers (equal to the injected carrier density) in this same region so as to maintain space charge neutrality in the bulk semiconductor. As the conductivity of the region between plane B and the base one contact increases, the effective resistance r_{B1} will decrease. The

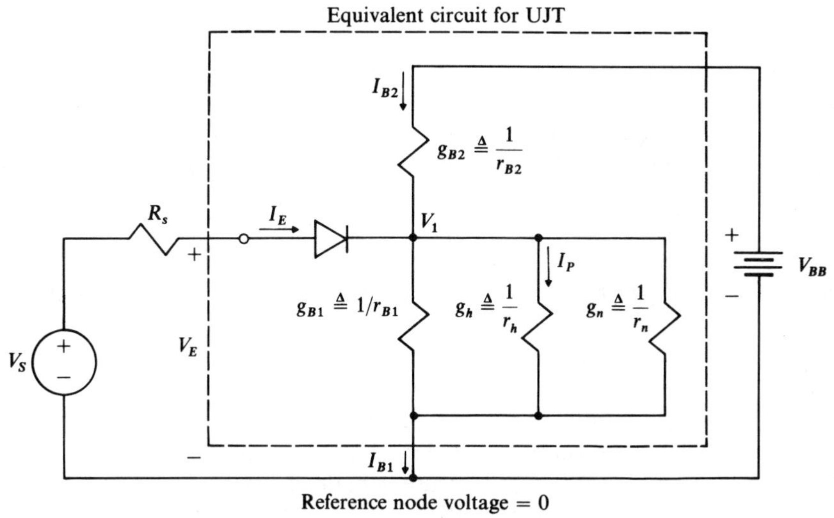

Fig. 10–3 An equivalent circuit for the unijunction transistor used to make some calculations.

external voltage source will put out *more* current because it sees a smaller effective resistance confronting it. The increased current implies more injected carriers and a further reduction in the resistance between the emitter and base one contact. Thus we expect to see a *decrease* in V_E with an *increase* in I_E. Hence, we see an incremental negative resistance region in $V_E - I_E$ characteristics. As the current I_E is increased still further, the V_E will continue to decrease until a stage is reached where a further increase in I_E leads to an *increase* in V_E. This position is merely the characteristics of the forward biased *p-n* junction in series with a low series resistance of the highly modulated portion of the device between plane *B* and the base one contact. This portion is operating under conditions of high injection. The negative resistance region, which makes the device interesting, is due to the conductivity modulation of the region between plane *B* and the base one contact. By conductivity modulation is meant the variation of conductivity of the semiconductor region due to high injection effects. It is clear that the negative resistance region would not exist unless the injected minority carriers were enough to change the conductivity of the region. This implies that the injected minority carrier density is at least 10 to 20 per cent of the majority carrier density. The peak voltage and current are designated by V_P and I_P. The valley voltage and current are designated by V_V and I_V. For currents less than I_V the characteristics of a *p-n* junction moved up by voltage V_{BO} results as shown in Fig. 10–2. Recall that the voltage V_{BO} is due to the battery V_{BB}. The resulting I-V characteristics are shown in Fig. 10–2. For currents less than I_P the device is said to be in the *cut-off* part of its characteristics, and for currents greater than I_V the device is said to be in the *saturated* part of its characteristics.

The equivalent circuit of Fig. 10–3 may be used to do some useful calculations such as the peak and valley currents and voltages. Using Ohm's law we have

$$I_{B2} = (V_{BB} - V_1)g_{B2}. \qquad (10\text{–}6)$$

Using Kirchhoff's current law applied to the node pair V_1 and the reference node, we obtain

$$I_{B2} + I_E = V_1(g_{B1} + g_h + g_n), \qquad (10\text{–}7)$$

where g_h and g_n = conductances due to the *excess* injected holes (into the region between plane B and the base one contact) and the consequent increase in majority carriers to maintain space charge neutrality.

Now in high injection

$$g_h = \mu_p(p_n + p_{\text{inj}})A/L_{AB1} \qquad (10\text{–}8)$$

$$g_n = q\mu_n(n_n + p_{\text{inj}})A/L_{AB1}, \qquad (10\text{–}9)$$

where p_{inj} = injected holes into the n-type bar,

A = cross-sectional area of the n-type bar,

L_{AB1} = distance between plane B and the base one contact.

In writing Eq. (10–9) we have used the fact that the *excess*† majority carrier density is equal to the *injected* minority carrier density.

At quite high injection level

$$p_{\text{inj}} \gg n_n \text{ and } p_n. \qquad (10\text{–}10)$$

Then from Eqs. (10–8) and (10–9) we have

$$\frac{g_h}{g_n} = \frac{\mu_p}{\mu_n} \qquad (10\text{–}11)$$

We may also define *injection efficiency for UJT* γ_u as follows

$$\gamma_u = \frac{I_p}{I_E} = \frac{V_1 g_p}{I_E}. \qquad (10\text{–}12)$$

† The word "excess" is used in the sense of over and above the thermal equilibrium value of the carrier density.

Using Eqs. (10–6) through (10–12) yields (see Problem 10–1)

$$V_1 = \mathscr{R}V_{BB} - \left(\frac{\mu_n}{\mu_p} + 1 - \frac{1}{\gamma_u}\right)\gamma_u I_E\left(\frac{r_{B1}r_{B2}}{r_{B1} + r_{B2}}\right). \tag{10–13}$$

The voltage V_E may be expressed using KVL

$$V_E = V_1 + V_D, \tag{10–14}$$

where V_D = voltage across the diode.

The voltage across the diode V_D may be calculated from the diode equation which is, for our case,

$$I_E = I_S(e^{qV_D/\eta kT} - 1) \tag{10–15}$$

where

$$I_E = I_S\, e^{qV_D/\eta kT} \qquad \text{if } V_D \gg \frac{\eta kT}{q}, \tag{10–16}$$

$$1 \leqslant \eta \leqslant 2. \tag{10–17}$$

The left hand limit is for low injection and the right hand limit is for very high injection. Here I_S is given either by Eq. (3–56) in the low injection case or by Eq. (12–158) in the very high injection case.

Equation (10–16) may be solved for V_D as follows

$$V_D = \eta\left(\frac{kT}{q}\right)\ell\text{n}\left(\frac{I_E}{I_S}\right). \tag{10–18}$$

Equations (10–18) and (10–13) may be substituted into Eq. (10–14) to obtain the following expression for V_E

$$V_E = \mathscr{R}V_{BB} - \left(\frac{\mu_n}{\mu_p} + 1 - \frac{1}{\gamma_u}\right)\left(\frac{r_{B1}r_{B2}\gamma_u I_E}{(r_{B1} + r_{B2})}\right) + \frac{\eta kT}{q}\ell\text{n}\left(\frac{I_E}{I_S}\right). \tag{10–19}$$

The coordinates of the *peak point* in Fig. 10–2 may be readily calculated by requiring that

$$\frac{dV_E}{dI_E} = 0. \tag{10–20}$$

The result is

$$I_P \doteq \left(\frac{\eta kT}{q\gamma_u}\right)\cdot\frac{\left(\dfrac{r_{B1} + r_{B2}}{r_{B1}r_{B2}}\right)}{\dfrac{\mu_n}{\mu_p} + 1 - \dfrac{1}{\gamma_u}} \tag{10–21}$$

$$V_P \doteq \mathscr{R}V_{BB} + \frac{\eta kT}{q}\ell\text{n}\,\frac{I_P}{I_S}. \tag{10–22}$$

Note that if I_P is sufficiently low

$$V_P \doteq \mathscr{R}V_{BB} \qquad (\text{low } I_p) \tag{10–23}$$

The valley coordinates may be obtained by observing that at the valley point a very large injected carrier density must occur in the region between the emitter and base-one contact. The conductivity of this region is expected to be high under these circumstances, and the voltage V_1 is expected to be quite low. We shall approximate this voltage by zero, i.e.

$$V_1 \doteq 0. \tag{10–24}$$

Equation (10–24) may be substituted into Eq. (10–13) and may be solved for I_E which is now to be recognized as the valley current,

$$0 \doteq \mathscr{R}V_{BB} - \left(\frac{\mu_n}{\mu_p} + 1 - \frac{1}{\gamma_u}\right)\left(\frac{r_{B1}r_{B2}}{r_{B1} + r_{B2}}\right)\gamma_u I_V \tag{10–25}$$

and

$$I_V \doteq \frac{\mathscr{R}V_{BB}\left(\dfrac{r_{B1}r_{B2}}{r_{B1} + r_{B2}}\right)^{-1}}{\gamma_u\left(\dfrac{\mu_n}{\mu_p} + 1 - \dfrac{1}{\gamma_u}\right)}. \tag{10–26}$$

Cancelling $\mathscr{R}$ with a part of the quantity in the bracket in the numerator of Eq. (10–26), the following simpler expression for I_V is obtained

$$I_V \doteq \left(\frac{V_{BB}}{r_{B2}\gamma_u}\right)\frac{1}{\left(\dfrac{\mu_n}{\mu_p} + 1 - \dfrac{1}{\gamma_u}\right)}. \tag{10–27}$$

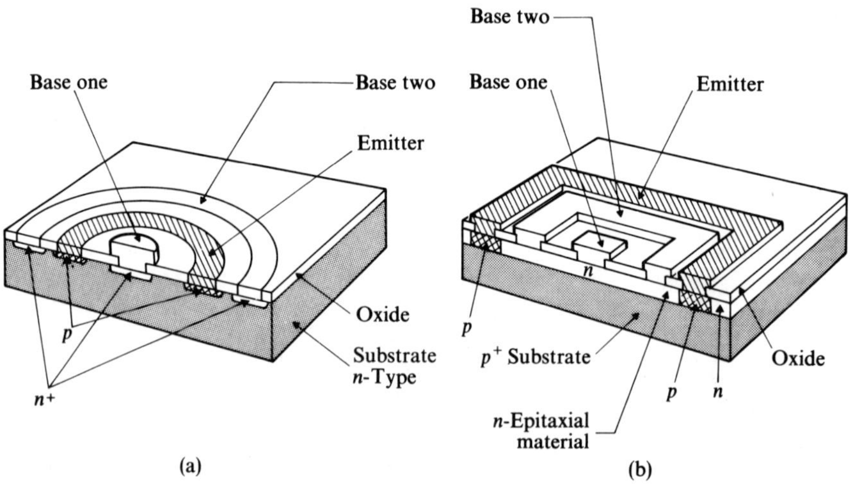

Fig. 10–4 Physical structure of unijunction transistor. (After Senhouse.) (a) Diffused planar structure. (b) Epitaxial planar structure.

The corresponding valley voltage is obtained by substituting Eq. (10–24) into Eq. (10–14)

$$V_V = \frac{\eta kT}{q} \ell n\left(\frac{I_V}{I_S}\right). \tag{10–28}$$

In the above calculations we have treated the injection efficiency γ_u as a constant. It is expected to diminish with increasing injection level (according to the arguments given in Chapter 12) as we move from the peak point to the valley point. It is possible to choose the proper device parameters so that I_P, V_P and I_V are insensitive to temperature variations (see Problems 10–4). These results have been found to be justified by experimental results.

Just like the tunnel diode, the UJT can be operated as a bistable device if the load line intersects the characteristics of Fig. 10–2 in three places, *including* the negative resistance region. It is possible to operate the device as an amplifier if the load line makes only one intersection and *only* in the negative resistance region. An oscillator can also be designed with the same load line with a suitable choice of *external* parameters. Two of the physical structures which are used in the fabrication of UJT are shown in Fig. 10–4.

10–3 TUNNEL DIODES

A practical and useful semiconductor device which uses the phenomenon of quantum mechanical tunneling is the *tunnel diode.* In the *p-n* junction the Zener breakdown phenomenon which is due to tunneling usually sets the upper bound on the range of reverse bias that these devices can be operated under.

In the case of the tunnel diodes the important and useful features of this device are *due* to tunneling. Here we wish to study the physics of the device.

L. Esaki was doing research in the 1950s on heavily doped germanium. He discovered, quite accidentally, that when a *p-n* junction is made using heavily doped germanium, the terminal current–voltage relationship which results is that shown in Fig. 10–8 with a prominent negative resistance region.

Esaki was able to give an explanation which still remains valid in many respects for the *I-V* characteristics in terms of quantum mechanical tunneling. The theory now has been extended beyond the early Esaki proposal.

As heavily doped semiconductors are needed to obtain tunnel diodes, let us see what the simple band structure of semiconductors looks like. There are three major changes in the band structure of heavily doped semiconductors compared to the lowly or moderately doped semiconductors. First, the donor states broaden out from what was a highly concentrated density of states function. The reason for this broadening out is the same, in most essential features, as the cause of the original development of the conduction and valence bands. That is, as the density of donor atoms gets high, at the high doping densities needed for tunnel junctions, the interactions between the donor atoms

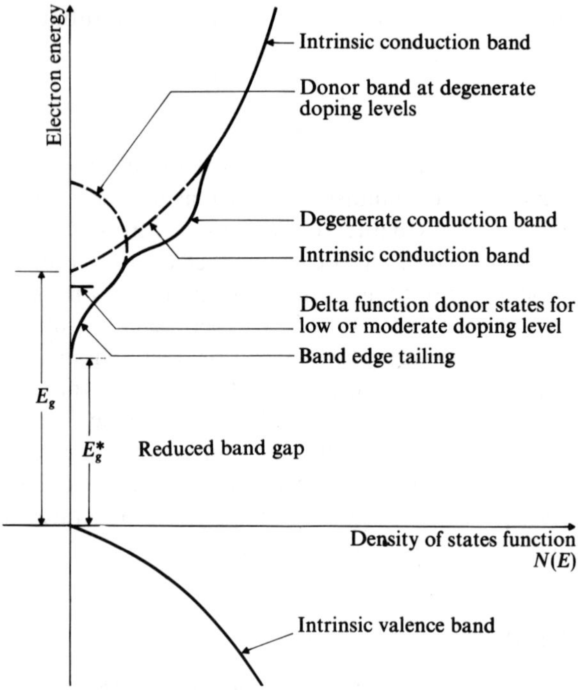

Fig. 10–5 The density of states function for a degenerate *n*-type semiconductor for a particular doping density.

themselves becomes non-negligible, and the Pauli exclusion principle leads to the broadening of the donor bands. Second, the donor bands in *n*-type semiconductors *merge* with the conduction band. Third, instead of the sharp edge of the intrinsic conduction and valence bands, which apply to lightly or moderately doped semiconductors, the combined conduction and donor bands trail off *gradually*, called *band edge tailing*, into the band gap to essentially zero† density of states function as shown in Fig. 10–5. Fourth, as a result of the band tail, the *effective band gap* E_g^* is significantly less than the intrinsic band gap E_g.

A fourth consequence of the heavy donor density is that the Fermi level in this case ends up within the conduction band. *We define a degenerate semiconductor as one in which the Fermi level is within or at the band edge of the conduction or the valence band.* Degenerate semiconductors are necessary in order to have tunnel diodes. The amount of degeneracy of a semiconductor is measured

† Actually the band tailing results in the density of states function having a nonzero, though small, value throughout the band gap. Thus the phrase "essentially zero" is meant as an approximation.

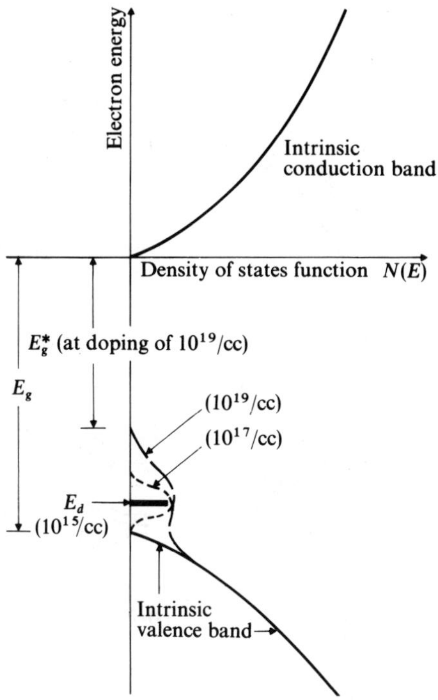

Fig. 10–6 Density of states function in *p*-type Ge as a function of doping density.

by the energy difference between the Fermi level and the band edge. Thus, if the Fermi level is 2.5 millielectron Volts (meV) above the band edge, the amount of degeneracy is 25 meV.

The density of states function is a function of the doping density in heavily doped semiconductors. Figure 10–6 shows schematically what happens in the *p*-type germanium. The density of states function in a degenerate *p*-type semiconductor is seen to be the mirror image of that in an *n*-type semiconductor if a comparison is made between Figs. 10–6 and 10–5. In germanium and silicon a doping density of 10^{19}/c.c. is sufficient to produce a degenerate semiconductor.

No adequately satisfactory analytic model exists which predicts or describes the density of states function at band edges and in the band tails at different doping levels needed to produce degeneracy. Because of this, it is extremely difficult to do precise calculations of the doping density necessary to obtain degeneracy. As the Fermi level is close to but inside the conduction or valence bands, the Boltzman approximation is *not valid* in these calculations. The more general formulations in terms of Fermi integrals of Eq. (2–64) must be used here. We shall assume for simplicity and convenience that the intrinsic density

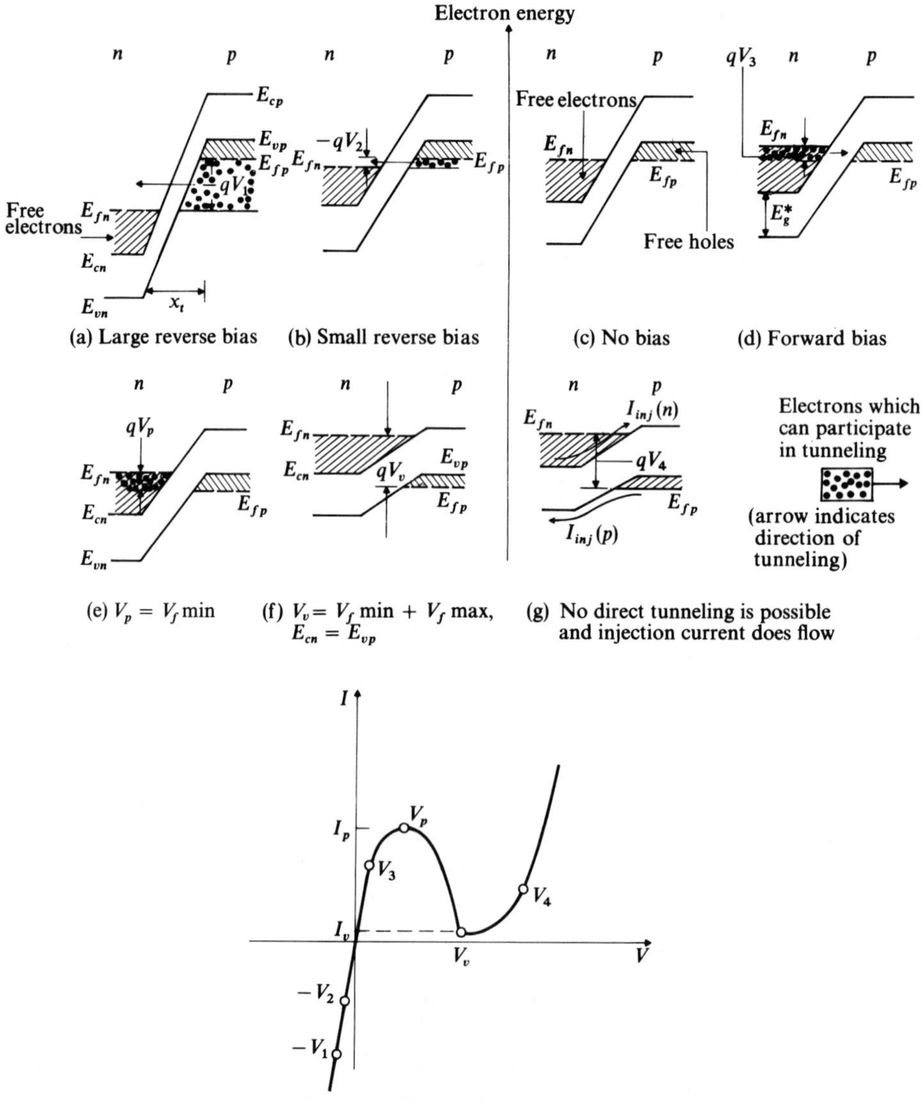

Fig. 10–7 Simple energy band pictures as a function of externally applied voltage at 0° K and the predicted *I-V* characteristics.

of states is what we assumed to be valid in Chapter 1. We shall further assume that even in this case the donor and acceptor levels are discrete and delta functions. We have already commented that this is *invalid* in degenerate semiconductors. Even so, making the above assumptions we may use Eq. (2–61) to calculate the donor density in a *pure* n-type semiconductor (that is, $N_a = 0$) to *just* make it degenerate, i.e.

$$\eta_f = \eta_c = 0 \tag{10–29}$$

which leads to

$$N_d = N_0 G_e F_{\frac{1}{2}}(0)(1 + e^{-\eta_d}). \tag{10–30}$$

We readily confess the inaccuracy of Eq. (10–30) in view of the above comments. It is nevertheless a useful result. If N_d predicted by Eq. (10–30) is used, it *ensures* degeneracy in the actual semiconductors.

Figure 10–7c shows the simple band diagram for a tunnel diode at no bias (terminals left open circuited). Note that as before we require that the Fermi levels, which are now within the main bands in a degenerate semiconductor, must line up. The condition which obtains only at sufficiently high reverse bias in ordinary *p-n* junctions, Fig. 3–11b, occur in tunnel *p-n* junctions both at reverse and part of forward bias. The four required conditions, stated in Section 3–6, for tunneling to occur are also required in tunnel diodes. These will be restated in slightly more general terms valid for tunnel diodes.

1. Electrons must exist at permissible energy states on one side of the tunneling barrier *from which* they could tunnel.
2. Permissible but empty energy states must exist on the other side of the transition region *into which* electrons can tunnel.
3. Overall energy and momentum conservation laws must be satisfied in the tunneling transition.
4. Tunneling probability for the carriers must not be too low.

The tunneling barrier is precisely the same as shown in Fig. 3–12. Note that the tunneling barrier width b_t is always less than the transition region width. That is

$$b_t < x_t. \tag{10–31}$$

Because very heavily doped semiconductors are needed to produce tunnel diodes, the transition region width is quite narrow, usually 20–100 Å. This is sufficiently narrow for tunneling probability to be adequate.

Figure 10–7 shows the simple one-dimensional band picture for the tunnel diode *at absolute zero degrees*. We choose this temperature because the Fermi–Dirac distribution function is sharp and we can say with certainty which electronic states are occupied by electrons and which are not. Furthermore, the tunnel diode characteristics are relatively insensitive to temperature changes and hence our conclusions on the behavior of tunnel diodes at absolute zero temperature will give us a good idea of how the device behaves. (See Fig. 10–7h.)

We are considering *direct tunneling* transitions in which the electrons tunnel from one side of the tunneling barrier to the other side without the intermediary of any other particles. Semiconductors in which the lowest conduction band valley is directly above the highest valence band peak (at the same value of k_c) are called *direct* semiconductors. Materials in which this fact is not true are called *indirect* semiconductors. In direct semiconductors the tunneling transition does not need to involve some other intermediary such as a phonon. In the indirect semiconductors the intermediary of other agents such as phonons or impurities or some others must be involved to satisfy the overall energy and momentum conserving requirements of the tunneling transition.

As the momentum conservation is automatically taken care of (because k_c before and after tunneling are the same) in tunnel diodes made from direct semiconductors, the energy conservation requirement will be satisfied if the energy of the electron before tunneling is the same as that after tunneling. Now consider Fig. 10–7c. No tunneling current can flow because when electrons exist there are no empty and available states on the other side of the barrier into which they can tunnel. Thus the origin is a point on the tunnel diode characteristics. If the reverse voltage of $-V_2$ is applied, the situation depicted in Fig. 10–7b is obtained. We see now that electrons existing in the valence band on the *p*-side can tunnel to the conduction band on the *n*-side. Recall that electrons moving from the *p*-side to the *n*-side constitute *conventional* current going from the *n*-side to the *p*-side. This is reverse current for the tunnel diode. The current flowing at the terminals is shown in Fig. 10–7 marked by $-V_2$. The voltage V_2 is the difference between the Fermi levels on the two sides of the junction. If the tunnel diode reverse bias is increased to $-V_1$, as can be seen from Fig. 10–7a, more filled electronic states on the *p*-side are exposed to empty states on the *n*-side at the same energy levels. Hence the reverse current increases. Fundamentally we can write the following equation for tunneling current density

$$\underset{\substack{\text{Tunneling}\\\text{current}\\\text{density}}}{1} = \begin{pmatrix} 2 \\ \text{Density of occupied} \\ \text{states on one side} \\ \text{of the barrier} \end{pmatrix} \times \begin{pmatrix} 3 \\ \text{Density of available} \\ \text{states on the other} \\ \text{side of the barrier} \end{pmatrix} \times \begin{pmatrix} 4 \\ \text{Tunneling} \\ \text{probability} \end{pmatrix} \tag{10–32}$$

In Fig. 10–7a items 2 and 3 are both higher than the same quantities for Fig. 10–7b. In Fig. 10–7h the resulting current is marked by the point $-V_1$. It is clear that if the *p-n* junction is further reverse biased, the current will continue to increase. When the *p*-side is biased positive with respect to the *n*-side by a voltage V_3, we see from Fig. 10–7d that there are now electrons in the conduction band on the *n*-side which are exposed opposite holes which are empty states in the valence band on the *p*-side. Thus electrons can tunnel from the *n*- to the *p*-side. This constitutes conventional current going from the *p*- to the *n*-side and is called *forward* current. The current flowing in this case is marked by V_3 in Fig. 10–7h. Let us define

$$\text{minimum of } V_{fn} \text{ and } V_{fp} \triangleq V_{f\,\text{min}} \tag{10–33}$$

$$\text{maximum of } V_{fn} \text{ and } V_{fp} \triangleq V_{f\,\text{max}}. \tag{10–34}$$

In Fig. 10–7e we see that when the externally applied voltage is equal to $V_{f\,\text{min}}$ the maximum number of filled states on the *n*-side are opposite the empty states on the *p*-side. Thus the maximum amount of forward tunneling current will now flow. This is the peak current I_p and the voltage is peak voltage V_p. If the forward voltage is increased beyond V_p, the current is expected to diminish for two reasons. First, fewer filled states on the *n*-side are exposed opposite the empty states on the *p*-side. Second, the tunneling probability decreases as the forward bias is increased. When the bands have *uncrossed,* i.e. when the conduction band edge on the *n*-side E_{cn} is lined up at the same energy as the top of the valence band edge on the *p*-side E_{vp}, no tunneling is possible (as seen in Fig. 10–7f) and the *direct* tunneling current ceases. This marks the *valley point.* The voltage at which this happens is

$$V = V_V = V_{f\,\text{min}} + V_{f\,\text{max}}. \tag{10–35}$$

The current at the valley is designated I_V. The reason why I_V is not zero is that even though the direct tunneling current reaches zero, *injection* current, though usually small at the valley, is *not* zero. There is also another contribution to current which we shall presently discuss. In Fig. 10–7h the ideal current for voltage greater than V_V is due to injection.

The negative resistance region between the peak and the valley point is primarily due only to the tunneling process. In the discussion above we have tacitly assumed that the density of states function is the governing factor which determines the negative resistance region. It is also possible for the tunneling probability factor to determine the negative resistance region.

It was discovered fairly early that more current was found to exist particularly in the valley region which was not accounted for by the theory just presented. Figure 10–8 shows the current expected from the theory presented so far as well as that which is experimentally measured in a tunnel diode. It is seen that the experimental *I-V* characteristics seem to show the presence of more current than predicted by the normal tunneling theory. This additional current has been appropriately called *excess current.*

The most successful theory which explains this excess current is a two step process only one of which is tunneling. Recall that at potentials where the excess current exists, the bands have *uncrossed.* Under these circumstances an electron from the conduction band on the *n*-side can go into an available and empty state in the valence band on the *p*-side along one of three possible paths shown in Fig. 10–9. The largely successful theory which explains most of the character-

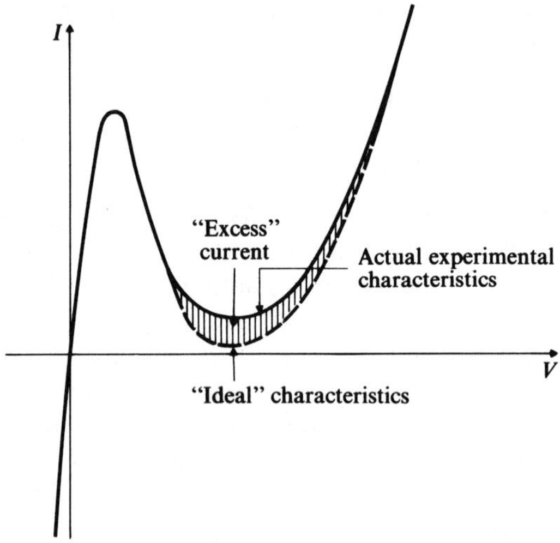

Fig. 10–8 Tunnel diode *I-V* characteristics showing "excess" current.

istics of excess current is based on transition II shown in Fig. 10–9. This is a two step process. Step *a* is a vertical motion of the electron which falls into a band gap state N_x. Step *b* is a tunneling transition from the band gap state to the valence band which is the *rate controlling step*. The most likely source of the band gap states is band-edge tailing. Any crystaline imperfection of chemical or mechanical origin usually will contribute to band gap density of states. The model which we shall use assumes that the density of states in the band gap is *uniformly distributed*.

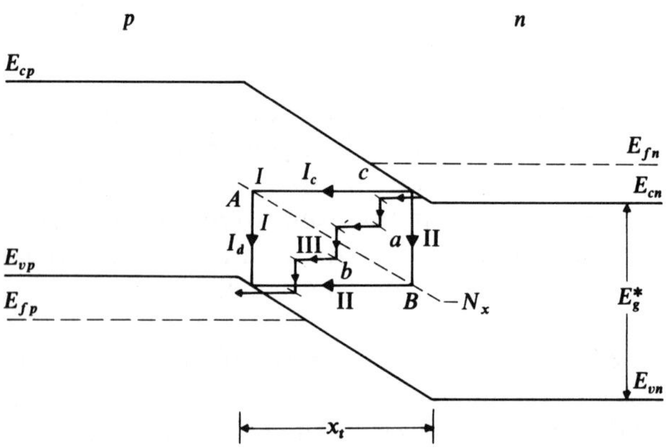

Fig. 10–9 A model to explain "excess" current. I, II and III are possible paths along which an electron can leave the conduction band on the *n*-side and reach the valence band on the *p*-side. Path *b* is rate controlling.

On the basis of path II in Fig. 10–9, which was not considered in Fig. 10–7, an expression for excess current I_x may be written as follows

$$I_x = CN_xP_x, \tag{10-36}$$

where C = constant,

N_x = density of occupied band gap states (e.g. at B) above the top of valence band,

P_x = probability of electrons tunneling from the band gap states N_x (e.g. B) to the valence band.

Note that in Eq. (10–36) N_x represents those band gap states which are participating in the tunneling process such as path b in Fig. 10–9. The tunneling probability P_x has been worked out by Franz (1956) and is given by

$$P_x = e^{-\alpha_x E_x^{3/2}/\mathscr{E}}, \tag{10-37}$$

where $\mathscr{E}$ = electric field in the transition region,

$$\alpha_x = \frac{4(2m_x)^{\frac{1}{2}}\theta}{3q\hbar}, \tag{10-38}$$

θ = numerical factor of the order of one,

E_x = energy gap through which the electron tunnels when it makes the transition along path b,

m_x = reduced effective mass for the tunneling electrons.

In writing Eq. (10–37) it was assumed that the electric field in the transition region is constant. The field in the transition region is not really constant. We shall, however, assume that this approximation is not too bad. For $\mathscr{E}$ we shall use its *maximum* value $\mathscr{E}_{max}$.

$$\mathscr{E} = \mathscr{E}_{max} = 2(V_{BI} - V_x)^{\frac{1}{2}}/w_1, \tag{10-39}$$

where V_{BI} = built-in voltage,

w_1 = width of the transition region when $(V_{BI} - V_x)$ is 1 Volt.

Also

$$E_x = E_g^* - qV_x + q(V_{fn} + V_{fp}) \tag{10-40}$$

$$(V_{BI} - V_x) = \frac{E_g^*}{q} + V_{fn} + V_{fp} - V_x, \tag{10-41}$$

where V_{fn}, V_{fp} = are the voltages corresponding to the Fermi energies on the n- and p-sides respectively.

Combining Eqs. (10–36) through (10–41) we finally obtain the following expression for excess current I_x,

$$I_x = CN_x e^{-(\alpha_x w_1 q^{1/2}/2)\{E_g^* + [(V_{fn}+V_{fp})q] - qV_x\}}. \tag{10–42}$$

Chynoweth (1961) et al have shown the validity of Eq. (10–42) at least as far as its dependence on w_1, E_g^*, N_x and V_x in silicon is concerned. Nanavati and de Andrade (1964) showed that GaAs tunnel diodes follow Eq. (10–42). Nanavati and Eisencraft (1966) demonstrated that Eq. (10–42) works well for GaSb tunnel diodes. They also showed that the temperature dependence of excess current is primarily due to the built-in voltage $V_{BI} = [(E_g^*/q) + (V_{fn} + V_{fp})]$ which was assumed to vary *linearly* with temperature

$$V_{BI}(T) = V_{BI}(0) - \beta_g T, \tag{10–43}$$

where β_g = constant.

Figures 10–10 and 10–11 show some experimental data on GaSb tunnel diodes. Furthermore, they show that the thermal current component for

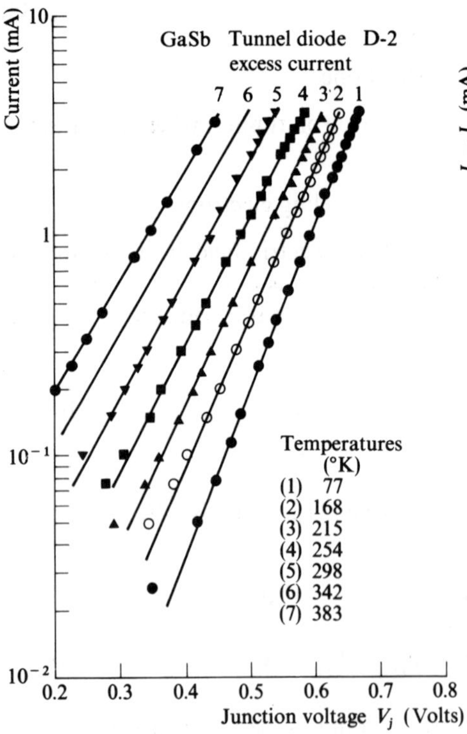

Fig. 10–10

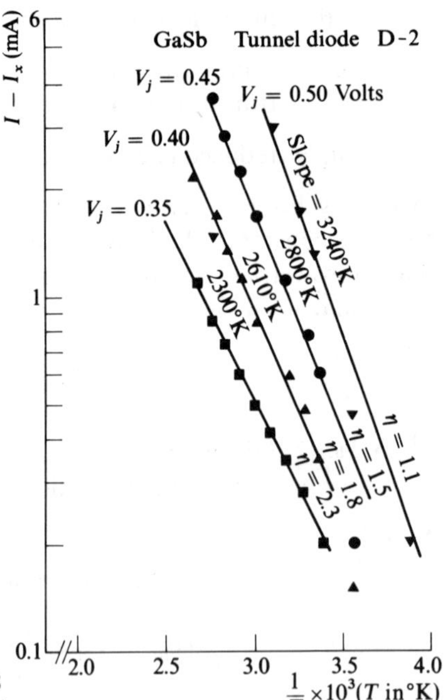

Fig. 10–11

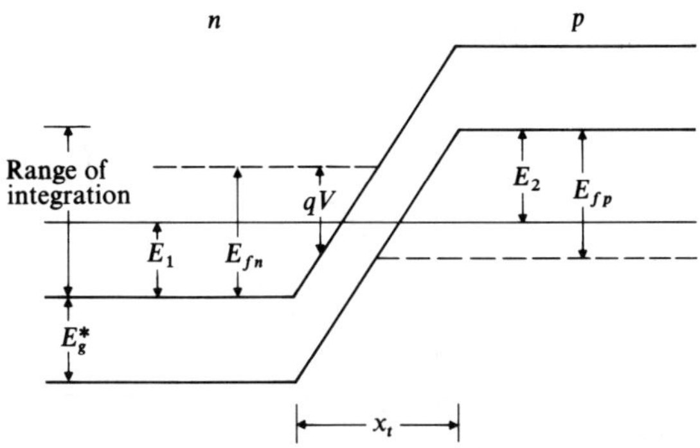

Fig. 10–12 Relavent definitions of symbols used in Eq. (10–43).

$V > V_V$ changes from recombination-generation through deep traps at low voltage to low injection current at higher voltages.

It is worthwhile pointing out that because the tunneling probability is exponentially dependent on the effective mass, the lighter mass electrons will be preferentially tunneled. The "heavy" electrons will determine the location of the Fermi levels, and the light electrons will determine the tunneling currents. We would expect in this device that the direction dependence of the effective mass may be important. The orientation of the crystal *has* been found to influence tunnel diode characteristics. The direct tunneling current is given below (Kane, 1961).

$$J_t = \frac{qm^*}{18\hbar^3} e^{-\pi m^{*1/2} E_g{}^{3/2}/(2\sqrt{2}\ E)} (\bar{E}_\perp/2) \int [f_n(E_1) - f_p(E_2)](1 - e^{-2E_s/E_\perp}) dE, \tag{10–44}$$

where F = force on the electron in the assumed constant electric field in the transition region,

$$\bar{E} \triangleq \frac{\sqrt{2}\hbar F}{\pi m^{*\frac{1}{2}} E_g^{*\frac{1}{2}}} \tag{10–45}$$

f_n, f_p = Fermi functions on the *n*- and *p*-sides,

E_s = smaller of E_1 and E_2.

The quantities E_1, E_2, E_{fn}, E_{fp} and the range of integration are defined in Fig. 10–12.

In tunnel diodes made from *indirect* band gap materials, of which silicon and germanium are examples, the tunneling process must necessarily be a co-operative effort between a tunneling electron and a *phonon* or some other scattering agent which could make up the difference in crystal momentum

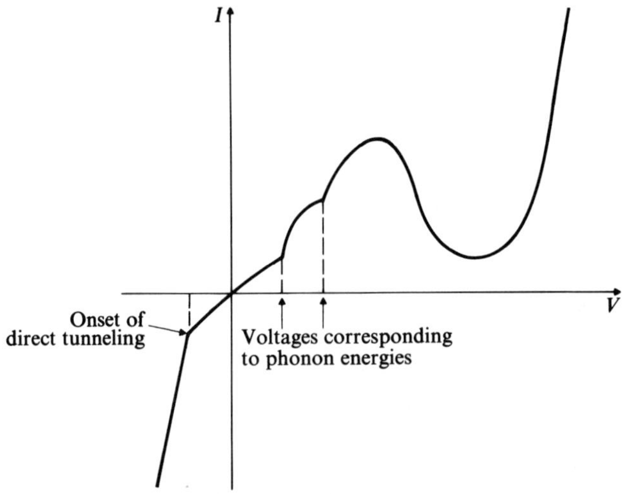

Fig. 10–13 Phonon assisted tunneling in Si.

between the initial and final values of momentum for the tunneling electron. If the phonon is the cooperating agent in the *indirect tunneling process*, it is termed *phonon assisted tunneling* (Holonyak, 1959; Hall et al, 1960). Figure 10–13 shows the *I-V* characteristics of Si at liquid helium temperature at which the phonon assisted tunneling is clearly visible; where the phonon energy and momentum couple properly, a sharp increase in current is observed. These cusps determine the phonon emission (or absorption) energies. Indeed the observing of tunneling current at low temperatures is a simple and quite accurate way of determining phonon energies in materials. While no analytic

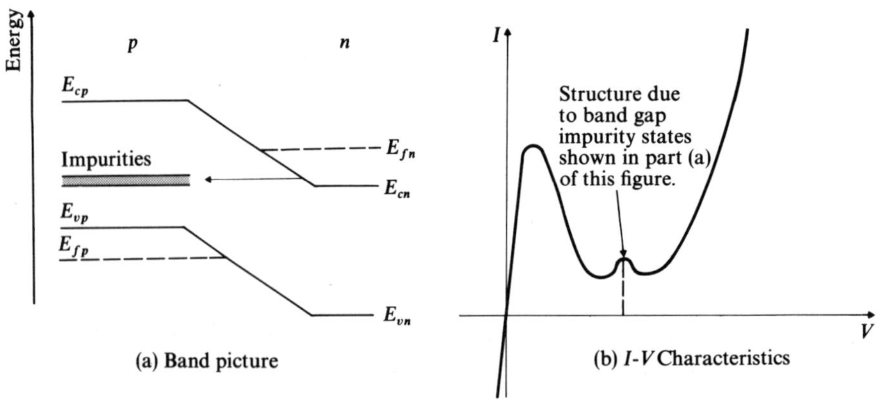

Fig. 10–14 An example of tunneling spectroscopic. (a) Band picture. (b) *I-V* Characteristics.

results have been stated here for indirect tunneling, the interested reader may look up Kane (1961) and Nanavati (1963).

The tunnel diode is one of the remarkable devices from whose terminal *I-V* characteristics it is possible to deduce some fundamental properties of the semiconductor material out of which the device is fabricated. It has been possible to measure the separation between the conduction band direct and indirect valleys (Morgan and Kane, 1959). (See Problem 10–5.) By observing the reverse *I-V* characteristics of GaSb tunnel diodes it has been possible to measure the energy separation of bands 1 and 2 from band 3 (the split-off band) in the valence band (Nanavati and Eisencroft, 1968).

If the material from which the tunnel diode is made has unwanted chemical impurities present, it is possible to detect them by an observation of the *I-V* characteristics at low temperatures (liquid nitrogen 77°K). Figure 10–14 shows how the presence of impurities is detected. If a catalog is made of energies at which the band gap states are located, it may be possible to do *tunneling spectroscopy.* Tunneling spectroscopy has been done in materials such as Bi.

It is perhaps in the applications such as the above mentioned that the tunnel diode has its greatest value. It is a device which can be operated over a very much wider range of temperatures than most other semiconductor devices. It is also a device which has very high cut-off frequency. In Ge the *intrinsic* cut-off frequency for tunnel diodes has been estimated at

$$f_{i\,\max} \doteq 10^{12} \text{ cps.} \tag{10–46}$$

The major drawbacks of the device are that it is a two terminal device, the *I-V* characteristics are *not stable,* and the negative resistance region often disappears when operated in a region of voltage significantly higher than V_V. Particularly vulnerable to this degradation are GaAs and GaSb tunnel diodes (Nanavati, 1962, 1968).

10–4 SOLAR CELLS AND PHOTODIODES

We shall look at the recombination-generation process due to thermal causes in some detail in Chapter 12. It is possible to generate electron-hole pairs by applying enough energy to the valence band electrons to excite them directly into the conduction band. If the energy is in the form of light of appropriate frequency, and hence adequate quantum energy $h\nu$, we can have either a *photodiode* or a *solar cell.* A photodiode is a *p-n* junction whose *I-V* characteristics are altered in the presence of light. Figure 10–15 shows the structure of a photodiode which is seen to consist of a *p-n* junction on which external light is focused with the help of a lens. Figure 10–16 shows the structure of a solar cell which is designed to convert solar light energy into electrical energy. Generally

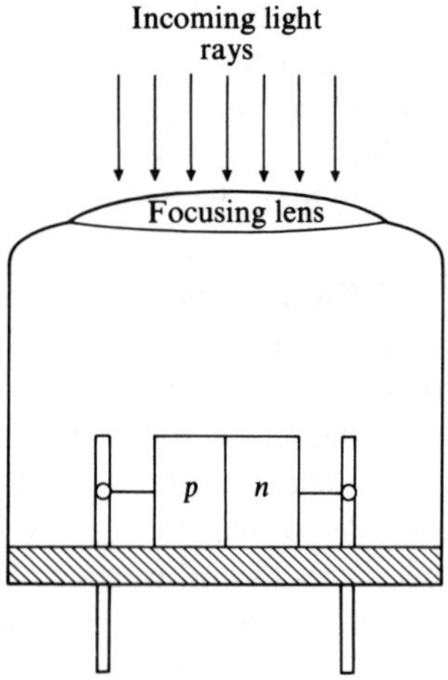

Fig. 10–15 Photo-diode.

the solar cell has a large junction area. The ring ohmic contact permits maximum exposure of the *p*-type material to sunlight.

The general continuity equations including the generation terms G_p and G_n will need to be considered to analyze photo-diodes or solar cells. These equations are repeated below for convenience

$$\nabla^2 p - \frac{p - p_n}{L_p^2} = \frac{1}{D_p}\frac{\partial p}{\partial t} - \frac{G_p}{D_p} \tag{10–47}$$

$$\nabla^2 n - \frac{n - n_p}{L_n^2} = \frac{1}{D_n}\frac{\partial n}{\partial t} - \frac{G_n}{D_n}. \tag{10–48}$$

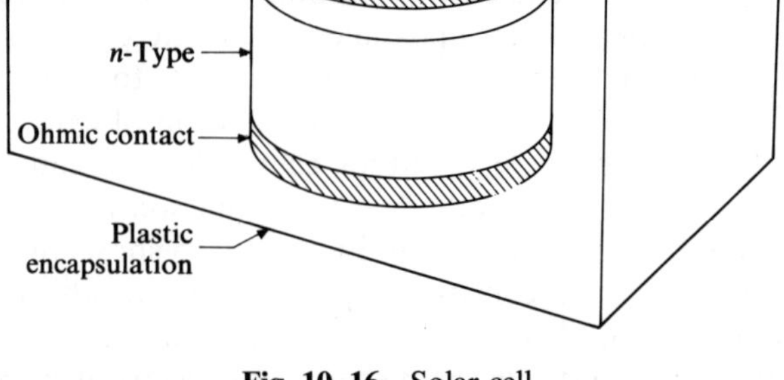

Fig. 10–16 Solar cell.

For the one-dimensional and d-c analysis the above equations become

$$\frac{d^2p}{dx^2} - \frac{p - p_n}{L_p^2} = -\frac{G_p}{D_p} \tag{10–49}$$

$$\frac{d^2n}{dx^2} - \frac{n - n_p}{L_n^2} = -\frac{G_n}{D_n}. \tag{10–50}$$

The following will be the basic assumptions of the analysis.

1. The incoming light, artificial or sunlight, falls uniformly on the semiconductor device.
2. The light generates electron hole pairs at the rate of

$$G_c = G_n = G_p \tag{10–51}$$

per unit volume per second.

3. Generation rate G_c is constant.
4. One-dimensional analysis is adequate.

In solving Eqs. (10–49) and (10–50) we note that with assumption 3 being true the only difference between the solution of these equations will be simply in the *particular* solution. The old particular solution for holes previously found in the d-c case was p_n. We can obtain the *new* particular solution by setting $\partial^2 p/\partial x^2$ equal to zero. This yields on the n-side.

$$p = p_n + G_p\tau_p \qquad \text{particular solution.} \tag{10–50}$$

Similarly on the p-side

$$n = n_p + G_n\tau_n \qquad \text{particular solution.} \tag{10–53}$$

When the particular and the homogeneous solutions are introduced into the transport equations and the current crossing the *junction* is solved for the contributions due to the generation terms *alone*, it will be *similar* to the term which resulted from the original particular terms p_n and n_p. The currents which resulted from particular solutions p_n and n_p are

$$I_p = -\left(\frac{qAD_p}{L_p}\right)p_n \qquad \text{(holes on } n\text{-side)} \tag{10–54}$$

$$I_n = -\left(\frac{qAD_n}{L_n}\right)n_p \qquad \text{(electrons on } p\text{-side).} \tag{10–55}$$

Thus the current contribution due to the generation terms will be as follows.

$$I_{gh} = -\left(\frac{qAD_p}{L_p}\right)G_c\tau_p \qquad \text{(holes on } n\text{-side)} \tag{10–56}$$

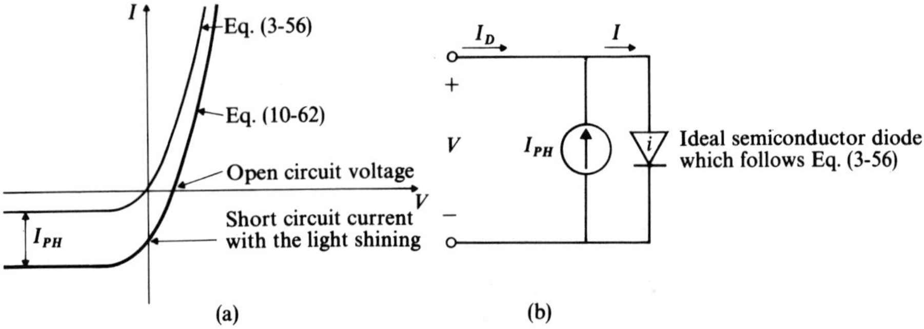

Fig. 10–17 Photo-diode model and characteristics. (a) *I-V* characteristics for photo-diode. (b) Model for the photo-diode.

$$I_{gn} = -\left(\frac{qAD_n}{L_n}\right) G_c\tau_n \qquad \text{(electrons on } p\text{-side).} \tag{10–57}$$

Note that the right hand side of Eqs. (10–56) and (10–57) are obtained by simply substituting the particular solution due to generation of carriers in place of p_n and n_p in Eqs. (10–54) and (10–55). The right hand side of Eqs. (10–56) and (10–57) may be modified to

$$I_{gh} = -qAL_pG_c \tag{10–58}$$

$$I_{gn} = -qAL_nG_c. \tag{10–59}$$

We may now write down the total *I-V* equation for *p-n* junctions on which light is incident by simply adding the contributions due to generation obtained above to the diode Eq. (3–56) as follows

$$I = I_S\,(e^{qV/kT} - 1) - qAG_c(L_p + L_n). \tag{10–60}$$

If we define

$$I_{PH} \triangleq qAG_c(L_p + L_n) \tag{10–61}$$

Equation (10–60) may be rewritten as

$$I_D = I_S(e^{qV/kT} - 1) - I_{PH} \tag{10–62}$$

As is clear from Eq. (10–62) that it is just the ideal diode curves lowered by I_{PH} as shown in Fig. 10–17a. A simple equivalent circuit for the device is shown in Fig. 10–17b. The short circuit current with *light on* is seen to be

$$I_{sc} = I_{PH}. \tag{10–63}$$

The open circuit voltage of either device of Figs. 10–15 and 10–16 is obtained by noticing that the ideal diode current is given by

$$I = I_{PH} = I_S(e^{qV_{oc}/kT} - 1) \tag{10–64}$$

if

$$V_{oc} \gg \frac{kT}{q} \tag{10–65}$$

$$I \doteq I_S(e^{qV_{oc}/kT}). \tag{10–66}$$

Finally,

$$V_{oc} = \frac{kT}{q} \ell n \left(\frac{I_{PH}}{I_S}\right). \tag{10–67}$$

In most photodiodes the photocurrent I_{PH} is proportional to light flux. If the photodiode is used in an application, the circuit needs to discriminate between the condition when the light is on and when the light is off. In such an application the ratio of light current to dark current is the important parameter. This is the ratio

$$I_{PH}/I_S$$

which can be made very large because I_S is extremely small in the present day devices.

10–5 LIGHT EMITTING *p-n* JUNCTIONS AND JUNCTION LASERS

We have seen that the recombination process can take place in semiconductors and can be either of the following two types.

A. Direct recombination of a conduction band electron with a valence band hole.

B. Recombination through the intermediary of the band gap trap or the recombination center.

When an electron-hole pair recombines, an energy roughly equal to E_g is lost. This energy can either appear as heat energy in the crystal or as light energy by way of photons being produced. In either case the energy E_g appears in some other form. It turns out that the recombination process *B* usually leads to the energy (being heat energy) being given up to the crystal. In the recombination process *A* the energy is given off in the form of photons of energy E_g.

We have seen in the diffusion theory of the *p-n* junction diode that recombination is an important semiconductor process and that the flow of current in the device is accompanied by recombination. In all real devices both the recombination processes *A* and *B* will be present simultaneously. We know that in *indirect materials* like Ge and Si, process *B* predominates. Hence these are not expected to be good light emitting diodes. In direct materials such as gallium arsenide and gallium phosphide (GaP) the radiative recombination process *A* predominates. Junction diodes made from such materials will give off light when they are forward biased and the recombination process proceeds steadily. These light emitting diodes radiate light at a frequency approximately

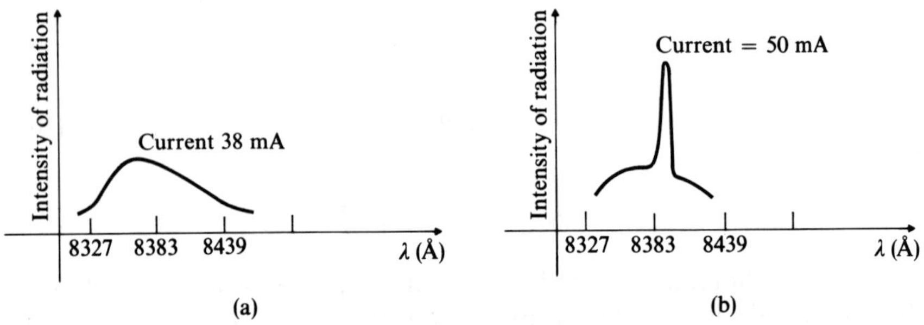

Fig. 10–18 Emission spectrum of a GaAs light emitting *p-n* junction at a temperature of 2°K (after W. E. Howard et al). (a) Light emitting diode spectrum (incoherent light). (b) Junction laser spectrum (coherent light).

centered at a photon energy of E_g as follows

$$E_g = h\nu \tag{10–68}$$

or

$$\nu = \frac{E_g}{h}. \tag{10–69}$$

The emission spectrum from GaAs is shown in Fig. 10–18a. The reasons why radiation slightly above and below that predicted by Eq. (10–69) is also observed are the following.

a) Acceptors and/or donors may participate in the radiative recombination process. These would contribute radiation at a frequency *less* than that of Eq. (10–69).

b) The carriers that participate in the radiative recombination process can have energy of the order of kT. These then will contribute photons of energy greater than E_g and hence they will radiate at a frequency *higher than* that given by Eq. (10–69).

These radiative recombinations are more efficient at lower temperatures and become less efficient as the temperature increases. Only a few diodes give any appreciable radiation at room temperatures. The light emitted in a spectrum such as that of Fig. 10–18a is *incoherent*. To define the term, consider the electric field vector of the following mathematical form

$$\mathscr{E}(t) = E \sin(2\pi\nu t + \phi). \tag{10–70}$$

If E and ϕ are *random functions of time* the corresponding radiation is said to be *incoherent* radiation. If E and ϕ are *constants*, the radiation is said to be *coherent*. It is clear from Eq. (10–70) that in coherent radiation the light intensity emitted along a given direction is predictable at any future time if its value is known at any given time. This is a very important and useful property in many applications such as communications systems.

Coherent light emitting *p-n* junctions are called *junction lasers.* In Fig. 10–18b is shown a GaAs junction laser which is the same diode as used in Fig. 10–18a. One difference is that this diode is operated at a *higher current level.* It turns out that a current exceeding a certain *threshold value* for a given device must be used before laser action is observed. A considerable sharpening in the emission spectrum is observed as seen in Fig. 10–18b. The reason for the change in the spectrum is because that under certain conditions the probability of a photon producing recombination depends on the presence of another photon of the same energy near the site of recombination. Under conditions where a large number of carriers are available for recombination and the diode surfaces are shaped precisely and polished and treated so as to retain many of the photons within the device, then photon assisted *stimulated emission* occurs. Useful coherent light is produced in such devices. These devices are called *injection lasers* or simply *junction* lasers.

Laser and light emitting *p-n* junctions can be formed by many fabrication methods. Most junctions including junction lasers are produced by the *diffusion method,* discussed in detail in Chapter 11. For example, GaAs laser can be formed by diffusing Zn acceptor atoms into an *n*-type substrate. Doping densities between about 5×10^{17} to 5×10^{19} cm^{-3} are necessary for junctions laser. Junction depth can be controlled by judicious choice of diffusion time and temperature.

Two other methods of fabrication have also been used. One is the *vapor regrowth method.* Here a vapor having the desired doping agent is passed through the substrate. Under appropriate temperature conditions epitaxial regrowth takes place to obtain the *p-n* junction. The other is the *solution regrowth method.* If we desired a GaAs *p-n* junction, we would prepare a solution of GaAs together with doping agents dissolved in it, for example Ga at or near its saturation temperature. This solution is permitted to set on a GaAs substrate of opposite doping. Subsequent cooling produces an epitaxial growth on the substrate, and a *p-n* junction is thus formed. Table 10–1 lists many of the semiconducing materials out of which lasers and light emitting diodes (LEDs) have been made along with the wavelength of the emitted light as well as photon energy. Note that in alloys such as $Ga(As_xP_{1-x})$ it is possible to get a continuum of photon energy and wavelength by a judicious variation in the alloy composition. Thus it is possible to fabricate a device which will yield a desired wavelength of light for a given application within certain limits.

It should be mentioned that in the last few years the development of LED has gained versatility. The cost per device has diminished dramatically while at the same time reliability has increased by several orders of magnitude. New materials are being discovered at an amazing rate. The compatibility of these devices with the integrated circuit technology holds out a promise of very great

Table 10–1 Semiconductor Laser and LED Materials

Material	Wavelength (microns)	Photon energy (eV)
GaAs	0.84	1.47
InP	0.90	1.37
GaSb	1.50	0.82
InSb	5.2	0.23
PbS	4.26	0.29
PbTe	6.5	0.19
PbSe	8.5	0.145
†GaP (Zn, 0-doped red)	0.698	1.76
†GaP (Zinc diffused green)	0.553	2.24
†$Ga_{1-x}Al_xAs$ (red) (band only)	0.688	1.78
$Ga(As_xP_{1-x})$	1.41–1.95	0.88–0.63

†LED only.

growth of the use of LED in the years to come. Pocket electronic calculators are a prime example of such an application.

10–6 SCHOTTKY BARRIER DIODES AND SOME ASSOCIATED DEVICES

We consider in this section rectifying metal-semiconductor contacts. These contacts are to be contrasted with the ohmic contacts, considered in the earlier chapters, as the latter are nonrectifying. Metal-semiconductor rectifying contacts were first studied in a systematic way by Braun (1874) who noticed the dependence of the total resistance of such a system on the polarity of the voltage applied to the system.

As a preparatory step to understanding metal-semiconductor contacts, we shall first examine the metal-vacuum system. The energy diagram of such a system is shown in Fig. 10–19. The energy necessary for an electron located at the Fermi level in a metal to escape into the vacuum is defined to be the *work function*. With no external fields applied, the energy needed by an electron at E_f to escape into the vacuum is seen, from Fig. 10–19, to be $q\phi_m$, where ϕ_m is in Volts. The value of work functions for metals is in the range 2 to 6 eV, and is quite sensitive to surface contamination. Values of work functions for clean surfaces of a few metals are given in Table 10–2.

Consider that a negative charge $-Q$ is located a distance x from the surface of a metal. This negative charge will induce a positive charge on the surface of the metal. The force of attraction between the negative charge $-Q$ outside the metal and the induced positive charge on the metal surface *as felt by the negative charge* remains unaltered if we replace the induced positive charge with a single

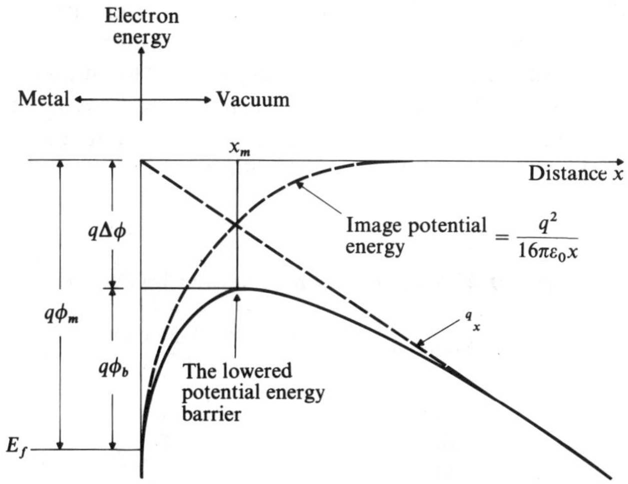

Fig. 10–19 The energy band picture for the Schottky effect, lowered energy barrier, in the metal-vacuum system.

point charge $+Q$ located a distance $-x$ from the surface of the metal opposite the charge $-Q$. The positive charge $+Q$ is called *image charge*. The attractive force is also referred to as *image force* and is given by

$$F = -\frac{Q^2}{16\pi\varepsilon_0 x^2}. \tag{10–70}$$

Eq. (10–70) is quite general in that it applies for any point charge $-Q$ at x. In particular if the charge happens to be an electron, Eq. (10–70) becomes

$$F = -\frac{q^2}{16\pi\varepsilon_0 x^2}, \tag{10–71}$$

where ε_0 = permittivity of free space,

q = magnitude of the electronic charge,

x = distance of the electron charge from the metal surface.

If an external electric field $\mathscr{E}$ is applied, the total potential energy PE as a function of distance is given by the following in electron-Volts

$$PE(x) = \frac{q^2}{16\pi\varepsilon_0 x} + q\mathscr{E}x. \tag{10–72}$$

The first term on the right hand side of Eq. (10–72) is the contribution due to the image charge, and the second term is the contribution due to the external electric field. Equation (10–72) is shown plotted as a solid curve on Fig. 10–19. We note that the potential energy barrier which an electron would need to surmount, in order to escape, is lowered from $q\phi_m$ to $q\phi_b$. The amount of lowering is referred to as *Schottky barrier lowering* or *image force lowering*. The lowering by $\Delta\phi$ of the metal work function because of both the image force and the external electric field is called the *Schottky effect*.

Table 10–2 Work Function Values for Clean Surfaces

Material	Work function (eV)
Ge	4.4
Si	4.8
GaAs	4.7
A	4.25
Na	2.3
Cr	4.6
Ni	4.5
Cu	4.4
Sn	4.3
Sb	4.1
Au	4.8
Ag	4.3

The location x_m and the amount of Schottky barrier lowering $\Delta\phi$ may be calculated from the requirement given below

$$\frac{d(PE)}{dx} \doteq 0. \tag{10–73}$$

The results are given as follows

$$x_m = \sqrt{\frac{q}{16\pi\varepsilon_0\mathscr{E}}} \quad \text{cm} \tag{10–74}$$

$$\Delta\phi = \sqrt{\frac{q\mathscr{E}}{4\pi\varepsilon_0}} = 2x_m\mathscr{E} \text{ Volts.} \tag{10–75}$$

The above results obtained for the metal-vacuum system apply, with relatively minor changes also, to the metal-semiconductor system. Table 10–3 shows transformations from the metal-vacuum system to the metal-semiconductor system relationship.

Table 10–3 Transformation Relationship Between thc Metal-Vacuum System and the Metal-Semiconductor System.

Metal-vacuum system	Metal-semiconductor system
ε_0 ⟷	ε_s
$\mathscr{E}$ ⟷	$\mathscr{E}_{\max}$ (maximum field at the metal–semiconductor interface)

Equations corresponding to Eqs. (10–74) and (10–75) are the following

$$x_m(m\text{-}s) = \sqrt{\frac{q}{16\pi\varepsilon_s\mathscr{E}_{\max}}} \tag{10–76}$$

$$\Delta\phi(m\text{-}s) = \sqrt{\frac{q\mathscr{E}_{\max}}{4\pi\varepsilon_s}} = 2x_m\mathscr{E}_{\max}. \tag{10–77}$$

Example 10–1. Calculate the Schottky barrier lowering and its location for the metal-silicon system when $\mathscr{E}_{\max} = 10^5$ Volts/cm.

Solution. Schottky barrier lowering

$$\Delta\phi = \sqrt{\frac{q\mathscr{E}_{\max}}{4\pi 11.8\varepsilon_0}} = \sqrt{\frac{1.6 \times 10^{-11} \times 10^5}{4\pi(11.8)8.854 \times 10^{-14}}} = 0.036 \text{ Volts.} \tag{1}$$

Equation (10–77) may now be used to obtain x_m

$$x_m = \frac{\Delta\phi}{2\mathscr{E}_{\max}} = \frac{0.036}{2 \times 10^5} = 18 \text{ Å}. \tag{2}$$

Although the barrier lowering is quite small it has a profound influence on the transport of carriers in the device.

In Fig. 10–20a is shown a metal and an n-type semiconductor *not* in thermal equilibrium. The energy needed to free an electron located at the bottom of the conduction band is $q\chi$ and is called *electron affinity.*

If a wire is connected between the metal and the n-type semiconductor, charge will flow in this case from the semiconductor to the metal until equilibrium is established when the *Fermi levels line up*, as discussed in Chapter 3, in the two materials as shown in Fig. 10–20b. As the distance w_g is diminished to the vanishing point shown in Fig. 10–20c, the barrier height which the electrons in the metal would need to surmount to reach the semiconductor is $q\phi_{bn}$. The magnitude of this barrier can be calculated by using the requirement that the Fermi levels in the two materials must line up. Thus from Figs. 10–20a and 10–20b we may conclude

$$q\chi + q\phi_{bn} = q\phi_m. \tag{10–78}$$

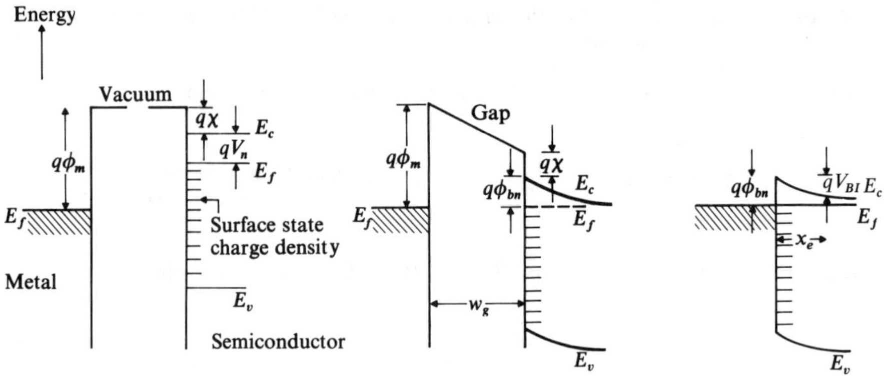

Fig. 10–20 Energy band diagram for metal-n-type semiconductor contact at thermal equilibrium.

The barrier height neglecting the Schottky lowering may now be solved for

$$q\phi_{bn} = q(\phi_m - \chi). \tag{10–79}$$

In a metal-p-type semiconductor contact, the barrier height is given by

$$q\phi_{bp} = E_g - q(\phi_m - \chi). \tag{10–80}$$

An intriguing result is obtained if Eqs. (10–79) and (10–80) are added together

$$q(\phi_{bn} + \phi_{bp}) = E_g. \tag{10–81}$$

Equation (10–81) says that for a given semiconductor the sum of the barrier heights of n-type and p-type substrates is the band gap energy of the semiconductor for any metals used.

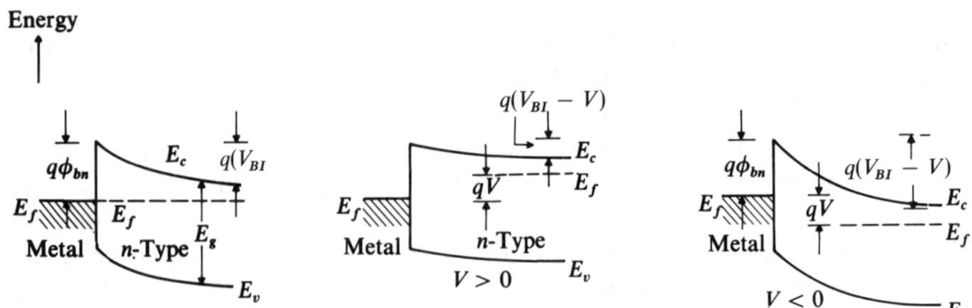

Fig. 10–21 Energy band diagram for metal–n-type semiconductor contact with external voltage applied.

When the n-type semiconductor–metal system is in thermal equilibrium, the Fermi level of the semiconductor drops in relation to that of the metal by the contact potential difference of

$$\text{Contact potential} = q\phi_m + q(\chi V_n). \tag{10–82}$$

As a result of the above an electric field will exist in the gap. We shall assume that the density of surface states shown in Fig. 10–20 is very high. So that they may accept additional surface charges as the gap width w_g is diminished without changing the Fermi level E_f remains at essentially the same level as before. Under these circumstances the barrier height is determined by the semiconductor surface and *not* by the metal work function.

We shall now obtain the current–voltage relationship governing Schottky barrier diodes by the use of Schottky's isothermal diffusion theory. Fig. 10–21 shows the band diagram for Schottky barrier diode under nonthermal equilibrium condition. The following assumptions will be used.

1. Barrier height is significantly greater than kT.
2. Carrier densities at $x = 0$, the interface, and at $x = x_t$ have their thermal equilibrium values and are not influenced by current flow.
3. The semiconductor is non-degenerate, i.e. the Fermi level is within the band gap.
4. Only one-dimensional current flow will be considered. The influence of electron collisions has not been neglected.

We shall begin with the transport equation for electrons, and use it first in the transition region.

$$J_x = J_n = qn(x)\mu_n\mathscr{E} + qD_n\frac{\partial n}{\partial x} \tag{10–83}$$

$$J_n = qD_n\left[qn(x)\frac{\mu_n}{D_n}\left(-\frac{\partial V(x)}{\partial x}\right) + \frac{\partial n}{\partial x}\right]. \tag{10–84}$$

But Einstein relationship permits us to write

$$\frac{\mu_n}{D_n} = \frac{q}{kT}. \tag{10–85}$$

Equation (10–85) may be substituted into Eq. (10–84) to obtain

$$J_n = qD_n\left[-\frac{qn(x)}{kT}\left(\frac{\partial V(x)}{\partial x}\right) + \frac{\partial n}{\partial x}\right]. \tag{10–86}$$

In steady state, the current density is independent of x. Equation (10–86) may be integrated with the help of the integrating factor $\exp[-qV(x)/kT]$. With J_n

treated as a constant with respect to x. The following boundary conditions may be used

$$qV(0) = -q(V_n + V_{BI}) = -q\phi_{bn} \tag{10–87}$$

$$qV(w_g) = -qV_n + qV \tag{10–88}$$

$$n(0) = N_c\, e^{-(E_c(0)-E_f)/(kT)} = N_c\, e^{-q\phi_{bn}/(kT)} \tag{10–89}$$

$$n(x_t) = N_c\, e^{qV_n/(kT)}, \tag{10–90}$$

where the origin of the x axis is at the interface of the semiconductor and the metal. The transport equation can be integrated subject to the above boundary conditions and with the aid of the integrating factor and the solution of Poisson's equation in the transition region to obtain the following result.

$$J_n = \left[\frac{q^2 D_n N_c}{kT}\left(\frac{q(V_{BI} - V)8\pi N_D}{\varepsilon_s}\right)^{\frac{1}{2}} e^{-q\phi_{bn}/(kT)}\right](e^{qV/(kT)} - 1) \tag{10–91}$$

or

$$I_n = J_{SS}(e^{qV/(kT)} - 1), \tag{10–92}$$

where J_n and I_n = current density and current respectively in metal-n-semiconductor Schottky barrier junctions, and where the bulk saturation current is given by

$$I_{SS} = \frac{q^2 D_n N_c}{kT}\left[\frac{q(V_{BI} - V)8\pi N_D}{\varepsilon_s}\right]^{\frac{1}{2}} e^{-q\phi_{bn}/(kT)}. \tag{10–93}$$

Equation (10–91) is valid for both forward bias $V > 0$ and for reverse bias $V < 0$. Equation (10–93) is not valid for large forward voltage. The ideal junction diode Eq. (3–56) and Eq. (10–93) are of the same form. Unlike the back saturation current I_S of Eq. (3–56), which is independent of voltage, Eq. (10–93) has an explicit voltage dependence which may be of some significance particularly under reverse bias conditions.

From the above development we may conclude that the Schottky barrier diode is a *majority carrier device.* Thus the limitations in speed of response due to minority carrier phenomena in ordinary *p-n* junctions and junction transistors are not present in Schottky barrier diodes. These devices are thus expected to have faster response times. In Fig. 10–22 we see a composite transistor which utilizes a Schottky diode between the collector and base terminals when the transistor is saturated, both the emitter and collector junctions are forward biased. If the forward voltage in the barrier diode has a smaller value than the forward biased collector junction, a vast majority of the *excess base* current will flow through the Schottky barrier diode in which the minority carrier storage effects are not present. For this reason the storage time in the composite transistor can be as little as 10 per cent that of the same transistor *without* the Schottky barrier diode. In experimental devices storage times less than a nanosecond have been measured.

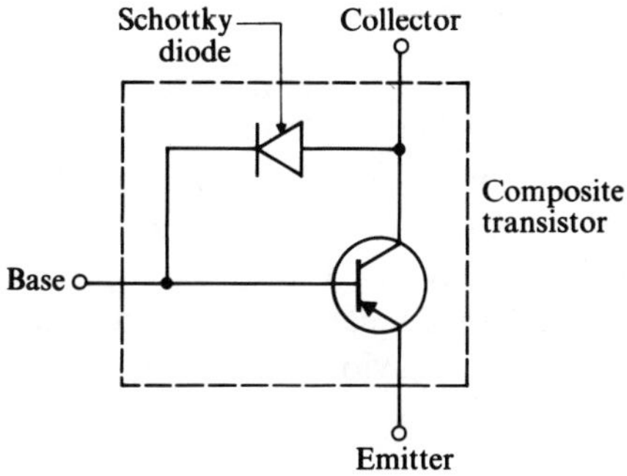

Fig. 10–22 A composite transistor employing a Schottky barrier diode between the collector and base terminals.

The Schottky barrier gate field effect transistor has the extremely simple geometry shown in Fig. 10–23 and permits the fabrication of the device at a significantly lower temperature than in the regular field effect transistor. The drain characteristics of these devices are quite similar to those of the regular field effect transistors.

10–7 COMPUTER AIDED DESIGN

The computer has revolutionized the world. Circuit analysis and device fabrication are not immune to this onslaught of the computer. Computer-aided circuit analysis and device modeling are being resorted to with greater frequency. A much more detailed and accurate analysis thus becomes possible. Computer-aided analysis and design has been used to study MOS integrated circuits, large signal MOS transistor models, bipolar transistor circuits, and a nonlinear JFET model. See Dov Frohman-Bentchkowsky (1969), Merckel et al (1972), McCalla and Howard (1971), Biehl (1971), and Calahan (1968). The last named source contains extensive bibliography on the subject. See the references at the end of Chapter 11 for computer aided fabrication of semiconductor devices.

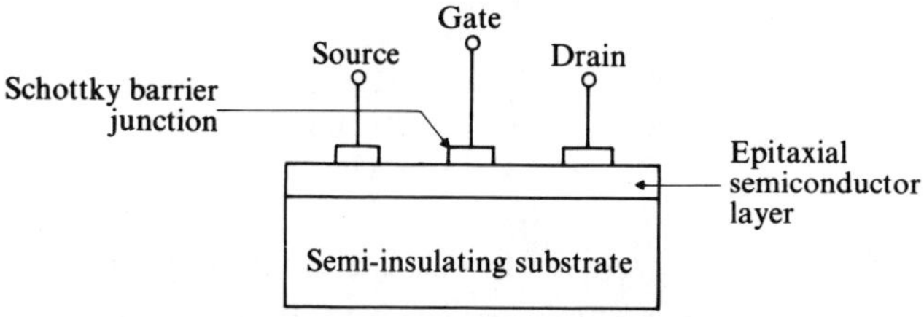

Fig. 10–23 Schottky barrier gate field effect transistor.

10–8 SUMMARY

In this chapter we examined a number of new devices which could be analyzed by the theory already presented in previous chapters. Many more devices could have been considered in a similar way. The purpose of this chapter was to show the generality of the theories presented so far, and to show how they could be applied to some present day devices. These should provide the reader with the confidence that as a new device appears he may be able to utilize the knowledge gained here to understand its behavior and to model it.

PROBLEMS

10– 1 Derive Eq. (10–13).

10– 2 Derive Eq. (10–21).

10– 3 Derive Eq. (10–22).

10– 4 Discuss how you would go about choosing the conductivity of the *n*-type base of the UJT so as to minimize the temperature variation of I_P, V_P, and I_V.

10– 5 The following information is available on a germanium tunnel diode. $E_{fn} = 0.050$ eV, $E_{fp} = 0.010$ eV, and the direct conduction band valley is 0.13 eV above the lowest conduction band indirect valley. Calculate (a) peak voltage, (b) valley voltage, (c) the voltage at which the current should show a change in $I(V)$ plot slope in the reverse current.

10– 6 Justify the statement "A semiconductor having a band gap E_g is transparent to incident radiation of $\nu < E_g/h$."

10– 7 Why are light emitting diodes made from direct gap semiconductors rather than indirect band gap semiconductors?

10– 8 Discuss how the excess current component can be experimentally separated from the normal injection current.

10– 9 You are given a tunnel diode for which the following constants are known $E_{fn} =$ 0.05 eV, $E_{fp} = 0.025$ eV, $E_g = 0.5$ eV. The *I-V* plot shows a kink due to impurities on the *p*-side (as shown in Fig. 10–14b) at $V = 0.35$ Volt. Calculate where the impurity band is located within the band gap on the *p*-side.

10–10 Consider a photo-diode the following of whose parameters are known

$$\mu_n = 2000 \text{ cm}^2/\text{Volt-sec.}$$
$$\mu_p = 1000 \text{ cm}^2/\text{Volt-sec.}$$
$$n_i^2 = 10^{24} \text{ cm}^{-6}$$
$$\sigma_p = 0.1\ (\Omega\text{-cm})^{-1}$$
$$\sigma_n = 0.2\ (\Omega\text{-cm})^{-1}$$
$$A = 10^{-4} \text{ cm}$$
$$I_{Ph} = 10 \text{ mA}$$

Calculate the open circuit voltage when the light is "on" to produce 10 mA.

10–11 Draw energy band diagrams, similar to Fig. 10–21, for the metal–*p*-type semiconductor junction.

REFERENCES

Biehl, B. L., "Nonlinear JFET Model for Computer-Aided Circuit Analysis," *IEEE J.* of Solid State Circuits Book Co., **SC-6,** No. 1, 55 (1971).

Bluhn, V. A., and Sylvan, T. P., "A High Performance Unijunction Transistor Using Conductivity Modulation of Spreading Resistance," *Solid State Design,* **5,** 26–31 (1964).

Braun, F., "Uber die Stromleitung durch Schwefelmetalle," *Ann. Physik Chem.,* **153,** 556 (1874).

Calahan, D. A., *Computer-Aided Network Design,* McGraw-Hill Book Co., New York (1968).

Chynoweth, A. G., et al, "Excess Tunnel Currents in Si Esaki Junctions," *Phys. Rev.,* **121,** No. 3, 684–694 (1961).

Clark, L. E., "Now New Unijunction Geometrics," *Electronics,* **38,** 93–97 (1965).

Clark, L. E., "Unijunction Transistor," Patent 3,325,705 (1967).

Esaki, L., "New Phenomenon in Narrow Ge *p-n* Junctions," *Phys. Rev.,* **109,** 603–604 (1958).

Franz, W., "Handblick der Physik," Vol. 17, p. 155 in S. Flugge (Ed.), Springer-Verlag OHG, Berlin (1956).

Frohman-Bentchkowsky, D., "Computer-Aided Design and Characterization of Digital MOS Integrated Circuits," *IEEE J. of Solid State Circuits,* **SC-4,** No. 2 (1969).

Hall, R. N., et al, "Direct Observation of Polorons and Phonons During Tunneling in Group 3–5 Semiconductor Junctions," *Phys. Rev. Letters,* **4,** No. 9, 456–458 (1960).

Holonyak, N. Jr., et al, "Direct Observation of Phonons during tunneling in Narrow Junction Diodes," *Phys. Rev. Letters,* **3,** No. 4, 167–168 (1959).

Kane, E. D., "Theory of Tunneling," *J. Appl. Phys.,* **32,** No. 1, 83–92, Jan. (1961).

Less, I. A., and Mathis, V. P., "The Double-Base Diode—A New Semiconductor Device," *IRE Conv. Rec.,* **Part 6,** 2 (1953).

McCalla, W. J., and Howard, W. G., "BIAS-3-A Program for the Nonlinear DC Analysis of Bipolar Transistor Circuits," *IEEE J. of Solid State Circuits,* **SC-6,** No. 1, 14 (1971).

Merckel, G. et al, "Am Accurate Large-Signal MOS Transistor Model for Use in Computer-Aided Design," *IEEE Trans. on Electron Devices,* **ED-19,** No. 5, 681 (1972).

Morgan, J. V., and Kane, E. O., "Observation of Direct Tunneling in Ge," *Phys. Rev. Letters,* **3,** No. 10, 466–468 (1959).

Nanavati, R. P., "Some Important Mechanisms Contributing to Tunnel Diode Failure," *Proc. of Symposium on the Physics of Failure in Electronics,* 214–223, Sept. (1962).

Nanavati, R. P., *An Introduction to Semiconductor Electronics,* Chapter 12, McGraw-Hill Book Co., New York (1963).

Nanavati, R. P., "GaSb Tunnel Diode Degredation and Lattice Defects," *Lattice Defects in Semiconductors,* Edited by R. R. Hasigrite, Published by *Univ. of Tokyo Press and Pennsylvania State Univ. Press,* 252–262 (1968).

Nanavati, R. P., and deAndrade, A. M., "Excess Current in GaAs Tunnel Diodes," *IRE Proc.,* **52,** No. 7, July (1964).

Nanavati, R. P., and Eisencraft, M., "Reverse Characteristic of GaSb Tunnel Diodes," *Proc. of the Int'l Conf. on the Physics of Semiconductors,* Journal of the Physical Societies of Japan, Vol. 21, Supplement, 603–608 (1966).

Nanavati, R. P., and Eisencraft, M., "On Thermal and Excess Currents in GaSb Tunnel Diodes," *IEEE Trans. on Electron Devices,* Oct. (1968).

Nelson, H., "Epitaxial Growth from the Liquid State and Its Application to the Fabrication of Tunnel and Laser Diodes," *RCA Rev.,* **24,** 603 (1963).

Pankove, J. L., "Optical Absorption by Degenerate Germanium," *Phys. Rev. Letters,* **4,** No. 9, 454–455 (1960).

Shockley, W., Pearson, G. L., and Haynes, J. R., "Hole Injection in Germanium-Quantitative Studies and Filamentary Transistors," *Bell System Tech. J.,* **28,** 344 (1949).

Sylvan, T. P., "The Unijuction Transistor Characteristics and Applications," *Application Note,* Semiconductor Products Department, General Electric Co., May (1965).

Trofimenkoff, F. M., and Huffs, G. J., "D-c Theory of the Unijunction Transistor," *Int. J. Electronics,* **20,** 217–225 (1966).

Winogradoff, N. N., and Kessler, H. K., "Light Emission and Electrical Characteristics of Epitaxial GaAs Lasers and Tunnel Diodes," *Solid State Commun.,* **2,** 119 (1964).

CHAPTER 11

STRUCTURE AND FABRICATION OF DISCRETE SEMICONDUCTOR DEVICES AND INTEGRATED CIRCUITS

11-1 INTRODUCTION

So far we have analyzed semiconductor devices and the internal mechanisms which govern their terminal performance. We have, however, already seen, for example in the case of junction capacitance, that the method of fabricating the junction influences the mathematical form of its dependence on total junction voltage V_t. The method of fabrication in turn depends on the semiconductor technology which is utilized. The structure of devices is also strongly influenced by the methods employed in their fabrication.

It is therefore imperative that we pay attention to both the semiconductor technology as well as to the methods of fabricating devices. Dramatic and rapid changes have occurred in semiconductor technology in a relatively short time. It is precisely because of the continuing rapid changes in the technology that we shall concentrate on only the *fundamental aspects of technology* which will stand us in good stead for a longer length of time, and not on the details of technology which change from month to month and which are zealously guarded secrets of the manufacturing corporations.

The dominant semiconductor material of the 1950s was germanium. Silicon rapidly eclipsed its use in the 1960s. While some other material, possibly GaAs, may replace Germanium and silicon in the 1970s we shall consider silicon in this chapter because of its widespread use and because the fundamental aspects which will be discussed will be probably applicable to any material (with some changes) which may emerge in the future. The group of techniques in widespread use relating to silicon device fabrication is called *planar diffusion technology*. With the help of this technology it has been found possible to produce on a single previously homogenous chip of silicon crystal, resistances, capacitances, *p-n* junctions, bipolar and monopolar MOSFET transistors, as well as means of interconnecting the various regions of the silicon chip. This technology also provides the means with which to protect the surface of the

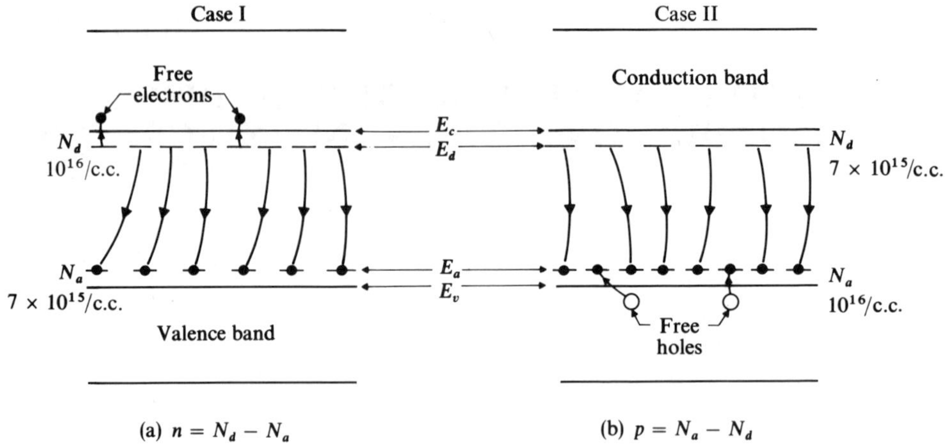

Fig. 11–1 Illustrative Example 11–1.

device with a protective layer of insulating oxide which helps to stabilize the semiconductor device characteristics as well as to materially reduce leakage currents.

We shall now examine the major techniques which constitute the planar diffusion technology.

11–2 SOLID STATE DIFFUSION AND COMPENSATION

We saw in Example 2–4 that the free carrier density and the conductivity in a semiconductor crystal depend on the *difference* between the donor density N_d and the acceptor density N_a, and not on N_d or N_a. To clarify this further consider the simple band picture of silicon in which the donor is phosphorous and the acceptor is boron. Consider two cases:

$$\text{Case I} \quad N_d = 10^{16}/cc, \quad N_a = 7 \times 10^{15}/cc$$
$$\text{Case II} \quad N_d = 7 \times 10^{15}/cc, \quad N_a = 10^{16}/cc$$

The acceptor energy level is very close to the top of the valence band and the donor energy level is very close to the bottom of the conduction band. Figure 11–1a and 11–1b show these two cases schematically. In case I we can see that the electrons which occupy the donor states will prefer to be in the available and empty acceptor states rather than go into the conduction band. Once all acceptor states are occupied, the remaining electrons in the donor states will easily go into the conduction band with the help of thermal energy, for example at room temperature. As all the acceptor states are now occupied, electrons in the valence band have no place to go because the thermal energy that they have at room temperature is too small to overcome the band gap E_g. Hence the free electron density is given in case I by

$$n = N_d - N_a = 10^{16} - 7 \times 10^{15} = 3 \times 10^{15} \text{ c.c.}; \quad N_d > N_a \quad (11\text{–}1)$$

and the material will behave like an n-type semiconductor. In Case II shown schematically in Fig. 11–1b again the electrons from the donor state will preferentially fall into the empty and available acceptor states. As virtually all the donor state electrons find room in the acceptor states in this case (because $N_a > N_d$) there are no donor state electrons in the conduction band. The valence band electrons can and will occupy the remaining acceptor states not already occupied by the electrons which came from the donor states. Free holes created in the valence band will hence be

$$p = N_a - N_d = 10^{16} - 7 \times 10^{15} = 3 \times 10^{15} \text{ c.c.}; \quad N_a > N_d. \quad (11\text{–}2)$$

The material will behave like a p-type semiconductor.

We see then that it would seem possible that a material which was initially n-type may be converted to p-type by simply doing it with a sufficiently high density of acceptors. Similarly an initially p-type crystal may be converted to an n-type crystal by doping it with a sufficiently high density of donors.

This method of changing the conductivity type of a crystal by means of introducing impurities into the crystal is called *compensation*. Compensation is of fundamental importance to the planar diffusion technology.

We have seen in previous chapters that a very narrow base width is desirable for good high frequency response of transistors. It is necessary to produce very narrow widths in a precisely controlled fashion. A process which meets this objective very well is *solid state diffusion*. This process allows the introduction of chemical impurities into the crystal lattice. *Solid solubility* determines the maximum density of chemical impurities which can be introduced into the crystal. Practically the range of maximum densities of chemical donors or acceptors which can be introduced into germanium or silicon is between 10^{18} to 10^{20} per cc. The impurities introduced in this manner into the crystal tend to enter the lattice *substitutionally*, i.e. they occupy lattice sites normally occupied by the host of atoms.

The motion of impurities *within* the crystal is by solid state diffusion. The equation which governs this process is essentially the same as the diffusion equation for minority carriers except that the recombination terms will be missing. The relevant *one-dimensional* diffusion equation for solid state diffusion is:

$$\frac{\partial^2 C(x, t)}{\partial x^2} = \left(\frac{1}{D}\right) \frac{\partial C(x, t)}{\partial t}, \quad (11.3)$$

where D = solid state diffusion coefficient,

C = volume density of chemical impurities (donors or acceptors).

The solution to Eq. (11–3) will depend on the boundary conditions. Two major sets of boundary conditions are worth considering.

Case A. Constant Surface Concentration

This circumstance is obtained as follows. The semiconductor to be doped is placed in a vapor containing a compound one of whose constituents is the doping impurity atoms. The compound is such that it reacts with the semiconductor surface resulting in the release of the doping atoms establishing the surface concentration C_s. The vapor can act as an "infinite" source of the doping atoms as the vapor goes past the surface. Under these circumstances a *constant surface* concentration C_s can result.

Thus,
$$C(0, t) = C_s. \tag{11–4}$$

The other boundary conditions is
$$C(\infty, t) = 0. \tag{11–5}$$

The solution to Eq. (11–3) subject to the above boundary condition is
$$C(x, t) = C_s \operatorname{erfc} \{x/[2(Dt)^{\frac{1}{2}}]\}. \qquad (11\text{–}6)\dagger$$

This is the *complementary error function* solution.

Case B. This is the *limited source* condition. In this case at temperatures lower than needed for solid state diffusion, the doping impurity deposited on the surface of the semiconductor. These impurity atoms will diffuse into the crystal when the temperature is raised sufficiently. In this case one of the boundary conditions is that the total number of impurity atoms during solid state diffusion is a constant S_{id}. The other boundary condition is the same as Eq. (11–5). The solution to Eq. (11–3) under these boundary conditions is the following:
$$C(x, t) = [S_{id}/(\pi Dt)^{\frac{1}{2}}]\, e^{-x^2/(4Dt)}. \tag{11–7}$$

The right hand side of Eq. (11–7) is called the *Gaussian function.* The Gaussian function and the complementary error function are shown plotted on a dimensionless set of coordinates in Fig. 11–2. For a fairly large number of impurities the doping profiles resulting from solid state diffusion are well approximated by either Eq. (11–6) or (11–7). There are, however, a number of impurities whose

† The *error function* is defined as
$$\operatorname{erf}(x) \triangleq \frac{2}{\sqrt{\pi}} \int_0^x e^{-a^2}\, da. \tag{11–6a}$$

The *error function complement* is defined by
$$\operatorname{erfc}(x) = 1 - \operatorname{erf}(x). \tag{11–6b}$$

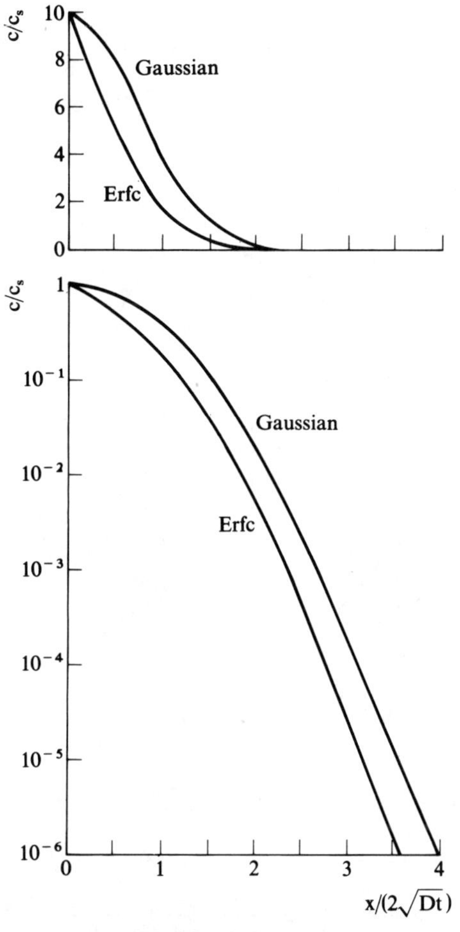

Fig. 11–2 Gaussian and complimentary error-function plots for normalized concentration vs. normalized distance. (After Carslaw and Jaeger[3].)

diffusion profiles are more complicated than those given by Eqs. (11–6) or (11–7).

The diffusion coefficient D for solid state diffusion is a strong function of temperature as can be seen in Fig. 11–3. It can be seen from this figure that in the case of silicon, if the crystal is heated to 1000°C or higher that the desired impurity profile can be obtained by the variation of *temperature* and the *length of time* for which a given temperature is maintained at a given value. It is to be noted that the diffusion constant increases as the temperature is increased, and decreases as the temperature is decreased. Thus, the impurity distribution can be "frozen in" by reducing the temperature below that needed for solid state diffusion.

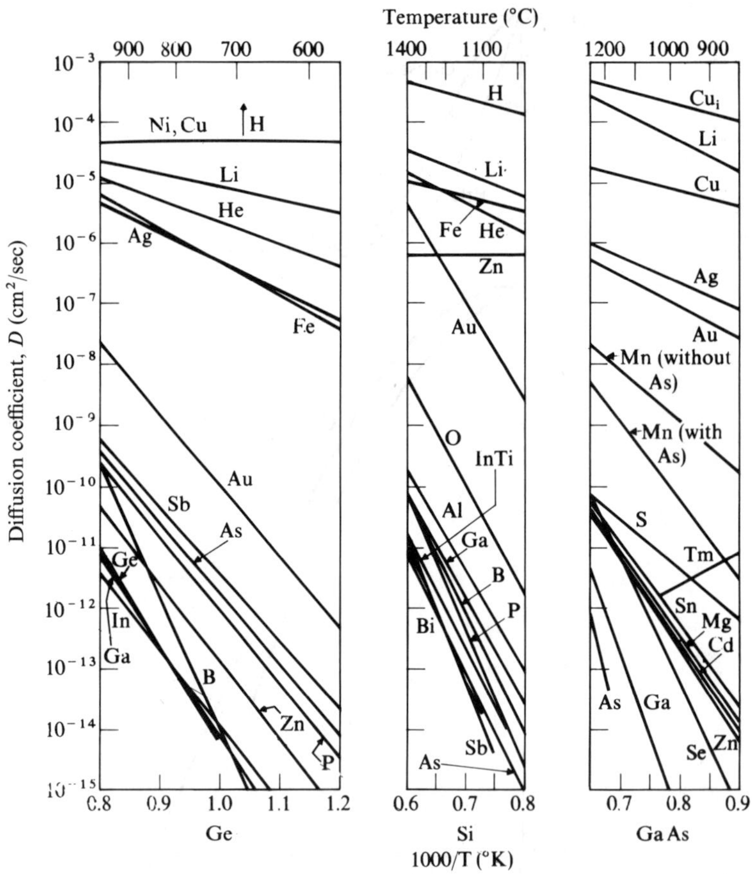

Fig. 11–3 Diffusion coefficients of impurities as a function of temperature. (After Burger and Donovan, and Kendall.)

The combination of the twin processes of compensation and solid state diffusion can be utilized to produce *p-n* junctions. Consider for example the *p*-type semiconductor having initially a uniform acceptor concentration N_a.

Solid state diffusion of donors according to Case A is now carried out in this semiconductor. Figure 11–4 shows the original acceptor concentration and the donor density introduced by solid state diffusion. The final effective chemical impurity density is also shown in Fig. 11–4 which is obtained by satisfying the following equation:

$$N_{eff}(x) = N_d(x) - N_a. \qquad (11\text{–}8)$$

Figure 11–4a shows shallow diffusion and Fig. 11–4c, shows deep diffusion.

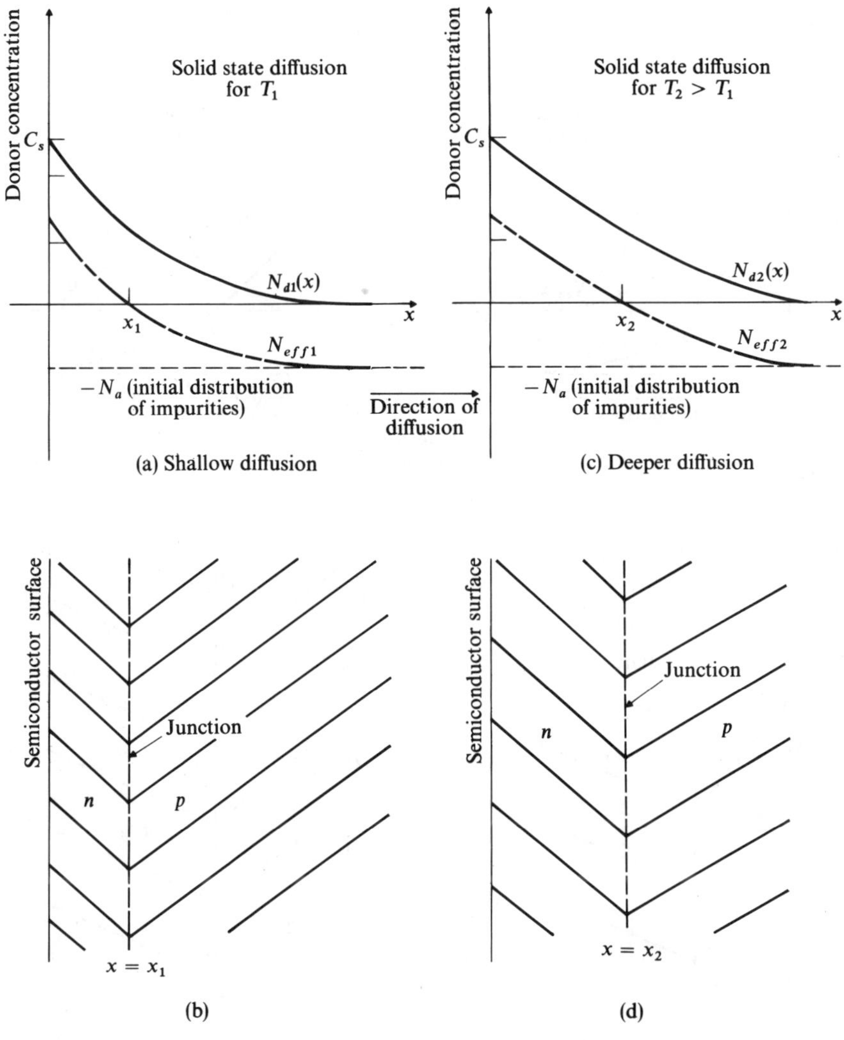

Fig. 11–4 Solid state diffusion into an initially p-type semiconductor under the case of constant surface concentration.

The latter is obtained by maintaining the higher temperature longer. If solid state diffusion takes place according to Case B, where the deposited donors are a fixed number, the area under the two curves $N_{d3}(x)$ and $N_{d4}(x)$ must be equal to each other in Figs. 11–5a and 11–5c. Note in this case that the surface concentration is not constant. Again a p-n junction is obtained. The surface concentration is less for deeper diffusions. Notice that when the solid state diffusion is

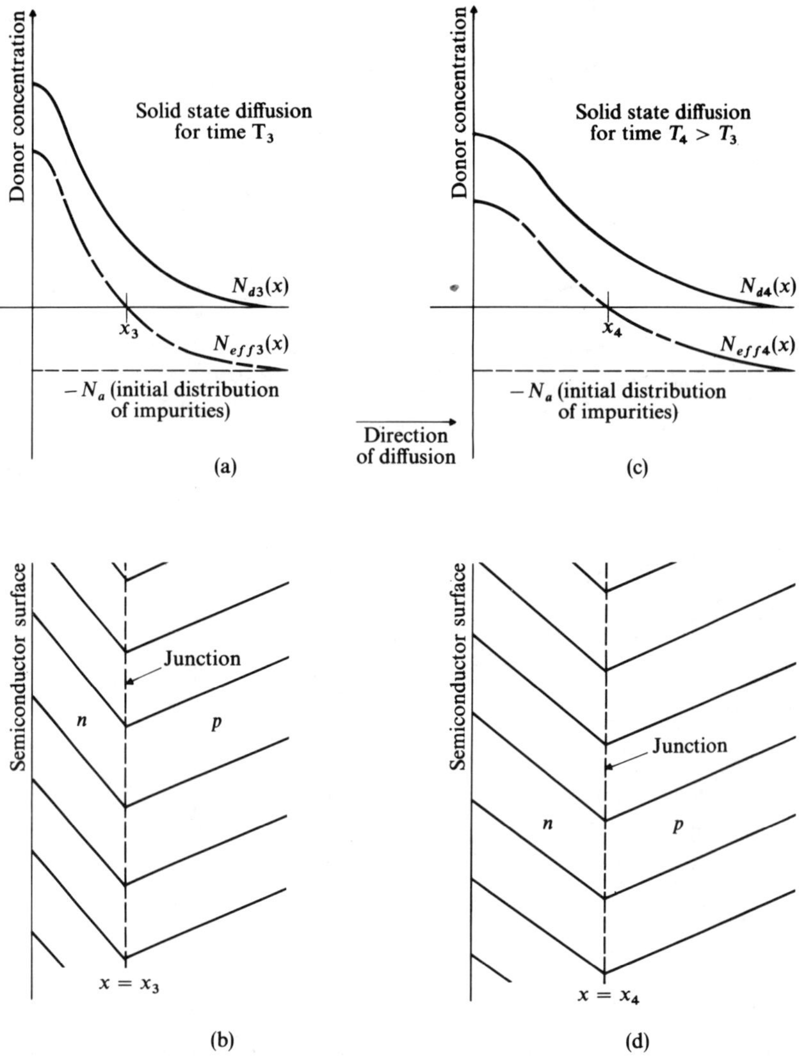

Fig. 11–5 Solid state diffusion into an initially *p*-type semiconductor under conditions of limited sources.

allowed to occur for a longer time, deeper diffusion profile is obtained and the *p-n* junction is obtained farther into the original semiconductor†.

Nothing in the above discussion limits the solid state diffusion technique from being applied more than once. Thus, two successive solid state diffusions

†A well defined junction results only if the gradient of $N_{eff}(x)$ is sufficiently large at $x = x_1, x_2, x_3$, and x_4 in Figs. 11–4, and 11–5.

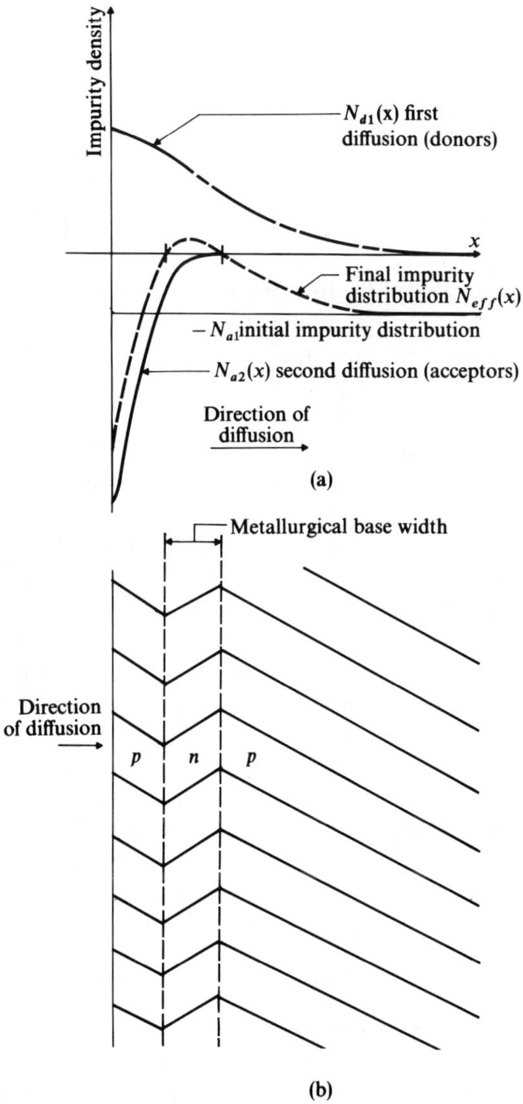

Fig. 11-6 Double diffusion to fabricate a *p-n-p* transistor.

into a *p-type* substrate of donors $N_{d1}(x)$ and of acceptors $N_{a2}(x)$ can produce a *p-n-p* transistor as shown in Fig. 11-6. Similarly *n-p-n* transistors may be fabricated in quite a similar way to that described for *p-n-p* transistors.

The net final chemical impurity density may be expressed as:

$$N_{\text{eff}}(x) = N_{d1}(x) - N_{a1} - N_{a2}(x). \tag{11-9}$$

Generally speaking, for practical reasons not more than *three* successive solid state diffusions should be carried out. These reasons are listed below.

1. There is an absolute upper bound set by the solid solubility at a given temperature on the *total* chemical impurity density (not the effective impurity density) at the surface of the semiconductor through which the diffusions are carried out.

   Even before the solid solubility limit is reached the following limitations set in:

2. Beyond a certain value of the total impurity density in the lattice, the mean free time of carriers will be diminished because, on the average, they will suffer more collision per unit time. Through Eqs. (2–43) and (2–44) we readily see that mobilities will be reduced due to this effect and so also will the diffusion constants for free carriers. As the high frequency response is critically dependent on these parameters and diminishes directly with D and μ, (see Problem 11–1) an increase in total impurity density becomes unacceptable.

   From Fig. 11–6 we can deduce that as the number of solid state diffusion steps increases to produce more regions of different conductivity types the surface concentration increases. Thus, a limit on the total impurity concentration at the surface puts very definite limits on the number of diffusion-compensation steps which can be employed. In practice usually *not more than three solid state diffusions are acceptable.*

   Solid state diffusion process allows us through the judicious control of the types of chemical impurities, and hence values of D, of temperature during diffusion, and of the duration of time for diffusion to control fairly precisely the metallurgical base width of a transistor. The impurity profile also can be properly controlled. This is one of the major reasons for the widespread use of solid state diffusions in the fabrication of semiconductor devices.

11–3 OXIDE LAYERS, PASSIVATION, MASKS, AND AREA SELECTION

In the previous section we saw how the combination of compensation and solid state diffusion permits the fabrication of *p-n* junctions. To obtain useful semiconductor devices either as discrete components or as integrated circuits (multiple circuits fabricated on a single chip of semiconductor crystal), it is necessary to allow the change of conductivity type to only specific areas on a semiconductor chip. In the early 1960s ways of growing SiO_2 (silicon dioxide) on silicon in a controlled way were found. The oxide serves three major useful functions. First, it can act as a *mask* which blocks the chemical impurities, donors and acceptors, from entering the silicon crystal. Second, it can act as a protective shield to prevent contamination of the silicon surface. Third, it can act as an electrical insulator.

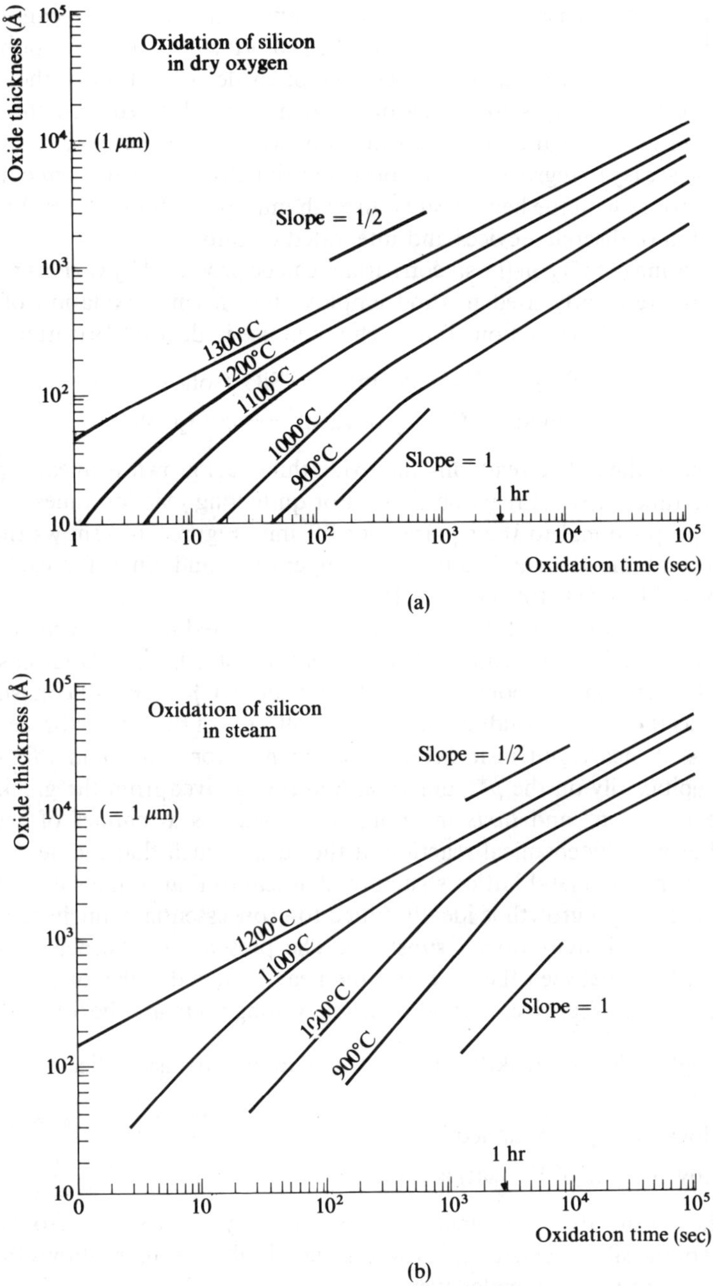

Fig. 11–7 Oxide thickness as a function of reaction time and temperature (experimental). (After Deal and Grove.)

It is relatively easy to thermally grow a thin layer of oxide, typically 1000 Å, on a silicon crystal. Through the use of the *photoresist technique* portions of the oxide can be removed leaving a network of oxide protection of the original crystal surface. Donors (or acceptors) can now be diffused into the crystal. These will go through *only* the openings or *windows* in the oxide coating. As the oxide is easy to regrow it is possible to continually protect the semiconductor crystal surface with the help of SiO_2 through multiple diffusion steps during the fabrication of discrete devices and integrated circuits.

A coating of SiO_2 on the silicon surface can be produced by various methods. The most frequently used method employs the thermal oxidation of silicon through a chemical reaction. Two such are in use in device fabrication

$$\text{Si(solid)} + 2\text{H}_2\text{O(steam)} \longrightarrow \text{SiO}_2\text{(solid)} + 2\text{H}_2 \qquad (11\text{–}10)$$

$$\text{Si(solid)} + \text{O}_2\text{(dry oxygen)} \longrightarrow \text{SiO}_2\text{(solid)}. \qquad (11\text{–}11)$$

By either of the above reactions the oxide thickness grown is linearly proportional to time for short reaction times. For quite long reaction times the thickness is proportional to the square root of time. Figure 11–7 shows the oxide thickness obtained as a function of temperature and time, for each of the reactions of Eqs. (11–10) and (11–11).

A specific example of how a transistor is fabricated will clarify many things. Suppose we wish to fabricate a silicon *p-n-p* transistor by the planar passivated method. Our starting point is a uniformly doped p^+-type silicon substrate which has the correct conductivity for the collector of the transistor we wish to fabricate. A nearly intrinsic layer of semiconductor (silicon in this case) is grown epitaxially on the p^+ substrate. *Epitaxy* (derived from the greek words *epi* meaning "on" and *taxis* meaning "arranged") is a manner of arranging crystal growth by chemical reaction at the surface such that the newly grown thin layer has a crystal lattice structure identical to that of the substrate. This method of crystal growth is ideally suited to grow essentially intrinsic material on top of the heavily doped substrate. The presence of the nearly intrinsic epitaxial layer between the base and the heavily doped collector substrate (of the order of 10^{19}/cc) gives several benefits. Among them are the following:

a) A high collector breakdown voltage and a low extrinsic collector resistance r_c' are obtained

b) A lower of C_{tc} is obtained

c) A lower value of V_{CE} (sat) is obtained.

By either of the above two methods a uniform layer of oxide is grown on the surface of the silicon crystal as shown in Fig. 11–9b (this figure shows the use of dry O_2 to produce the oxide layer.)

The photoresist technique can be used to remove the oxide layer in selected areas as shown by the various steps summarized in Fig. 11–8. Only through

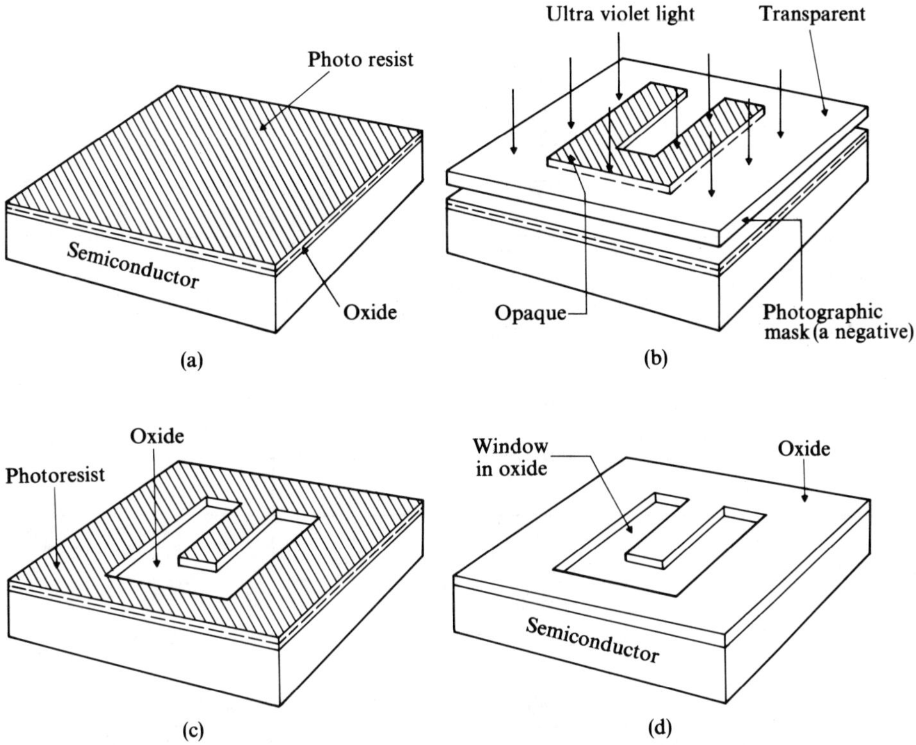

Fig. 11–8 The photoresist technique to produce the desired design of windows in the oxide layer. (a) A layer of photoresist covers the oxide on top of the semiconductor chip. (b) Ultraviolet light shines through a precision photograph mask on to the photoresist resulting in a hardening of photoresist. (c) Chemical etching with hydrofloric acid (HF) removes selected oxide region. (U-shape) the hardened photoresist acts as a mask. (d) The photoresist is finally dissolved with an organic solvent in the oxide leaving the desired opening.

these windows of the oxide can solid state diffusion be carried out. Thus, a donor diffusion produces the n-type region as shown in Fig. 11–9c. Note that the donor diffusion into the epitaxial region also results in some side diffusion into the epitaxial region also results in some side diffusion of donors. Therefore the junction ends up being *under* the oxide and remains protected during further processing. Thus, silicon dioxide, which is a form of glass, serves an additional function besides that of a mask. The oxide layer also protects the surface of the semiconductor during and after fabrication of the device because it stops a vast number of contaminants which could alter the characteristics of the device. It is this function which is the reason for the word *passivated* to describe devices made with the help of this process. The presence of oxide passivates the device.

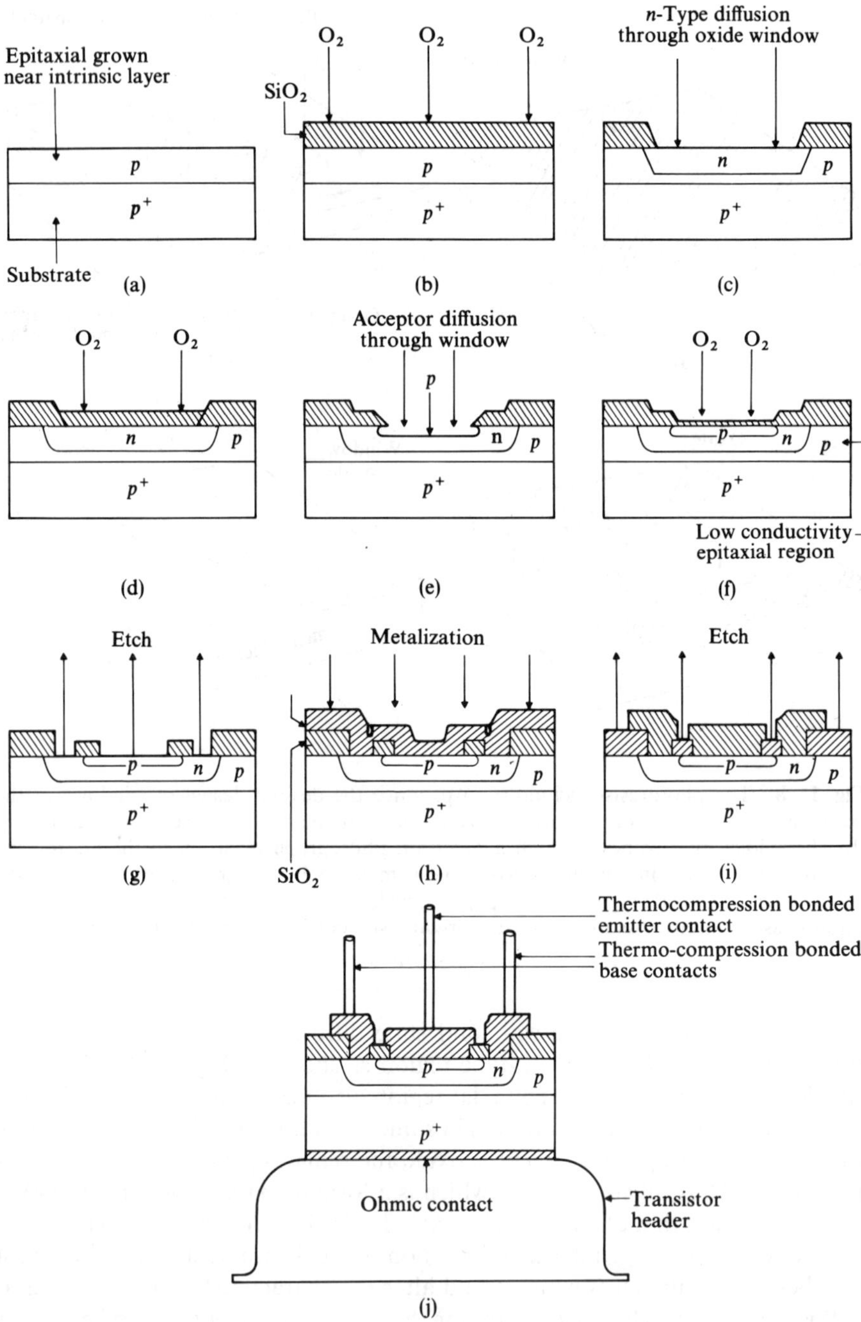

Fig. 11–9 Fabrication of a *p-n-p* transistor by the planar passivation technique.

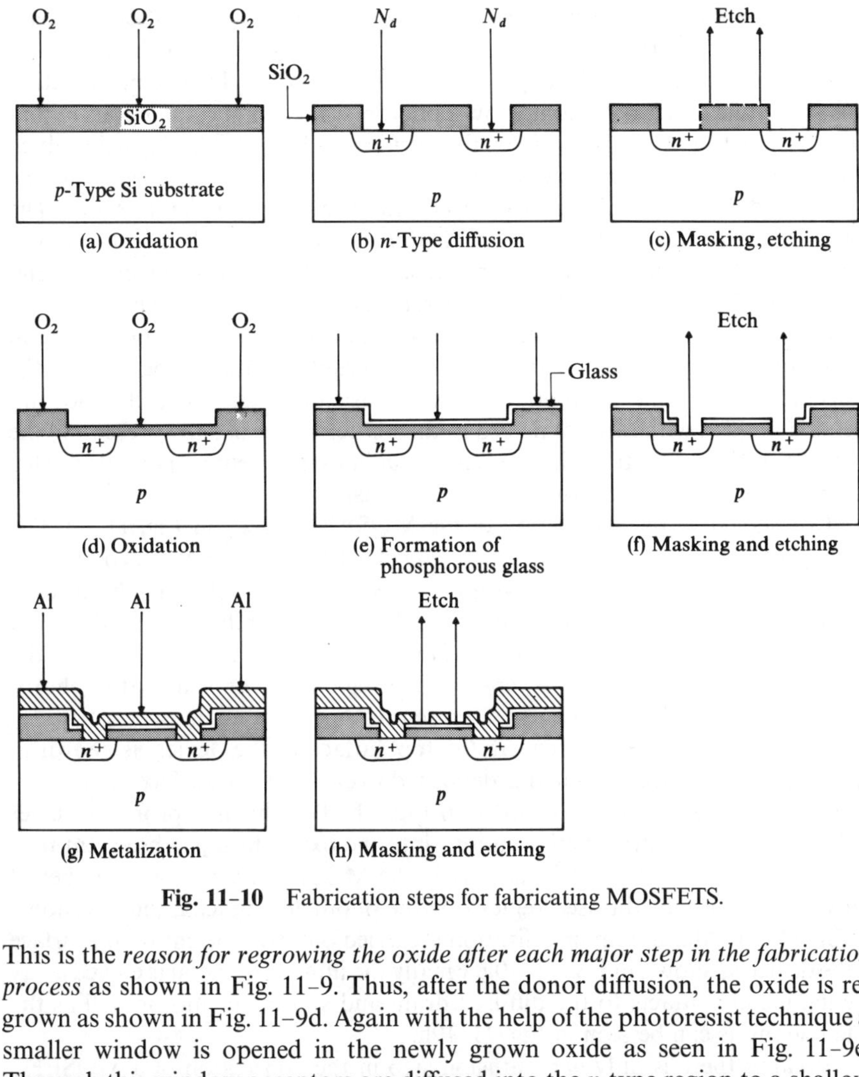

Fig. 11–10 Fabrication steps for fabricating MOSFETS.

This is the *reason for regrowing the oxide after each major step in the fabrication process* as shown in Fig. 11–9. Thus, after the donor diffusion, the oxide is regrown as shown in Fig. 11–9d. Again with the help of the photoresist technique a smaller window is opened in the newly grown oxide as seen in Fig. 11–9e. Through this window acceptors are diffused into the *n*-type region to a shallow depth. Compensation produces the *p*-type emitter region. Again note that the side diffusion results in the *p-n* junction being *under* the oxide. The oxide is regrown as shown in Fig. 11–9f.

Note that all the junctions make their appearance on the top plane surface of the semiconductor. It is this feature which makes it possible to do all the processing *on the same plane surface*—hence the adjective *planar*. Again photomasking permits the selective removal of oxides as shown in Fig. 11–9g. The oxide is still intact and completely covers the emitter and collector junctions.

Now electrical contacts would need to be made to the emitter and base regions. For this purpose a layer of aluminium is deposited over the entire top surface as in Fig. 11–9h. Either gold or aluminium would do. The metalization is usually done under a vacuum. Now a photoresist technique analogous to that discussed in Fig. 11–8 is used on the surface of Fig. 11–9h. An acid which attacks aluminium but *not* the oxide is used to remove unwanted metal. Phosphoric acid or hydrochloric acid satisfy the requirements for this job. The appropriate metalization pattern to make the necessary electrical contacts remains as shown in Fig. 11–9i. Finally, wires are fastened to the emitter and base regions by the process of *thermocompression bonding* as shown in Fig. 11–9j. Also shown in this figure is the ohmic contact made on the entire bottom surface of the original semiconductor chip to permit a contact to be made to the collector region. This is connected to a transistor *header*. The thermocompression bonded wires are connected to thicker terminals usually present on the transistor header. By utilization an exactly similar set of steps an *n-p-n* transistor can be fabricated starting with an n^+-type substrate.

Consider now the fabrication of the MOSFET shown in Fig. 11–10. The starting point is a *p*-type substrate. In Fig. 11–10a is shown the formation of SiO_2 on the surface. Two windows are opened in the oxide by the photo resist technique already discussed. Solid State diffusion permits n^+-type diffusion through the windows in the oxide as shown in Fig. 11–10b, thus forming the drain and source regions. The center oxide region is removed by the photoetching technique discussed earlier. The entire surface is again exposed to dry oxygen and hence the SiO_2 covers the top surface of the device as shown in Fig. 11–10d. Phosphorous glass is deposited over the entire surface completely covering the oxide surface as shown in Fig. 11–10e. This is a protective layer which prevents the introduction of ions into the oxide during subsequent processing which would lead to instability in the MOSFET. Windows are opened above the two *n*-type diffused regions by the photoetching technique as shown in Fig. 11–10f. Aluminium metalization is carried out over the entire top surface of the device as shown in Fig. 11–10g. Finally the unwanted metal is etched away leaving metal contacts to the diffused drain and source regions as well as the gate regions as can be seen in Fig. 11–10h.

Because there is only *one* diffusion step in the fabrication of a MOSFET compared to three required for the bipolar junction transistor, the former is made more efficiently and less expensively. This is what makes MOSFET integrated circuits attractive.

11–4 SIMPLE SEMICONDUCTOR CIRCUIT ELEMENTS

The planar technology can be used to produce additional electronic components as we shall see below.

a) *Resistors.* A resistor can be produced in a silicon crystal in the following way. First an n-type silicon crystal chip can form the starting point. Silicon dioxide is formed on the surface by passing dry O_2 over the surface of Si. The photoetching can be used to produce a long winding window in SiO_2. Through this window acceptors are diffused by solid state diffusion. At the ends of the newly diffused p-type region metalization is carried out. Between these metal contacts a diffused resistance is obtained. The turn around points may also be metalized as shown in Fig. 11–11 to prevent current crowding and to permit easy calculation of the resistance value. The p-n junction formed between the n-type substrate and the p-diffused channel must be reverse biased to use the resistor as a resistor. This is easily accomplished by ensuring that the substrate potential is more negative than the potential at both terminals of the resistor.

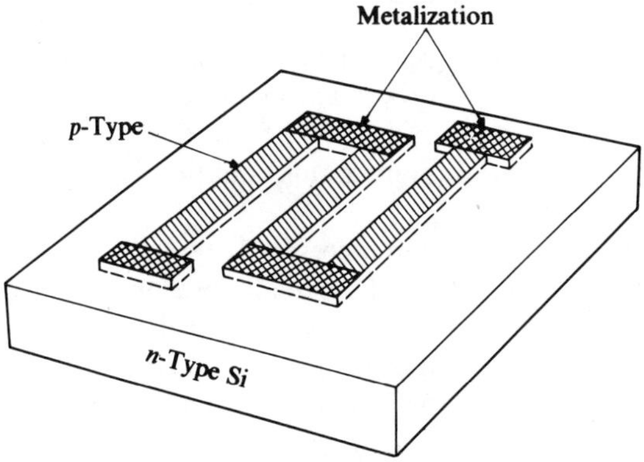

Fig. 11–11 Diffused resistance.

The resistance values so obtained will depend on the width and depth of diffusion, density of donors in the resistive channel, and the length of the channel. Resistance values in the range of

$$1 < R_{\text{diffused}} < 5 \times 10^4\ \Omega$$

are obtainable.

The major disadvantages of such diffused resistances are (i) their typical precision is $\pm 20\%$, (ii) poor temperature coefficient, i.e. a large variation in its value for a given change in temperature typically 0.1 per cent per degree centigrade, (iii) there is a distributed junction capacitance associated with the n-type channel and the p-type substrate.

Major advantages are that (i) several resistances diffused simultaneously on the same semiconductor chip have relative values within about 2% tolerances, (ii) they can be fabricated at the same time as other semiconductor components on the same substrate.

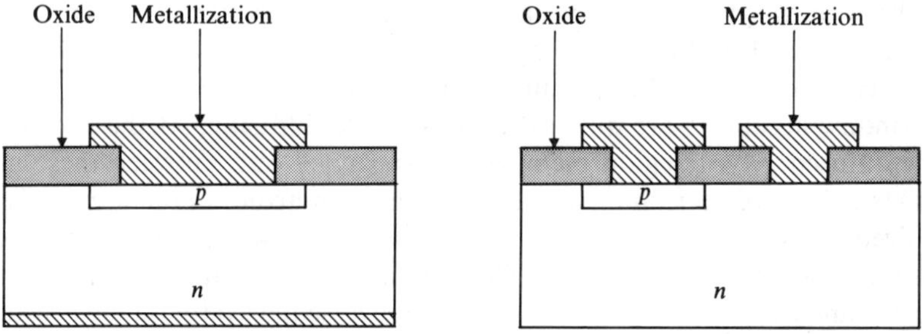

(a) The bottom contact made to the header

(b) Both contacts on the top surface

Fig. 11–12 Junction diode.

b) *Junction Diode.* Only a single *p*-type diffusion through an oxide window on an *n*-type silicon chip results in a *p-n* junction diode. Photoetching and metalization yields the diode shown in Fig. 11–12. Such a diode follows the same *I-V* equation as Eq. (3–56). Deviation from this equation is predictable on the basis of the material of Chapter 12.

c) *Capacitors.* As a *p-n* junction necessarily has the junction capacitance associated with it, the fabrication of a capacitance on a semiconductor chip is precisely the same as that of fabricating the *p-n* junction diode explained in part b. It will serve as a reasonable capacitance if the junction is reverse biased and the voltage variation on the junction is small. The value of the capacitance depends on the junction area, impurity profile, and bias. Capacitor values in the range of 20 to 5000 pF/mm^2 are not too difficult to obtain.

The major disadvantages of such capacitances are (i) the somewhat limited range of values, (ii) presence of leakage current, (iii) they require relatively fixed voltage. The major advantage is that it can be fabricated on the same semiconductor chip along with other electronic components.

The MOS structure of Fig. 9–19 can be easily fabricated by the use of planar technology to obtain capacitors. Only the formation of the oxide layer and a subsequent metalization obtains the MOS capacitor. This capacitor is a lot less leaky compared with the *p-n* junction capacitance. The capacitance, however, is not only voltage dependent but also is frequency dependent.

No satisfactory way has yet been found to fabricate discrete inductance as an integrated circuit element. Some attempts have been made but these proved to be too costly to be practical. For this reason the inductances are either avoided or the combination of feedback, gain, and capacitance are used together to simulate an equivalent inductor.

11–5 INTEGRATED CIRCUITS

In the previous sections we have seen how each of the active as well as passive discrete electronic elements can be fabricated. The planar passivation technology allows us to obtain all the significant active and passive elements needed to make complete electronic circuits on a single semiconductor crystal.

The processing of the device consisting of oxide growing, photo-etching, solid state diffusion, epitaxial crystal growing, metalization, and area selection are adequate to fabricate quite complex integrated circuits of the small scale, medium scales or the large scale (LSI) variety. All of these steps have been examined in detail in the previous sections of this chapter. Only those features which are different from those already discussed previously will be considered in this section.

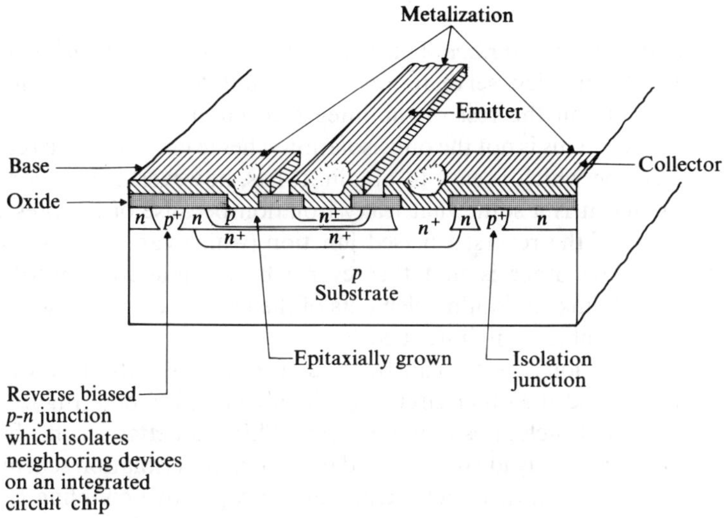

Fig. 11–13 A bipolar transistor as a component in the integrated circuit with junction isolation.

The bipolar junction transistor in an integrated circuit differs from the discrete device of Fig. 11–9 in that in the former case the collector terminal must be brought out through the top crystal surface. Such a structure produced with the help of planar technology is shown in Fig. 11–13. Several things are noteworthy in this figure. First, the collector contact is indeed brought out through the top surface. Because the integrated transistor has its collector contact brought out on top, the path length along which the collector current flows will be longer than is the case for discrete transistors. There are two undesirable consequences due to this fact. The series resistance r_c' is bigger and so also is the voltage across the transistor in saturation $V_{CE,\text{sat}}$. The elongated high conductivity n^+ regions, also called the "buried layers," in Figs. 11–13 and 11–14 help to reduce both r_c' and $V_{CE,\text{sat}}$. Second, the last etching process leaves the metaliza-

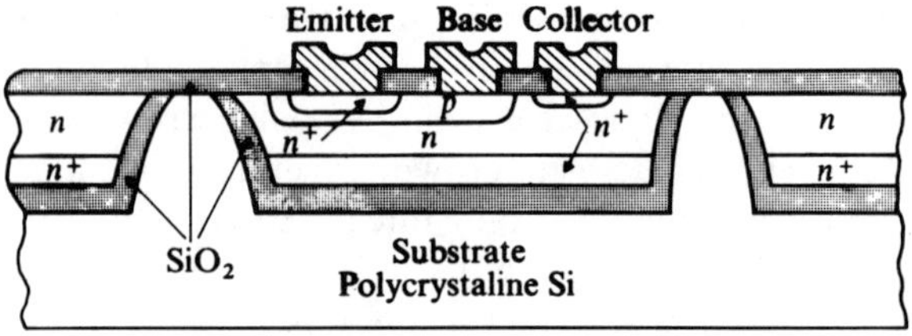

Fig. 11–14 Cross section of a bipolar transistor as a component in the integrated circuit with dielectric isolation.

tion pattern *all on the top surface*. Third, and most important, the p^+-region on the extreme right and left serve to isolate the middle transistor from other adjoining elements on the rest of the integrated circuit by way of a reverse biased *p-n* junction. This is not the only isolation scheme used. This is a relatively inexpensive scheme which is easy to obtain. It has, however, two major drawbacks. One is that it is a somewhat leaky isolation because of the presence of leakage currents of the reversed biased junction (which can be kept small by proper design). The other is that the reverse biased junction contributes a capacitance which links adjoining elements of the integrated circuit which may prove to be undesirable at high frequencies.

Dielectric isolation scheme can be used to obtain isolation between the bipolar transistor and the other circuits of the integrated circuit. Figure 11–14 shows an example of such an isolation scheme. This is a better scheme because the isolation is more nearly ideal compared to the junction isolation scheme. The leakage currents are almost nonexistent, and the capacitive coupling between adjoining circuits is quite small.

The transistor as an integrated circuit component shown in Figs. 11–13 and 11–14 is seen to be an *n-p-n* transistor. Indeed, the *n-p-n* transistor is the most normal form of the integrated circuit transistor. Let us see the reason for this. Because the collector is subjected to repeated heating during the base and emitter diffusions, it is important to choose the collector doping impurities which have a small thermal diffusion coefficient. Such a choice will prevent the movement of collector junctions during successive diffusions. The donor impurities are seen from Fig. 11–3 to have lower diffusion coefficients than acceptor impurities. It is, therefore, seen to be advantageous to have *n*-type collectors. Furthermore, the solid solubilities of some donor impurities are higher than almost any acceptor impurities, which fact permits heavier doping of the n^+ regions, such as the emitter, for better injection efficiency and other regions under metallic contacts. For these reasons it is easier to fabricate *n-p-n* transistors as integrated circuit elements as compared with *p-n-p* transistors.

There are important applications in which both *p-n-p* and *n-p-n* transistors are used in close proximity as for example in complementary symmetry circuits.

Such circuits prove to be economical in the use of electronic circuit components mentary bipolar transistor structure obtained by the Stanford (1973) complementary process which utilizes only one more processing step than the standard required to perform a given circuit function. In Fig. 11–15 is shown the complebipolar transistor process. The diffusion steps are the following: (1) n^+ buried layer (B.L.) is diffused for the *n-p-n* collector; (2) p^+ layer is diffused for the *p-n-p* and a *p*-isolation followed by an *n*-epitaxial growth; (3) a *p*-diffusion for the *n-p-n* base, *p*-isolation and collector contact for *p-n-p* transistor; (4) a p^+ emitter is diffused; and finally (5) an n^+ emitter is diffused.

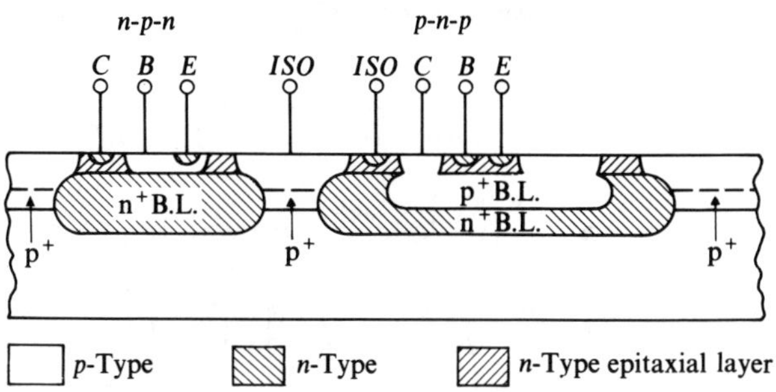

Fig. 11–15 The Stanford complementary process.

The impurities for the buried layers are phosphorous for n^+ region and boron for p^+ region. The base of the *p-n-p* transistor is the original uncompensated epitaxial layer. Both transistors have highly doped collector regions for low $V_{CE,\,\mathrm{sat}}$ and low r_c'. Such processing which involves only one extra diffusion does result in quite good complementary transistors on a single chip. However, because the MOSFET technology requires fewer processing steps for the fabrication of the active element it should be possible to fabricate these devices at less cost and higher density of circuits than the corresponding bipolar transistor technology. A bipolar junction transistor as a component of an integrated circuit may take up roughly 50 mil^2 of area, a MOSFET requires approximately only one-tenth of this area. There are however three obstacles. First, MOS device characteristics were found to be unstable. The source of the instability is widely believed to be *migrating ions* in the oxide under gate metalization. Second, the currently available MOSFETS are as yet not as good in high frequency performance as the bipolar transistors. Third, because the MOSFETS circuits cannot generally be interconnected to bipolar circuits directly, the necessary interfacing circuits between the two introduce additional costs.

The recently developed silicon-gate technology seems to remove several of the above pitfalls of the conventional MOS technology. Figure 11–16 summarizes this technology. There are two major differences between the device of

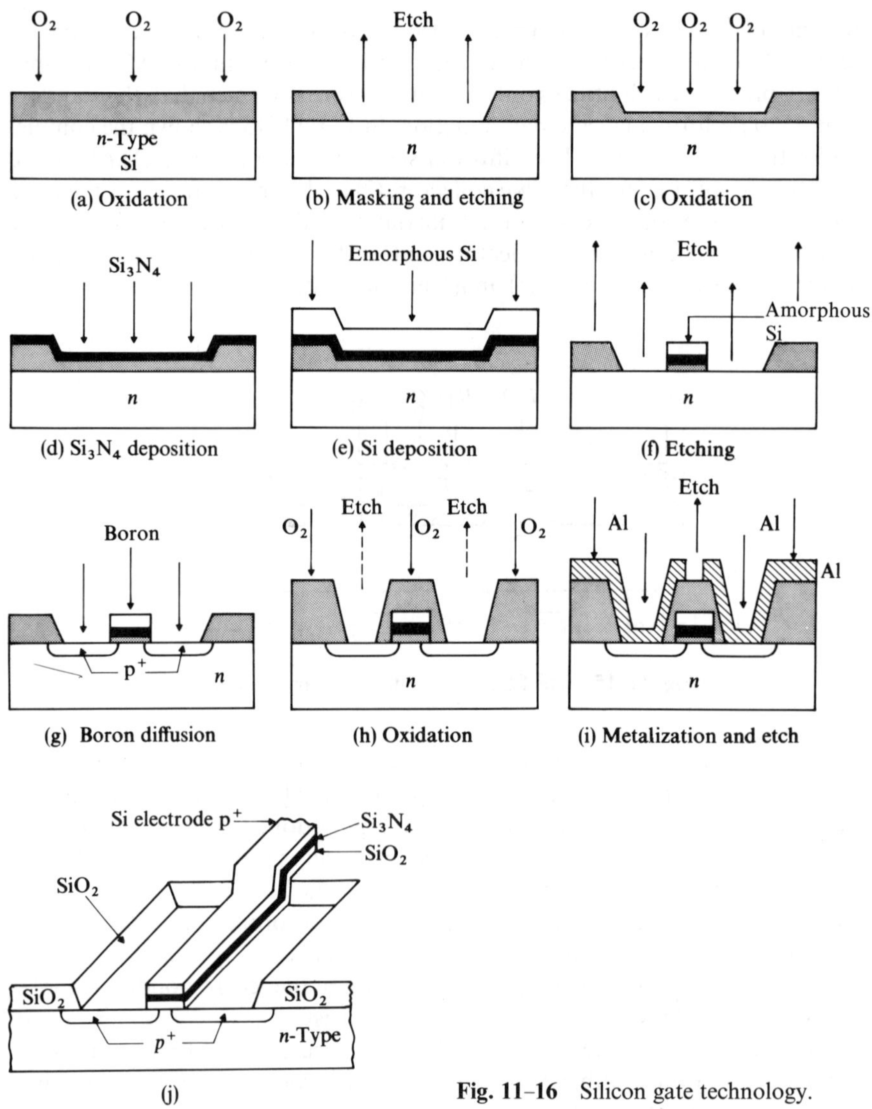

Fig. 11–16 Silicon gate technology.

Fig. 11–16i and the conventional MOSFET. First, the gate electrode is not a metal but is heavily doped silicon. Second, under this silicon gate are two dielectric layers—silicon nitride Si_3N_4 and SiO_2. This system has several advantages in integrated circuits over the conventional MOSFET structure. They are:

1. It turns out that Si_3N_4 is a very effective barrier to migrating ions believed to be the major cause of instability of MOSFETS.

2. As this is an induced type device the threshold voltage V_T needed to establish a conducting channel is important. Lower threshold voltage is desirable because it facilitates the interfacing of MOSFETS circuits to the bipolar circuits and it also improves circuit performance. The device of Fig. 11–16i has V_T which is lower than that of the corresponding conventional MOSFET structure because the combination of Si_3N_4 and SiO_2 results in *an increase in the effective dielectric constant* of the combination being higher than that of SiO_2. The reason for the use of SiO_2 under Si_3N_4 is that in the absence of SiO_2 it is found that due to the presence of unwanted spurious charges the V_T actually rises. The threshold voltage is also influenced by the material out of which the gate electrode is made. The use of p^+-type Si gate electrode for the *p*-channel devices, which we are considering results in V_T being approximately 1 Volt lower than in comparable devices made from aluminium gate MOSFETS. The combination of p^+-type gate Si_3N_4–SiO_2 sandwich for *p*-channel devices results in V_T being *typically about 2 Volts.*[2]
3. The structure of Fig. 11–16i results in what is known as "self-aligning" gate structure. The *p*-type diffused regions and the gate electrode are "self-aligned" because the dielectric under the gate electrode is the diffusion mask which defines the edges of the drain and source regions. They are, therefore, automatically aligned. In the conventional MOSFETS integrated circuit this result is obtained only by two sequential masking procedures which are inherently inaccurate. To take this inaccuracy into account space is wasted and relatively larger electrodes are used. Thus, the self aligned scheme leads to reduced gate capacitance, reduced gate to drain capacitance C_{dg} (which is the very important Miller capacitance† in common source amplifier configuration), and reduced drain junction capacitance because now the drain metal electrode can overlap the gate regions as these are separated by an insulating layer of SiO_2.
4. Higher functional density can be achieved on the silicon chip because the individual devices are smaller on account of the above reasons.
5. Because the silicon gate electrode and aluminium electrodes are separated by an insulator, we have two separate networks of conductors on top of the silicon wafer allowing the aluminum pattern to cross the silicon electrode pattern. Additional diffused conductors can be laid down on the substrate, thus three separate conducting patterns seem possible. These can handle

---

† The most commonly used MOSFET amplifier is the common source configuration in which the *source* terminal is commonly shared by both the input terminals (gate and source) and the output terminals (drain and source). According to Miller's theorem the capacitance which links the output and input terminals of an amplifier having a gain of $-A_v$ (having a 180° phase shift from input to output) appears as seen from the input terminals to be bigger by a factor of $(A_v + 1)$. This is the reason for keeping C_{dg} capacitance as low as possible.

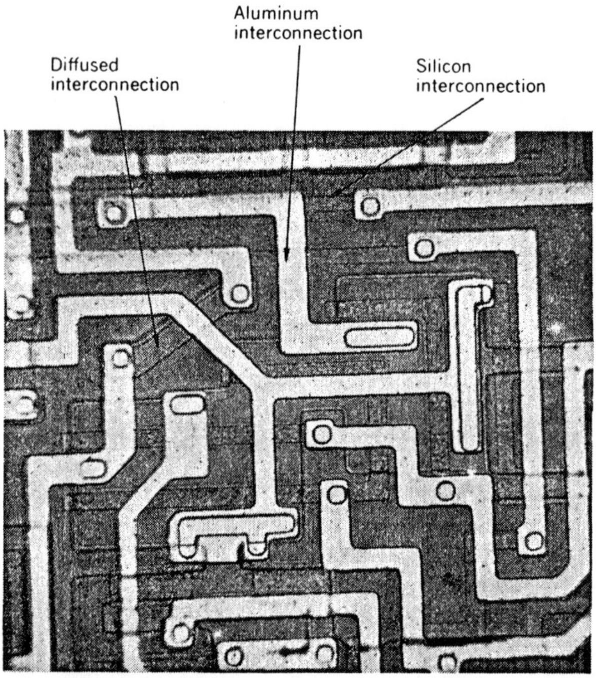

Fig. 11–17 Photomicrograph of a small region of Intel's i-1101 256-bit RAM, showing the three possible modes of interconnection: diffused regions within the wafer, deposited silicon, and metal interconnecting lines. (After Vadasz et al. 1969).

quite complex and overlapping efficient interconnections of complex functional blocks within the integrated circuit. Figure 11–17 shows a portion of an integrated circuit which efficiently utilizes the three conductor scheme just mentioned.

The MOS integrated scheme usually uses only the induced type active MOSFETS *because the isolation* of one active element from another is automatic—a feature not present in the bipolar integrated circuits. Furthermore, MOS integrated circuits do not have any parasitic transistors. Unfortunately the bipolar junction transistor integrated circuits do result in parasitic transistors between two adjoining transistors (such as the *p-n-p* parasitic transistor seen in Fig. 11–13) in the junction isolation scheme. The MOS technology can provide integrated circuits up to perhaps ten times more complex than the bipolar transistor integrated circuits. Thus, the MOS technology would seem to have a great future. The major drawback of MOS integrated circuits is that the frequency range over which useful amplification can be obtained is less than that in the corresponding bipolar transistor integrated circuits.

Recently *ion implantation* as a method of doping semiconductor wafers is used where devices of very small dimensions are to be fabricated. Let us suppose that we wish to make a MOSFET having surface dimensions of the order of a micron. Thermal-diffusion technique will not readily allow us to fabricate such a device. The technique of implanting accelerated ions into say silicon does allow us to make such a device. Because this technique allows us to control the number of implanted ions as well as their depth it is likely to challenge the thermal-diffusion technique of fabrication whenever extremely small dimension devices are to be produced. As methods improve for producing more devices per unit area, it appears to be within the possibility of this technology to be able to produce a few hundred billion devices in a year from a single machine. Currently the total world production is about a trillion devices per year (F. F. Morehead and B. L. Crowder, 1973). MOSFETS have been fabricated by the use of this method.

11–6 INTEGRATED CIRCUIT DIODES

Doping or impurity profiles are chosen in integrated circuit design to optimize the most important component, the transistor. Usually additional diffusions are avoided where possible because of the added cost. For this reason if diodes are needed in integrated circuits, they are obtained by connecting a transistor appropriately to obtain the diode function. Five possible diode configurations, obtainable in this way, are shown in Fig. 11–18. In devices A, B, and E the breakdown voltage corresponds to that of the emitter junction and hence is low, typically about 7 Volts. In devices C and D the breakdown voltage corresponds to that of the collector junction and is hence relatively high, about 40 Volts or higher. In devices C and D the leakage current is that corresponding to the collector junction which is lower than the other three devices. The storage time for given ON and OFF drives is expected to be higher in device E and lowest in device B. (Do you know why? See Problem 11–6.) Device E also has the highest junction capacitance, about twice that of the others.

11–7 TWO SPECIFIC EXAMPLES OF INTEGRATED CIRCUITS

Consider the following two specific examples of integrated circuits obtained by using the fabrication principles discussed above.

A. The Darlington Amplifier

The Darlington common emitter amplifier shown in Fig. 11–19a is a widely used amplifier configuration. If the circuit is provided an external load resistance R_L (going to a power supply V_{CC}) in the collector circuit, the amplifier offers a moderately high input resistance and a very high current, voltage and power gains. This configuration of Fig. 11–19a can also be used as a Darlington

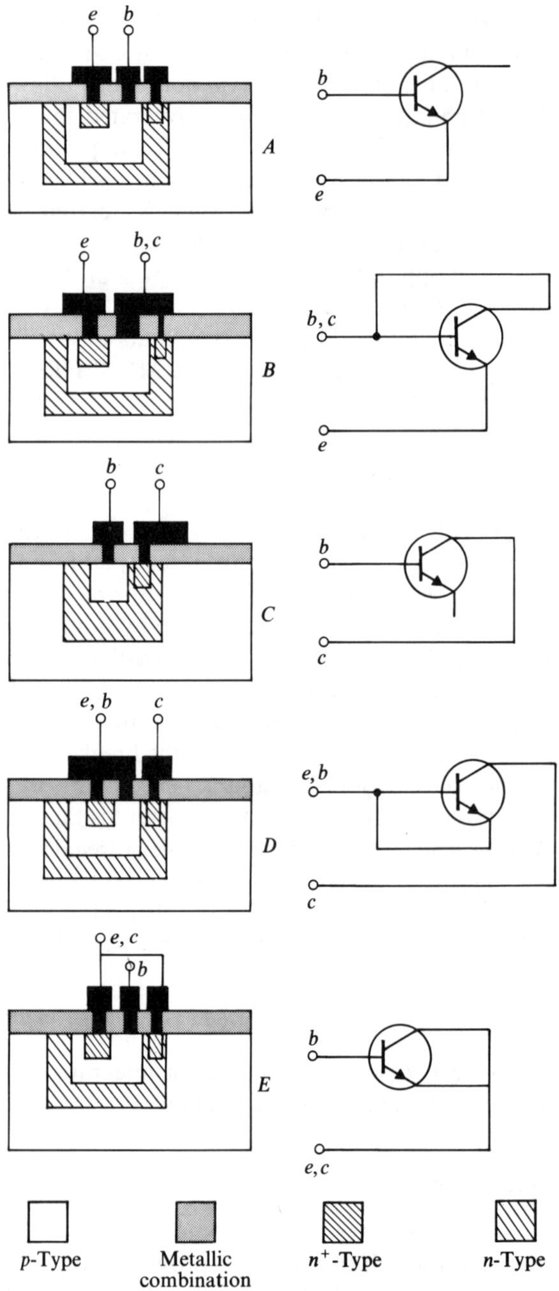

Fig. 11–18 Integrated circuit diodes obtained from transistor terminal interconnections; cross-sectional view and circuit representation.

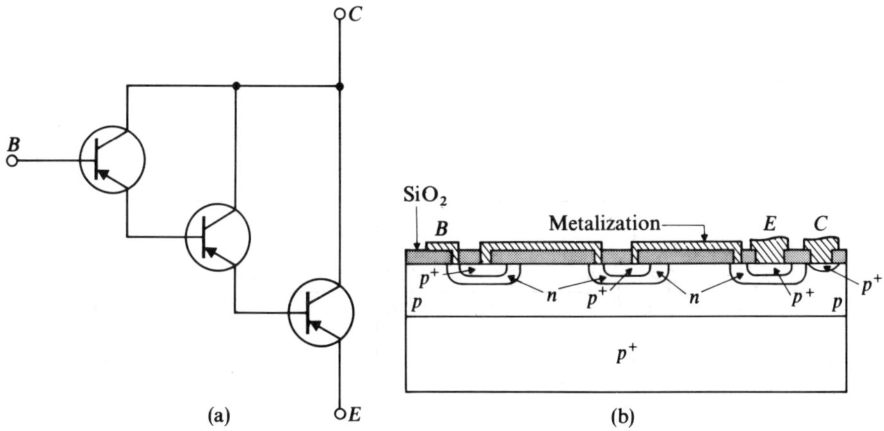

Fig. 11–19 Integrated circuit realization of a Darlington Amplifier. (a) Darlington amplifier. (b) Cross section of an integrated circuit Darlington amplifier.

connected common collector (emitter follower) giving a quite high input resistance and a very high current gain. In this case the load resistance would need to be in the emitter circuit.

The Darlington amplifier of Fig. 11–19b can be fabricated as a single integrated circuit. Figure 11–19a shows a p^+-type substrate with an epitaxially grown p region on top. Two solid state diffusions—first of donors and then acceptors brings the three (or more if needed) transistors into existence with a common collector for all the transistors. The metalization and photo-etching results in connecting the emitter of the first transistor to the base of the second, and the emitter of the second transistor to the base of the third transistor. Figure 11–19 shows the cross section of such an integrated circuit as well as its circuit diagram.

B. The NOR Gate

As the second specific example we consider the important digital logic circuit the NOR gate† shown in Fig. 11–20a. This is called resistor transistor logic and is abbreviated RTL. Figure 11–20b shows the top view of the integrated circuit chip schematic. The three transistors can be fabricated with the same simultaneous diffusions. The collector is common for the three transistors. Note that all the four diffused resistors are on island *A*, and the three transistors are fabricated on island *B*. Successive masking, photo-etching, and metalization produce patterns shown in the top view of Fig. 11–20b.

Many more circuits similar to those just considered can be fabricated as integrated circuits, It is also possible to make several hundred identical circuits,

†The NOR gate produces an output, logic **1**, if and only if *all* its inputs are OFF (logic **O**).

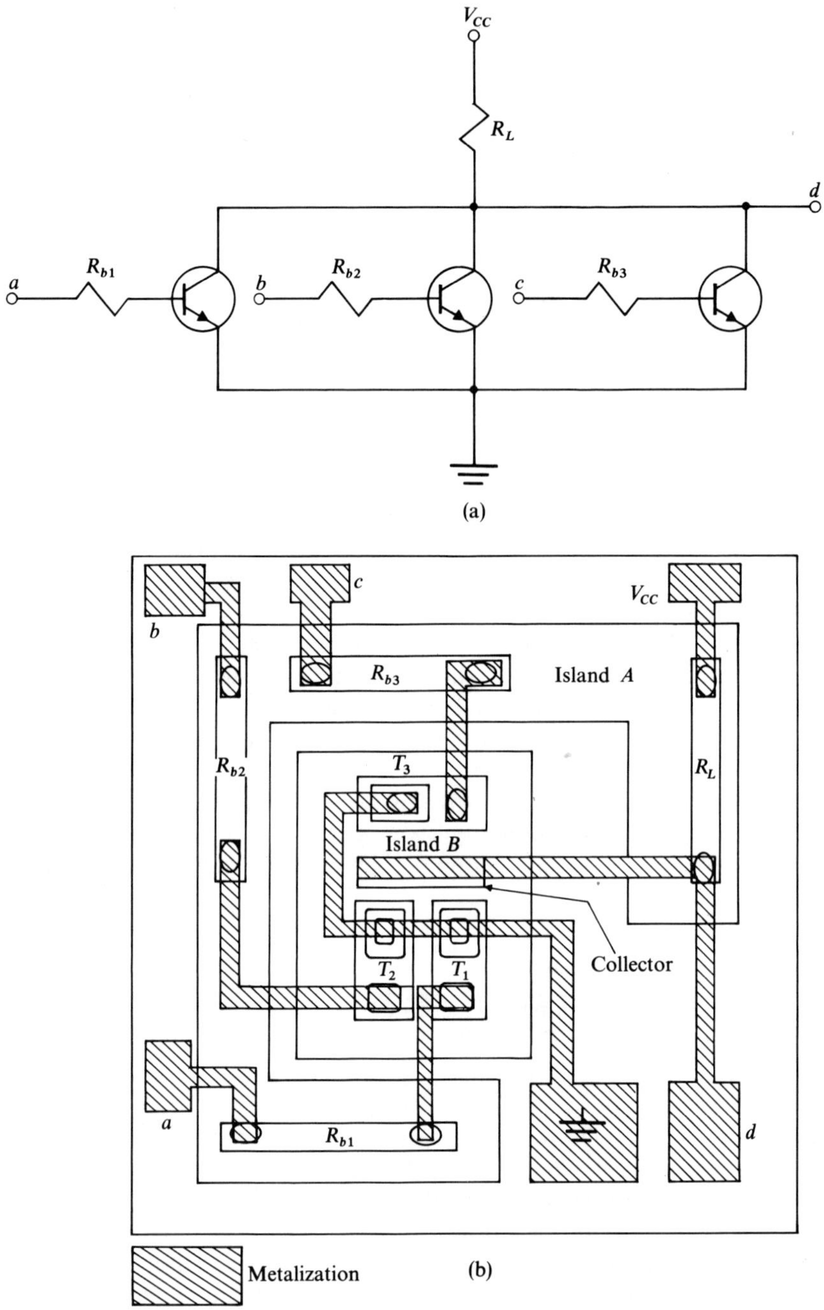

Fig. 11-20 RTL NOR gate.

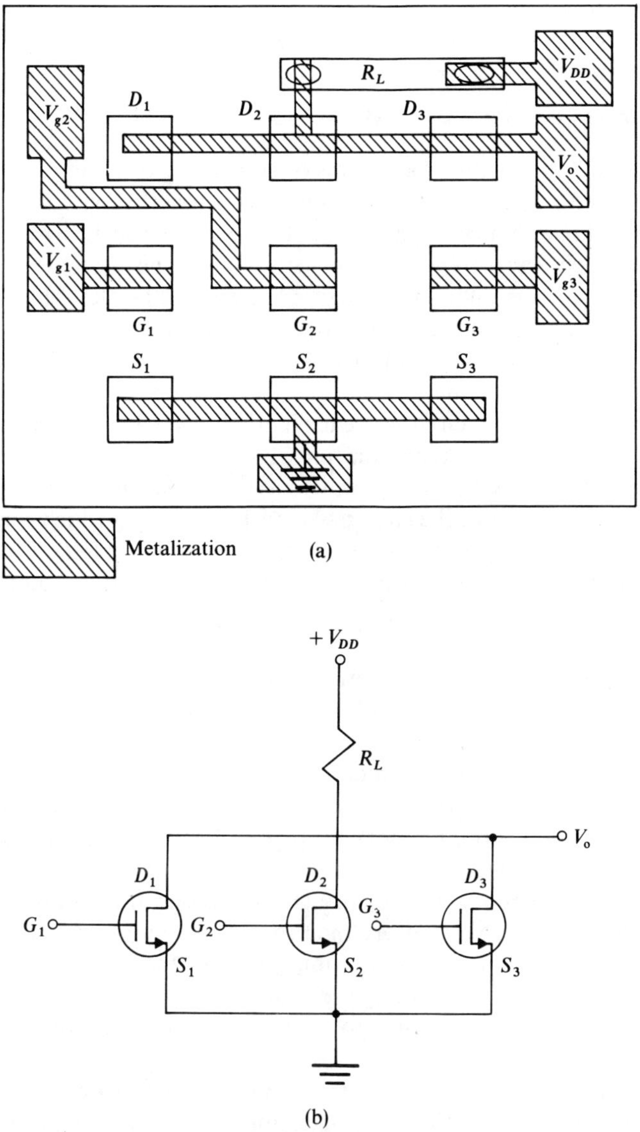

Fig. 11-21 MOS integrated circuit realization of a NOR circuit. (a) Top view of the integrated circuit realization of the NOR gate using MOS technology. (b) NOR circuit.

such as, for example, that of Fig. 11-20b, simultaneously on a single silicon crystal. This is the reason why the cost per circuit has now come down dramatically in the last few years. Because precision masks of considerable complexity can be made, large numbers of circuits can be fabricated on a single silicon crystal chip. Circuits of the previous few figures can also be realized where the MOSFETS are the active elements. Integrated circuit realization of the NOR circuit through MOS technology is shown in schematic form in Fig. 11–21.

11–8 LARGE SCALE INTEGRATION

Large scale integration (LSI) is a process of fabricating complete subsystems or systems containing a large number of components together with their interconnections on a single semiconductor crystal chip. The term LSI is reserved for those chips which have 1000 or more components on a single chip. If the number of components on a single chip is between 100 to less than 1000, it is said to be *medium scale integration* (MSI). *Small scale integration* is a term reserved for the case where the number of components on a single chip is fewer than 100 but greater than one.

Large scale integrated circuits are currently being fabricated and sold. It is possible to build a single silicon chip on which there are 1000 logic gates (as many as 4000 gates can be put on a Si wafer having a 1.5 inches diameter). An LSI memory module which uses the silicon gate technology and which can hold 256 "words" each of which is 32 bits‡ in length is shown in Fig. 11–22. Figure 11–22 shows the photo-micrograph of a 256-bit (random access) memory element along with circuits needed to select a particular bit from an eight bit binary coded address.

With the help of a suitable microwiring scheme, 25 to 50 such LSI memory modules can be interconnected and fitted into a cube having four inch sides. It seems possible that 100 LSI chips can constitute the central processor of a high capacity computer. Computers which utilize such a central processor can be made so that it is readily portable. It seems well within the realm of possibility to replace all magnetic core memory units with LSI units in computers. Such memory units, as currently conceived, must have all gates within it operate properly. The LSI memory will require shorter access time than the conventional core memory. Perhaps before long the LSI memory modules will completely replace the core memory in computers. As the memory module is a basic component of the computer, the market for LSI memory modules is extensive. This is already leading to a rapid development of LSI memory modules.

If a fabrication procedure involves say 30 processing steps and if only a 2 per cent loss of circuits results per processing step, only 53.3 per cent of the circuits will be left working at the end of thirty steps! Thus, to have a high percentage of circuits working at the end of a number of processing steps, extremely low loss per step must be achieved. One way of getting around this problem is to simply have many more circuits on an LSI chip than are needed for a given application. Usually each of the circuits on a LSI chip can be tested

---

†A bit is defined to be a binary digit which may either have the value of **0** or **1**.

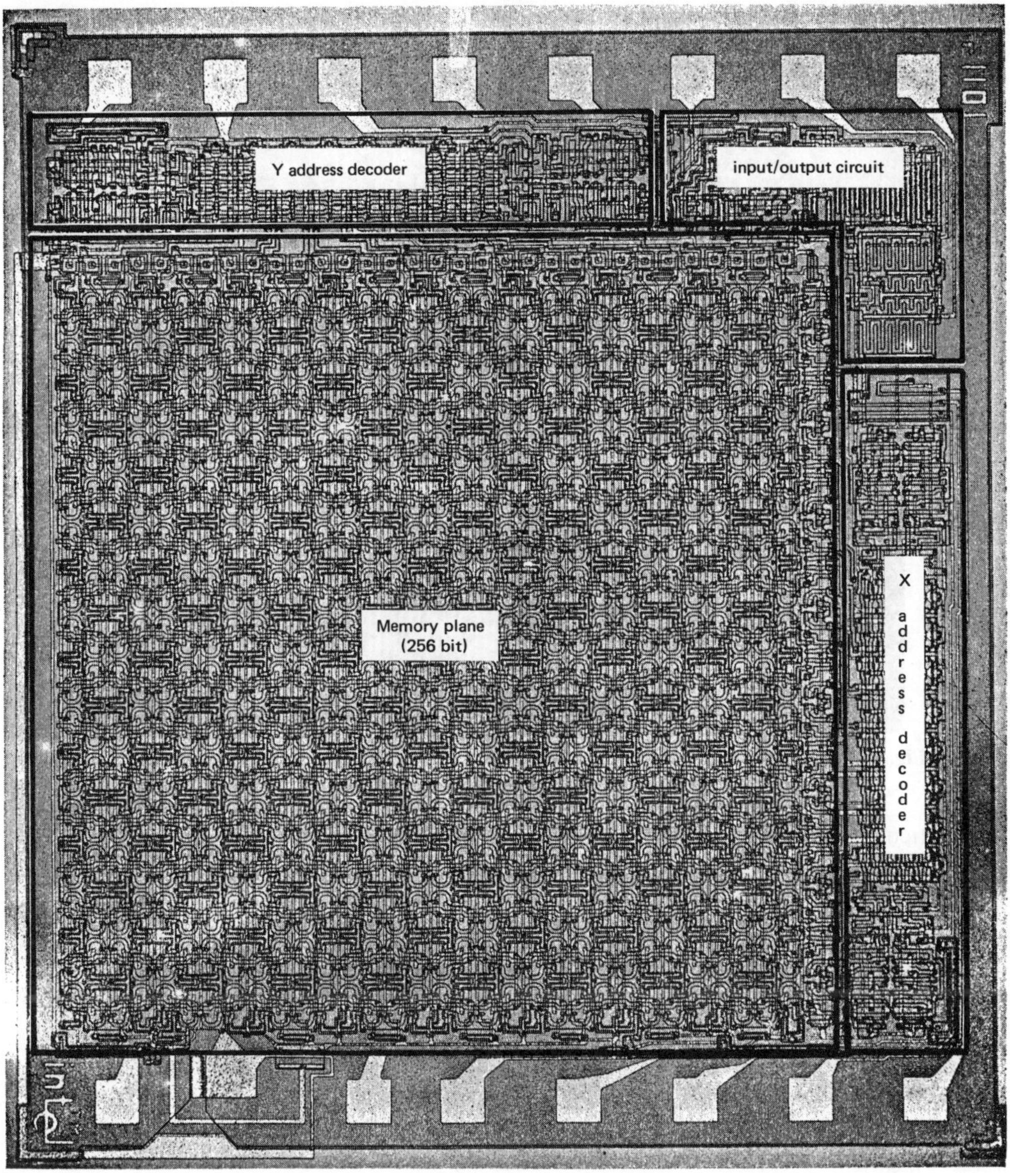

Fig. 11–22 Photo-micrograph of an integrated circuit (chip size 3.10 mm × 2.8 mm) 256 bit random-access memory element utilizing silicon gate technology. Block diagram is superimposed. (After Vadas et al.)

by a computer aided testing procedure before the final interconnection metalization pattern is deposited on to the chip. Thus, with the good and the faulty circuits identified, the interconnection pattern simply circumvents the defective circuits. This is called the *discretionary wiring scheme.* It is possible to do computer aided tests on individual circuits on an LSI chip with the information about good and defective circuits stored in a computer memory bank. For any par-

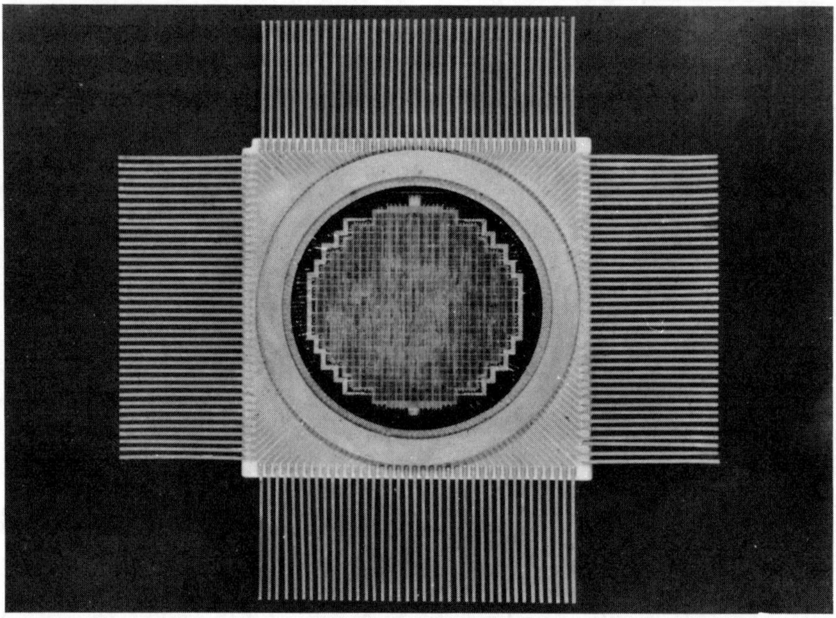

Fig. 11–23 An integrated circuit LSI which employs the discretionary wiring scheme in a complete circuit package. (Courtesy of Texas Instruments Inc.)

ticular use the computer can be programmed to figure out the most efficient wiring plan to interconnect as many good devices as needed for a particular application. A cathode ray tube can separately generate vertical and horizontal wiring patterns; when these are put together the complete wiring scheme is obtained. A custom design is thus produced for a given application. Texas Instruments Company uses the scheme discussed above. Figure 11–23 shows the wiring pattern as well as a complete circuit package. A portion of the LSI semiconductor chip used in connection with discretionary wiring is shown in Fig. 11–24.

Figure 11–25 shows 64 chips employing both MOS and bipolar circuits having 40,960 transistors, diodes, and resistors all of which are accommodated within an area of 0.6×0.75 square inches thus having a density of 9×10^4 devices per square inch.

The LSI may well incorporate light emitting diodes as well as photocells. A wristwatch television set, a flat TV set may well become possible with the LSI technology. Perhaps in the not too distant future it may be possible to duplicate the function of the eye by the direct connection of large numbers of photocells to the appropriate receptors in the human head. For many human health disorders which can be helped by implanted active devices, e.g. pacemakers for the heart and many other conditions, the LSI and integrated circuits may well help to prolong life.

People are now beginning to use computer controlled microfabrication schemes. By these schemes submicron pattern dimensions are now feasible. Precision microcircuit lithography, electron probe resist exposure, etc., are

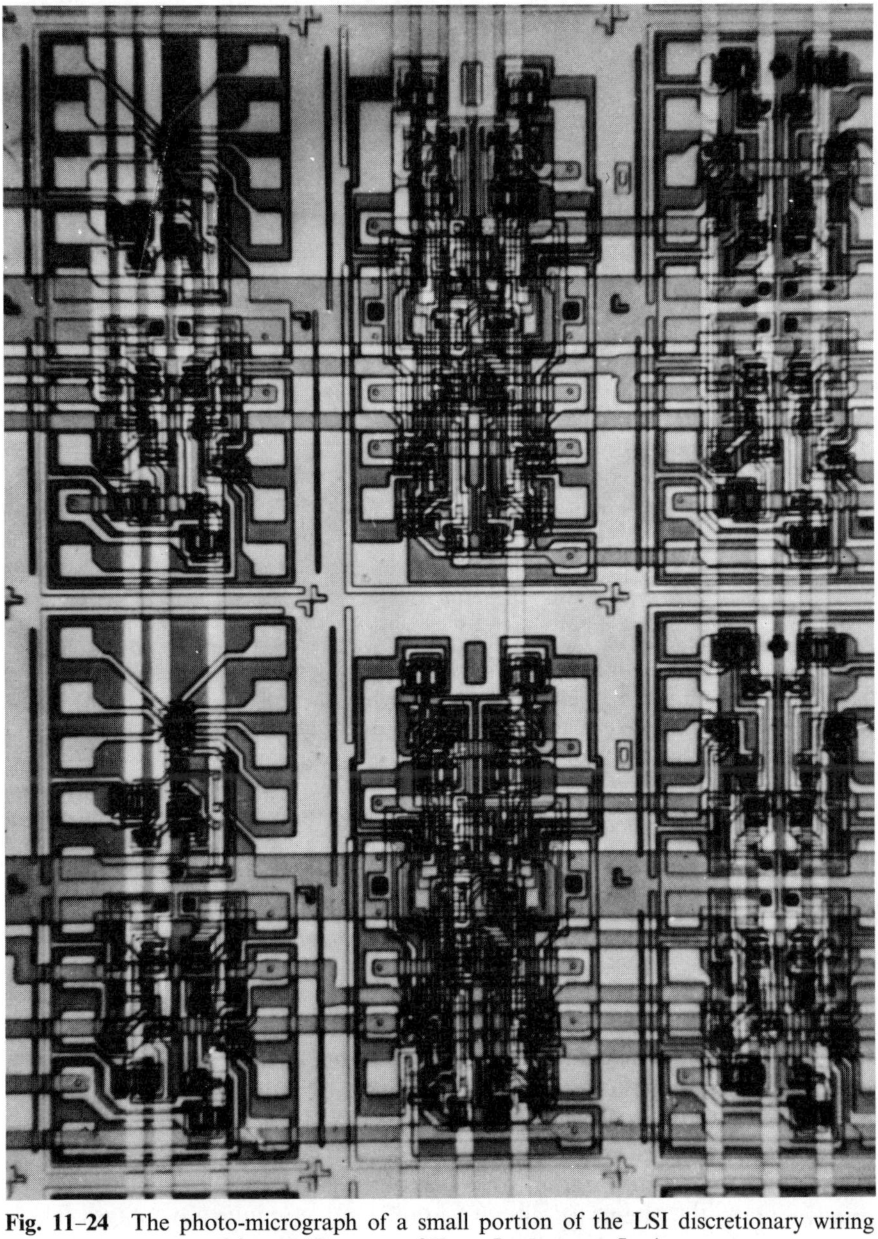

Fig. 11–24 The photo-micrograph of a small portion of the LSI discretionary wiring scheme. (Courtesy of Texas Instruments Inc.)

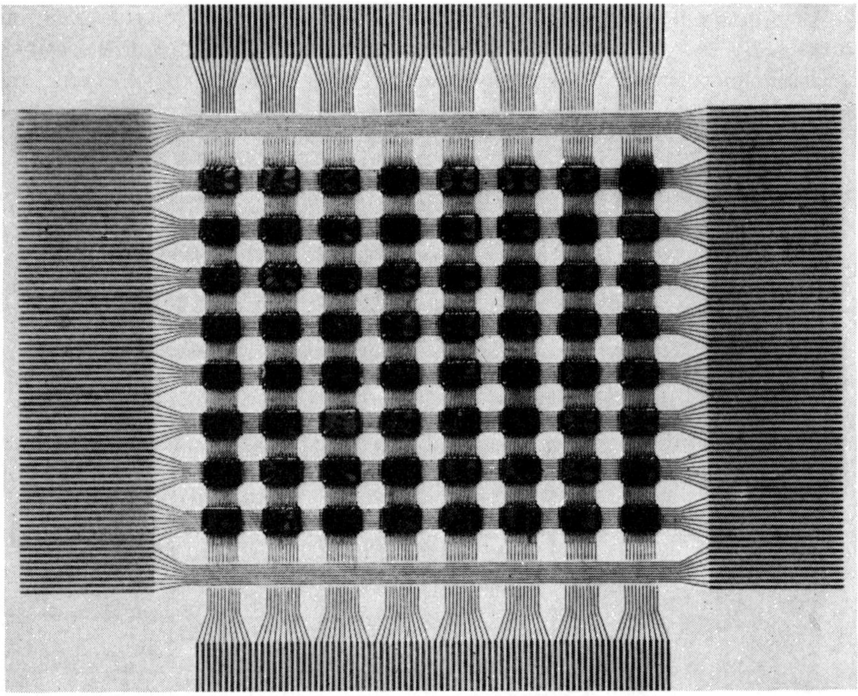

Fig. 11–25 Right-scale integration, or RSI, is the way the Bell Telephone Laboratories describes a technology that permits the mixing of various kinds of circuits, such as bipolar and MOS, each at an appropriate state of development. In the photograph 64 bipolar memory chips, each 0.046 by 0.063 inch, are mounted on a ceramic wafer and connected by "beam leads" into a large memory array. The 64 chips fit within an area 0.6 by 0.75 inch and contain 40,960 transistors, diodes and resistors. This equals 90,000 devices per square inch. (After F. G. Heath.)

now possible. The third, fifth, seventh, eighth and tenth references at the end of this chapter will be found to be useful in this connection.

11–9 SPECIAL DESIGN CONSIDERATIONS FOR INTEGRATED CIRCUITS

In this section we look at some problems and some unique advantages of integrated circuits. The major problems are the following. It is highly impractical to obtain inductors, transformers, and large value capacitors in integrated circuits. Furthermore, variations of most parameters with temperature are higher in integrated circuits than with standard discrete components. For example, a diffused resistor temperature coefficient is an order of magnitude higher than that of a carbon resistor, and the resistor value of the former may vary by a factor of three or four times that of the latter. Even in the face of these disadvantages, the integrated circuits do offer some unique advantages. These advantages arise from the following fundamental facts. First, active devices

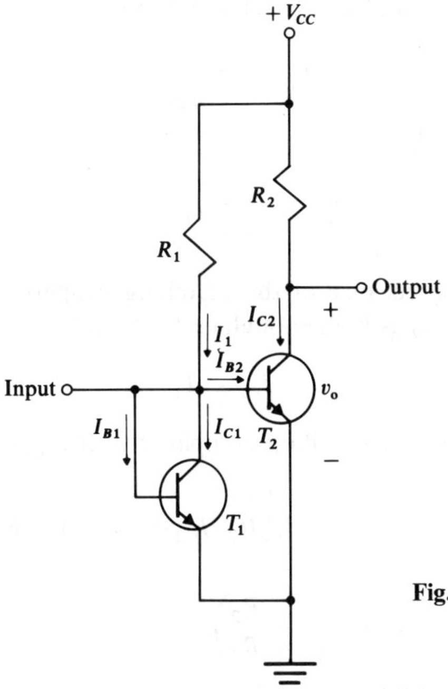

Fig. 11–26 Diode biasing.

are available in large numbers at relatively little additional cost. Second, because of their close proximity on the semiconductor chip and because of their being fabricated at the same time, the integrated circuit components match and track much better than discrete components. The typical performance of integrated components is summarized in Table 11–1 which justifies the last statement. To make these observations more concrete, we shall consider a series of specific examples of circuits which exploit the unique advantages of integrated circuit technology.

Diode Biasing

Consider the biasing circuits of Fig. 11–26 which utilizes two identical transistors T_1 and T_2. Despite the fact that T_1 is connected as a diode (see Section 11–6), with a sufficiently high collector supply voltage, both transistors will be in their normal active regions (defined in Chapter 6). Because of the close proximity of T_1 and T_2 we shall assume that their emitter junction characteristics are identical. We shall also assume that the d-c collector current is related to the corresponding d-c base current through the d-c h_{FE}. Then

$$I_{B1} = I_{B2} \triangleq I_B, \tag{11–12}$$

$$I_{C2} = h_{FE2} I_{B2}, \tag{11–13}$$

$$I_1 = I_{B1} + I_{B2} + I_{C1} = I_{B1} + I_{B2} + h_{FE1} I_{B1}. \tag{11–14}$$

Solving for I_{B2} and using Eq. (11–12) we obtain

$$I_B = \frac{(I_1 - 2I_B)}{h_{FE1}}. \tag{11–15}$$

From Eq. (11–13) we have

$$I_{C2} = \frac{h_{FE2}}{h_{FE1}}(I_1 - 2I_B). \tag{11–16}$$

If $h_{FE2} \approx h_{FE1} \triangleq h_{FE}$ because of the matching property of integrated circuit components, and if h_{FE} is high enough (>50) then

$$I_{C2} \doteq I_1. \tag{11–17}$$

Neglecting the emitter-base voltage compared with V_{CC}, the d-c output voltage is seen to be

$$V_{C2} = V_{CC} - I_{C2}R_2; I_{C2} = I_1 = V_{CC}/R_1,$$

hence

$$V_{C2} = V_{CC}\left(1 - \frac{R_2}{R_1}\right). \tag{11–18}$$

Note that V_{C2} depends on a resistance ratio which is highly predictable as Table 11–1 shows. Thus, despite the variations of R_1 and R_2, V_{C2} is relatively stable.

In discrete transistor amplifiers the temperature stabilization of collector current and voltage is usually achieved by the introduction of a resistor R_e in the emitter circuit. This scheme provides not only temperature stabilization but also provides stability with respect to the normal spread in the value of β_0 for transistors of even the same type when one transistor is replaced by another. Normally the resistor R_e is bypassed by a capacitor C_e, usually of large value, so that the a-c voltage gain does not diminish on account of the presence of R_e.

A transistor amplifier in an integrated-circuit form cannot make use of the above scheme; it is virtually impossible to include the needed large value capacitor C_e. Two or more transistors of the same semiconductor crystal and made at the same time turn out to have not only the same electrical characteristics but the same changes, called *tracking*, in characteristics with changes in temperature. This feature together with the fact that the cost of including an extra transistor in an integrated circuit is only two cents or less, results in the circuit of Fig. 11–27. This circuit is widely used in integrated circuits to provide stability of the quiescent operating point with respect to temperature and the variation in the value of β_0.

In Fig. 11–27 the amplifier network N_2 needs to be stabilized. The circuit of the active network N_1 provides the necessary stabilization as can be seen

Table 11–1 Typical performance of Integrated Components†

	Symbol	Typical Value	Tolerance	Temperature Coefficient
n-p-n Transistors and Diodes				
Current amplification factor	h_{FE}	30–100	+50, −30 per cent	+0.5 per cent/°C
Matching of h_{FE} between identical transistors in close proximity	Δh_{FE}	—	±10 per cent	+10.5 per cent/°C
Base—emitter diode forward voltage drop (low current)	V_{BE}	0.7 Volts	±3 per cent	−2 mV/°C
Matching of V_{BE} between identical transistors in close proximity	ΔV_{BE}	—	±2 mV	±10 μV/°C
Base—emitter diode reverse breakdown voltage	BV_{EBO}	6–8 Volts	±5 per cent	+3 mV/°C
Collector—base breakdown voltage	BV_{CBO}	<45 Volts	±30 per cent	—
Collector—substrate breakdown voltage	BV_{CS}	<60 Volts	—	—
Lateral p-n-p Transistors				
Current amplification factor	h_{FE}	0.5–20	+200, −50 per cent	+0.5 per cent/°C
Base—collector breakdown voltage	BV_{CBO}	<45 volts	—	—
Resistors				
Resistance of diffused resistors (base layer)	R	100 Ω–20 kΩ	±20 per cent	±0.15 per cent/°C
Resistance of deposited resistors	R	100 Ω–100 kΩ	±18 per cent	±0.10 per cent/°C
Matching between identical resistors in close proximity	ΔR	—	±3 per cent	±0.005 per cent/°C
Pinch Resistors				
Sheet resistance (small applied voltage)	R_s	5–10 kΩ/□	+100, −50 per cent	+0.3 – 0.5 per cent/°C
Matching of identical pinch resistors in close proximity	ΔR	—	±5 per cent	—

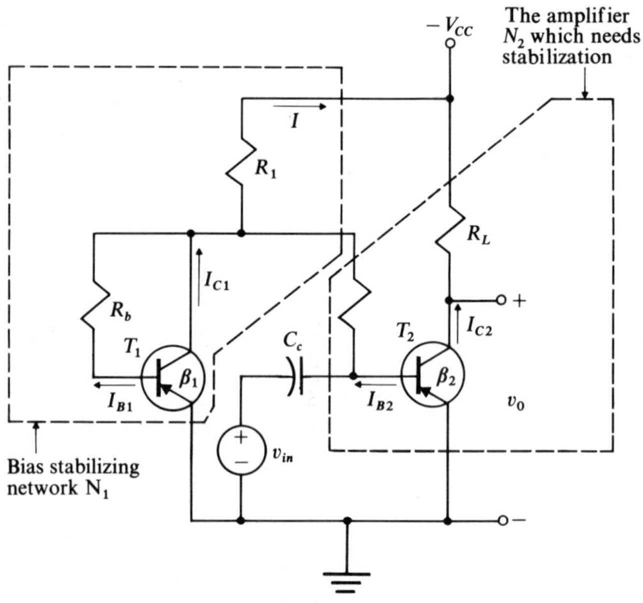

Fig. 11–27 The bias stabilization in an integrated circuit transistor amplifier.

from the following reasoning. Suppose the temperature of the integrated circuit increases. The collector current I_{C2} will try to increase because of the higher temperature. Transistor T_1 will draw more current with increasing temperature. The combination of V_{CC} and R_1 behaves like a current source which puts out essentially a fixed current I. Thus, as the current drawn by T_1 is increased with increasing temperature, the current in the base circuit I_{B2} of T_2 is reduced. The reduction of I_{B2} will compensate for the increase in I_{C2} which would otherwise occur. Because R_b is directly connected between the base and collector terminals of T_1, in effect T_1 can be replaced by a simple diode. The transistor T_1 is used instead of a diode for better matching between N_1 and N_2 and for better *tracking*. If the current gains β_1 and β_2 are equal to β and large, it can be shown that I_{C2} is independent of both temperature and β. Note too that no capacitors were needed or used in this bias stabilizing scheme. The resistance R_b is needed so that T_1 will not load down the signal source v_{in}. The capacitor C_c is there merely to isolate the d-c currents from the source. The circuit of Fig. 11–27 or its simple modifications find widespread use in integrated circuit bias stabilization in transistor amplifiers.

Compound Devices

Because the gain-bandwidth product of an amplifier stage is significantly improved if a compound device (more than one active device per amplifying

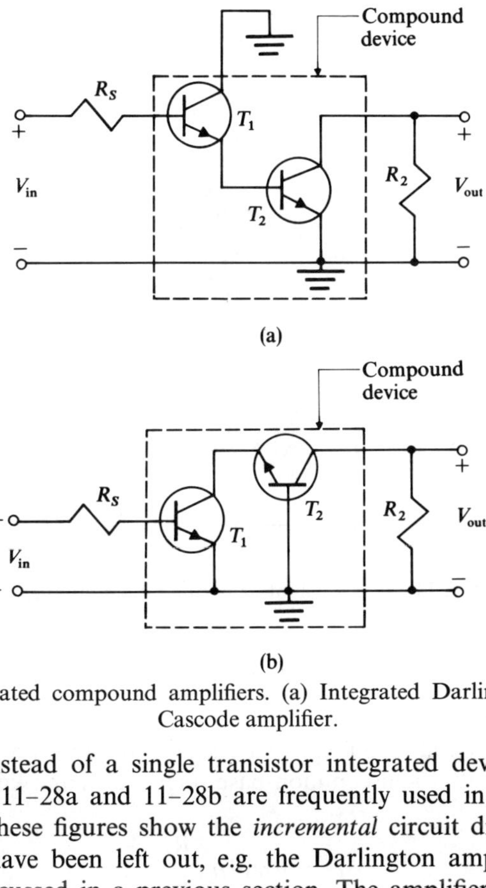

Fig. 11–28 Integrated compound amplifiers. (a) Integrated Darlington amplifier. (b) Cascode amplifier.

stage) is used instead of a single transistor integrated device, such devices as those of Fig. 11–28a and 11–28b are frequently used in *linear* integrated circuit design. These figures show the *incremental* circuit diagrams in which biasing details have been left out, e.g. the Darlington amplifier, which has already been discussed in a previous section. The amplifier T_1 may be used as a common collector (or emitter follower) which is a good isolation amplifier stage with an essentially close to unity voltage gain. The input stage effectively isolates the input from the output. Thus the Miller capacitance of T_2 has little effect at the input of T_1. Using the fact that the *incremental* emitter to base voltages are negligibly small for both transistors and the collector current is related to the base current by

$$i_c = h_{fe} i_b, \tag{11–19}$$

we may write

$$A_{v1} = (h_{fe1} + 1) h_{fe2} \frac{R_L}{R_s} \doteq \frac{h_{fe1} h_{fe2} R_L}{R_s}. \qquad (11\text{–}20)\dagger$$

Again note that the gain depends on a *resistance ratio* which is easily controlled.

† Note that $h_{fe} = \beta$ and $h_{fb} = \alpha$.

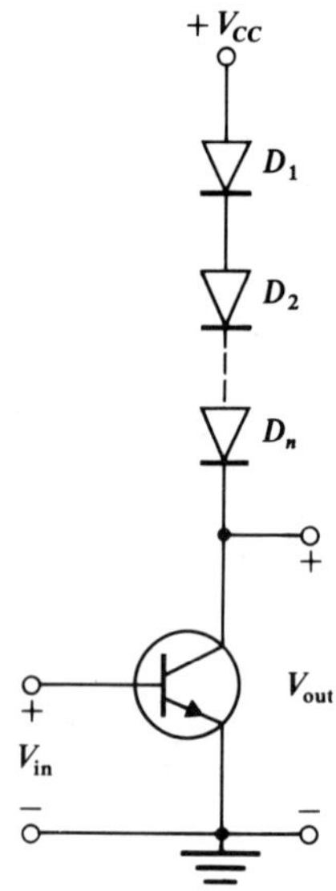

Fig. 11–29 Constant gain integrated amplifier.

The circuit of Fig. 11–28b is called a cascode amplifier which is a combination of common-emitter and common-base stages. This amplifier has a voltage gain of

$$A_{v2} = h_{fe1}h_{fb2}\frac{R_L}{R_s}. \tag{11–21}†$$

This compound amplifier offers a high frequency performance greatly improved over a single transistor amplifying stage. The output of T_1 looks into the low input impedance of T_2.

In both the compound devices the reverse transmission is markedly reduced because of the high mismatch in the output impedance of T_1 and the input impedance of T_2. We must note, however, that the compound device which utilizes more than one transistor results in higher amount of phase shift from input to output of the compound device. Hence great care should be exercised in employing overall feedback around the compound device.

Constant Gain Stages

The matching and tracking properties of integrated elements can be utilized to obtain a *constant gain* over a wide range of parameter and temperature changes. Figure 11–29 shows such an amplifier. Each diode in the collector circuit offers

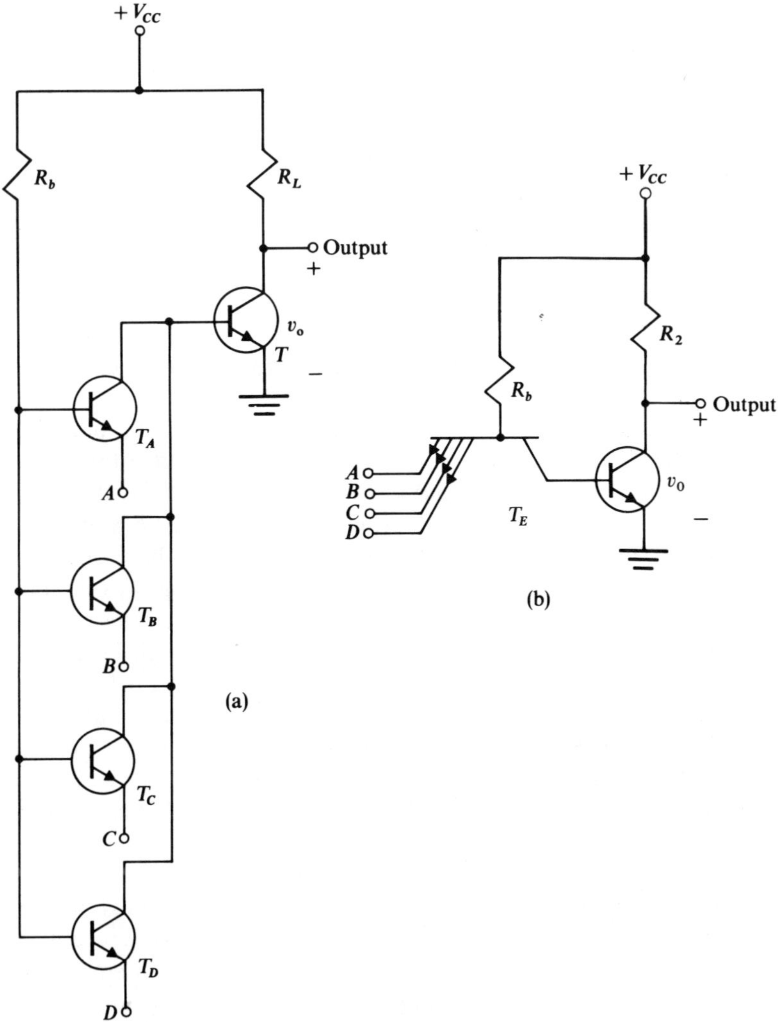

Fig. 11–30 Transistor-transistor-logic circuit. (a) Discrete circuit. (b) Integrated circuit version.

the incremental resistance given by

$$r_{d1} = \frac{kT}{qI_C} \tag{11–22}$$

for n-identical diodes in the collector circuit offer an effective load resistance of

$$R_{L\text{eff}} = n\frac{kT}{qI_C}. \tag{11–23}$$

But the input voltage is given by

$$V_{\text{in}} = \left(\frac{kT}{qI_E}\right) I_C. \tag{11-24}$$

Finally, the overall gain is given by

$$A_{v3} = \frac{V_{\text{out}}}{V_{\text{in}}} = -\frac{R_{L\text{eff}} I_C}{V_{\text{in}}} \doteq -n. \tag{11-25}$$

The voltage gain is seen to depend on only the number of diodes in the collector circuit!

Finally, consider the transistor-transistor-logic (TTL) circuit of Fig. 11–30. With the emitters of transistors T_A, through T_D open, these transistors merely act as diodes feeding current to the base of T which is hence ON. If at least one (say T_B) of the inputs is connected to ground, it now has *both* of its junctions forward biased. The transistor is saturated and forces the base of T to be at ground potential thus cutting off this transistor. The voltage V_0 will approach V_{CC}. Thus, if the ground potential input is assigned the logic binary number **0** and the high collector voltage at the output is assigned the logic binary number **1**, then the circuit produces a **1** when at least one of the inputs is **0**.

Because the bases of T_A, T_B and T_C are connected together and so also are the collectors, it is possible to produce the function of these three devices by a single integrated transistor with multiple emitters as shown in Fig. 11–30b (see Problem 11–7). The circuit of Fig. 11–30b is economical on chip area utilization.

11–10 SUMMARY

We studied the fabrication of single active devices as well as integrated circuits. The impact of integrated circuits and LSI is likely to be very far reaching and profound. The electronics engineer of the future will need to simultaneously understand circuit analysis and synthesis as well as device theory. *The knowledge of technology is important in that it limits what devices are possible to be made.*

The fundamental steps which we found to be of enormous usefulness are plannar passivation, masking, photo-etching, solid state diffusion, and metalization. The details of this technology change almost everyday and hence are not covered in this chapter or this book. The emphasis has been on fundamental processing steps.

The complete analysis of integrated circuits needs to involve computer aided analysis. The distributed circuit analysis may also be necessary. For many purposes, however, the integrated circuits can be analyzed as functional blocks each being considered to be made up of discrete elements.

This modern technology has it within its capacity to help to cure some diseases, to help to sustain life not otherwise possible, and to make some glorious dreams come true such as a blind man being able to feel the sensation of sight.

PROBLEMS

11–1 Discuss how the most important transistor parameters are affected by the reduction in μ and D.

11–2 Show that impedance Z of the circuit of Fig. P11–2 is inductive in nature. Use only the low frequency model for the transistor.

11–3 Describe in detail the steps to be followed to realize the circuit of Fig. P11–3 as a bipolar integrated circuit.

11–4 Show a circuit diagram for a more realistic model of the integrated circuit of Problem 11–3 including parasitic elements.

11–5 You are given a rectangular thin sheet of uniform homogeneous conducting material of width W, length L, and thickness T and having resistivity ρ.

a) Calculate the resistance along the *length* of the rectangular sheet as a function of the given parameters.

b) Show that if the thin sheet is square in shape that its resistance along the length is given by

$$R = 8/T.$$

This is called sheet resistance and is designated Ohms/square or graphically designated by $\Omega/\square$.

c) If a thin rectangular sheet has sheet resistance of $2000\Omega/\square$, $L = 2$ mm, and $W = 1$ mm, what is the resistance of this sheet along its length?

11–6 Noting that minority carriers can be stored in the base and the collector show that the expected storage time in the devices of Fig. 11–18 are the least in device B and the most in device E.

11–7 Show the top view of the integrated circuit layout of transistor T_E of Fig. 11–30b.

REFERENCES

Camenzind, H. R., *Circuit Design for Integrated Electronics*, Addison-Wesley, Reading, Mass., 1968.

Camenzind, H. R., and Grebene, A. B., "An Outline of Design Techniques for Linear Integrated Circuits," I.E.E.E. Journal of Solid-State Circuits, Vol. SC-4, No. 3, June 1969.

Chang, T. H. P., and Wallman, B. A., "A Computer-Controlled Electron-Beam Machine for Microcircuit Fabrication," IEEE Trans. on Electron Devices, Vol. ED—19, No. 5, p. 629, May 1972.

Chirlian, P. M., *Integrated and Active Network Analysis and Synthesis*, Ch. 2, Prentice-Hall Inc., Englewood Cliffs, N.J., 1967.

Dix, C. et al., "Paper-Tape-Controlled Electron Probe Resist Exposure and Direct Metallic Deposition," IEEE Trans. on Electron Devices, Vol. ED-19, No. 5, p. 641, May 1972.
Gray, P. E., and Searle, C. L., *Electronic Principles*, Ch. 5, John Wiley and Sons Inc., New York, 1969.
Herzog, R. F. et al., "Computer Controlled Resist Exposure in the Scanning Electron Microscope," IEEE Trans. on Electron Devices, Vol. ED-19, No. 5, p. 635, May 1972.
Liveay, W. R., and Fritz, R. B., IEEE Trans. on Electron Devices, Vol. ED-19, No. 5, p. 647, May 1972.
Lynn, D. L., Meyer, C. S., and Hamilton, D. J., *Analysis and Design of Integrated Circuits*, Motorola Semiconductors Inc., McGraw Hill, New York, 1967.
Ozdemir, F. S., et al., "Computer-Controlled Scanning Electron Microscope System for High-Resolution Microelectronic Pattern Fabrication," IEEE Trans. on Electron Devices, Vol. ED-19, No. 5, p. 624, May 1972.
Schilling, D. L., and Belove, C., *Electronic Circuits: Discrete and Integrated*, Ch. 9, McGraw-Hill Book Co., New York, 1968.
Vadasz, L. L., Grove, A. S., Rowe, T. A., and Moore, C. T., "Silicon Technology," IEEE Spectrum, pp. 28–35, Oct. 1969.
Warner, R. M., and Fordemwalt, J. N., *Integrated Circuits*, Chs. 11–15, McGraw-Hill Book Co., New York, 1965.

CHAPTER 12

RECOMBINATION-GENERATION IN THE TRANSITION REGION OF *p-n* JUNCTIONS AND HIGH INJECTION

12-1 INTRODUCTION

In this chapter we consider removing one of the key assumptions used in obtaining terminal current–voltage characteristics of junction diodes and transistors. For example, the following previously described equation

$$I = I_S(e^{qV/kT} - 1) \tag{12–1}$$

assumes, among other things, that recombination or generation of carriers in the transition region is negligibly small. Some experimentally observed $I(V)$ characteristics of *p-n* junction diodes deviate significantly both from Eq. (12–1) and from that predicted by the complete model of the diode including series bulk resistance and leakage resistance. We shall, in this chapter, take account of *recombination and generation of carriers in the transition region* and see under what conditions it significantly influences the $I(V)$ characteristics. We shall also consider the effect of high injection on *p-n* junction $I(V)$ characteristics. High injection effects are important when the injected minority carrier densities are not negligible compared to the normally present majority carrier density in the region of interest. The consideration of both the recombination–generation of carriers in the transition region, and high injection remove the two most important restrictions on Eq. (12–1).

12-2 RECOMBINATION-GENERATION STATISTICS

We wish to consider the general recombination–generation process. An expression for the recombination rate which is more general than the expressions obtained in Section 2–11 will be obtained. The major assumptions of the analysis are:

1 recombination–generation centers, also known as *traps*, are located at a *single* energy level somewhere in the band gap, and
2 these centers are *uniformly* distributed throughout the material.

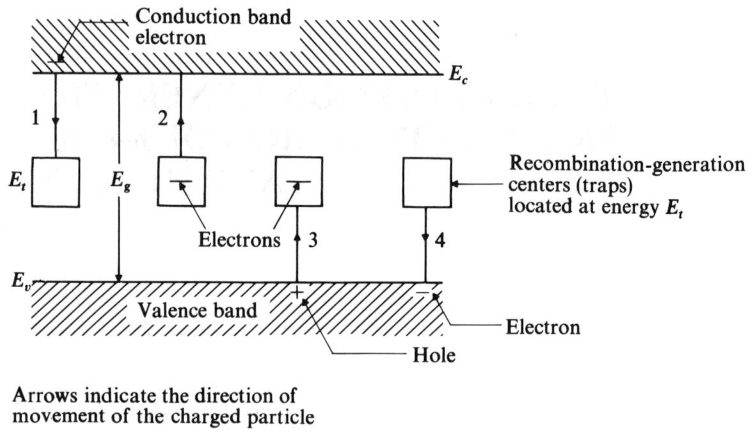

Fig. 12–1 Electron and hole emission and capture by trap. 1. Electron capture by trap. 2. Electron emission by trap. 3. Hole capture by trap. 4. Hole emission by trap.

Figure 12–1 shows the four different processes which are involved in the recombination-generation of carriers. Two of these are the *electron capture* and *emission.* The remaining two are *hole capture* and *emission.* Here the words *emission* and *capture* are *with reference to the traps.* For example, an electron emission implies the trap is emitting an electron, whereas a hole capture implies the hole is being captured by the trap. It is useful to note that when the trap or recombination-generation center has an electron, it can give it up either to the conduction band or to the valence band. The former is termed electron emission. The latter is hole capture. Note that a trap occupied by a hole does not have an electron in it. The capture of a hole comes about by an electron which leaves the trap and recombines with a hole in the valence band. Now, as the trap is empty, it is considered to be occupied by a hole.

An analogous situation occurs when the trap is empty. Now a conduction band electron can fall into the empty and available trap, or a valence band electron can make the transition to the trap. When a conduction band electron falls into the trap we call this *electron capture.* When a valence band electron makes the transition to the trap, it is useful to consider this as the transition of the hole from the trap to the valence band. That is, when a valence band electron goes into the trap, the trap is said to have emitted the hole. In Fig. 12–1 the arrow refers to the motion of the carrier of interest for both electrons and holes.

It is reasonable to expect that the electron capture rate should be proportional to the density of electrons in the conduction band from where they come. The capture rate should also be proportional to the fraction of traps which are *not* already occupied by electrons (or occupied by holes) and hence are available for electron occupancy. Thus the capture rate U_{ct} of conduction band electrons by available traps is

$$U_{ct} = \frac{nf_{tp}}{\tau_{n0}}, \tag{12–2†}$$

where n = free electron density in conduction band,

f_{tp} = fraction of traps occupied by holes (i.e. not occupied by electrons),

τ_{n0} = limiting lifetime of injected electrons into a highly p-type specimen.

Equation (12–2) shows that the proportionality constant is the reciprocal of the limiting lifetime of electrons.

Electron emission rate from trap to conduction band U_{tc} is expected to be proportional to the available electronic states in the conduction band, the fraction of traps occupied by electrons, probability of electron emission, and trap density. It is, however, easier to obtain U_{tc} by requiring that *under thermal equilibrium* the electron capture rate must equal the electron emission rate. This requirement leads to (Sah, et al, 1957)

$$U_{tc} = \frac{n_1 f_t}{\tau_{n0}}, \tag{12–3}$$

where U_{tc} = rate of electron emission from traps to conduction band,

f_t = fraction of traps occupied by electrons,

n_1 = density of electrons in the conduction band when the Fermi level is at the trap level,

τ_{n0} = limiting lifetime of electrons in highly p-type semiconductor.

Because each trap is occupied either by an electron or by a hole, it follows that

$$f_t = 1 - f_{tp}. \tag{12–4}$$

The definition of n_1 allows us to write it explicitly as

$$n_1 = n_i e^{(E_t - E_i)/kT}, \tag{12–5}$$

where E_t = trap energy,

E_i = intrinsic energy,

n_i = intrinsic carrier density.

Here E_t has been substituted for E_f in Eq. (3–13) as required by the definition.

At *thermal equilibrium*

$$U_{ct} = U_{tc}. \tag{12–6}$$

† The order of the subscripts on U tells where the carrier of interest originates and where it ends up. Thus, for example, U_{ct} is the capture rate of electrons going from the *conduction band* to the *trap*.

When external sources such as batteries are used in a circuit containing the *p-n* junction, conditions of thermal equilibrium do not exist. Under these circumstances the steady net rate of electron capture is given by

$$U_{cn} = U_{ct} - U_{tc} = (nf_{tp} - n_1 f_t)/\tau_{n0}. \tag{12-7}$$

Hole capture rate is expected to be proportional to the hole density p in the valence band and to the fraction of traps occupied by electrons. Thus the rate at which the holes in the valence band arrive at the traps is given by

$$U_{vt} = \frac{pf_t}{\tau_{p0}}, \tag{12-8}$$

where τ_{p0} = limiting lifetime of holes in highly *n*-type semiconductor.

Analogous to Eq. (12–3), the rate at which the traps emit holes to the valence band is given by

$$U_{tv} = \frac{p_1 f_{tp}}{\tau_{p0}}, \tag{12-9}$$

where p_1 = hole density in the valence band when the Fermi level is at the trap level.

With the direct use of the above definition, an explicit expression can be written for p_1 as

$$p_1 = n_i e^{(E_i - E_t)/kT}. \tag{12-10}$$

At thermal equilibrium

$$U_{vt} = U_{tv}. \tag{12-11}$$

At *steady state* but *thermal nonequilibrium* the net rate of hole capture by the traps is given by

$$U_{cp} = (pf_t - p_1 f_{tp})/\tau_{p0}. \tag{12-12}$$

We seek an expression for the recombination rate U for *thermal nonequilibrium* though *steady-state condition.* Furthermore, such an expression should be preferably independent of f_t or f_{tp}. We shall therefore first express f_{tp} in terms of f_t by using Eq. (12–4). This allows us to write Eqs. (12–7) and (12–12) in terms of carrier densities and only f_t. Then finally f_t itself can be eliminated by invoking the condition of steady state though thermal nonequilibrium, namely that the net rates of capture of electrons and holes must be the same. Mathematically it may be stated as

$$U_{cn} = U_{cp}. \tag{12-13}$$

Now that the strategy is clear we may proceed by rewriting Eq. (12–7) to eliminate f_{tp}.

$$U_{cn} = [n(1 - f_t) - n_1 f_t]/\tau_{n0} \tag{12-14}$$

or

$$U_{cn} = [n - f_t(n + n_1)]/\tau_{n0}. \tag{12–15}$$

Similarly we may rewrite Eq. (12–12) to eliminate f_{tp} from it

$$U_{cp} = [pf_t - p_1(1 - f_t)]/\tau_{p0} \tag{12–16}$$

or

$$U_{cp} = [f_t(p + p_1) - p_1]/\tau_{p0}. \tag{12–17}$$

Equation (12–13) may now be used to solve for f_t as follows by direct substitution

$$[n - f_t(n + n_1)]/\tau_{n0} = [f_t(p + p_1) - p_1]/\tau_{p0} \tag{12–18}$$

or

$$\frac{n}{\tau_{n0}} - f_t\frac{(n + n_1)}{\tau_{n0}} = f_t\frac{(p + p_1)}{\tau_{p0}} - \frac{p_1}{\tau_{p0}}. \tag{12–19}$$

Rearranging we have

$$\frac{n}{\tau_{n0}} + \frac{p_1}{\tau_{p0}} = f_t\left[\frac{p + p_1}{\tau_{p0}} + \frac{n + n_1}{\tau_{n0}}\right] \tag{12–20}$$

or

$$\frac{n\tau_{p0} + p_1\tau_{n0}}{\tau_{n0}\tau_{p0}} = \frac{f_t}{\tau_{p0}\tau_{n0}}[(p + p_1)\tau_{n0} + (n + n_1)\tau_{p0}]. \tag{12–21}$$

Thus f_t may be expressed as

$$f_t = \frac{n\tau_{p0} + p_1\tau_{n0}}{[(p + p_1)\tau_{n0} + (n + n_1)\tau_{p0}]}. \tag{12–22}$$

We may substitute this expression for f_t into Eq. (12–17) to obtain

$$U_{cp} = \left(\frac{1}{\tau_{p0}}\right)\left[\frac{(n\tau_{p0} + p_1\tau_{n0})(p + p_1)}{(p + p_1)\tau_{n0} + (n + n_1)\tau_{p0}} - p_1\right]. \tag{12–23}$$

Clearing fractions in the square brackets we obtain

$$+np_1\tau_{p0} + p_1^2\tau_{n0}$$

$$U_{cp} = \left(\frac{1}{\tau_{p0}}\right)\frac{pn\tau_{p0} + pp_1\tau_{n0} - p_1p\tau_{n0} - p_1^2\tau_{n0} - p_1n\tau_{p0} - p_1n_1\tau_{p0}}{(p + p_1)\tau_{n0} + (n + n_1)\tau_{p0}}$$

or

$$U_{cp} = \frac{\tau_{p0}(pn - p_1n_1)}{\tau_{p0}[(p + p_1)\tau_{n0} + (n + n_1)\tau_{p0}]}. \tag{12–24}$$

Finally

$$U_{cp} = U_{cn} = U = \frac{pn - p_1n_1}{(p + p_1)\tau_{n0} + (n + n_1)\tau_{p0}}. \tag{12–25}$$

Equations (12–5) and (12–10) allow us to write

$$p_1n_1 = (n_i\,e^{(E_i - E_t)/kT})(n_i\,e^{(E_t - E_i)/kT}) = n_i^2. \tag{12–26}$$

Thus our desired expression for the *steady state nonthermal equilibrium recombination rate for electrons or holes is*

$$U = \frac{pn - n_i^2}{(p + p_1)\tau_{n0} + (n + n_1)\tau_{p0}}. \tag{12–27}$$

Equation (12–27) *is a very important result* which is general within the framework of the assumptions which were our starting point. We may interpret U to be the *recombination rate* when it is positive and the *generation rate* when it is negative.

Let us see how the more general Eq. (12–27) is related to the simple expression for the recombination rate of electrons and holes given in Eqs. (2–106) and (2–107). Consider an n-type semiconductor, so that

$$n = n_n \gg p_p n_i. \tag{12–28}$$

Also

$$n_n \gg n_1, p_1, p. \tag{12–29}$$

Now if we assume that τ_{p0} and τ_{n0} *are of the same order of magnitude*, then Eq. (12–27) simplifies considerably to

$$U = \frac{pn - p_n n}{(\cancel{p} + \cancel{p_1})\tau_{n0} + (n + \cancel{n_1})\tau_{p0}} \tag{12–30†}$$

$$U = \frac{pn - p_n n}{n\tau_{p0}} = \frac{p - p_n}{\tau_{p0}}. \tag{12–31}$$

It is seen that Eq. (12–31) is identical to Eq. (2–106). Use was made of $n = n_n$ from Eq. (12–28) in arriving at Eq. (12–31).

Next we consider a p-type semiconductor so that

$$p = p_p \gg n_n \gg n_i. \tag{12–32}$$

Also

$$p_p \gg p_1, n_1, n. \tag{12–33}$$

Again if we assume that τ_{p0} and τ_{n0} *are the same order of magnitude*, then Eq. (12–27) simplifies to

$$U = \frac{pn - pn_p}{(p + \cancel{p_1})\tau_{n0} + (\cancel{n} + \cancel{n_1})\tau_{p0}} \tag{12–34}$$

$$U = \frac{pn - pn_p}{p\tau_{n0}} = \frac{n - n_p}{\tau_{n0}}. \tag{12–35}$$

We see that Eq. (12–35) is the same as Eq. (2–107). Thus Eq. (12–27) does indeed simplify to the simple expressions obtained earlier. Eq. (12–27), however, is a

---

† A line through a quantity signifies that it can be neglected.

more general result, and the approximations involved in obtaining Eqs. (12–31) and (12–35) are explicitly stated.

12-3 SYMMETRIC JUNCTIONS AND SIMPLE CALCULATIONS

In this section we shall do some simple calculations based on Eq. (12–27) as applied to a *symmetric junction.* This will enable us to develop a feel for the meaning of Eq. (12–27). We shall also assume that the traps are located at intrinsic energy E_i for simplicity. Thus for this section we shall assume

$$p_1 = n_1 = n_i \tag{12–36}$$

$$\tau_p = \tau_n \triangleq \tau \tag{12–37}$$

$$\mu_p = \mu_n \triangleq \mu \tag{12–38}$$

$$D_p = D_n \triangleq D \tag{12–39}$$

$$n_p = p_n \tag{12–40}$$

$$p_p = n_n \tag{12–41}$$

$$L_p = L_n \triangleq L \tag{12–42}$$

Case A. Reverse-Biased Junction

In considering a reverse biased *p-n* junction, if we focus our attention on the *transition region,* it is clear that because of the presence of a rather high electric field in this region, the free carriers will be swept out of the transition region rather rapidly. The recombination rate U in this case can be obtained from the general Eq. (12–27), where

$$n, p \ll n_i, n_1, p_1. \tag{12–43}$$

Equation (12–27) then becomes

$$U = \frac{\not{p}\not{n} - n_i^2}{(\not{p} + p_1)\tau_{n0} + (\not{n} + n_1)\tau_{p0}}. \tag{12–44}$$

Using Eqs. (12–43), (12–36), and (12–37) we have

$$U = -\frac{n_i^2}{2n_i\tau} = -\frac{n_i}{2\tau}. \tag{12–45}$$

Note that the negative sign in Eq. (12–45) indicates that carriers are being *generated* in the transition region.

In general, to obtain the current density resulting from recombination-generation of carriers, the following integration is required for the one-dimensional device

$$J_{RG} = q\int U\,dx, \tag{12–46}$$

where q = magnitude of the electronic charge.

By substituting Eq. (12–45) into Eq. (12–46), noting that U in this case is *not* a function of x, and by evaluating the integral over the transition region width x_t, we obtain

$$J_{RG} = -\frac{qn_i x_t}{2\tau}. \tag{12–47}$$

The corresponding recombination-generation current i_{rg} is obtained simply as

$$I_{RG} = -\frac{qn_i x_t A}{2\tau}. \tag{12–48}$$

It should be noted that Eqs. (12–47) and (12–48) relate to terminal current flow because of recombination-generation of carriers in the transition region. This component of current was assumed to be zero in obtaining the ideal *p-n* junction diode equation. If both diffusion and recombination-generation currents are considered, the terminal current, exclusive of ohmic leakage currents, is given by

$$I = I_{RG} + I_D, \tag{12–49}$$

where I_{RG} = recombination-generation current,

I_D = current due to diffusion in the bulk region.

Current I_D for a reverse-biased *p-n* junction obtained from Eqs. (3–56) and (3–56a) is given below

$$I_D = -qA\left[\frac{p_n D_p}{L_p} + \frac{n_p D_n}{L_n}\right]. \tag{12–50}$$

The use of symmetry conditions given by Eqs. (12–39), (12–40), and (12–42) allow us to simplify Eq. (12–50) still further

$$I_D = -2q\frac{A n_p D}{L}. \tag{12–51}$$

As

$$L^2 \triangleq D\tau. \tag{12–52}$$

Equation (12–51) may be rewritten as

$$I_D = -2q\frac{An_p D\tau}{L\tau} = -2q\frac{An_p L}{\tau}. \tag{12–53}$$

We can now obtain the ratio I_{RG}/I_D as follows

$$\frac{I_{RG}}{I_D} = \frac{-qn_i x_t A/2\tau}{-2qAn_p L/\tau} = \frac{1}{4}\left(\frac{n_i}{n_p}\right)\left(\frac{x_t}{L}\right). \tag{12–54}$$

Another form of Eq. (12–54) is the following

$$\frac{I_{RG}}{I_D} = \frac{1}{4}\left(\frac{p_p}{n_i}\right)\left(\frac{x_t}{L}\right). \tag{12–55}$$

Equation (12–55) indicates that I_{RG} will tend to be more important in those materials which have low n_i and/or low L. Generally speaking all other things being equal the wider band gap materials will have a lower n_i. *Thus we may say that in* p-n *junctions made out of materials having wider band gap and/or lower minority carrier diffusion length,* I_{RG} *is expected to be important.* This observation is found to be true in practice. For example, in silicon usually the recombination–generation current is significantly more important than in germanium particularly at low current density, as we shall see by the following numerical calculations in the above two materials.

Example 12–1. Consider germanium and silicon *p-n* junctions whose *n* and *p* regions have conductivity of 1 Ohm-cm, a microsecond lifetime, and 1μ transition region width.

Calculate the ratio of recombination–generation current to diffusion current in both these devices at room temperature.

Solution We may rewrite Eq. (12–55) which is true for reverse-biased *p-n* junctions, in terms of the above known quantities

$$\frac{I_{RG}}{I_D} = \frac{1}{4}\left(\frac{\sigma_p}{q\mu_p n_i}\right)\left(\frac{x_t}{L}\right). \tag{12–56}$$

If we are to be consistent with the assumptions made in this section, it would be advisable to choose values of μ_p and D_p to be the average between the corresponding quantities for electrons and holes. Thus for silicon,

$$\mu = \frac{\mu_p + \mu_n}{2} = \frac{480 + 1350}{2} = 915 \text{ cm}^2/\text{Volt-sec.}$$

$$D = \frac{D_p + D_n}{2} = \frac{12.5 + 31}{2} = 21.75 \text{ cm}^2/\text{sec.}$$

So

$$L = \sqrt{D\tau} = \sqrt{21.75 \times 10^6} = 4.66 \times 10^{-3} \text{ cm.}$$

For germanium,

$$\mu = \frac{1900 + 3900}{2} = 2900 \text{ cm}^2/\text{Volt-sec.}$$

$$D = \frac{D_p + D_n}{2} = \frac{44 + 93}{2} = 68.5 \text{ cm}^2/\text{sec.}$$

So
$$L = \sqrt{D\tau} = \sqrt{68.5 \times 10^{-6}} = 8.3 \times 10^{-3} \text{ cm}.$$

We are now ready to compute the ratio of recombination–generation current to diffusion current for silicon and germanium by using Eq. (12–56) as follows.

Silicon

$$\frac{I_{RG}}{I_D} = \frac{(1)(10^{-4})}{4(1.6 \times 10^{-19})(915)(1.38 \times 10^{10})(4.66 \times 10^{-3})}$$

$$\frac{I_{RG}}{I_D} = \underline{2650.}$$

Germanium

$$\frac{I_{RG}}{I_D} = \frac{1}{4}\frac{\sigma_p x_t}{q\mu n_i L} = \frac{(1)(10^{-4})}{4(1.6 \times 10^{-19})(2900)(2.35 \times 10^{13})(8.3 \times 10^{-3})}$$

$$\frac{I_{RG}}{I_D} = \underline{0.277.}$$

We see from the above numerical example that the recombination–generation current is indeed dominant in reverse-biased silicon *p-n* junctions, and not quite so important in germanium *p-n* junctions. The major difference between silicon and germanium is due to the large difference in the value of n_i which is importantly determined by the band gap. In the above example the specific numbers are not quite so important as the order of magnitude of the calculated ratios.

Considering *only the intrinsic p-n* junction characteristics, we have already noted in Section 3–4 that for reverse bias greater than a few kT/q, the incremental resistance is extremely large. In this range of reverse bias, $i(v)$ is a perfectly flat curve because

$$I = -I_S \tag{12–57}$$

for

$$V \ll -q/kT. \tag{12–58}$$

If we include the contribution due to recombination–generation current, it makes the incremental resistance, even in reverse bias, satisfying Eq. (12–1), finite.

In *step junctions* I_{RG} becomes

$$I_{RG} = q\frac{n_i A x_t}{2\tau} = -q\frac{n_i A}{2\tau} k_a (V_{BI} - V)^{\frac{1}{2}}. \tag{12–59}$$

The incremental resistance of the *p-n* junction is simply obtained by computing dV/dI from Eq. (12–59). Thus

$$\frac{dI_{RG}}{dV} = +\frac{qn_i Ak_a}{2\tau}(V_{BI} - V)^{-\frac{1}{2}} \tag{12–60}$$

and
$$\frac{dV}{dI_{RG}} \triangleq r_{rg} = \frac{2\tau(V_{BI} - V)^{\frac{1}{2}}}{qn_iAk_a}. \tag{12-61}$$

If the *p-n* junction is *linearly graded*, the recombination-generation current is given by the following

$$I_{RG} = -q\frac{n_iAx_t}{2\tau} = -q\frac{n_iAK_\ell}{2\tau}(V_{BI} - V)^{\frac{1}{3}}. \tag{12-62}$$

Then

$$\frac{1}{r_{rg}} = \frac{dI_{RG}}{dV} = \frac{qn_iAk_\ell(V_{BI} - V)^{-\frac{2}{3}}}{6\tau}. \tag{12-63}$$

Finally

$$\frac{dV}{dI_{RG}} = r_{rg} = \frac{6\tau(V_{BI} - V)^{\frac{2}{3}}}{qn_iAk_\ell}. \tag{12-64}$$

Whether or not the recombination-generation current proves to be the limiting incremental resistance of a reverse-biased *p-n* junction diode depends on whether the magnitude of r_{rg} as given by Eqs. (12–61) or (12–64) is less than or greater than the leakage resistance r'_ℓ respectively. Note also that r_{rg}, owing to recombination-generation in the transition region, is a function of reverse voltage, whereas r'_ℓ is not. This may make it possible to distinguish between r_{rg} and r'_ℓ experimentally.

Case B. Forward-Biased Junction

We wish to investigate how the $I(V)$ characteristics of a forward-biased junction change when the contribution due to recombination-generation of carriers *in the transition* region is taken into account.

Because of the symmetry of the device assumed in this section it stands to reason that the electron and hole densities are precisely equal to each other at the center of the transition region. Use of Eqs. (3–33) and (3–34) which are repeated below allows us to write for electron and hole densities.

$$p = n_i\, e^{q(\varphi_p - \psi)/kT} \tag{12-65}$$

$$n = n_i\, e^{q(\psi - \varphi_n)/kT}. \tag{12-66}$$

Thus the product *pn* evaulated at the *center* of the transition region is

$$p_{\text{center}}n_{\text{center}} = n_i^2\, e^{(q/kT)(\varphi_p - \varphi_n)} = n_i^2\, e^{qV/kT}.$$

Then

$$p_{\text{center}} = n_{\text{center}} = n_i\, e^{qV/2kT}. \tag{12-67}$$

Let us return to Eq. (12–27) to obtain the appropriate expression for the recombination rate U for carriers in the transition region of a forward-biased

p-n junction as a function of carrier densities. Because of Eq. (12–67), and assuming that

$$V > \text{several } kT/q \tag{12–68}$$

the following inequalities are true

$$p, n \gg n_i, n_1, p_1. \tag{12–69}$$

This allows us to write Eq. (12–27) as follows

$$U = \frac{pn - \not{n}_i^2}{(p + \not{p}_1)\tau_{n0} + (n + \not{n}_1)\tau_{p0}}. \tag{12–70}$$

Thus with $\tau_{p0} = \tau_{n0} = \tau$

$$U = \frac{pn}{2n\tau} = \frac{p}{2\tau} = \frac{n}{2\tau}. \tag{12–71}$$

At the center of the transition region

$$U = \frac{n_i\, e^{qV/2kT}}{2\tau}, \tag{12–72}$$

where use of Eq. (12–67) was made to arrive at Eq. (12–72). The recombination rate U is expected to decrease exponentially with a characteristic length of $kT/q\mathscr{E}$ as we move away from the center of the transition region in either direction. This is because the minority carrier density diminishes as we move away from the center of the transition region. For example, hole density diminishes as we move into the n-region of the transition region, just as electron density diminishes as we move away from the center of the transition region into the p-region. Figure 12–2 shows how U changes with distance.

If we designate the origin of the coordinate system to be the center of the transition region, then in the n-region

$$x > 0,$$

$$U(x) = \frac{n_i\, e^{qV/2kT}}{2\tau}\, e^{-x/[kT/(q\mathscr{E})]}, \tag{12–73}$$

where $\mathscr{E}$ = electric field, *assumed* to be constant, in the transition region.

For $x \leqslant 0$

$$U(x) = \frac{n_i\, e^{qV/2kT}}{2\tau}\, e^{x/[kT/(q\mathscr{E})]}. \tag{12–74}$$

We shall take advantage of the symmetry of the problem as seen in Eqs. (12–73) and (12–74), and integrate Eq. (12–73) over half the transition region and then double the result. Strictly speaking the range of integration in this case should

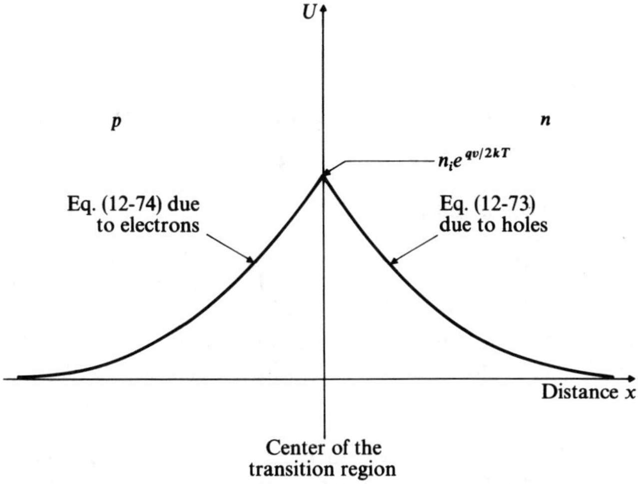

Fig. 12–2 U_{po} in the transition region of a *p-n* junction.

be $x = 0$ to $x = x_t/2$. For mathematical simplicity, however, the integration will be from $x = 0$ to $x = \infty$.

$$I_{RG} = 2qA \int_0^\infty \frac{n_i}{2\tau} e^{qV/2kT} e^{-x/[kT/(q\mathscr{E})]} dx \tag{12–75}$$

$$= \frac{qAn_i}{\tau} e^{qV/2kT} \left(-\frac{kT}{q\mathscr{E}}\right) [e^{-x/[kT/q\mathscr{E})]}]_0^\infty . \tag{12–76}$$

finally

$$I_{RG} = \frac{qAn_i\, e^{qV/2kT}}{\tau} \left(\frac{kT}{q\mathscr{E}}\right) \tag{12–77}$$

or

$$I_{RG} = \frac{qAn_i}{\tau} \left(\frac{kT}{q\mathscr{E}}\right) e^{qV/2kT}.$$

The diffusion component of current I_D for a *symmetric device* can be obtained from Eq. (3–55)

$$I_d = \frac{2qAn_pL}{\tau} e^{qV/kT}. \tag{12–79}$$

Equation (12–53) was utilized in arriving at Eq. (12–79).

We are now ready to compute the ratio I_{RG}/I_D as follows

$$\frac{I_{RG}}{I_D} = \frac{(qAn_i/\tau)(kT/q\mathscr{E})\, e^{qV/2kT}}{(2qAn_pL/\tau)\, e^{qV}/^{kT}}. \tag{12–80}$$

Finally
$$\frac{I_{RG}}{I_D} = \left(\frac{n_i}{n_p}\right)\left(\frac{kT}{q\mathscr{E}}\right)\left(\frac{1}{2L}\right)e^{-qV/2kT}. \tag{12–81}$$

If a linear approximation of the electric field in the transition region is used
$$\mathscr{E} = (V_{bi} - V)/x_t \tag{12–82}$$
then Eq. (12–81) becomes finally
$$\frac{I_{RG}}{I_D} = \left(\frac{n_i}{n_p}\right)\left(\frac{kT}{q(V_{BI} - V)}\right)\left(\frac{x_t}{2L}\right)e^{-qV/2kT}. \tag{12–83}$$

We shall again do a numerical calculation for germanium, silicon, and gallium arsenide to develop a feel for the relative magnitude of currents i_{rg} and i_d in different materials.

Example 12–2. Again we shall consider a *p-n* junction whose *p* and *n* regions have conductivity of 1 Ohm-cm, minority carrier lifetime of 10^{-6} sec, transition region width of 0.1 μ at a forward externally applied voltage of half the built-in voltage (of the junction) which is assumed to be half the band gap voltage.

Solution. It is useful to rewrite the ratio (n_i/n_p) of Eq. (12–83) in terms of known quantities as follows
$$\left(\frac{n_i}{n_p}\right) = \frac{\sigma_p}{q\mu_p n_i} \tag{12–84}$$
The ratio of Eq. (12–83) can be now rewritten utilizing Eq. (12–84) as
$$\frac{I_{RG}}{I_D} = \left(\frac{\sigma_p}{q\mu_p n_i}\right)\left(\frac{kT}{q(V_{BI} - V)}\right)\left(\frac{x_t}{2L}\right)e^{-qV/2kT}. \tag{12–85}$$

If we consider the external voltage to be half the built-in voltage, we can compute the ratio of I_{RG}/I_D for the three different materials as follows at room temperature ($kT/q = 0.026$ Volt). *For germanium*, $V_{BI} = 0.33$ Volt, $V = 0.165$ Volt. The ratio given by Eq. (12–85) may now be evaluated as follows.

$$\frac{I_{RG}}{I_D} = \left(\frac{\sigma_p}{q\mu_p n_i}\right)\left(\frac{kT}{q(V_{BI} - V)}\right)\left(\frac{x_t}{2L}\right)e^{-qV/2kT}$$
$$= \left(\frac{(1)}{(1.6\times 10^{-19})(2900)(2.35\times 10^{13})}\right)\left(\frac{26}{165}\right)\left(\frac{(10^{-5})}{2\times 8.3\times 10^{-3}}\right)e^{-3.17}$$
$$\frac{I_{RG}}{I_D} = 8.68\times 10^{-3}\times e^{-3.17} = \underline{3.65\times 10^{-2}}$$

For Si,
$$V_{BI} = 0.55 \text{ Volt}, V = 0.275 \text{ Volt}$$

$$\frac{I_{RG}}{I_D} \left(\frac{(1)}{(1.6 \times 10^{-19})(915)(1.38 \times 10^{10})}\right)\left(\frac{26}{275}\right)\left(\frac{10^{-5}}{2 \times 4.66 \times 10^{-3}}\right) e^{-5.29}$$

$$\frac{I_{RG}}{I_D} = 5.0 \times 10^2 \, e^{-5.29} = 2.5.$$

For GaAs,

$$V_{BI} = 0.72 \text{ Volt}, \; V = 0.36$$

$$\mu = \frac{\mu_p + \mu_n}{2} = \frac{400 + 7000}{2} = 3700$$

$$D_p = \frac{kT}{q} \mu_p = (0.026)(400) = 10.4 \text{ cm}^2/\text{sec}$$

$$D_n = \frac{kT}{q} \mu_n = (0.026)(7000) = 182 \text{ cm}^2/\text{sec}$$

Thus

$$D = \frac{D_p + D_n}{2} = \frac{10.4 + 182}{2} = 96.2 \text{ cm}^2/\text{sec}.$$

Finally

$$L = \sqrt{D\tau} = \sqrt{96.2 \times 10^{-6}} = 9.78 \times 10^{-3} \text{ cm}.$$

We are now ready to compute I_{RG}/I_D as follows

$$\frac{I_{RG}}{I_D} = \left(\frac{(1)}{(1.6 \times 10^{-19})(3700)(3.5 \times 10^{6})}\right)\left(\frac{26}{360}\right)\left(\frac{10^{-5}}{2 \times 9.78 \times 10^{-3}}\right) e^{-6.92}$$

$$\frac{I_{RG}}{I_D} = 1.78 \times 10^4 \times e^{-6.92} = \underline{17.8.}$$

Again we see that the recombination–generation current is dominant in silicon and gallium arsenide and not in germanium at a forward voltage across the transition region of a quarter of the band-gap voltage. Thus we see that all other things being approximately equal, I_{RG} is more important in the wider band-gap materials. Because of the negative exponential factor $e^{-qV/2kT}$, at higher forward bias voltage the diffusion component is again expected to be dominant. Usually then, the recombination–generation current is expected to be quite important at low currents in devices of almost all materials. The recombination–generation is expected to diminish just as injection current is expected to increase as forward bias is increased. In *p-n* junctions made from wider band gap and low minority carrier diffusion length material, the range of bias over which the recombination–generation current is important is wider than that of those devices made from narrower band gap and higher minority carrier diffusion length material.

It is important to note that the above discussion *does not* say that we can neglect recombination–generation current in narrower band gap material like germanium, nor does it say that in silicon and gallium arsenide the recombination–generation current is always dominant. At sufficiently low forward bias recombination–generation current is probably important in all *p-n* junctions. As forward bias is increased, the injection current usually will begin to dominate in all devices.

12–4 MORE ACCURATE AND GENERAL EXPRESSIONS FOR THE RECOMBINATION RATE OF U

So far we have considered only the *symmetrical p-n* junction for which Eqs. (12–36) through (12–42) are valid. A more general formulation is necessary if some or all of these equations are not valid. In this section the more general problem is considered.

We begin by noticing that Eq. (12–27) is quite general and does not require symmetry. For the general junction the following equations are valid.

$$p = n_i\, e^{(q/kT)(\varphi_p - \psi)} \tag{12–86}$$

$$n = n_i\, e^{(q/kT)(\psi - \varphi_n)} \tag{12–87}$$

$$p_1 = n_i\, e^{(E_i - E_t)/kT} \tag{12–88}$$

$$n_1 = n_i\, e^{(E_t - E_i)/kT} \tag{12–89}$$

$$-q\psi \triangleq E_i = [E_c + E_v - kT\,\ell\text{n}\,(N_c/N_v)]/2. \tag{12–90}$$

Equations (12–86), (12–87), and (12–89) have been established in the earlier chapters. Equations (12–88) and (12–89) follow from the definitions of n_1, and p_1. The quasi Fermi potential for electrons, and holes φ_n, φ_p, and the intrinsic electrostatic potential ψ have all been introduced in the earlier chapters.

When Eqs. (12–86) through (12–90) are substituted into Eq. (12–27), the following expression for the recombination rate U is obtained

$$U = \frac{n_i}{\sqrt{\tau_{p0}\tau_{n0}}} \cdot \frac{\sinh[(q/2kT)(\varphi_p - \varphi_n)]}{\cosh\{(q/kT)[\psi - (\varphi_p + \varphi_n)/2] + \ell\text{n}\,(\tau_{p0}/\tau_{n0})^{\frac{1}{2}}\} + e^{-(q/2kT)(\varphi_p - \varphi_n)} \cosh[(E_t - E_i)/kT + \ell\text{n}\,(\tau_{p0}/\tau_{n0})^{\frac{1}{2}}]}. \tag{12–91}$$

In principle the current density due to recombination–generation in the transition region J_{rg} may be found by substituting Eq. (12–91) into (12–46). In practice, however, it is not easy to do this calculation simply because the *x*-dependence of quasi Fermi levels and the electrostatic potential are quite difficult to obtain since simultaneous non-linear partial differential equations would need to be solved first. In an attempt to obtain analytical results we shall choose to make some judicious approximations. It will be useful to consider the *I-V* characteristics in three separate regions.

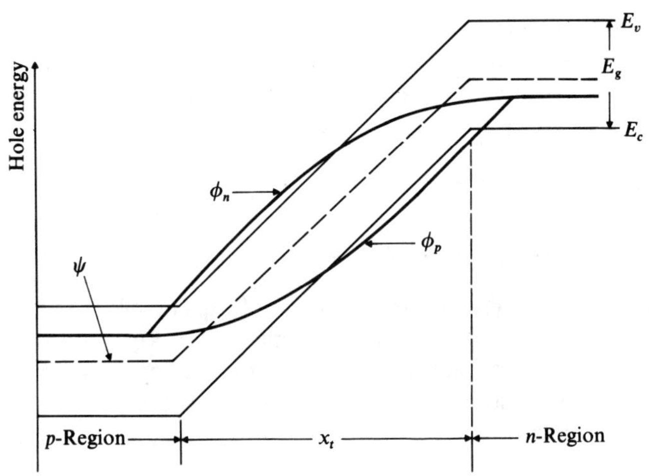

Fig. 12–3 Quasi-Fermi levels for electrons and holes in a *p-n* junction under high reverse bias.

A. Large Reverse Bias

It can be shown (Sah et al, 1957) that unlike the case of forward bias and small forward and reverse bias, where φ_n, φ_p are essentially constant† in the transition region, in the case of large reverse bias the φ_n and φ_p *change the most in the transition region* as shown in Fig. 12–3. Furthermore, $(\varphi_n - \varphi_p)$ is positive and several (kT/q) along almost the entire width of the transition region. Thus, for this case we may say

$$(\varphi_n - \varphi_p) \gg kT/q. \qquad (12\text{–}92)$$

The magnitude relationship (12–92) allows us to neglect the first term in comparison with the second term in the denominator of Eq. (12–91). With this in mind and using the definition of sinh x we may rewrite Eq. (12–91) as follows

$$U = \frac{n_i}{\sqrt{\tau_{p0}\tau_{n0}}} \cdot \frac{(1/2)(e^{(q/2kT)(\varphi_p - \varphi_n)} - e^{-(q/2kT)(\varphi_p - \varphi_n)})}{e^{-(q/2kT)(\varphi_p - \varphi_n)} \cosh[(E_t - E_i)/kT + \ell n\,(\tau_{p0}/\tau_{n0})^{\frac{1}{2}}]}. \qquad (12\text{–}93)$$

Because of (12–92), the first term on the right hand side of Eq. (12–93) may be neglected in comparison with the second term in the numerator. After neglecting this first term and cancelling the common terms in the numerator and the denominator of Eq. (12–93) we obtain

$$U_r = \frac{n_i}{2\sqrt{\tau_{p0}\tau_{n0}} \cosh[(E_t - E_i)/kT + \ell n\,(\tau_{p0}/\tau_{n0})^{\frac{1}{2}}]}. \qquad (12\text{–}94)$$

† This was shown in Section 3–2.

Note that Eq. (12–94) *is independent of the quasi-Fermi levels.* Furthermore the term $(E_t - E_i)$ is independent of x. Hence it may be concluded that U_r is constant with x in the transition region. With the help of Eq. (12–46) we may write

$$I_{RG} = J_{RG}A = qA \int_{-x_t/2}^{x_t/2} U_r dx = qAU_r x_t, \tag{12–95}$$

where U_r is given by Eq. (12–94).

Equation (12–95) is a more accurate version of Eq. (12–48). The latter assumes symmetry which the former does not assume. In fact U_r is not quite constant but diminishes close to the edges of the transition region. Thus the assumption of constant U_r in the transition region overestimates I_{RG}.

For abrupt junctions

$$I_{RG} = -qAU_r k_a (V_{BI} - V)^{\frac{1}{2}}. \tag{12–96}$$

For linearly graded junctions

$$I_{RG} = -qAU_r K_\ell (V_{BI} - V)^{\frac{1}{3}}. \tag{12–97}$$

Case B. Forward Bias

In contrast with the situation for large reverse bias, in forward bias the quasi-Fermi levels are essentially constant in the transition region. And it may be stated to a very good approximation that

$$\varphi_p - \varphi_n = V. \tag{12–98}$$

As noted earlier, in principle Eq. (12–27) gives the recombination rate. In general, however, it cannot be introduced into Eq. (12–46) to obtain the current–voltage relationship because the quasi-Fermi levels and the electrostatic potentials are not in general known as a function of the position coordinates. To get around this practical difficulty some judicious simplifications will be introduced. Equation (12–98) is one of these which has been proved earlier. If the electrostatic potential $\psi(x)$ is assumed to be *linear* throughout the transition region we may write the following

$$\psi(x) = (V_{BI} - V)\frac{x}{x_t}, \tag{12–99}$$

where V_{BI} = built-in voltage.

Equation (12–99) is written by first noting that the net voltage across the transition region is $(V_{BI} - V)$. Using Eq. (12–98), Eq. (12–99) may be rewritten

$$\psi(x) = [\psi_{BI} - (\varphi_p - \varphi_n)]\frac{x}{x_t} \tag{12–100}$$

for

$$-(x_t/2) < x < (x_t/2) \tag{12–101}$$

where $\psi_{BI} = V_{BI}$.

Note that a linear distribution of the electrostatic potential implies a constant field in the transition region. This approximation together with Eq. (12–98) allows us to integrate U according to Eq. (12–46) with the following result.

$$J_{rg} = \frac{2qn_i x_t}{\sqrt{\tau_{p0}\tau_{n0}}} \cdot \frac{\sinh(qV/2kT)}{(V_{BI} - V)q/kT} \cdot f(b), \tag{12–102}$$

where

$$f(b) \triangleq \int_{z_1}^{z_2} \frac{dz}{z^2 + 2bz + 1} \tag{12–103}$$

$$b \triangleq e^{-qV/2kT} \cosh\left[\frac{E_t - E_i}{kT} + \frac{1}{2}\ell n\,(\tau_{p0}/\tau_{n0})\right] \tag{12–104}$$

$$z_1, z_2 \triangleq (\tau_{p0}/\tau_{n0})^{\frac{1}{2}}\, e^{\pm(V_{BI} - V)q/2kT}. \tag{12–105}$$

Equation (12–102) was derived under the condition that the integration indicated in (12–46) was carried out over the entire device. Strickly speaking, the integration should be carried out over only the transition region. If this more accurate calculation is carried out, the following results are obtained.

$$J_{rg} = \frac{qn_i}{\sqrt{\tau_{p0}\,\tau_{n0}}}\left(\frac{x_t}{\alpha}\right) f(b) \sinh\left(\frac{qV}{2kT}\right), \tag{12–106}$$

where

$$f(b) = \frac{1}{\sqrt{1 - b^2}} \tan^{-1}\left[\frac{2\sqrt{1 - b^2}\sinh\alpha_r}{\sqrt{\tau_{p0}/\tau_{n0}} + \sqrt{\tau_{n0}/\tau_{p0}} + 2b\cosh\alpha_r}\right], \tag{12–107}$$

where b is still given by Eq. (12–104) and α_r is defined by

$$(q/kT)[\psi - (\varphi_p + \varphi_n)/2] = \alpha_r\left(\frac{x}{x_t/2}\right) \tag{12–108}$$

a) Very Low Voltage

For low voltages

$$V < \frac{kT}{q}. \tag{12–109}$$

We may rewrite Eq. (12–102) as follows:

$$J_{rg} = \frac{qn_i x_t(0)}{\sqrt{\tau_{p0}\tau_{n0}}} \cdot \frac{2(qV/2kT)}{(qV_{BI}/kT)} f(b) \tag{12–110}$$

$$i_{rg} = J_{rg}A = \frac{qAn_i x_t(0)V}{\sqrt{\tau_{p0}\tau_{n0}}\, V_{BI}} f(b) \tag{12–111}$$

where $x_t(0)$ = transition region width when $V = 0$.

From Eq. (12–111) we may calculate the resistance of the *p-n* junction due to recombination-generation as follows:

$$r_{rg}(\text{origin}) = \frac{V}{I_{RG}} = \frac{V_{BI}\sqrt{\tau_{p0}\tau_{n0}}}{qAn_i f(b) x_t(0)}. \tag{12–112}$$

b) Medium Forward Bias and Deep Recombination–Generation Centers

The recombination-generation centers are located fairly near the intrinsic energy level E_i. For this case Eqs. (12–102) through (12–105) lead to the conclusion that

$$I_{RG} \sim e^{qV/2kT}. \tag{12–113}$$

c) Small or Medium Forward Bias and Shallow Recombination–Generation Centers

By shallow recombination–generation centers is meant centers which are not too far from the band edge or centers which are many kTs away from the intrinsic energy level E_i. For this case Eqs. (12–102) through (12–105) yield

$$I_{RG} \sim e^{qV/kT}. \tag{12–114}$$

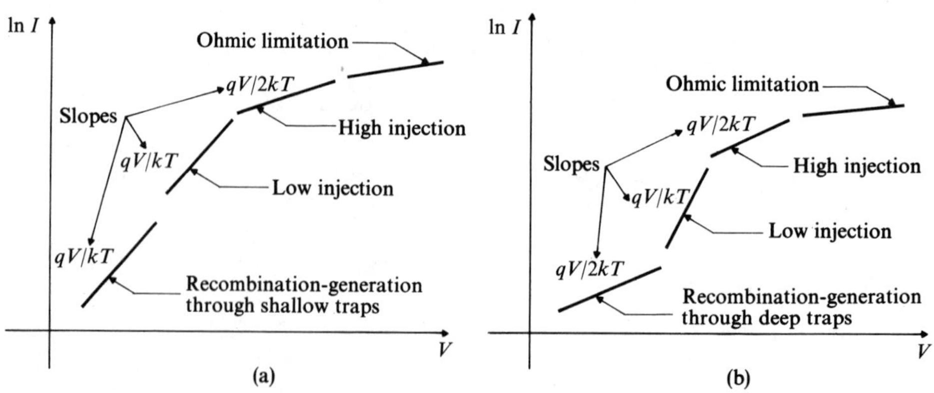

Fig. 12–4 Current voltage relationships in *p-n* junctions.

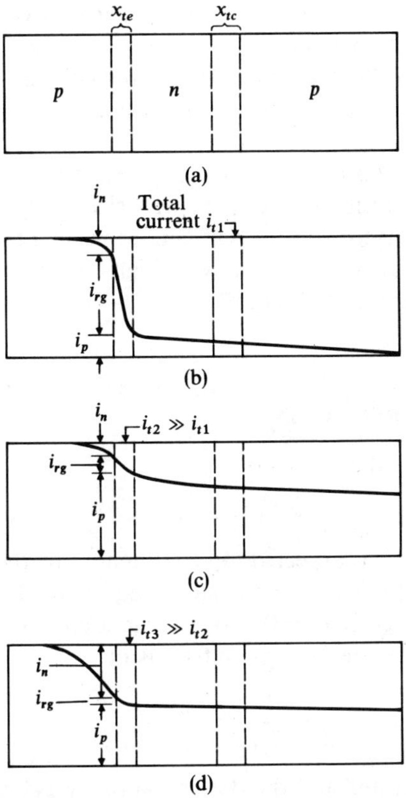

Fig. 12–5 Current distributions including recombination–generation and injection currents in a *p-n-p* transistor.

It is important to distinguish between injection current and current due to recombination–generation in the transition region. Low injection current is proportional to exp (qV/kT) just as I_{RG} through shallow traps is also. The former current is expected to occur at higher voltages than the latter. High injection current will be shown to be proportional to exp $(qV/2kT)$ just as I_{RG} through deep traps is also. The constant of proportionality is different in each case. The recombination–generation current is expected to be important at a lower voltage than that at which the high injection current is important. Figures 12–4a and 12–4b summarize the above conclusions. In both cases the ultimate limitation is due to ohmic *IR* voltage drop in the *bulk* of the device outside the transition region.

Figure 12–4 is a schematic drawing which shows each region if it were to exist separately. If the recombination–generation is sufficiently dominant it may swamp out the immediately adjoining region which would hence not be easily detectable. It is possible, for example, that in a GaAs *p-n* junction the recombination–generation region may swamp out the low injection region altogether thus exp $(qV/2kT)$ behavior due to recombination–generation in the transition region may merge with the high injection region of the same voltage dependence.

Based on calculations done so far it is possible to plot the different components of the total current flowing in the transistor. Figures 12–5 (a) through (d) show a series of plots for various bias conditions in a *p-n-p* transistor. It will be shown later that under conditions of very high injection the current is carried equally by electrons as it is by holes.

12–5 INFLUENCE OF RECOMBINATION–GENERATION CURRENT ON SHORT-CIRCUIT CURRENT GAIN

We have already noted that

$$\alpha = \gamma\beta^*\alpha^*. \tag{12–115}$$

The injection efficiency γ is expected to be influenced by the presence of recombination–generation in the emitter transition region. By definition γ is the ratio of the minority carrier current to the total current crossing the junction. Using this definition we may write for a *p-n-p* transistor

$$\gamma = \frac{I_{pE}}{I_{pE} + I_{nE} + I_{RG}}. \tag{12–116}$$

By dividing the numerator and the denominator by I_{PE} Eq. (12–116) may be expressed as

$$\gamma = \frac{1}{1 + \dfrac{I_{RG} + I_{nE}}{I_{pE}}}$$

or

$$\gamma = \frac{1}{1 + (I_{RG}/I_{pE}) + (I_{nE}/I_{pE})}. \tag{12–118}$$

From the discussion of the last section at *low currents*

$$I_{RG} \gg I_{nE}, I_{pE} \tag{12–119}$$

so that

$$\gamma = \frac{1}{1 + (I_{RG}/I_{pE})} \tag{12–120}$$

which can be quite low at sufficiently low currents.

As the emitter forward bias is increased the diffusion current I_{pE} will dominate I_{RG} and γ will tend to be quite close to unity. At still higher forward bias γ

diminishes again because I_{nE} increases as will be shown later. Thus the short circuit current gains α and β are expected to diminish with *decreasing* emitter current at low currents due to I_{RG} in the emitter transition region. There are other high injection effects as a result of which we also expect a diminuation of α and β with decreasing currents. This decrease due to I_{RG} is expected to be more important in transistors made out of wider band gap materials if all other things are equal.

12-6 HIGH INJECTION CONSIDERATIONS

In all of our previous considerations we have analyzed junction devices under the *low injection assumption.* According to this assumption the minority carrier density injected into a semiconductor is significantly less than the thermal equilibrium value of the majority carrier density in the semiconductor. Thus in p-type semiconductors:

$$n_{\text{inj}} \ll p_p, \tag{12-121}$$

where n_{inj} = injected electrons,

p_p = thermal equilibrium density of holes in the p-type semiconductor.

In n-type semiconductors

$$p_{\text{inj}} \ll n_n, \tag{12-122}$$

where p_{inj} = injected holes,

n_n = thermal equilibrium density of electrons in the n-type semiconductor.

Whenever either or both of the magnitude relationships (12-121) and (12-122) are not valid, the *low injection assumption is not* valid and high injection conditions must be considered.

High injection effects are not easy to analytically calculate because the problem involves *three* coupled partial differential nonlinear equations! Closed form solutions to these equations have not been obtained. The problem is therefore attacked in one of the following ways. In some cases a limiting form of the solution under extreme conditions is obtained. In other cases the attention is focused on a particular parameter of the device. Solutions pertaining to only this parameter are obtained. We shall consider an example of the former by finding the limiting form of I-V characteristics of the p-n junction diode at very high injection level. An example of the latter will be considered in obtaining the common emitter current gain β as a function of injection level.

In both approaches approximations are made to obtain analytical solutions. The computer can be and has been used to find numerical solutions to devices operating in the high injection region.

12-7 CURRENT-VOLTAGE CHARACTERISTICS OF p-n JUNCTION AT EXTREMELY HIGH INJECTION LEVEL

We wish to examine in this section the behavior of *p-n* junctions *at very high injection levels.* The following assumptions will be made in the analysis which follows:

1. Only a one-dimensional analysis will be carried out.
2. $\sigma_p \gg \sigma_n$. (12–123)
3. The *p*-side and *n*-side are uniformally doped.
4. The boundary conditions of Eqs. (3–38) and (3–39) will be assumed to be valid.
5. No *IR* drops in the bulk regions will be considered.
6. No leakage current (currents which do not go through junctions) will be considered.
7. No recombination or generation of carriers takes place in the transition region.
8. Electric fields in the bulk *n* and *p* regions are significantly less than the breakdown field for the semiconductor material used in the device.

Assumption 2 invites the following two conclusions. First, most of the current crossing the *p-n* junction is made up of holes (see Problem 3–2). Second, Eq. (12–123) implies that when μ_n and μ_p are the same order of magnitude,

$$p_p \gg n_n. \qquad (12\text{–}124)$$

Hence the injected electron density would need to be significantly higher on the *p*-side than the injected hole density on the *n*-side before the high injection effects need to be considered. Thus, assumption 2 permits us to neglect high injection effects on the *p*-side. We shall hence focus attention on the high injection effects on the *n*-side.

It can be shown through quantum theory arguments that a departure from space charge neutrality in bulk semiconductor material can give rise to enormous electric fields such as have never been experimentally measured in the bulk material. The conclusion from the above is that space charge neutrality is valid in the bulk region. We have already seen in Chapter 3 that the minority carrier density diminishes exponentially with distance as we move away from the junction transition region.

Figure 12–6 shows the minority carrier density as a function of distance on the *n*-side. In the same figure is also shown the majority carrier density. Note that the *excess* majority and minority carrier densities have the same total charge and charge density distribution. Recall that in the bulk semiconductor in thermal equilibrium having thermal equilibrium minority and majority carrier densities, there is no space charge. Thus, only the *excess* carrier densities must be equal to

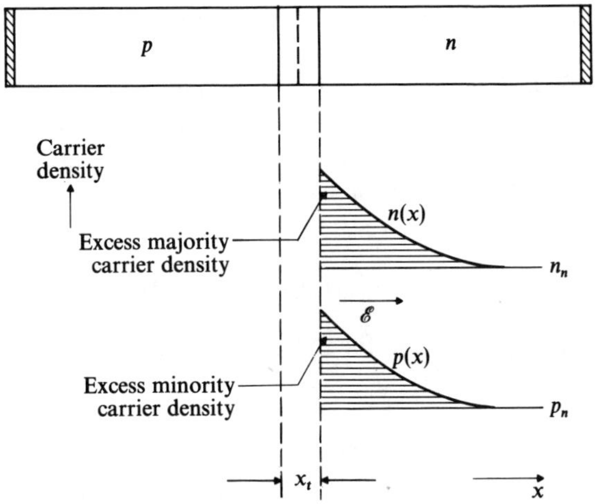

Fig. 12–6 Majority and minority carrier density distribution consistent with the requirements of space charge neutrality.

satisfy the space charge neutrality. These arguments lead to the following conclusion on the *n*-side

$$\nabla p = \nabla n \tag{12–125}$$

and

$$p - p_n = n - n_n \tag{12–126}$$

First we shall see how the high injection effects affect the transport equation for minority carriers. We shall then look at what influence high injection has on the diffusion equation solution. According to assumption 2, and the results of Problem 3–2 and the above discussion, the electron current density component is negligibly small.

We know that the gradient of holes is what leads to the transport of holes to the right due to the process of diffusion. The transport process of diffusion is remarkably impartial to the kind of particles being transported. The presence of the gradient of electrons will tend to transport the electrons to the right in Fig. 12–6 as well. When holes get transported to the right in Fig. 12–6, there are plenty of holes coming from the *p*-side to take their place. If, however, the electrons get transported to the right, there are very few electrons on the *p*-side to replenish the ones on the *n*-side that are moving away on account of the gradient of majority carriers. Under these conditions if the electrons, majority carriers, were to move to the right, space charge neutrality would be destroyed which is not permitted because extremely large fields would come into existence which have never been detected. The inescapable conclusions is that the majority carriers cannot move. What can hold these carriers (electrons) in place? An electric field pointed towards the positive x-axis can prevent the motion of majority carriers. We shall presently calculate this field. This same field, however, will aid the flow of minority carriers (holes) to the right. What is the cause of this field? Space charge *must* exist in the bulk region to cause the field to exist. This space charge needed to produce the field can be and usually is quite small so that our original assumption of space charge neutrality is still valid to a good approximation.

$$\mathbf{j}_n \doteq 0 = q\mu_n n\mathscr{E} + qD_n\nabla n \tag{12–127}$$

Equation (12–127) may be solved for the electric field

$$\mathscr{E} = -\left(\frac{D_n}{\mu_n}\right)\frac{\nabla n}{n} = -\frac{kT}{q}\frac{\nabla n}{n}. \tag{12–128}$$

The extreme right hand side of Eq. (12–128) is obtained by the use of the Einstein relationship Eq. (3–27). The expression for the electric field can be introduced into Eq. (2–112) to obtain

$$\mathbf{j}_p = -\frac{q\mu_p pkT\nabla p}{qn} - qD_p\nabla p. \tag{12–129}$$

From Eq. (3–26)

$$\frac{\mu_p kT}{q} = D_p. \tag{12–130}$$

Eq. (12–129) may be rewritten with the help of Eq. (12–130) as follows:

$$\mathbf{j}_p = -qD_p\frac{p}{n}\nabla p - qD_p\nabla p. \tag{12–131}$$

Thus

$$\mathbf{j}_p = -qD_p\left[1 + \frac{p}{n}\right]\nabla p. \tag{12–132}$$

Let us define

$$D_{p\text{eff}} \triangleq D_p\left[1 + \frac{p}{n}\right]. \tag{12–133}$$

With the use of Eq. (12–126), we may rewrite Eq. (12–133) as follows:

$$D_{p\text{eff}} = D_p\left[1 + \frac{p}{n_n + p - p_n}\right]. \tag{12–134}$$

In terms of the effective diffusion constant defined in Eq. (12–134) the transport equation takes the same form as though the diffusion process was the only mechanism of transport. The effective diffusion constant replaces the ordinary diffusion constant. The effective diffusion constant has hidden in it the influence of the electric field due to high injection.

At very high injection level

$$p \gg n_n, p_n \tag{12–135}$$

and Eq. (12–134) takes the following form

$$D_{p\text{eff}} = 2D_p. \tag{12–136}$$

We see that the effective diffusion constant at very high injection level is twice the low injection diffusion constant.

Let us now consider the diffusion equation solution under conditions of high injection. First the high injection electric field will result in a voltage V_{D1} in the bulk region. This voltage can be calculated by integrating the field in the semi-infinite n-region as follows in one dimension.

$$V_{D1} = \int_{xtn}^{\infty} \mathscr{E}\, dx = -\frac{kT}{q}\int_{xtn}^{\infty} \frac{1}{n}\frac{dn}{dx}\, dx = \int_{n_{xtn}}^{n(\infty)=n_n} \frac{dn}{n}. \tag{12–137}$$

Carrying out the integration we obtain

$$V_{D1} = -\frac{kT}{q}\ell\text{n}\left(\frac{n_n}{n_{xtn}}\right) = \frac{kT}{q}\ell\text{n}\left(\frac{n_{xtn}}{n_n}\right). \tag{12–138}$$

The terminal voltage V_0 is the sum total of the voltage across the transition region, V due to external source and V_{D1}. Thus,

$$V_0 = V + V_{D1} \tag{12–139}$$

It is worth noting that under low injection condition

$$n_{xtn} \doteq n_n \qquad \text{(low injection)} \tag{12–140}$$

and

$$V_{D1} \doteq 0, \mathscr{E} = 0 \qquad \text{(low injection).} \tag{12–141}$$

At very high injection the magnitude relationship (12–135) reduces Eq. (12–126) to the following

$$p_{xtn} \doteq n_{xtn}. \tag{12–142}$$

But

$$p_{xtn} = p_n e^{qV/kT} = n_{xtn} \tag{12–143}$$

and

$$n_n = n_i^2/p_n. \tag{12–144}$$

With the use of Eqs. (12–142) through (12–144) we may rewrite Eq. (12–138) as follows:

$$V_{D1} = \frac{kT}{q}\ell\text{n}\left(\frac{p_n^2 e^{qV/kT}}{n_i^2}\right) \tag{12–145}$$

$$V_{D1} = \frac{kT}{q} \ell\text{n} \left(\frac{p_n}{n_i}\right)^2 + \ell\text{n}\, e^{qV/kT} \tag{12–146}$$

$$V_{D1} = \frac{2kT}{q} \ell\text{n} \left(\frac{p_n}{n_i}\right) + V. \tag{12–147}$$

Equation (12–147), when substituted into Eq. (12–139), gives

$$V_0 = V + V_{D1} = \frac{2kT}{q} \ell\text{n} \left(\frac{p_n}{n_i}\right) + 2V, \tag{12–148}$$

because

$$\ell\text{n}\, e^x = x \tag{12–149}$$

or

$$V = \frac{V_0}{2} - \frac{kT}{q} \ell\text{n} \left(\frac{p_n}{n_i}\right). \tag{12–150}$$

The expression for V may be substituted into Eq. (3–46) which is repeated here: for $x > x_{tn}$

$$p(x) = p_n + p_n(e^{qV/kT} - 1)\, e^{-(x - x_{tn})/L_p}. \tag{12–151}$$

Using Eq. (12–150) we may rewrite Eq. (12–151) as follows:

$$p(x) = p_n + p_n\left(\exp\left\{\frac{q}{kT}\left[\frac{V_0}{2} + \frac{kT}{q}\ell\text{n}\,(n_i/p_n)\right]\right\} - 1\right) e^{-(x - x_{tn})/Lp} \tag{12–152}$$

$$p(x) = p_n + p_n\,[e^{qV_0/(2kT)} \cdot e^{\ell\text{n}(n_i/p_n)} - 1]\, e^{-(x - x_{tn})/Lp}. \tag{12–153}$$

The transport equation in terms of the effective diffusion coefficient is written as follows:

$$\mathbf{j}_p = -qD_{peff}\, \nabla p. \tag{12–154}$$

At very high injection

$$\mathbf{j}_p = -q(2D_p)\nabla p. \tag{12–155}$$

Equation (12–153) when substituted into Eq. (12–155) gives

$$\text{j}_p(x) = \frac{2qD_p p_n}{L_p}\left[\frac{n_i}{p_n} e^{qV_0/(2kT)} - 1\right] e^{-(x - x_{tn})/L_p} \tag{12–156}$$

$$i_p = i_p(x_{tn}) = \text{j}_p(x_{tn})\,A = \frac{2qD_p n_i A}{L_p}\left[e^{qV_0/(2kT)} - \frac{p_n}{n_i}\right]. \tag{12–157}$$

Furthermore because of Eq. (12–127)

$$i_p \doteq i = \frac{2qAD_p n_i}{L_p}\left[e^{qV_0/(2kT)} - \frac{p_n}{n_i}\right]. \tag{12–158}$$

It is worthwhile to compare Eq. (12–158) with a comparable expression for low injection case for a junction diode in which the electron component of current is negligible. Such an expression can be easily obtained from Eq. (3–55). The result is

$$I \doteq \frac{qAD_p p_n}{L_p}(e^{qV/(kT)} - 1). \qquad (12\text{–}159)$$

Three differences are observable from a close comparison of Eqs. (12–158) with Eq. (12–159). These are summarized in Table 12–1.

Table 12–1 Low Injection–Very High Injection Transformations

	Low injection	Very high injection
1.	D_p →	$2D_p$
2.	p_n →	n_i
3.	V →	$V_0/2$

A fourth difference is that the 1 inside the bracket on the right hand side of Eq. (12–159) is replaced by p_n/n_i which obeys

$$\frac{p_n}{n_i} \ll 1. \qquad (12\text{–}160)$$

As the exponential terms on the right hand side of Eqs. (12–159) and (12–160) are very much greater than 1 at high injection level, the above difference is negligible. We see now where the factor of 2 in the exponent of the slope of $\ell n\, I$ versus V observed at high current density in Fig. 12–4 comes from. The most significant change is item 2 in Table 12–1 because of the magnitude relationship (12–160). Item 3 in Table 12–1 is used under certain circumstances to detect when a device is in very high injection.

Example 12–3. You are given a *p-n* junction Si diode having the following parameters:

$\sigma_p = 0.1\ (\Omega\text{-cm})^{-1}$, $\sigma_n = 0.001\ (\Omega\text{-cm})^{-1}$, $\mu_n = 1350\ \text{cm}^2/(\text{Volt-sec})$,
$L_p = 10^{-3}$ cm, $D_p = 12\ \text{cm}^2/\text{sec}$, $A = 10^{-2}\ \text{cm}^2$.

Estimate the current level, in this device, at which the injected minority carrier density at x_{tn} equals the thermal equilibrium value of majority carrier density on the *n*-side. (At this level the high injection effects would need to be considered.)

Solution. Using the condition that

$$p_{xtn} - p_n = n_n \tag{12–161}$$

or

$$p_{xtn} \doteq n_n \tag{12–162}$$

$$D_{peff} = D_p\left[1 + \frac{p_{xtn}}{n_n + p_{xtn}}\right]. \tag{12–163}$$

With the use of Eq. (12–162) the above equation becomes

$$D_{peff} = 1.5\, D_p \tag{12–164}$$

$$I \doteq \frac{qA1.5D_p}{\sqrt{\tau_p 1.5D_p}}(p_n\, e^{qV/kT}). \tag{12–165}$$

But

$$p_n\, e^{qV/kT} = p_{xtn} = n_n \tag{12–166}$$

thus Eq. (12–166) may be substituted into Eq. (12–165) to obtain

$$I = \frac{qA1.5D_p n_n}{\sqrt{\tau_p 1.5D_p}}. \tag{12–167}$$

But

$$n_n = \frac{\sigma_n}{q\mu_n}. \tag{12–168}$$

Thus Eq. (12–167) may be rewritten as follows:

$$I = \frac{A1.5D_p\sigma_n}{\mu_n L_p\sqrt{1.5}}. \tag{12–169}$$

The numerical value is obtained as follows:

$$I = \frac{10^{-2}(1.5)(12)10^{-3}}{1.35 \times 10^3\, 10^{-3}\sqrt{1.5}} = \underline{1.089 \times 10^{-4}\text{ Amp.}} \tag{1}$$

It is to be noted that the smaller the area of the device, the higher the current density, and the lower the current at which the high injection effects become important.

High injection effects are fundamentally dependent on the current density rather than current.

12–8 COMMON EMITTER LOW FREQUENCY INCREMENTAL SHORT CIRCUIT CURRENT GAIN AS A FUNCTION OF EMITTER CURRENT

In this section we shall concentrate on the incremental common emitter short circuit current gain β_0 (also designated h_{fe}). We wish to study how this parameter varies with emitter current. The experimental variation of β_0 with I_E is

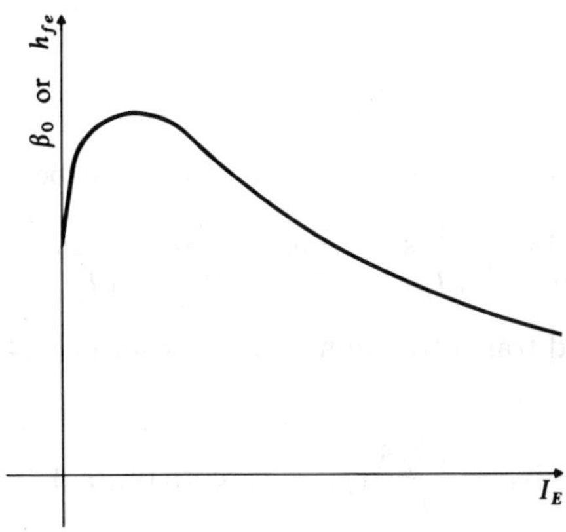

Fig. 12–7 Variation of short-circuit common emitter current gain with emitter current.

shown in Fig. 12–7. We shall attempt to explain this curve. A number of judicious approximations will be made to obtain an analytic expression.

It turns out that the analytical calculations are more straight forward if we consider the reciprocal β_0. Consider the base current and KCL,

$$I_B = I_E - I_C. \tag{12–170}$$

Equation (12–170) may be rewritten as follows:

$$I_B = I_{Ep} + I_{En} + I_{RG} - I_{Cp} - I_{Cn} \tag{12–171}$$

$$= (I_{Ep} - I_{Cp}) + I_{RG} + I_{En} - I_{Cn}, \tag{12–172}$$

where I_{RG} is the recombination–generation current in the emitter transition region.

The term within the bracket

$$I_{Ep} - I_{Cp} \triangleq I_R \tag{12–173}$$

is recognized as the *recombination current* which exists only when I_{Cp} and I_{Ep} are not equal. We have noted in Chapter 2 that there is *bulk* or *volume* recombination and also surface recombination. Thus,

$$I_R = I_{VR} + I_{SR}. \tag{12–174}$$

Equation (12–172) substituted into Eq. (12–170) gives

$$I_B = I_{VR} + I_{SR} + I_{En} + I_{RG} - I_{Cn}. \tag{12–175}$$

Differentiation gives

$$\frac{\partial I_B}{\partial I_E} = 1 - \alpha_0 \doteq \frac{1 - \alpha_0}{\alpha_0} = \frac{1}{\beta_0}. \tag{12–176}$$

Also we assume

$$\frac{\partial I_B}{\partial I_E} \doteq \frac{\partial I_B}{\partial I_{Ep}}. \tag{12-177}$$

With the use of Eq. (12–175), reciprocal gain β_0 may be expressed as follows:

$$\frac{1}{\beta_0} \doteq \frac{\partial I_B}{\partial I_{Ep}} = \frac{\partial I_{VR}}{\partial I_{Ep}} + \frac{\partial I_{SR}}{\partial I_{Ep}} + \frac{\partial I_{En}}{\partial I_{Ep}} + \frac{\partial I_{RG}}{\partial I_{Ep}} - \frac{\partial I_{Cn}}{\partial I_{Ep}}. \tag{12-178}$$

For a normally biased transistor we may write from Eqs. (4–23) and (4–24) the following:

$$I_{Ep} \doteq \frac{qD_pA}{L_p}(p_E - p_n)\coth(w_0/L_p) \tag{12-179}$$

$$I_{Cn} \doteq \frac{qD_nAn_{pc}}{L_{nc}}. \tag{12-180}$$

It is clear that I_{Ep} and I_{Cn} are independent of each other. Therefore,

$$\frac{\partial I_{Cn}}{\partial I_{Ep}} = 0. \tag{12-181}$$

In 1951 Shockley calculated the volume and surface recombination terms given below:

$$\frac{\partial I_{VR}}{\partial I_{Ep}} = \left(\frac{1}{2}\right)\left(\frac{w_0}{L_p}\right)^2 = \frac{w_0^2}{2D_p\tau_p} \tag{12-182}$$

$$\frac{\partial I_{SR}}{\partial I_{Ep}} = \frac{A_s w_0 s}{AD_p}. \tag{12-183}$$

See problem 12–2 for the derivation of Eq. (12–182). Consider now a quantity (I_{En}/I_{Ep}) closely related to $\partial I_{En}/\partial I_{Ep}$ given below:

$$\frac{I_{En}}{I_{Ep}} = \frac{1}{\gamma_0} - 1. \tag{12-184}$$

There is therefore, an intimate relationship between the left hand side of Eq. (12–184) and the injection effeciency γ_0. Also

$$\gamma_0 = \frac{1}{1 + (I_{En}/I_{Ep})} \tag{12-185}$$

From a comparison of Eq. (12–185) with Eq. (4–74) we conclude:

$$\frac{I_{En}}{I_{Ep}} = \left(\frac{\sigma_B}{\sigma_E}\right)\left(\frac{w_0}{L_{ne}}\right). \tag{12-186}$$

We shall return to Eq. (12–186) later to calculate $\partial I_{En}/\partial I_{Ep}$.

Now from Eqs. (12–80), (12–84), and (12–181) we obtain

$$\frac{I_{RG}}{I_{Ep}} = \frac{\dfrac{qAn_i}{\tau} \cdot \dfrac{kT\,x_{te}}{q(V_{BI} - V_E)}\, e^{qV_E/(2kT)}}{\dfrac{qD_pA}{L_p}\, p_E \coth\,(w_0/L_p)}. \tag{12–187}$$

If a one-term Taylor series expansion is carried out of the hyperbolic cotangent in the denominator and the explicit expression for p_E is written in terms of V_E, the following is obtained:

$$\frac{I_{RG}}{I_{Ep}} = \left(\frac{n_i}{p_n}\right)\left(\frac{kT}{q(V_{BIE} - V_E)}\right)\left(\frac{x_{te}w_0}{D_p\tau}\right) e^{-qV_E/(2kT)}, \tag{12–188}$$

where V_{bie} = built-in voltage of the emitter junction.

The lifetime τ in Eq. (12–188) comes from the consideration of recombination-generation current in the emitter transition region under the conditions of the emitter junction being symmetric in all respects. The emitter base junction is most likely *not* symmetric. Equation (12–106) which refers to asymmetric junctions seems to indicate that the following transformation may be made

$$\tau \longrightarrow \sqrt{\tau_{p0}\,\tau_{n0}}. \tag{12–189}$$

Furthermore

$$p_n = n_i^2/n_n. \tag{12–190}$$

Equation (12–188) may be rewritten with the help of transformation (12–189) and Eq. (12–190) as follows:

$$\frac{I_{RG}}{I_{Ep}} = \left(\frac{n_n}{n_i}\right)\left[\frac{kT}{q(V_{BIE} - V_E)}\right]\frac{x_{te}w_0\, e^{-qV_E/(2kT)}}{D_p(\tau_{p0}\,\tau_{n0})^{\frac{1}{2}}}. \tag{12–191}$$

We note that none of the right hand side of Eqs. (12–182), (12–183), and (12–186) show any explicit dependence on I_{Ep}. They cannot therefore explain the curve of Fig. 12–7. These above equations are really the low injection values of the respective parameters. Our present objective is to devise simple rules by which the low injection expressions may be converted to ones which are valid at any injection level.

Two parameters will now be defined which will aid us in the above objective. The first of these is the injection parameter Z, which is obtained from an integration of the transport equation. All the arguments presented in connection with the transport equation (12–132) are also valid generally and hence usable also in the case of the transistor. Equation (12–126), which expresses the requirements of space charge neutrality in the bulk semiconductor, is also valid in the base region of the transistor. Thus we may rewrite Eq. (12–132) with the help of Eq. (12–126).

$$\mathbf{j}_p = -qD_p\left[1 + \frac{p}{n_n + p - p_n}\right]\nabla p. \tag{12–192}$$

If all donors are assumed to be ionized in the base of a *p-n-p* transistor

$$n_n \doteq N_d. \tag{12–193}$$

Equation (12–192) may be rewritten as follows:

$$i_p = -qAD_pN_d\left[1 + \frac{p/N_d}{1 + (p/N_d) - (p_n/N_d)}\right]\frac{d(p/N_d)}{dx}. \tag{12–194}$$

Now

$$p_n/N_d \ll 1. \tag{12–195}$$

Thus, Eq. (12–194) may be simplified to

$$i_p = -qAD_pN_d\left[1 + \frac{p/N_d}{1 + (p/N_d)}\right]\frac{d(p/N_d)}{dx}. \tag{12–196}$$

Equation (12–196) may be integrated throughout the base region. To a good approximation in this integration we may assume that

$$i_p \doteq I_{Ep} \tag{12–197}$$

which is valid if

$$(w_0/L_p) \ll 1. \tag{12–198}$$

With the use of Eq. (12–197) we may carry out the integration as follows across the base region shown in Fig. 12–8.

$$\int_0^{w_0} \frac{-I_{Ep}}{qAD_pN_d}\,dx = \int_{p_E}^{0} d(p/N_d) + \int_{p_E}^{0} \frac{(p/N_d)\,d(p/N_d)}{1 + (p/N_d)}. \tag{12–199}$$

Carrying out the above integration we obtain

$$-\frac{I_{Ep}\,w_0}{qAD_pN_d} = -\frac{p_E}{N_d} + \left[-\frac{p_E}{N_d} + \ell\text{n}\left(1 + \frac{p_E}{N_d}\right)\right] \tag{12–200}$$

or

$$Z_p \triangleq \frac{I_{Ep}\,w_0}{qAD_pN_d} = \frac{2p_E}{N_d} - \ell\text{n}\left(1 + \frac{p_E}{N_d}\right) \tag{12–201}$$

Note the two expressions obtained for Z_p under limiting conditions given below. If

$$p_E/N_d \ll 1 \tag{12–202}$$

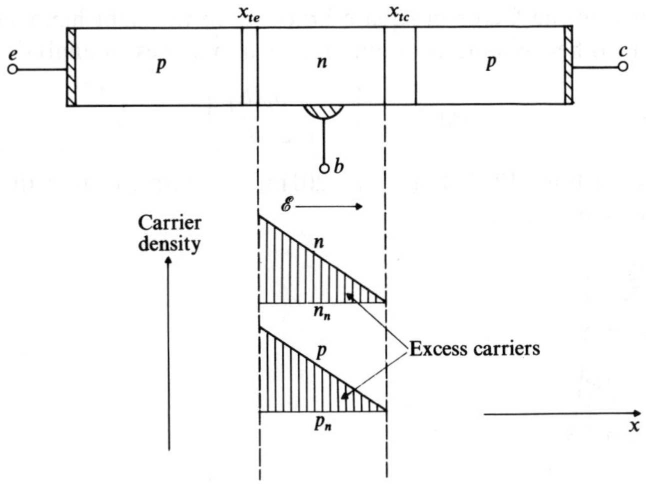

Fig. 12–8 Carrier densities in the base region for assumed space charge neutrality in the base.

$$Z_p \doteq 2p_E/N_d - p_E/N_d = p_E/N_d \qquad \text{(low injection).} \tag{12–203}$$

If

$$p_E/N_d \gg 1$$

$$Z_p \doteq 2p_E/N_d \qquad \text{(very high injection).} \tag{12–204}$$

The other parameter important to our analysis is associated with the effective diffusion constant. Equation (12–161) specialized to the transistor case obtains the following:

$$D_{peff} = D_p\left[1 + \frac{p_E}{n_n + p_E}\right] \tag{12–205}$$

$$D_{peff} = D_p\left[\frac{n_n + 2p_E}{n_n + p_E}\right]. \tag{12–206}$$

Also in the base region of a *p-n-p* transistor

$$n_n \doteq N_d. \tag{12–207}$$

Thus, Eq. (12–206) may be rewritten

$$D_{peff} = D_p\left[\frac{1 + 2p_E/N_d}{1 + p_E/N_d}\right]. \tag{12–208}$$

From Eq. (12–201) we see that

$$Z_p = f(p_E/N_d). \tag{12–209}$$

The reciprocal of the factor in square brackets on the right hand side of Eq. (12–208) is the other parameter of importance in our present analysis.

$$g_p(Z_p) = \left[\frac{1 + p_E/N_d}{1 + 2p_E/N_d}\right]. \tag{12–210}$$

With the use of Eqs. (12–210) and (12–201) it is possible to plot the function $g_p(Z_p)$ as shown in Fig. 12–9.

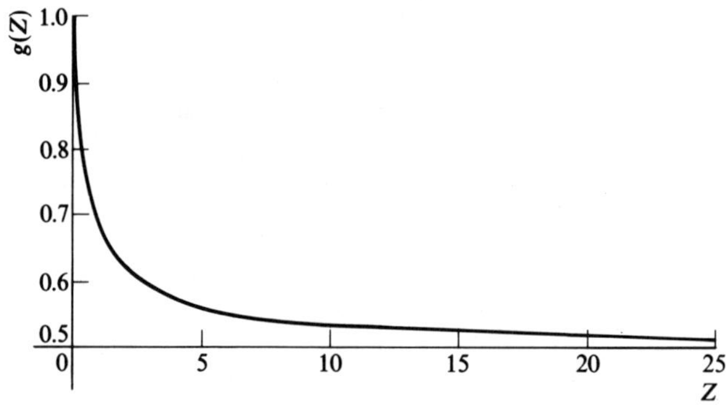

Fig. 12–9 $g(Z)$. Notice that $g(Z) = g_p(Z_p$ or $g_n(Z_n)$.

From the earlier discussion of the effective diffusion constant and Eq. (12–210) we may observe that the high injection field may be adequately taken care of in all equations including transport by using the transformations

$$D_p \longrightarrow D_p/[g_p(Z_p)] \qquad (p\text{-}n\text{-}p) \tag{12–211}$$

and

$$D_n \longrightarrow D_n/[g_n(Z_n)] \qquad (n\text{-}p\text{-}n), \tag{12–212}$$

where

$$g_n(Z_n) = \left[\frac{1 + n_E/N_a}{1 + 2n_E/N_a}\right] \tag{12–213}$$

and

$$Z_n = \frac{I_{En} w_0}{qAD_n N_a} = 2(n_E/N_a) - \ell n\,[1 + n_E/N_a], \tag{12–214}$$

where N_a = acceptor density in the *p*-type base of an *n-p-n* transistor.

We shall now return to I_{En}/I_{Ep}. First note that as the injected minority carrier density into the base becomes comparable to the thermal equilibrium value of the majority carrier density in the base region, the new value of the latter is given by

$$\text{New majority carrier density} = n_n + p_E. \tag{12–215}$$

Hence all transistor parameters which were proportional to n_n at low injection values should be multiplied by the ratio

Multiply all terms proportional to n_n by $\dfrac{n_n + p_E}{n_n} \doteq \dfrac{N_d + p_E}{N_d} = 1 + \dfrac{Z_p}{2}$.

$$(12\text{–}216)$$

The last equality is obtained by utilization of Eq. (12–204) for high injection. It is not hard to see from the above that the conductivity of the base region will be increased because of the increased carrier densities at moderate and high injection level. For this reason this change in base conductivity is referred to as *conductivity modulation.*

We see from Eq. (12–186) that here is a quantity that is proportional, at low injection, to n_n. Hence it should be modified by the factor of Eq. (12–216), thus

$$\frac{I_{En}}{I_{Ep}} = \left(\frac{\sigma_B}{\sigma_E}\right)\left(\frac{w_0}{L_{ne}}\right)\left(1 + \frac{Z_p}{2}\right). \tag{12–217}$$

The injection parameter Z_p can be written out explicitly as follows:

$$\frac{I_{En}}{I_{Ep}} = \left(\frac{\sigma_B}{\sigma_E}\right)\left(\frac{w_0}{L_{ne}}\right)\left[1 + \frac{I_{Ep}w_0}{2qAD_pN_d}\right]. \tag{12–218}$$

Thus

$$I_{En} = \left(\frac{\sigma_B}{\sigma_E}\right)\left(\frac{w_0}{L_{ne}}\right)\left[I_{Ep} + \frac{I_{Ep}^2 w_0}{2qAD_pN_d}\right]. \tag{12–219}$$

Differentiation yields:

$$\frac{\partial I_{En}}{\partial I_{Ep}} = \left(\frac{\sigma_B}{\sigma_E}\right)\left(\frac{w_0}{L_{ne}}\right)\left[1 + \frac{I_{Ep}w_0}{qAD_pN_d}\right]. \tag{12–220}$$

By the use of Eq. (12–201) we may rewrite Eq. (12–220) as follows:

$$\frac{\partial I_{En}}{\partial I_{Ep}} = \left(\frac{\sigma_B}{\sigma_E}\right)\left(\frac{w_0}{L_{ne}}\right)(1 + Z_p). \tag{12–221}$$

From the above calculations we draw the following useful transformation laws which will allow us to convert low injection expressions to the corresponding general injection level–expressions valid also at high injection level.

Applying the transformation 1 and 4 of Table 12–2 to Eq. (12–191) we obtain

$$\frac{\partial I_{RG}}{\partial I_{Ep}} = \left(\frac{n_n}{n_i}\right)\left[\frac{kT}{q(V_{BIE} - V_E)}\right]\frac{x_{te}w_0\, e^{-qV_E/(2kT)}}{D_p(\tau_{p0}\tau_{n0})^{\frac{1}{2}}}\, g_p(Z_p)(1 + Z_p). \tag{12–222}$$

Table 12–2 Transformations to Convert Low Injection Expressions to Those Valid at General Injection Level

Low injection expression	High injection expression
1. D_p →	$D_p/[g_p(Z_p)]$ (p-n-p)
2. D_n →	$D_n/[g_n(Z_n)]$ (n-p-n)
3. Expressions proportional to n_n (base) →	should be multiplied by: $\left(1 + \frac{Z_p}{2}\right)$ (p-n-p)
4. Expressions proportional to n_n (base) and involving $\frac{\partial}{\partial I_{Ep}}$ →	should be multiplied by: $(1 + Z_p)$ (p-n-p)
5. Expressions proportional to p_p (base) →	should be multiplied by: $\left(1 + \frac{Z_n}{2}\right)$ (n-p-n)
6. Expressions proportional to p_p (base) and involving $\frac{\partial}{\partial I_{En}}$ →	should be multiplied by: $(1 + Z_n)$ (n-p-n)

With the guide lines of Table 12–2 we may now deduce the general expressions for reciprocal β_0

$$\frac{1}{\beta_0} = \frac{w_0 s A_s}{A D_p} g_p(Z_p) + \frac{\sigma_B w_0}{\sigma_E L_{ne}}(1 + Z_p) + \frac{1}{2}\left(\frac{w_0}{L_p}\right)^2 (1 + Z_p) g_p(Z_p)$$

$$+ \left(\frac{n_n}{n_i}\right)\left[\frac{kT}{q(V_{BIE} - V_E)}\right]\frac{x_{te} w_0 \, e^{-qV_E/(2kT)}}{D_p(\tau_{p0}\tau_{n0})^{\frac{1}{2}}} g_p(Z_p)(1 + Z_p) \quad (p\text{-}n\text{-}p). \qquad (12\text{–}223)$$

Similarly,

$$\frac{1}{\beta_0} = \frac{w_0 s A_s}{A D_n} g_n(Z_n) + \frac{\sigma_B w_0}{\sigma_E L_{pe}}(1 + Z_n) + \frac{1}{2}\left(\frac{w_0}{L_n}\right)^2 (1 + Z_n) g_n(Z_n)$$

$$+ \left(\frac{p_p}{n_i}\right)\left[\frac{kT}{q(V_{BIE} - V_E)}\right]\frac{x_{te} w_0 \, e^{-qV_E/(2kT)}}{D_n(\tau_{p0}\tau_{n0})^{\frac{1}{2}}} g_n(Z_n)(1 + Z_n) \quad (n\text{-}p\text{-}n). \qquad (12\text{–}224)$$

The reason why the third term on the right hand side of Eq. (12–223) has both the factors present in it is that $L_p^2 = D_p \tau_p$, and τ_p is *inversely* proportional to n_n.

From Eqs. (12–223) and (12–224) we may conclude the following. The function $g_p(Z_p)$ is a monotonically decreasing function which decreases by 80 per cent of total possible change when the injection parameter varies between 0 and 3. Those terms in $(1/\beta_0)$ which are multiplied by $g_p(Z_p)$ will decrease with injection level. Hence β_0 will increase with injection level. We see from Eqs. (12–223) and (12–224) that the surface and volume recombination terms and the recombination–generation terms will result in an initial rise in β_0 with injection level. The recombination–generation term here is expected to be dominant in those devices in which it is important. This term will increase with injection level not only because of $g_p(Z_p)$, but also because of the exponential term. All other things being equal we would expect that this component may be of greater relative importance in wider band gap semiconductors such as Si and GaAs.

Terms in $1/\beta_0$ which are proportional to $(1 + Z_p)$ are expected to increase with injection level. Thus, β_0 will diminish with injection level on account of the factor $(1 + Z_p)$ contained in terms in Eq. (12–221) associated with injection efficiency γ_0, volume recombination, and recombination in the emitter transition region.

Equations (12–223) and (12–224) are adequate to explain the curve of Fig. 12–7. One contributing element that explains the experimental $\beta_0(I_E)$, which is left out of the above analysis, is the so called *self-bias effect*. At increasing currents the lateral flow of base current *tends to forward bias that part of the emitter junction closest to the base contact heavier than the regions farther away*. The net effect of this phenomenon is for the emission of minority carriers into the base region not to be uniform over the emitter junction area and to concentrate it over a smaller area close to the base contact. Thus for a given current level, the current density is higher. As the high injection phenomenon is fundamentally dependent on current density rather than on current, the self bias effect tends to accentuate the *high injection effects*. This effect is quite difficult to reduce to analytical analysis, and hence will not be discussed further here. We shall simply note its existence. More recently it has been shown that emitter heavy doping effects are the predominantly important cause of limiting injection efficiency and hence the transistor current gain. We have already noted that emitter doping is usually high to obtain a reasonable value of injection efficiency. At these heavy doping levels (a) the band gap diminishes, (b) donor (or acceptor) energy levels broaden into bands, and (c) band edge tailing occurs. These effects have been discussed in Sec. 10–3 in some detail. It turns out that in most present day plannar transistors, these heavy doping effects are the major cause limiting transistor current gain rather than minority carrier lifetime in the emitter. The interested reader is referred to Martens et al. *I.E.E.E. Trans. Electron Devices* Sept. 1973.

It turns out that the injection parameter can be rewritten as follows:

$$Z_p = \frac{qI_{Ep}w_0}{A\sigma_B kT}\left(\frac{\mu_n}{\mu_p}\right) \quad (p\text{-}n\text{-}p) \qquad (12\text{–}225)$$

$$Z_n = \frac{qI_{En}w_0}{A\sigma_B kT}\left(\frac{\mu_p}{\mu_n}\right) \quad (n\text{-}p\text{-}n). \qquad (12\text{–}226)$$

Thus, if all parameters of *p-n-p* and *n-p-n* transistors are identical, for the same current level the injection parameter is lower in *n-p-n* transistors than in *p-n-p* transistors because usually

$$\mu_n > \mu_p. \tag{12–227}$$

Thus, we expect *n-p-n* transistors to be less sensitive to high injection effects than *p-n-p* transistors. This advantage of *n-p-n* transistors over *p-n-p* transistors diminishes with increasing doping level in the base because μ_n and μ_p approach each other as doping level is increased.

As ω_α is proportional to D_p we expect that ω_α should double in the limit as injection level increases. This has been observed in diffusion transistors.

12–9 HIGH INJECTION EFFECTS IN TRANSISTORS HAVING NON-UNIFORMLY DOPED BASE REGIONS

Earlier in the present chapter we obtained the following equation:

$$\mathbf{j}_p = \frac{-q\mu_p D_n p[\nabla p + \nabla N_d(x)]}{\mu_n[p + N_d(x)]} - qD_p \nabla p. \tag{12–228}$$

Dividing both sides of the above equation by (qp) we obtain

$$\frac{\mathbf{j}_p}{(qp)} = u = -\frac{\mu_p D_n[\nabla p + \nabla N_d(x)]}{\mu_n[p + N_d(x)]} - D_p \frac{\nabla p}{p}, \tag{12–229}$$

where u = average velocity with which the minority carriers traverse the base region.

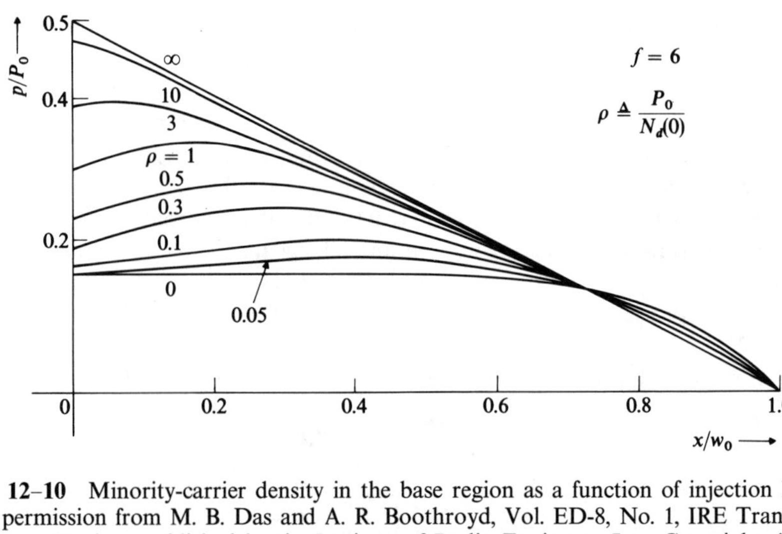

Fig. 12–10 Minority-carrier density in the base region as a function of injection level. (By permission from M. B. Das and A. R. Boothroyd, Vol. ED-8, No. 1, IRE Trans. on Electron Devices, published by the Institute of Radio Engineers, Inc., Copyright, 1961.)

The Einstein relationship has been used in arriving at Eq. (12–228), i.e. the ratio of mobility to diffusion constant is identical for holes and electrons. In the limit of very high injection level such that

$$p \gg N_d \tag{12–230}$$

$$\nabla p \gg \nabla N_d \tag{12–231}$$

Equation (12–229) takes the following form:

$$u = -\frac{2D_p \nabla p}{p}. \tag{12–232}$$

For uniformly doped base region the value of u can be shown to be that given by Eq. (12–232) (see Problem 12–3). We see thus that the beneficial effects of the built-in field of a graded base transistor is reduced to zero if the transistor is operated at very high injection level when it behaves just like a diffusion transistor!

Figure 12–10 shows how the minority carrier density distribution varies as a function of injection level parameter ρ for a specific field factor f

$$\rho \triangleq \frac{P_0}{N_d(0)}. \tag{12–233}$$

In epitaxial diffused base transistors another phenomenon called the *base-push-out* occurs under high injection conditions. We shall explain this phenomenon as follows. When current density in the device is low the current in the epitaxial region is primarily due to drift. In *n-p-n* transistors

$$\mathscr{E}_e = \frac{J_c}{q\mu_n N_d}, \tag{12–234}$$

where N_d = doping density in the epitaxial region,
J_c = collector current which equals the current in the epitaxial region,
$\mathscr{E}_e$ = electric field in the epitaxial region.

If the width of the epitaxial region is w_e

$$\mathscr{E}_e = \frac{V_{CB} - V_{BIC}}{w_e}. \tag{12–235}$$

Let us define the *critical* current density J_c as follows:

$$J_c \triangleq \frac{q\mu_n(\mathscr{E})N_d}{w_e}(V_{CB} - V_{BIC}), \tag{12–236}$$

where $\mu_n(\mathscr{E})$ = mobility dependence on the epitaxial region field $\mathscr{E}_e$.

Current density J_c would result in all the current being carried by drift current and all donors being ionized. If current density exceeds J_c, the current everywhere cannot be sustained by the drift of electron density equal to N_d. Excess electrons will be introduced into a part of the epitaxial region where excess holes will

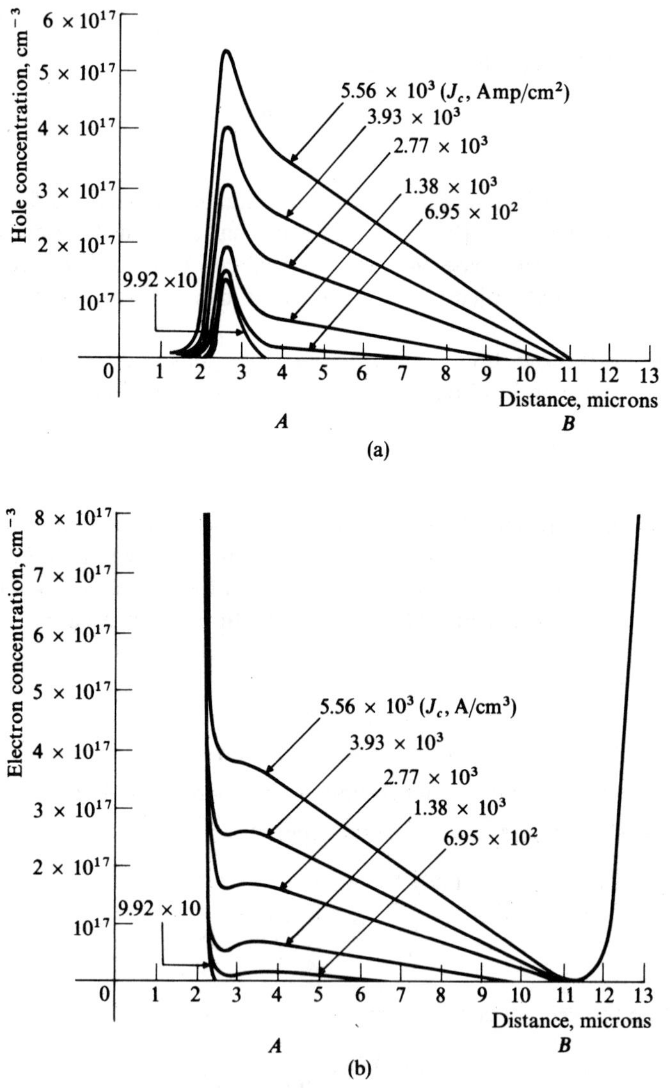

Fig. 12–11 Hole and electron densities in epitaxial transistors for various collector currents. (a) Hole concentration as a function of distance for various collector currents. (b) Electron concentration as a function of distance for various collector currents. (After Poon, et al., 1969)

neutralize the excess electrons introduced. Thus a kind of neutral region close to the base will exist in the epitaxial region. This amounts to an extension of the neutral base region into the epitaxial region. This effect will progress as the current density is increased until the "neutral" region extends throughout the epitaxial region. This is the *base-push-out phenomenon.* The effective base width w^* increases due to this phenomenon with a resulting increasing of the (effective) base transit time of minority carriers and a consequent decrease in ω_α. When the effective base width envelopes both the original base width and the epitaxial region width,

$$J_c \gg qv_mN_d, \tag{12-237}$$

where v_m = scattering limited velocity.

Then the effective base width is given by:

$$w^* = w + w_e - \left[\frac{2\varepsilon v_m(V_{BIC} - V_{CB} - V_w)}{J_c - qv_mN_d}\right]^{\frac{1}{2}}, \tag{12-238}$$

where

$$V_w = \int_0^{w^*} \mathscr{E}dx. \tag{12-239}$$

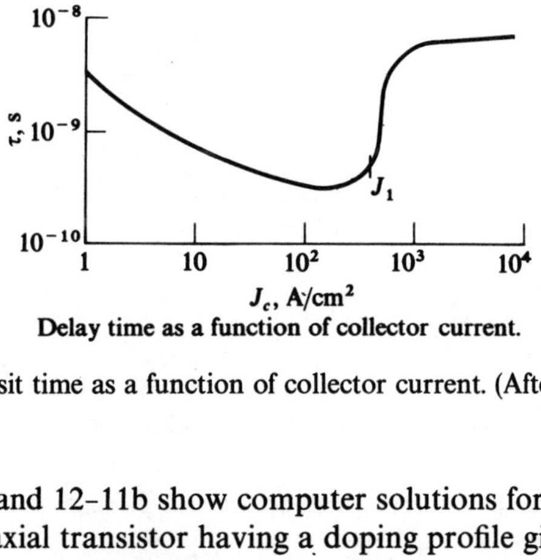

Delay time as a function of collector current.

Fig. 12–12 Base transit time as a function of collector current. (After Poon, et al., 1969)

Figure 12–11a and 12–11b show computer solutions for hole and electron densities in an epitaxial transistor having a doping profile given in Fig. 7–10b. It can be readily seen that the region of esentially a constant gradient extends farther into the epitaxial region with increasing current density. The effective base width is seen to widen. Figure 12–12 shows how the base transit time increases when current density is increased beyond J_1. These figures justify the conclusions drawn from the above analysis. It is to be noted that in the calculations of Figs. 12–11 and 12–12, *the base-push-out occurs in a relatively narrow range of current densities.*

12–10 SUMMARY

We extended the *p-n* junction theory in this chapter by removing two of the most important simplifying assumptions of the simple theory. We considered both recombination-generation in the transition region and the effect of high injection. Both of these effects were found to have a profound influence on transistor performance. The recombination-generation currents are likely to be more important in devices fabricated utilizing wider band-gap semiconductors. It also influences the variation of β_0 with current density.

The high injection effects aids the transport of minority carriers in the base region of diffusion transistors. The graded base transistors are worse off for high injection effects because the beneficial effect of the built-in field is neglected. Indeed the diffused base transistors behave about the same as diffusion transistors at very high injection levels.

Expitaxial graded base transistors suffer from the base-push-out phenomenon which decreases the cut-off frequency and increases transient times at high injection level.

PROBLEMS

12–1 You are given Si *p-n* junction the following of whose parameters are known

$$\sigma_n = 0.1\ (\Omega\text{-cm})^{-1}$$
$$\sigma_p = 0.001\ (\Omega\text{-cm})^{-1}$$
$$L_n = 10^{-4}\ \text{cm}$$
$$A = 10^{-1}\ \text{cm}^2.$$

(a) Estimate the current in this device at which the electrons injected into the *p*-side is 50 per cent of the thermal equilibrium density of holes on this side. (b) What is the value of V_{D1}?

12–2 Using Eqs. (12–23), (12–24), the conditions of *normal* bias, and using Taylor series expansion, prove Eq. (12–182).

12–3 Show that Eq. (12–232) is true also for uniformly doped transistors. *Hint:* Use the fact that the transistor is normally biased and use one term Taylor series expansion of hyperbolic functions.

12–4 Plot each of the four terms which comprise $1/\beta_0$ separately as a function of I_E for *p-n-p* transistors.

12–5 You are the manufacturer of transistors whose fundamental device constants such as mobilities, lifetimes, diffusion lengths, base widths, junction areas, etc., are known to you. A customer of yours wants to know at what current levels he would need to start taking into account the high-injection effects. Derive an analytical expression in terms of the fundamental device parameters to answer your customer's question.

12–6 How would you expect the shape of $\beta_0(I_E)$ to change as the surface-recombination velocity changes from a moderate to an extremely large value in a *p-n-p* transistor?

12–7 A *p-n-p* common-emitter circuit has a collector circuit which consists of a battery in series with a variable load resistance and which has the base current fed by a constant-current source. In such a circuit β_0 will vary not only because of Z_p and $g_p(Z_p)$ but also because of a change in the collector transition region. What single parameter is affected by the latter change? How would you expect $\beta_0(I_E)$ to change because of this effect?

12–8 Discuss at least one way in which the actual $\beta_0(I_E)$ curve, as obtained in the circuit of Problem 12–7, is expected to be different for *p-n-p* and *n-p-n* transistors.

REFERENCES

Bowler, D. L., and Lindholm, F. A., "High Current Regimes in Transistor Collector Regions," *IEEE Trans. Electron Devices,* **ED-20,** 257 (1973).

Das, M. B., and Boothroyd, A. R., "Determination of Physical Parameters of Diffusion and Drift Transistors," *IRE Trans. Electron Devices,* **ED-8,** No. 1 (1961).

Iwerson, J. E., et al, "Low Current Alpha in Si Transistors," *IRE Trans. Electron Devices,* **ED-9,** 74 (1962).

Kannam, P. J., "Effect of Emitter Doping on Device Characteristics," *IEEE Trans. Electron Devices,* **ED-20,** 845 (1973).

Kirk, C. T., "A Theory of Transistor Cut-off Frequency (f_T) Fall-off at High Current Density," *IEEE Trans. Electron Devices,* **ED-9,** 164 (1962).

Mertens, R. P., De Man, H. J., Van Overstraeten, R. J., "Calculation of the Emitter Efficiency of Bipolar Transistors," *IEEE Trans. Electron Devices,* **ED-20,** 772 (1973).

Misawa, R., "Emitter Efficiency of Junction Transistor," *J. Phys. Soc. Japan,* **10,** 362, May (1955).

Nanavati, R. P., *An Introduction to Semiconductor Electronics,* Chapters 3, 6, McGraw-Hill Book Co., New York (1963).

Poon, H. C., et al, "High Injection in Epitaxial Transistors," *IEEE Trans. Electron Devices,* May, (1969).

Rittner, E. S. "Extension of the Theory of Junction Transistor," *Phys. Rev.,* **94,** 1161 (1954).

Sah, C. T., Noyce, R. N., and Shockley, W., "Carrier Generation and Recombination in *p-n* Junction and *p-n* Junction Characteristics," *Proc. IRE,* **45,** 1228 (1957).

Webster, W. M., "On the Variation of Junction Transistor Current Amplification Factor with Emitter Currents," *Proc. IRE,* **42,** 914 (1954).

APPENDIX A

THE PERIODIC TABLE

IA	IIA	IIIB	IVB	VB	VIB	VIIB	VIII			IB	IIB	IIIA	IVA	VA	VIA	VIIA	0	Shell structure	Shell number
1 H 1.0																	2 He 4.0	2	1
3 Li 6.9	4 Be 9.0							Atomic number Symbol Atomic weight				5 B 10.8	6 C 12.0	7 N 14.0	8 O 16.0	9 F 19.0	10 Ne 20.2	2 8	1 2
11 Na 23.0	12 Mg 24.3											13 Al 27.0	14 Si 28.1	15 P 31.0	16 S 32.1	17 Cl 35.4	18 Ar 39.9	2 8 8	1 2 3
19 K 39.1	20 Ca 40.1	21 Sc 45.1	22 Ti 48.0	23 V 51.0	24 Cr 52.0	25 Mn 54.9	26 Fe 55.8	27 Co 58.9	28 Ni 58.7	29 Cu 63.5	30 Zn 65.4	31 Ga 69.7	32 Ge 72.6	33 As 75.0	34 Sc 79.0	35 Br 80.0	36 Kr 83.7	2 8 18 8	1 2 3 4
37 Rb 85.5	38 Sr 87.6	39 Y 88.9	40 Zr 91.2	41 Nb 92.9	42 Mo 96.0	43 Tc (99)	44 Ru 101.7	45 Rh 102.9	46 Pd 106.7	47 Ag 107.9	48 Cd 112.4	49 In 114.8	50 Sn 118.7	51 Sb 121.8	52 Te 127.6	53 I 126.9	54 Xe 131.3	2 8 18 18 8	1 2 3 4 5

IA	IIA	IIIB	IVB	VB	VIB	VIIB	VIII			IB	IIB	IIIA	IVA	VA	VIA	VIIA	0	Shell structure	Shell number
55 Cs 132.9	56 Ba 137.4	57 La 138.9	72 Hf 178.6	73 Ta 180.9	74 W 183.9	75 Re 186.3	76 Os 190.2	77 Ir 193.1	78 Pt 195.2	79 Au 197.2	80 Hg 200.6	81 Tl 204.4	82 Pb 207.2	83 Bi 209.0	84 Po (210)	85 At (211)	86 Ra (222)	2 8 18 32 18 8	1 2 3 4 5 6
87 Fr (223)	88 Ra 226.0	89 Ac 227.0																	
			Rare earths	58 Ce 140.1	59 Pr 140.9	60 Nd 144.3	61 Pm (147)	62 Sm 150.4	63 Eu 152	64 Gd 156.9	65 Tb 159.2	66 Dy 162.5	67 Ho 164.9	68 Er 167.2	69 Tm 169.4	70 Yb 173.0	71 Lu 175.0	2 8 18 32 9 2	1 2 3 4 5 6
			Actinide series	90 Th 232.1	91 Pa 231.0	92 U 238.1	93 Np (237)	94 Pu (242)	95 Am (243)	96 Cm (245)	97 Bk (245)	98 Cf (248)	99 E (255)	100 Fm (252)	101 Mv (256)			2 8 18 32 9 2	1 2 3 4 5 6 7

APPENDIX B

PHYSICAL CONSTANTS

Quantity	Symbol	Value	Unit
Angström	Å	10^{-10}	meter
Avogadro's number	N_{AVO}	6.025×10^{23}	molecules/(g-mole)
Bohr radius (first)	...	5.29172×10^{-11}	meter
Boltzmann's constant	k	1.38044×10^{-23}	joule/°K
		8.62×10^{-5}	eV/°K
Electronic charge	q	1.602×10^{-19}	coulomb
Electron rest mass	m	9.1083×10^{-31}	kg
Electron-Volt	eV	1.602×10^{-19}	joul
		23.1	k cal
Micron	μ	10^{-6}	meter
Neutron rest mass	...	1.67470×10^{-27}	kg
Permittivity of free space	ε_0	8.854×10^{-14}	Farad/cm
Permeability of free space	μ_0	1.257×10^{-8}	Henry(cm
Planck's constant	h	6.62517×10^{-34}	joul-sec
Proton rest mass	...	1.67239×10^{-27}	kg
Reduced Planck's constant	$\hbar$	1.054×10^{-34}	joul-sec
The Rydberg	R_H	10,973,730.9	m^{-1}
Stefan–Boltzman constant	σ_s	0.56687×10^{-7}	watt/((m^2) (°k)^4)
Standard atmospheric pressure	...	1.033×10^4	kg/m^2
		760	mmHg(torr)
Thermal Voltage	kT/q	0.0259	Volt
Thermal energy	kT	0.0259	eV
Velocity of light	c	2.997930×10^8	m/sec
Wavelength corresponding to 1 eV	λ	1.239×10^{-6}	meter

APPENDIX C

VALUES (AT 300°K) OF PARAMETERS OF GERMANIUM, SILICON, AND GALLIUM ARSENIDE

Parameter	Unit	Ge	Si	GaAs
Atom density	atom/cm^3	4.42×10^{22}	5.8×10^{22}	2.21×10^{22}
Atomic number		32	14	
Atomic weight		72.6	28.08	144.63
Breakdown electric field	Volt/cm	$\sim 10^5$	$\sim 3 \times 10^5$	$\sim 4 \times 10^5$
Density	g/cm^3	5.3267	2.328	5.32
Dielectric constant		16	11.8	10.9
Diffusion constant (electrons)	cm^2/sec	93	31	224
Diffusion constants (holes)	cm^2/sec	44	12.5	10.4
Effective density of states in the conduction band, N_c	cm^{-3}	1.04×10^{19}	2.8×10^{19}	4.7×10^{17}
Effective density of states in the valence band, N_v	cm^{-3}	6.1×10^{18}	1.02×10^{19}	7.0×10^{18}
Effective mass ratio m^*/m				
Electron m_ℓ^*		1.6	0.97	0.068
Electron m_t^*		0.082	0.19	
Holes m_{h1}^*		0.3	0.5	0.5
Holes m_{h2}^*		0.04	0.16	0.12
Electron affinity	χ	4.0	4.05	4.07
Energy gap indirect	eV	0.67	1.10	1.77 (100) direction
Energy gap direct	eV	0.804		1.43
Intrinsic carrier density, n_i	cm^{-3}	2.37×10^{13}	1.5×10^{10}	9×10^6
Lattice constant	Å	5.65748	5.43086	5.6534
Linear coefficient of thermal expansion $\Delta L/(L\Delta T)$	°C^{-1}	5.8×10^{-6}	2.6×10^{-6}	5.9×10^{-6}
Linear temperature coefficient for band gap energy β_g	eV/°K	4.0×10^{-4}	4.4×10^{-4}	4.9×10^{-4}
Melting point	°C	937	1420	1238
Mobility (electrons)	cm^2/(Volt-sec)	3900	1350	8600
Mobility (holes)	cm^2/(Volt-sec)	1900	480	400

MEASURED ACCEPTOR AND DONOR ENERGY LEVELS IN Ge, Si, AND GaAs

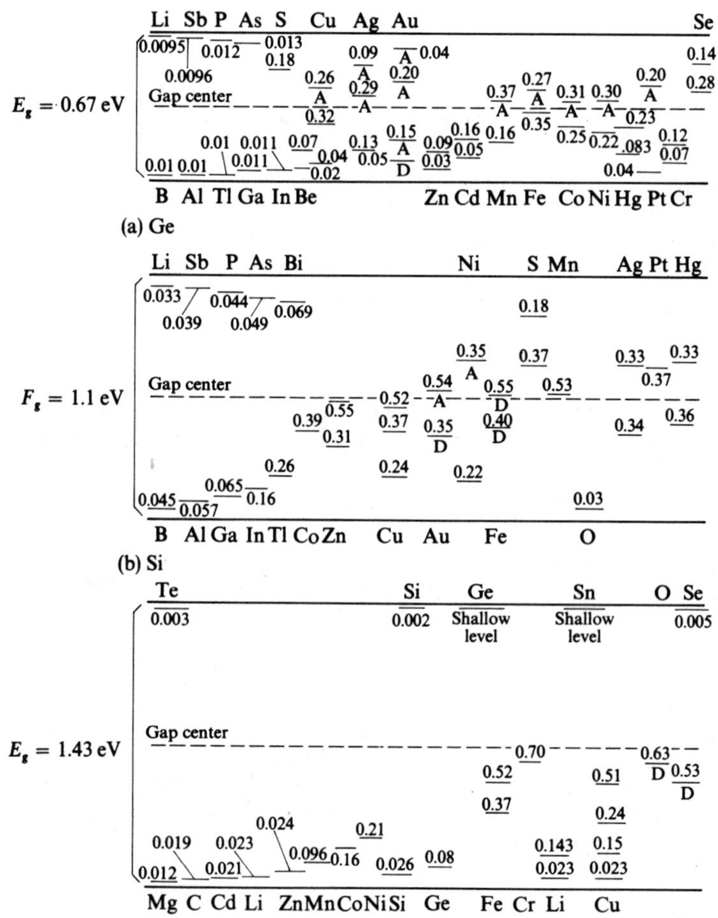

Fig. D–1

Levels below the band gap center are *acceptor levels* except when marked with a D in which case they are donors. Similarly, levels above the band gap center are *donor levels* except when marked with an A in which case they are acceptors. (After Conwell, and Sze and Irwin.)

APPENDIX E

VACUUM TUBE PHYSICS

E–1 INTRODUCTION

In this appendix we shall derive the ideal current–voltage relationship for vacuum diodes and triodes. First the emission characteristics of a cathode held at a fixed elevated temperature will be examined. Next, Poisson's equation will be solved for potential distribution between the cathode and the plate. The current–voltage equation will be then obtained under *space-charge limited* conditions. This result will be finally generalized to the multi-electrode vacuum tubes.

E–2 EMISSION LIMITED AND SPACE-CHARGE LIMITED FLOW IN VACUUM DIODES

The vacuum diode device consists of three elements (1) the cathode is a metal sleeve within which is (2) the *heater* element called the *filament* (insulated resistance wire), and (3) the *plate*, also called the *anode*. These three elements are housed within an evacuated glass envelope. These elements are schematically shown in Fig. E–1c.

The cathode is the emitter of electrons. When it is raised to a sufficiently high temperature it will emit electrons in large numbers. While any metal will emit electrons at elevated temperatures, cathodes coated with alkaline-earth oxides are found to emit electrons quite generously at moderate temperatures—between 700 and 800°C. The process by which the electrons are emitted from a cathode held at elevated temperatures is called *thermionic emission*. The vacuum within the envelope must be sufficiently good such that the device performance is not influenced by interactions between the molecules of any residual gas inevitably present within the vacuum and the emitted electrons.

Consider the situation when the cathode is heated to a reasonable temperature, but the terminals of the vacuum diode are left open-circuited. The

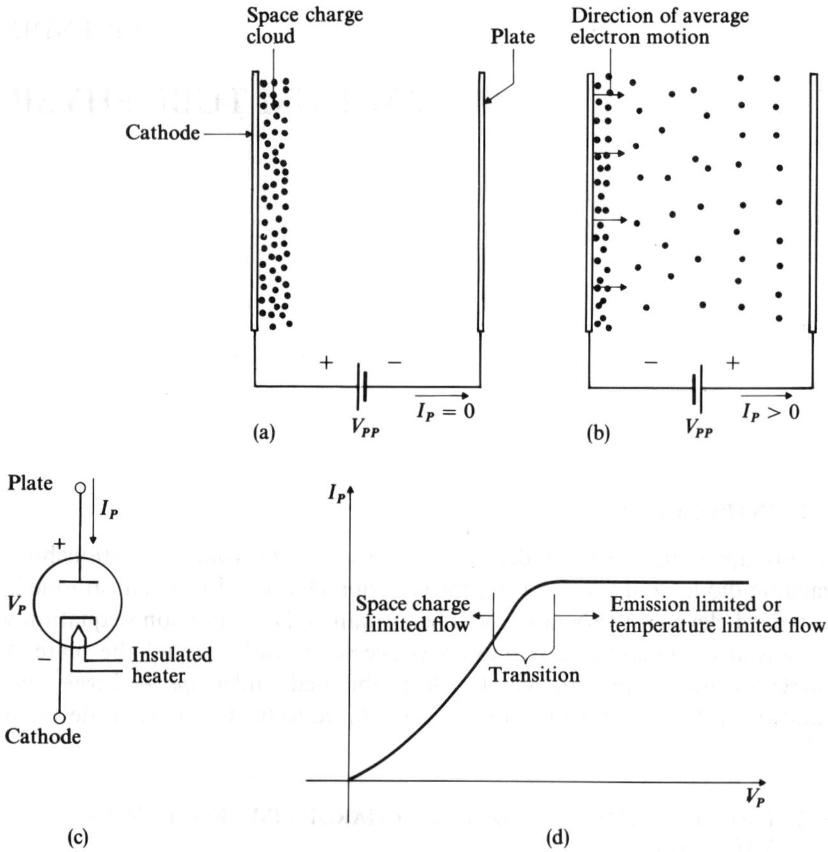

Fig. E–1 Vacuum tube diode, charge in the interelectrode space and I_P versus V_P.

heated cathode will emit electrons into the interelectrode space. The cathode material acquires a net positive charge because of the electrons it has lost. As the electrons are constantly "boiled off" the cathode, many of these will fall back into the cathode attracted by the net positive charge due to the coulombic force. A space charge cloud will form close to the cathode surface facing the interelectrode space. A dynamic equilibrium is set up where electrons are being exchanged between the cathode and the space charge cloud. The coulombic forces keep the electrons from wandering too far from the cathode. Thus virtually no electrons reach the plate. Hence the plate current is zero. This is consistent with the open-circuited terminals. If the plate has a negative potential with respect to the cathode, this negative potential will tend to move the space charge cloud of electrons even closer to the cathode than was the case with terminals left open-circuited. Again no plate current will flow as no electrons reach the plate. This situation is shown in Fig. E–1a.

Consider now the case where the plate is biased positively with respect to the cathode. The electrons which comprise the space charge cloud will be attracted by the positively biased plate. The electrons which make it to the plate electrode and enter the external circuit will constitute *conventional current from plate to cathode. Note that the direction of electron motion is opposite to the flow of conventional current due to the electron motion.* If the positive plate voltage is increased the space charge cloud will be reduced.

More electrons will reach the plate per unit time and the plate current I_P will increase as shown in Fig. E–1b. Generally, many more electrons are emitted in a unit time by the cathode into the space charge than are able to make it from the space charge to the plate. Many electrons fall back into the cathode from the space charge cloud. Thus a dynamic equilibrium exists. We see then that the space charge cloud exercises the controlling influence in this manner of current flow. The flow of current in the device is said to be *space-charge limited.* If the plate potential is increased sufficiently the space charge is thinned out so much that it no longer controls the current flow. Under these circumstances every electron emitted by the cathode reaches the plate. The rate at which electrons are emitted is also now the rate at which they are collected by the plate. This plate current is then expected to be independent of plate voltage. The current is importantly controlled by the cathode temperature and hence the cathode emission. For this reason the flow under these conditions is said to be *emission limited* or *temperature limited.* These characteristics can be seen in Fig. E–1d. Also shown on these characteristics is region due to space charge limited flow, as well as the *transition* between the two kinds of flow.

Figure E–2d shows the plate current as a function of the emission current from the cathode. To understand this curve we may note that when the cathode temperature is low, almost no electrons exist in the region between cathode and plate. Curves *a* in Figs. E–2a, E–2b and E–2c correspond to this situation. There is no space-charge formed yet as shown by curve *a* in Fig. E–2c. Hence as Fig. E–2a, curve *a* shows, the voltage distribution is linear when the plate to cathode voltage is V_P Volts. With a linear potential distribution, the electrostatic field $\mathscr{E}$ is constant as can be seen from the following relationship between electrostatic field and potential V

$$\mathscr{E} = -\nabla V. \tag{E–1}$$

For the case at hand

$$\mathscr{E} = -\frac{V_P}{L}. \tag{E–2}$$

which is depicted by curve *a* in Fig. E–2b.

The rate of emission of electrons from the cathode will be equal to the rate of collection by the plate because there is no space charge cloud formed yet.

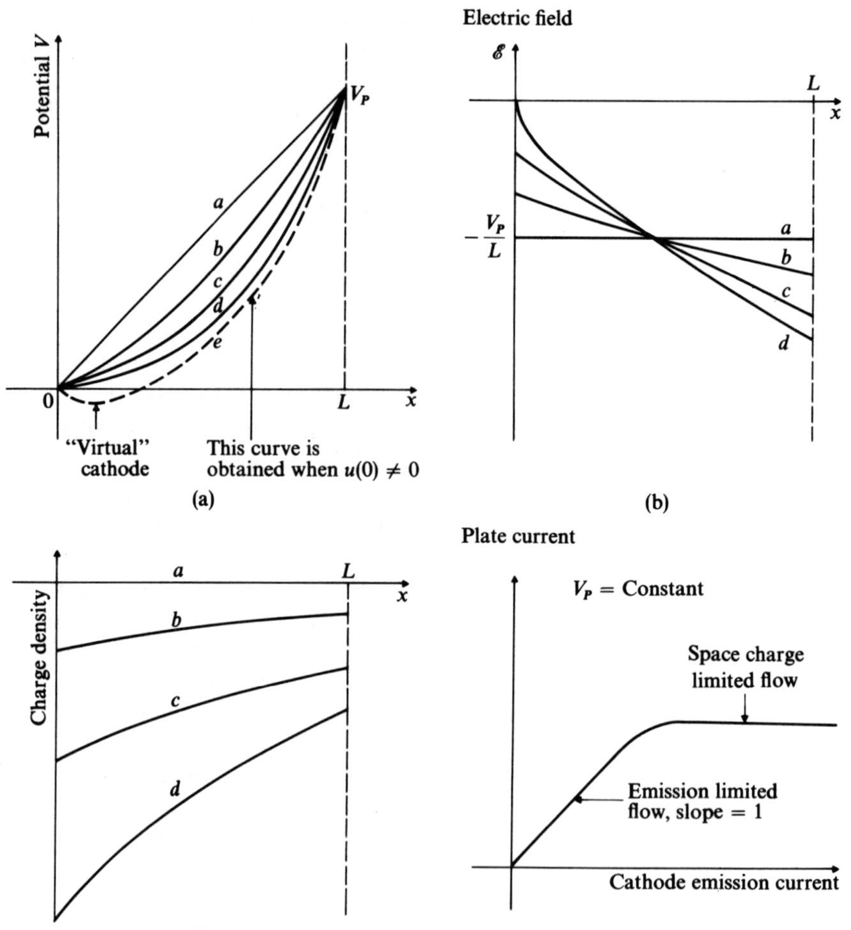

Fig. E–2 Potential, electric field, and charge density distribution as a function of position in a one-dimensional vacuum diode. Also shown is the plate current as a function of the cathode emission current.
Curve *a* represents no emission.
Curve *b* cathode emission rate is small (all emitted electrons reach the plate).
Curve *c* Cathode emission rate is greater than that for curves *b*.
Curve *d* Cathode emission rate is big enough so that the field is zero at the cathode.

The unity slope close to the origin in Fig. E–2d depicts this case which is *emission limited flow.*

As the cathode emission rate is increased an electron space charge cloud will form not far from the cathode as explained earlier. The charge density is seen to be negative. Its magnitude is biggest close to the cathode where the electron velocity is expected to be low. As the electrons travel towards the plate, they speed up due to the plate to cathode voltage V_P, and hence the charge density is the smallest close to the plate where the electron velocity is the highest. The emission rate is small for curve *b*. This corresponds to the emission limited flow of electrons, where the electron emission rate equals the collection rate. Curve *c* corresponds to the transition region between the emission limited flow and the space-charge limited flow. Note that in Fig. E–2a, the potential distribution is below the curves labeled *a* and *b* of this curve. The electric field is small at the cathode and large in magnitude at the plate. If the emission rate is increased still further, the field at the cathode is reduced to zero corresponding to the zero *slope* of the potential at the cathode. Under this condition the space charge cloud completely shields the cathode from the plate. The electrons emitted by the cathode will be unaware of the presence of the plate. Thus, the plate current is *not* emission limited. It is space charge limited as shown in Fig. E–2d.

The plate current I_P is generally a function of both temperature and the plate to cathode potential V_P. Figures E–3a and E–3b show both of these dependencies. Because cathode temperatures are hard to control precisely, and because temperature is the key parameter for emission limited current flow, this is not a desirable manner of operating the device. From Fig. E–3b we see that when the device is operated under conditions of space-charge limited current flow, the plate current is independent of the cathode temperature. Thus we see that it is desirable to operate the device under space-charge limited flow conditions. Almost all practical vacuum tubes, for this reason, are operated under space-charge limited conditions.

E–3 CURRENT-VOLTAGE EQUATION FOR VACUUM TUBE DIODE

So far we have looked at the performance of the vacuum diode in largely qualitative terms. Now we wish to obtain quantitative relationships between plate current I_P and plate to cathode voltage, abbreviated simply plate voltage V_P.

The starting assumptions of the ensuing analysis are the following.

1. The electrons are available at the cathode in large numbers.
2. The initial velocity with which electrons leave the cathode is zero.†

$$u(0) = 0 \tag{E-3}$$

†Actually the electrons are emitted by the cathode with a nonzero velocity. The average thermal energy of these electrons is less than 2 eV. Curve *e* in Fig. E–2a shows what happens to the potential distribution when the initial velocity is not zero. The dip in potential produces a retarding field between the cathode and the potential minimum. The initial velocity together with the retarding field produce zero velocity at the potential minimum which is called the "virtual" cathode. The neglect of initial velocity leads to no appreciable errors unless the plate voltage is a few Volts or less.

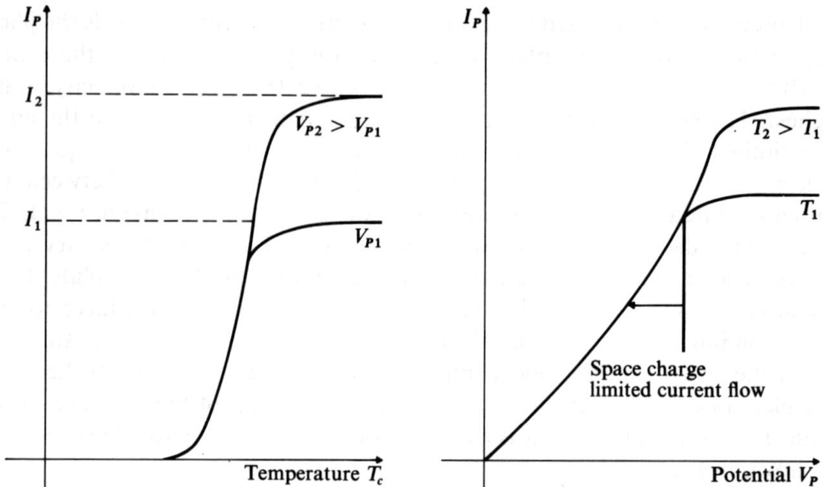

(a) Plate current as a function of cathode temperature for fixed plate to cathode voltages.

(b) Plate current as a function of plate to cathode voltage for fixed cathode temperatures.

Fig. E.3 Plate current as a function of plate voltage and cathode temperature.

where u = velocity of the electrons in the direction of cathode to plate. Origin of the x-coordinate is zero at the cathode.

3. Electron density in the interelectrode region is small enough so that collisions between electrons may be neglected.
4. Residual gas in the vacuum is small enough to be neglected.
5. The device is being operated under space-charge limited flow conditions.
6. Electron velocities in the interelectrode region are much less than the velocity of light so that relativistic effects may be neglected.
7. Only a one-dimensional flow of current is considered.
8. Plate voltage exceeds a few Volts.
9. Fringing fields may be neglected.

We shall first try to find the potential distribution as a function of distance. We shall need to solve Poisson's equation which is repeated below

$$\frac{d^2V(x)}{dx^2} = -\frac{\rho(x)}{\varepsilon_0}, \tag{E-4}$$

where ε_0 = vacuum permittivity.

The electron charge density $\rho(x)$ may be related to current density and the electron velocity $u(x)$ as follows

$$J = \rho(x)\, u(x), \tag{E–5}$$

where J = flow of electron current density of

$$\rho(x) = \frac{J}{u(x)}, \tag{E–6}$$

If we can relate the velocity to the potential function $V(x)$ we can rewrite Eq. (E–4) in terms of current and voltage. With this end in mind we may notice that with the assumptions made for the present analysis the law of conservation of energy applies to the motion of electrons from the cathode to the plate through the interelectrode space.

We may arbitrarily designate the cathode potential as zero, thus

$$V(0) = 0 \tag{E–7}$$

$$\begin{bmatrix}\text{Change in kinetic energy}\\ \text{between the cathode and point } x\end{bmatrix} = \begin{bmatrix}\text{Change in potential energy}\\ \text{between the cathode and point } x\end{bmatrix} \tag{E–8}$$

$$(1/2)mu^2(x) - (1/2)mu^2(0) = qV(x) - qV(0), \tag{E–9}$$

where m = mass of electron,

q = charge of an electron.

But

$$V(0) = 0 = u(0). \tag{E–10}$$

Thus

$$(1/2)mu^2(x) = qV(x). \tag{E–11}$$

We may solve for $u(x)$ from Eq. (E–11) to obtain

$$u(x) = \left(\frac{2q}{m}\right)^{\frac{1}{2}} V^{\frac{1}{2}}(x). \tag{E–12}$$

When Eq. (E–12) is substituted into Eq. (E–6) and the result is substituted into Eq. (E–4) the following is obtained:

$$\frac{d^2V(x)}{d^2x} = -\frac{J}{\varepsilon_0}\left(\frac{m}{2q}\right)^{\frac{1}{2}} V^{-\frac{1}{2}}(x). \tag{E–13}$$

If we use

$$J = -\frac{I_P}{A}. \tag{E–14}$$

Note that the negative sign in Eq. (E–14) is because the electron flow is opposite to the conventional current flow. Equation (E–13) takes the following form:

$$\frac{d^2V(x)}{dx^2} = \frac{I_p}{\varepsilon_0 A}\left(\frac{m}{2q}\right)^{\frac{1}{2}} V^{\frac{1}{2}}(x). \tag{E–15}$$

This equation is to be solved subject to the following boundary conditions for the vacuum tube diode:

$$V(0) = 0 \tag{E–16}$$

$$\frac{dV(x)}{dx} = 0 \quad \text{at} \quad x = 0 \tag{E–17}$$

$$V = V_P \quad \text{when} \quad x = L. \tag{E–18}$$

Equation (E–15) is a nonlinear equation. It can, however, be solved by integrating both sides of the equation. Before the integration is carried out, for convenience we shall multiply both sides of Eq. (E–15) by $(dV/dx)dx$. Thus,

$$\int \frac{dV(x)}{dx}\left(\frac{d^2V(x)}{dx^2}\right) dx = \frac{I_P}{\varepsilon_0 A}\left(\frac{m}{2q}\right)^{\frac{1}{2}} V^{-\frac{1}{2}}(x)\frac{dV(x)}{dx}dx. \tag{E–19}$$

Now the following identity may be used

$$\frac{dV(x)}{dx}\left(\frac{d^2V(x)}{dx^2}\right) = \frac{1}{2}\frac{d}{dx}\left(\frac{dV(x)}{dx}\right)^2. \tag{E–20}$$

Using the identity of (E–20) on the left hand side of Eq. (E–19) and cancelling dx's on the right hand side of Eq. (E–19) we obtain:

$$1/2 \int d\left[\left(\frac{dV(x)}{dx}\right)^2\right] = \int \frac{I_P}{\varepsilon_0 A}\left(\frac{m}{2q}\right)^{\frac{1}{2}} V^{-\frac{1}{2}}(x)dV(x). \tag{E–21}$$

Carrying out the integration we obtain

$$1/2\left[\frac{dV(x)}{dx}\right]^2 = \frac{2I_P}{\varepsilon_0 A}\left(\frac{m}{2q}\right)^{\frac{1}{2}} V^{-\frac{1}{2}}(x) + C_1, \tag{E–22}$$

where C_1 = constant of integration.

Utilizing Eqs. (E–16) and (E–17) to evaluate C_1 we get

$$(1/2)\,(0) = 2\frac{I_P}{\varepsilon_0 A}\left(\frac{m}{2q}\right)^{\frac{1}{2}} (0) + C_1 \tag{E–23}$$

or

$$C_1 = 0. \tag{E–24}$$

Thus Eq. (E–23) is simply

$$\left[\frac{dV(x)}{dx}\right]^2 = \frac{4I_P}{\varepsilon_0 A}\left(\frac{m}{2q}\right)^{\frac{1}{2}} V^{\frac{1}{2}}(x). \tag{E–25}$$

Let us take the square root of both sides of Eq. (E–25) to obtain

$$\frac{dV}{dx} = 2\left[\frac{I_P}{\varepsilon_0 A}\left(\frac{m}{2q}\right)^{\frac{1}{2}}\right]^{\frac{1}{2}} V^{\frac{1}{4}}. \qquad \text{(E–26)}$$

Equation (E–26) may be rewritten as follows and integrated

$$\int \frac{dV}{V^{\frac{1}{4}}} = 2\int\left[\frac{I_P}{\varepsilon_0 A}\left(\frac{m}{2q}\right)^{\frac{1}{2}}\right]^{\frac{1}{2}} dx. \qquad \text{(E–27)}$$

Carrying out the integration we obtain

$$(4/3)\, V^{3/4} = 2\left[\frac{I_P}{\varepsilon_0 A}\left(\frac{m}{2q}\right)^{\frac{1}{2}}\right]^{\frac{1}{2}} x + C_2, \qquad \text{(E–28)}$$

where C_2 = constant of integration.

Applying the boundary condition Eq. (E–16) to Eq. (E–28) we get

$$(4/3)\,(0) = 0 + C_2$$

or

$$C_2 = 0. \qquad \text{(E–29)}$$

Thus, Eq. (E–28) becomes

$$V^{3/4} = (6/4)\left[\frac{I_P}{\varepsilon_0 A}\left(\frac{m}{2q}\right)^{\frac{1}{2}}\right]^{\frac{1}{2}} x. \qquad \text{(E–30)}$$

Equation (E–18) may be applied to Eq. (E–30) to obtain

$$V_P = \frac{3L}{2}\left[\frac{1}{\varepsilon_0 A}\left(\frac{m}{2q}\right)^{\frac{1}{2}}\right]^{\frac{1}{2}} I_P^{\frac{1}{2}}. \qquad \text{(E–31)}$$

Both sides of Eq. (E–31) may be squared to obtain the following for I_p as a function of V_p.

$$I_P = \frac{4}{9L^2}\left[\varepsilon_0 A\left(\frac{2q}{m}\right)^{\frac{1}{2}}\right] V_P^{3/2} \qquad \text{(E–32)}$$

or

$$I_P = \left(\frac{4\varepsilon_0 A}{9L^2}\right)\left(\frac{2q}{m}\right)^{\frac{1}{2}} V_P^{3/2},\ V_P \geqslant 0 \qquad \text{(E–33)}$$

$$I_P = 0 \qquad \text{for } V_P < 0. \qquad \text{(E–34)}$$

Equation (E–33) is the *three-halves power law.* It is also known as the *Child–Langmuir Law*[1,2].

The above current–voltage law was derived for a planar device. Other geometries of the vacuum diode also lead to the plate current being proportional to the three-halves power of the plate voltage, though the constant of proportionality is different. For example, for concentric cylinders of length ℓ

where the cathode is the inner cylinder of radius R_c and the plate the outer cylinder of radius R_p, the three halves power law is as follows[3]

$$I_P = \frac{8\pi\varepsilon_0\ell}{9R_p\phi}\sqrt{\frac{2q}{m}}\,V_P^{\frac{3}{2}} \qquad \text{for } V_P \geqslant 0 \tag{E–35}$$

$$I_P = 0,\ V_P < 0, \tag{E–36}$$

where ϕ = a parameter which is a function of (R_p/R_c).

VACUUM TUBE TRIODES

The vacuum tube triode has a grid between the cathode and the plate. The function of the grid is to control the flow of electrons from the cathode to the plate. The triodes are operated under space-charge limited conditions. The grid electrode is ideally supposed to completely control the space charge in the electrode space between the cathode and the grid. This space charge can cause the electric field to be zero at the cathode. The electrons emitted by the cathode experience no force at the surface of the cathode. Only the ones which move through the space charge and go through the wide open spaces in the physical structure of the grid go on to the plate, whose positive potential entices the electron to move towards it, and results in the flow of plate current. The grid

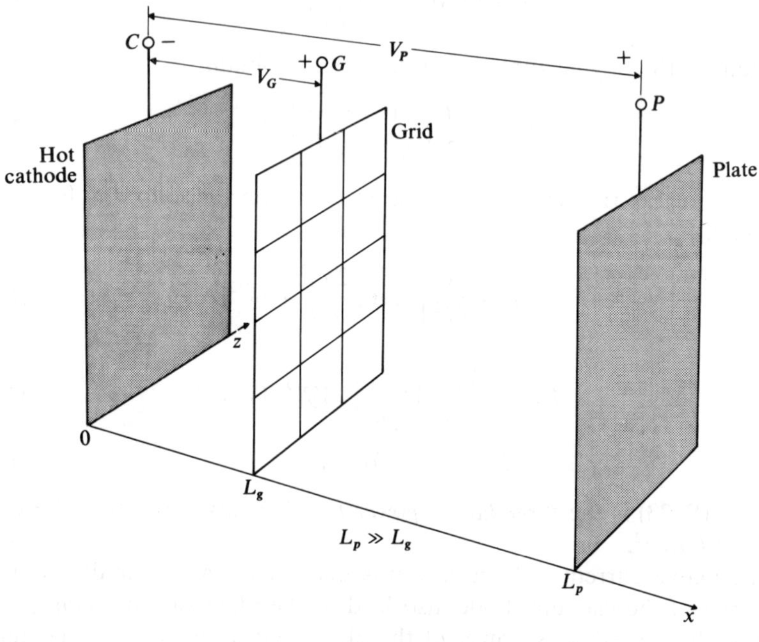

Fig. E–4 Structure of plane parallel vacuum triode.

structure is like a metallic screen with wide spacing so that it appears almost transparent to electrons. An ideal plane parallel structure of the triode is shown in Fig. E–4. The actual structure of a real triode is more likely to be concentric cylinders. For simplicity, however, we shall consider the plane parallel triode structure shown in Fig. E–4.

Usually the grid is operated at a negative potential with respect to cathode, and the plate is operated at a positive potential with respect to the cathode. The negative grid potential as well as its open physical structure leads to the grid collecting hardly any electrons and affords the very high input resistance which the grid offers when it is biased negatively with respect to the cathode.

Consider now the potential along the grid plane when the grid potential is negative ($-V_G$) and the plate potential is positive (V_P) with respect to the cathode. If a three-dimensional plot is made of the potential for the grid plane, it is clear that the potential at the wires of the grid must of necessity be negative ($-V_G$) if the grid is biased negatively with respect to the cathode. In the space away from the grid wires the potential will be less than V_P but will rise well above zero. The potential in the plane of the grid electrode is thus seen to be a complicated function.

An approximate way of dealing with this situation is to replace the grid with an equivalent equipotential surface whose potential is chosen as a weighted average of the plate and grid potentials combined linearly as follows

$$V_{EQ} = V_G + \frac{V_P}{\mu}, \tag{E–37}$$

where V_G and V_P are defined on Fig. E–4.

Now μ is called the amplification factor. Here we see that if μ is high, the relative importance of the grid voltage is enhanced over the plate voltage. We expect that because the grid is generally much closer to the cathode than is the plate, we anticipate the grid should exercise a greater controlling role than the plate. This implies that μ must be a high number typically between 10 and 250. As this single parameter determines the relative importance of the grid and the plate, it is expected to be a function of the structure of the grid itself as well as the relative spacing of the grid and the plate with respect to the cathode.

With the use of the equivalent grid potential the diode equation Eq. (E–33) may be applied to the triode with the use of the transformations in Table E–1.

Thus, the triode device equation is obtained readily by the use of the transformations of Table E–1 applied to Eq. (E–33) to get

$$I_P = K_T\left[V_G + \frac{V_P}{\mu}\right]^{\frac{3}{2}}, \tag{E–38}$$

when

$$\left(V_G + \frac{V_P}{\mu}\right) > 0, \tag{E–39}$$

and

$$V_P > 0, \tag{E–40}$$

where

$$K_T \triangleq \frac{4\varepsilon_0 A}{9L_g^2}\left(\frac{2q}{m}\right)^{\frac{1}{2}} \tag{E–41}$$

$$I_P = 0, \tag{E–42}$$

when

$$\left(V_G + \frac{V_P}{\mu}\right) \leqslant 0, \tag{E–43}$$

or

$$V_P < 0. \tag{E–44}$$

The above analysis assumes that V_G is negative. Under these conditions the cathode current and the plate current are identical.* If the grid becomes positively biased, the grid current will flow from the grid into the device. The cathode current will be the sum total of the plate current and the grid current. The situation is more complicated than the above analysis takes account of.

Table E–1 Transformations which convert the diode *I-V* Equation to the Corresponding Triode Equations

Eq. (E–33)	Eq. (E–38)
V_P ——	—— V_{EQ}
L ——	—— L_g

The other multi-electrode vacuum tubes such as pentodes also are operated in the space charge limited mode. These too follow the generalized three-halves power law in current voltage relationships. Experimental data on triodes and pentodes indicate that the models discussed above are quite good for vacuum tubes operated in the space charge limited mode.

Example E–1. You are given a vacuum tube diode whose filament is driven by a voltage of 5 Volts. When the plate voltage is 200 Volts, the plate current is 10 mA. When the filament voltage is increased to 6 Volts the plate current remains at 10 mA. Calculate the current when the plate voltage is 50 Volts and the filament voltage is 6.25 Volts.

† Except when the grid is greater than a few tenths of a Volt negative.

Solution. The fact that the plate current did not change when the filament voltage was increased from 5 to 6 Volts (i.e., the cathode temperature was increased) indicated that the tube was being operated in the space-charge limited conditions. A further small increase in the filament voltage, and hence cathode temperatures will not alter the plate current by itself. The new plate current can be calculated by the use of the three halves power law. Equation (E–33) applies, hence

$$\frac{I_P\,(\text{new})}{I_P\,(\text{old})} = \frac{K_D V_P^{\frac{3}{2}}\,(\text{new})}{K_D V_P^{\frac{3}{2}}\,(\text{old})} \tag{1}$$

$$I_P\,(\text{new}) = I_P\,(\text{old})\left[\frac{V_P\,(\text{new})}{V_P\,(\text{old})}\right]^{\frac{3}{2}} \tag{2}$$

$$I_P\,(\text{new}) = (10)\left(\frac{50}{200}\right)^{\frac{3}{2}} = 10(1/3)^{\frac{3}{2}} = \frac{10}{8} = \underline{1.25}\ \text{mA}. \tag{4}$$

REFERENCES

Child, D.C., "Discharge from Hot CaO", *Phys. Rev.*, **32,** 492–511, May (1911).

Langmuir, J., "The Effect of Space Charge and Residual Gases on Thermionic Currents in High Vacuum," *Phys. Rev.*, Ser. 2, **2,** 450–486, Dec. (1913).

Langmuir, J., "Currents Limited by Space Charge Between Coaxial Cylinders," *Phys. Rev.*, Ser. 2, **22,** 347–356, Oct. (1923).

INDEX

79 80 81 82 8 7 6 5 4 3 2